Algebraic Codes on Lines, Planes, and Curves

The past few years have witnessed significant advances in the field of algebraic coding theory. This book provides an advanced treatment of the subject from an engineering point of view, covering the basic principles of codes and their decoders. With the classical algebraic codes referred to as codes defined on the line, this book studies, in turn, codes on the line, on the plane, and on curves. The core ideas are presented using the ideas of commutative algebra and computational algebraic geometry, made accessible to the nonspecialist by using the Fourier transform.

Starting with codes defined on a line, a background framework is established upon which the later chapters concerning codes on planes, and on curves, are developed. Example codes include cyclic, bicyclic, and epicyclic codes, and the Buchberger algorithm and Sakata algorithm are also presented as generalizations to two dimensions of the Sugiyama algorithm and the Berlekamp–Massey algorithm. The decoding algorithms are developed using the standard engineering approach as applied to two dimensional Reed–Solomon codes, enabling the decoders to be evaluated against practical applications.

Integrating recent developments in the field into the classical treatment of algebraic coding, this is an invaluable resource for graduate students and researchers in telecommunications and applied mathematics.

RICHARD E. BLAHUT is Head of the Department of Electrical and Computer Engineering at the University of Illinois, Urbana Champaign, where he is also a Professor. He is a Fellow of the IEEE and the recipient of many awards, including the IEEE Alexander Graham Bell Medal (1998) and Claude E. Shannon Award (2005), the Tau Beta Pi Daniel C. Drucker Eminent Faculty Award, and the IEEE Millennium Medal. He was named Fellow of the IBM Corporation in 1980, where he worked for over 30 years, and was elected to the National Academy of Engineering in 1990.

Algebraic Codes on Lines, Planes, and Curves

Richard E. Blahut

CAMBRIDGE UNIVERSITY PRESS
Cambridge, New York, Melbourne, Madrid, Cape Town, Singapore, São Paulo, Delhi

Cambridge University Press
The Edinburgh Building, Cambridge CB2 8RU, UK

Published in the United States of America by Cambridge
University Press, New York

www.cambridge.org

Information on this title: www.cambridge.org/9780521771948

First published 2008

Printed in the United Kingdom at the University Press, Cambridge

A catalogue record for this publication is available from the British Library

ISBN 978-0-521-77194-8 hardback

In loving memory of
Lauren Elizabeth Kelley
– who always held the right thought

April 23, 1992 – January 2, 2007

Contents

3   The Many Decoding Algorithms for Reed–Solomon Codes   137

4   Within or Beyond the Packing Radius   190

Figures

Tables

Preface

This book began as notes for a collection of lectures given as a graduate course in the summer semester (April to July) of 1993 at the Swiss Federal Institute of Technology (ETH), Zurich, building on a talk that I gave in Brazil in 1992. Subsequently, in the fall of 1995 and again in the spring of 1998, the course notes were extensively revised and expanded for an advanced topics course in the Department of Electrical and Computer Engineering at the University of Illinois, from which course has evolved the final form of the book that appears here. These lectures were also given in various forms at Eindhoven University, Michigan Technological University, Binghamton University, Washington University, and the Technical University of Vienna. The candid reactions of some who attended these lectures helped me greatly in developing the unique (perhaps idiosyncratic) point of view that has evolved, a view that insists on integrating recent developments in the subject of algebraic codes on curves into the classical engineering framework and terminology of the subject of error-control codes. Many classes of error-control codes and their decoding algorithms can be described in the language of the Fourier transform. This approach merges much of the theory of error-control codes with the subject of signal processing, and makes the central ideas more readily accessible to the engineer.

The theme of the book is algebraic codes developed on the line, on the plane, and on curves. Codes defined on the line, usually in terms of the one-dimensional Fourier transform, are studied in Chapters 2, 3, and 4. These chapters provide a background and framework against which later chapters are developed. The codes themselves are defined in Chapter 2, while the decoding algorithms and the performance of the codes are studied in Chapters 3 and 4. Codes defined on the plane, usually in terms of the two-dimensional Fourier transform, are studied in Chapters 5 and 6. Codes defined on curves, again in terms of the two-dimensional Fourier transform, are studied in Chapters 10, 11, and 12. The exemplar codes under the three headings are the cyclic codes, the bicyclic codes, and the epicyclic codes. In addition, Chapters 7, 8, and 9 deal with some topics of mathematics, primarily computational algebraic geometry, in preparation for the discussion of codes on curves and their decoding algorithms. Readers who want to get quickly to the "algebraic geometry codes without algebraic geometry" may be put off by the digressions in the early chapters. My intention, however, is to assemble as much as I can about the role of the Fourier transform in coding theory.

The book is a companion to *Algebraic Codes for Data Transmission* (Cambridge University Press, 2003), but it is not a sequel to that book. The two books are independent and written for different audiences: that book written for the newcomer to the subject and this book for a reader with a more mathematical background and less need for instant relevance. The material in these books is not quite engineering and not quite mathematics. It belongs to an emerging field midway between these two subjects that is now sometimes called "infomatics."

I have three goals in preparing this book. The first goal is to present certain recent developments in algebraic coding theory seamlessly integrated into the classical engineering presentation of the subject. I especially want to develop the theory of codes on curves in a direct way while using as little of the difficult subject of algebraic geometry as possible. I have avoided most of the deep theory of algebraic geometry and also its arcane terminology and notation, replacing much of this with that favorite tool of the engineer, the Fourier transform. I hope that this makes the material accessible to a larger audience, though this perhaps makes it unattractive to the algebraic geometer. The second goal is to develop the decoding algorithms for these codes with a terminology and pedagogy that is compatible and integrated with the usual engineering approach to the decoding algorithms of Reed–Solomon codes. For the most useful of the algebraic geometry codes – the hermitian codes – the ideas of computational algebraic geometry have been completely restructured by the engineers so as to develop practical computational algorithms for decoding. This formulation will make the ideas accessible to engineers wanting to evaluate these codes against practical applications or desiring to design encoders and decoders, and perhaps will provide fresh insights to mathematicians. My final goal is to extract some of the ideas implicit in the decoding algorithms and to present these ideas distilled into independent mathematical facts in a manner that might be absorbed into the rapidly developing topic of computational commutative algebra. I do believe that the now active interface between the topics of algebraic coding and algebraic geometry forms an open doorway through which ideas can and should pass in both directions.

The book has been strengthened by my conversations many years ago with Doctor Ruud Pellikaan and Professor Tom Høholdt, and the book is probably a consequence of those conversations. Professor Ralf Koetter and Professor Judy L. Walker helped me to understand what little algebraic geometry I know. The manuscript has benefited from the excellent comments and criticism of Professor Ralf Koetter, Professor Tom Høholdt, Professor Nigel Boston, Professor Douglas Leonard, Doctor Ruud Pellikaan, Doctor Thomas Mittelholzer, Professor Ian Blake, Professor Iwan Duursma, Professor Rudiger Urbanke, Doctor William Weeks IV, Doctor Weishi Feng, Doctor

T. V. Selvakumaran, and Doctor Gregory L. Silvus. I attribute the good things in this book to the help I received from these friends and critics; the remaining faults in the book are due to me. The quality of the book has much to do with the composition and editing skills of Mrs. Francie Bridges and Mrs Helen Metzinger. And, as always, Barbara made it possible. Finally, Jeffrey shared the dream.

"The chief cause of problems is solutions."

– ERIC SEVAREID

1 Sequences and the One-Dimensional Fourier Transform

An *alphabet* is a set of symbols. Some alphabets are infinite, such as the set of real numbers or the set of complex numbers. Usually, we will be interested in finite alphabets. A *sequence* is a string of symbols from a given alphabet. A sequence may be of infinite length. An infinite sequence may be *periodic* or *aperiodic*; infinite aperiodic sequences may become periodic after some initial segment. Any infinite sequence that we will consider has a fixed beginning, but is unending. It is possible, however, that an infinite sequence has neither a beginning nor an end.

A *finite sequence* is a string of symbols of finite length from the given alphabet. The *blocklength* of the sequence, denoted n, is the number of symbols in the sequence. Sometimes the blocklength is not explicitly specified, but is known implicitly only by counting the number of symbols in the sequence after that specific sequence is given. In other situations, the blocklength n is explicitly specified, and only sequences of blocklength n are under consideration.

There are a great many aspects to the study of sequences. One may study the structure and repetition of various subpatterns within a given sequence of symbols. Such studies do not need to presuppose any algebraic or arithmetic structure on the alphabet of the sequence. This, however, is not the aspect of the study of sequences that we shall pursue. We are interested mainly in sequences – usually of finite blocklength – over alphabets that have a special arithmetic structure, the structure known as an *algebraic field*. In such a case, a sequence of a fixed finite blocklength will also be called a *vector*.

We can treat sequences over fields by using algebraic methods. We shall study such sequences by using the ideas of the linear recursion, the cyclic convolution, and the Fourier transform. We shall study here only the structure of individual sequences (and only those whose symbol alphabet is an algebraic field – usually a *finite field*), sets of sequences of finite blocklength n (called *codes*), and the componentwise difference of pairs of sequences (now called *codewords*) from a given code.

An important property of an individual vector over a field is its *Hamming weight* (or *weight*), which is defined as the number of components at which the vector is nonzero. An important property of a pair of vectors over a field is the *Hamming distance* (or *distance*) between them, which is defined as the number of coordinates in which the two vectors differ. We shall devote much effort to determining the weights of vectors and the distances between pairs of vectors.

1.1 Fields

Loosely, an algebraic field is any arithmetic system in which one can add, subtract, multiply, or divide such that the usual arithmetic properties of *associativity*, *commutativity*, and *distributivity* are satisfied. The fields familiar to most of us are: the *rational field*, which is denoted Q and consists of all numbers of the form a/b where a and b are integers, b not equal to zero; the *real field*, which is denoted R and consists of all finite or infinite decimals; and the *complex field*, which is denoted C and consists of all numbers of the form $a + ib$ where a and b are real numbers. The rules of addition, subtraction, multiplication, and division are well known in each of these fields.

Some familiar arithmetic systems are not fields. The set of integers $\{\ldots, -3, -2, -1, 0, 1, 2, 3, \ldots\}$, which is denoted Z, is not a field under ordinary addition and multiplication. Likewise, the set of natural numbers $\{0, 1, 2, \ldots\}$, which is denoted N, is not a field.

There are many other examples of fields, some with an infinite number of elements and some with a finite number of elements. Fields with a finite number of elements are called *finite fields* or *Galois fields*. The Galois field with q elements is denoted $GF(q)$, or F_q. The set of nonzero elements of a finite field is denoted $GF(q)^*$. "The" Galois field $GF(q)$ exists only if q equals a prime p or a prime power p^m, with m an integer larger than one. For other values of the integer q, no definition of addition and multiplication will satisfy the formal axioms of a field.

We may define the field F as a set that has two operations defined on pairs of elements of F; these operations are called "addition" and "multiplication," and the following properties must be satisfied.

(1) **Addition axioms.** The field F is closed under addition, and addition is associative and commutative,

$$a + (b + c) = (a + b) + c,$$
$$a + b = b + a.$$

There is a unique element called *zero*, denoted 0, such that $a + 0 = a$, and for every element a there is a unique element called the *negative* of a and denoted $-a$ such that $a + (-a) = 0$. *Subtraction* $a - b$ is defined as $a + (-b)$.

(2) **Multiplication axioms.** The field F is closed under multiplication, and multiplication is associative and commutative

$$a(bc) = (ab)c,$$
$$ab = ba.$$

There is a unique element not equal to zero called *one*, denoted 1, such that $1a = a$, and for every element a except zero, there is a unique element called the *inverse* of a and denoted a^{-1} such that $aa^{-1} = 1$. *Division* $a \div b$ (or a/b) is defined as ab^{-1}.

(3) **Joint axiom.** The *distributive law*

$$(a + b)c = ac + bc$$

holds for all elements a, b, and c in the field F.

The structure of the finite field $GF(q)$ is simple to describe if q is equal to a prime p. Then

$$GF(p) = \{0, 1, 2, \ldots, p - 1\},$$

and addition and multiplication are modulo-p addition and modulo-p multiplication. This is all the specification needed to determine $GF(p)$ completely; all of the field axioms can be verified to hold under this definition. Any other attempt to define a field with p elements may produce a structure that appears to be different, but is actually this same structure defined from a different point of view or with a different notation. Thus for every prime p, the finite field $GF(p)$ is unique but for notation. In this sense, only one field exists with p elements. A similar remark could be made for the field $GF(p^m)$ for any prime p and integer m larger than 1.

We can easily write down addition and multiplication tables for $GF(2)$, $GF(3)$, and $GF(5)$; see Table 1.1.

The field $GF(4)$ can *not* have this modulo-p structure because $2 \times 2 = 0$ modulo 4, and 2 does not have an inverse under multiplication modulo 4. We will construct $GF(4)$ in a different way as an *extension* of $GF(2)$. In general, any field that contains the field F is called an *extension field* of F. In such a discussion, F itself is sometimes called the *ground field*. A field of the form $GF(p^m)$ is formed as an extension of $GF(p)$ by means of a simple polynomial construction akin to the procedure used to construct the complex field from the real field. Eventually, we want to describe the general form of this construction, but first we shall construct the complex field C as an extension of the real field R in the manner of the general construction.

The extension field will consist of pairs of real numbers to which we attach a definition of addition and of multiplication. We will temporarily refer to this extension field using the notation $R^{(2)} = \{(a, b) \mid a \in R, b \in R\}$. The extension field $R^{(2)}$ must not be confused with the vector space R^2. We also remark that there may be more than one way of defining addition and multiplication on $R^{(2)}$. To define the arithmetic for the extension field $R^{(2)}$, we represent the elements of the extension field by polynomials. We will use the symbol z to construct polynomials for such purposes, leaving the symbol

Table 1.1. *Arithmetic tables for some small fields*

GF(2)

+	0	1
0	0	1
1	1	0

·	0	1
0	0	0
1	0	1

GF(3)

+	0	1	2
0	0	1	2
1	1	2	0
2	2	0	1

·	0	1	2
0	0	0	0
1	0	1	2
2	0	2	1

GF(5)

+	0	1	2	3	4
0	0	1	2	3	4
1	1	2	3	4	0
2	2	3	4	0	1
3	3	4	0	1	2
4	4	0	1	2	3

·	0	1	2	3	4
0	0	0	0	0	0
1	0	1	2	3	4
2	0	2	4	1	3
3	0	3	1	4	2
4	0	4	3	2	1

x for other things. Thus redefine the extension field as follows:

$$\boldsymbol{R}^{(2)} = \{a + bz \mid a \in \boldsymbol{R}, b \in \boldsymbol{R}\},$$

where $a + bz$ is a new and useful name for (a, b). Next, find a polynomial of degree 2 over $\boldsymbol{R}$ that cannot be factored over $\boldsymbol{R}$. The polynomial

$$p(z) = z^2 + 1$$

cannot be factored over $\boldsymbol{R}$. Although there are many other polynomials of degree 2 that also cannot be factored over $\boldsymbol{R}$ (e.g., $z^2 + z + 1$), this $p(z)$ is the usual choice because of its extreme simplicity. Define the extension field as the set of polynomials with degrees smaller than the degree of $p(z)$ and with coefficients in $\boldsymbol{R}$. Addition and multiplication in $\boldsymbol{R}^{(2)}$ are defined as addition and multiplication of polynomials modulo[1] the polynomial $p(z)$. Thus

$$(a + bz) + (c + dz) = (a + c) + (b + d)z$$

and

$$(a + bz)(c + dz) = ac + (ad + bc)z + bdz^2 \quad (\bmod z^2 + 1)$$
$$= (ac - bd) + (ad + bc)z.$$

---

[1] The phrase "modulo $p(z)$," abbreviated $(\bmod p(z))$, means to take the remainder resulting from the usual polynomial division operation with $p(z)$ as the divisor.

This is exactly the form of the usual multiplication of complex numbers if the conventional symbol i $= \sqrt{-1}$ is used in place of z because dividing by $z^2 + 1$ and keeping the remainder is equivalent to replacing z^2 by -1. The extension field that we have constructed is actually the complex field C. Moreover, it can be shown that any other construction that forms such an extension field $R^{(2)}$ also gives an alternative representation of the complex field C, but for notation.

Similarly, to extend the field $GF(2)$ to the field $GF(4)$, choose the polynomial

$$p(z) = z^2 + z + 1.$$

This polynomial cannot be factored over $GF(2)$, as can be verified by noting that z and $z + 1$ are the only polynomials of degree 1 over $GF(2)$ and neither is a factor of $z^2 + z + 1$. Then

$$GF(4) = \{a + bz \mid a \in GF(2), b \in GF(2)\}.$$

The field $GF(4)$ has four elements. Addition and multiplication in $GF(4)$ are defined as addition and multiplication of polynomials modulo $p(z)$. Thus

$$(a + bz) + (c + dz) = (a + c) + (b + d)z$$

and

$$(a + bz)(c + dz) = ac + (ad + bc)z + bdz^2 \quad (\bmod\ z^2 + z + 1)$$
$$= (ac + bd) + (ad + bc + bd)z$$

(using the fact that "$-$" and "$+$" are the same operation in $GF(2)$). Denoting the four elements $0, 1, z$, and $z + 1$ of $GF(4)$ by 0, 1, 2, and 3, the addition and multiplication tables of $GF(4)$ now can be written as in Table 1.2.

The notation used here may cause confusion because, for example, with this notation $1 + 1 = 0$ and $2 + 3 = 1$ in this field. It is a commonly used notation, however, in engineering applications.

To extend any field F to a field $F^{(m)}$, first find any polynomial $p(z)$ of degree m over F that cannot be factored in F. Such a polynomial is called an *irreducible polynomial* over F. An irreducible polynomial $p(z)$ of degree m need not exist over the field F (e.g., there is no irreducible cubic polynomial over R). Then $F^{(m)}$ does not exist. For a finite field $GF(q)$, however, an irreducible polynomial of degree m does exist for every positive integer m. If more than one such irreducible polynomial of degree m exists, then there may be more than one such extension field. Over finite fields, all such extension fields formed from irreducible polynomials of degree m are the same, except for notation. They are said to be *isomorphic* copies of the same field.

Table 1.2. *Arithmetic table for GF (4)*

+	0	1	2	3		·	0	1	2	3
	0	0	1	2	3		0	0	0	0	0
GF(4)	1	1	0	3	2		1	0	1	2	3
	2	2	3	0	1		2	0	2	3	1
	3	3	2	1	0		3	0	3	1	2

Write the set of polynomials of degree smaller than m as

$$F^{(m)} = \{a_{m-1}z^{m-1} + a_{m-2}z^{m-2} + \cdots + a_1 z + a_0 \mid a_i \in F\}.$$

The symbol z can be thought of as a kind of place marker that is useful to facilitate the definition of multiplication. Addition in $F^{(m)}$ is defined as addition of polynomials. Multiplication in $F^{(m)}$ is defined as multiplication of polynomials modulo $p(z)$.

The construction makes it evident that if F is $GF(q)$, the finite field with q elements, then the extension field is also a finite field and has q^m elements. Thus it is the field $GF(q^m)$, which is unique up to notation. Every finite field $GF(q)$ can be constructed in this way as $GF(p^\ell)$ for some prime p and some positive integer ℓ. The prime p is called the *characteristic* of $GF(q)$.

For example, to construct $GF(16)$ as an extension of $GF(2)$, choose[2] $p(z) = z^4 + z + 1$. This polynomial is an irreducible polynomial over $GF(2)$, and it has an even more important property as follows. If $p(z)$ is used to construct $GF(16)$, then the polynomial z represents a field element that has order 15 under the multiplication operation. (The *order* of an element γ is the smallest positive integer n such that $\gamma^n = 1$.) Because the order of the polynomial z is equal to the number of nonzero elements of $GF(16)$, every nonzero element of $GF(16)$ must be a power of z.

Any polynomial $p(z)$ over the ground field $GF(q)$ for which the order of z modulo $p(z)$ is equal to $q^m - 1$ is called a *primitive polynomial* over $GF(q)$, and the element z is called a *primitive element* of the extension field $GF(q^m)$. The reason for using a primitive polynomial to construct $GF(q)$ can be seen by writing the fifteen nonzero field elements of $GF(16)$, $\{1, z, z+1, z^2, z^2+1, z^2+z, z^2+z+1, z^3, z^3+1, z^3+z, z^3+z+1, z^3+z^2, z^3+z^2+1, z^3+z^2+z, z^3+z^2+z+1\}$, as powers of the field element z. In this role, a primitive element z *generates* the field because all fifteen nonzero field elements are powers of z. When we wish to emphasize its role as a primitive element, we shall denote z by α. We may regard α as the abstract field element, and z as the polynomial representation of α. In $GF(16)$, the nonzero field elements are expressed as powers of α (or of z) as follows:

$$\alpha^1 = z,$$

$$\alpha^2 = z^2,$$

---

[2] The use of p both for a prime and to designate a polynomial should not cause confusion.

$$\alpha^3 = z^3,$$

$$\alpha^4 = z + 1, \qquad \text{(because } z^4 = z + 1 \ (\text{mod } z^4 + z + 1)),$$

$$\alpha^5 = z^2 + z,$$

$$\alpha^6 = z^3 + z^2,$$

$$\alpha^7 = z^3 + z + 1,$$

$$\alpha^8 = z^2 + 1,$$

$$\alpha^9 = z^3 + z,$$

$$\alpha^{10} = z^2 + z + 1,$$

$$\alpha^{11} = z^3 + z^2 + z,$$

$$\alpha^{12} = z^3 + z^2 + z + 1,$$

$$\alpha^{13} = z^3 + z^2 + 1,$$

$$\alpha^{14} = z^3 + 1,$$

$$\alpha^{15} = 1 = \alpha^0.$$

The field arithmetic of $GF(16)$ works as follows. To add the field elements $z^3 + z^2$ and $z^2 + z + 1$, add them as polynomials with coefficients added modulo 2. (Writing only the coefficients, this can be expressed as $1100 + 0111 = 1011$.) To multiply 1100 by 0111 (here 1100 and 0111 are abbreviations for the field elements denoted previously as $z^3 + z^2$ and $z^2 + z + 1$), write

$$(1100)(0111) = (z^3 + z^2)(z^2 + z + 1) = \alpha^6 \cdot \alpha^{10} = \alpha^{16} = \alpha \cdot \alpha^{15}$$

$$= \alpha \cdot 1 = \alpha = z$$

$$= (0010).$$

To divide 1100 by 0111, write

$$(1100)/(0111) = (z^3 + z^2)/(z^2 + z + 1) = \alpha^6/\alpha^{10} = \alpha^6 \alpha^5$$

$$= \alpha^{11} = z^3 + z^2 + z$$

$$= (1110)$$

(using the fact that $1/\alpha^{10} = \alpha^5$ because $\alpha^5 \cdot \alpha^{10} = 1$).

The field $GF(256)$ is constructed in the same way, now using the irreducible polynomial

$$p(z) = z^8 + z^4 + z^3 + z^2 + 1$$

(which, in fact, is a primitive polynomial) or any other irreducible polynomial over $GF(2)$ of degree 8.

In any field, most of the methods of elementary algebra, including matrix algebra and the theory of vector spaces, are valid. In particular, the Fourier transform of blocklength n is defined in any field F, providing that F contains an element of order n. The finite field $GF(q)$ contains an element of order n for every n that divides $q-1$, because $GF(q)$ always has a primitive element α, which has order $q-1$. Every nonzero element of the field is a power of α, so there is always a power of α that has order n if n divides $q-1$. If n does not divide $q-1$, there is no element of order n.

One reason for using a finite field (rather than the real field) in an engineering problem is to eliminate problems of round-off error and overflow from computations. However, the arithmetic of a finite field is not well matched to everyday computations. This is why finite fields are most frequently found in those engineering applications in which the computations are introduced artificially as a way of manipulating bits for some purpose such as error control or cryptography.

1.2   The Fourier transform

The (*discrete*) *Fourier transform*, when defined in the complex field, is a fundamental tool in the subject of signal processing; its rich set of properties is part of the engineer's workaday intuition. The Fourier transform exists in any field. Since most of the properties of the Fourier transform follow from the abstract properties of a field, but not from the specific structure of a particular field, most of the familiar properties of the Fourier transform hold in any field.

The Fourier transform is defined on the vector space of n-tuples, denoted F^n. A *vector* v in the *vector space F^n* consists of a block of n elements of the field F, written as

$$v = [v_0, v_1, \ldots, v_{n-1}].$$

The vector v is multiplied by the element γ of the field F by multiplying each component of v by γ. Thus

$$\gamma v = [\gamma v_0, \gamma v_1, \ldots, \gamma v_{n-1}].$$

Here the field element γ is called a *scalar*. Two vectors v and u are added by adding components

$$v + u = [v_0 + u_0, v_1 + u_1, \ldots, v_{n-1} + u_{n-1}].$$

Definition 1.2.1 *Let v be a vector of blocklength n over the field F. Let ω be an element of F of order n. The Fourier transform of v is another vector V of blocklength* *n over the field F whose components are given by*

$$V_j = \sum_{i=0}^{n-1} \omega^{ij} v_i \qquad j = 0, \ldots, n-1.$$

The vector V is also called the *spectrum* of v, and the components of V are called *spectral components*. The components of the Fourier transform of a vector will always be indexed by j, whereas the components of the original vector v will be indexed by i. Of course, V is itself a vector so this indexing convention presumes that it is clear which vector is the original vector and which is the spectrum. The Fourier transform relationship is sometimes denoted by $v \leftrightarrow V$.

The Fourier transform can also be understood as the evaluation of a polynomial. The *polynomial representation* of the vector $v = [v_i \mid i = 0, \ldots, n-1]$ is the polynomial

$$v(x) = \sum_{i=0}^{n-1} v_i x^i.$$

The *evaluation* of the polynomial $v(x)$ at β is the field element $v(\beta)$, where

$$v(\beta) = \sum_{i=0}^{n-1} v_i \beta^i.$$

The Fourier transform, then, is the evaluation of the polynomial $v(x)$ on the n powers of ω, an element of order n. Thus component V_j equals $v(\omega^j)$ for $j = 0, \ldots, n-1$. If F is the finite field $GF(q)$ and ω is a primitive element, then the Fourier transform evaluates $v(x)$ at all $q-1$ nonzero elements of the field.

The Fourier transform has a number of useful properties, making it one of the strongest tools in our toolbox. Its many properties are summarized in Section 1, 3. We conclude this section with a lengthy list of examples of the Fourier transform.

(1) Q or R: $\omega = +1$ has order 1, and $\omega = -1$ has order 2. For no other n is there an ω in Q or R of order n. Hence only trivial Fourier transforms exist in Q or R. To obtain a Fourier transform over R of blocklength larger than 2, one must regard R as embedded into C.

There is, however, a *multidimensional Fourier transform* over Q or R with 2^m elements. It uses $\omega = -1$ and a Fourier transform of length 2 on each dimension of a two by two by ... by two m-dimensional array, and it is a nontrivial example of a multidimensional Fourier transform in the fields Q and R. (This transform is more commonly expressed in a form known as the (one-dimensional) *Walsh–Hadamard transform* by viewing any vector of length 2^m over R as an m-dimensional two by two by $\cdots$ by two array.)

(2) C: $\omega = \mathrm{e}^{-\mathrm{i}2\pi/n}$ has order n, where $\mathrm{i} = \sqrt{-1}$. A Fourier transform exists in C for any blocklength n. There are unconventional choices for ω that work also. For example, $\omega = (\mathrm{e}^{-\mathrm{i}2\pi/n})^3$ works if n is not a multiple of 3.

(3) $GF(5)$: $\omega = 2$ has order 4. Therefore

$$V_j = \sum_{i=0}^{3} 2^{ij} v_i \quad j = 0, \ldots, 3$$

is a Fourier transform of blocklength 4 in $GF(5)$.

(4) $GF(31)$: $\omega = 2$ has order 5. Therefore

$$V_j = \sum_{i=0}^{4} 2^{ij} v_i \quad j = 0, \ldots, 4$$

is a Fourier transform of blocklength 5 in $GF(31)$. Also $\omega = 3$ has order 30 in $GF(31)$. Therefore

$$V_j = \sum_{i=0}^{29} 3^{ij} v_i \quad j = 0, \ldots, 29$$

is a Fourier transform of blocklength 30 in $GF(31)$.

(5) $GF(2^{16} + 1)$. Because $2^{16} + 1$ is prime, an element ω of order n exists if n divides $2^{16} + 1 - 1$. Thus elements of order 2^{ℓ} exist for $\ell = 1, \ldots, 16$. Hence for each power of 2 up to 2^{16}, $GF(2^{16} + 1)$ contains a Fourier transform of blocklength n equal to that power of 2.

(6) $GF((2^{17} - 1)^2)$. This field is constructed as an extension of $GF(2^{17} - 1)$, using a polynomial of degree 2 that is irreducible over $GF(2^{17} - 1)$. An element ω of order n exists in the extension field if n divides $(2^{17} - 1)^2 - 1 = 2^{18}(2^{16} - 1)$. In particular, for each power of 2 up to 2^{18}, $GF((2^{17} - 1)^2)$ contains a Fourier transform of blocklength equal to that power of 2.

(7) $GF(16)$. If $GF(16)$ is constructed with the primitive polynomial $p(z) = z^4 + z + 1$, then z has order 15. Thus $\omega = z$ is an element of order 15, so we have the 15-point Fourier transform

$$V_j = \sum_{i=0}^{14} z^{ij} v_i \quad j = 0, \ldots, 14.$$

The components v_i (and V_j), as elements of $GF(16)$, can be represented as polynomials of degree at most 3 over $GF(2)$, with polynomial multiplication reduced by $z^4 = z + 1$.

Alternatively, $\omega = z^3$ is an element of order 5 in $GF(16)$, so we have the five-point Fourier transform

$$
\begin{bmatrix}
V_0(z) \\
V_1(z) \\
V_2(z) \\
V_3(z) \\
V_4(z)
\end{bmatrix}
=
\begin{bmatrix}
1 & 1 & 1 & 1 & 1 \\
1 & z^3 & z^6 & z^9 & z^{12} \\
1 & z^6 & z^{12} & z^{18} & z^{24} \\
1 & z^9 & z^{18} & z^{27} & z^{36} \\
1 & z^{12} & z^{24} & z^{36} & z^{48}
\end{bmatrix}
\begin{bmatrix}
v_0(z) \\
v_1(z) \\
v_2(z) \\
v_3(z) \\
v_4(z)
\end{bmatrix}.
$$

The components of $\boldsymbol{v}$ and of $\boldsymbol{V}$ have been written here in the notation of polynomials to emphasize that elements of the field $GF(16)$ are represented as polynomials. All powers of z larger than the third power are to be reduced by using $z^4 = z + 1$.

(8) $GF(256)$. An element ω of order n exists if n divides 255. If the primitive polynomial $p(z) = z^8 + z^4 + z^3 + z^2 + 1$ is used to construct the field $GF(256)$, then z has order 255. Thus

$$
\begin{bmatrix}
V_0(z) \\
V_1(z) \\
\vdots \\
V_{254}(z)
\end{bmatrix}
=
\begin{bmatrix}
1 & 1 & \cdots \\
1 & & \\
\vdots & z^{ij} & \\
& &
\end{bmatrix}
\begin{bmatrix}
v_0(z) \\
v_1(z) \\
\vdots \\
v_{254}(z)
\end{bmatrix}
$$

is a 255-point Fourier transform over $GF(256)$. Each component consists of eight bits, represented as a polynomial over $GF(2)$, and powers of z are reduced by using $z^8 = z^4 + z^3 + z^2 + 1$.

(9) $\boldsymbol{Q}^{(16)}$. The polynomial $p(z) = z^{16} + 1$ is irreducible over $\boldsymbol{Q}$. Modulo $z^{16} + 1$ multiplication is reduced by setting $z^{16} = -1$. An element of $\boldsymbol{Q}^{(16)}$ may be thought of as a "supercomplex" rational with sixteen parts (instead of two parts). To emphasize this analogy, the symbol z might be replaced by the symbol i. Whereas a complex rational is $a_0 + a_1 \mathrm{i}$, with $a_0, a_1 \in \boldsymbol{Q}$ and here $\mathrm{i}^2 = -1$, the "supercomplex" rational is given by

$$
a_0 + a_1 \mathrm{i} + a_2 \mathrm{i}^2 + a_3 \mathrm{i}^3 + a_4 \mathrm{i}^4 + \cdots + a_{14} \mathrm{i}^{14} + a_{15} \mathrm{i}^{15},
$$

with $a_\ell \in \boldsymbol{Q}$ for $\ell = 1, \ldots, 15$ and here $\mathrm{i}^{16} = -1$.

There is a Fourier transform of blocklength 32 in the field $\boldsymbol{Q}^{(16)}$. This is because $z^{16} = -1 \pmod{z^{16} + 1}$, so the element z has order 32. This Fourier transform takes a vector of length 32 into another vector of length 32. Components of the vector are polynomials of degree 15 over $\boldsymbol{Q}$. The Fourier transform has the form

$$
V_j(z) = \sum_{i=0}^{31} z^{ij} v_i(z) \pmod{z^{16} + 1}.
$$

We can think of this as an operation on a 32 by 16 array of rational numbers to produce another 32 by 16 array of rational numbers. Because multiplication by z can by implemented as an indexing operation, the Fourier transform in $Q^{(16)}$ can be computed with no multiplications in Q.

1.3   Properties of the Fourier transform

The Fourier transform is important because of its many useful properties. Accordingly, we will list a number of properties, many of which will be useful to us. A sketch of the derivation of most of these properties will follow the list.

If $V = [V_j]$ is the Fourier transform of $v = [v_i]$, then the following properties hold.

(1) Linearity:

$$\lambda v + \mu v' \leftrightarrow \lambda V + \mu V'.$$

(2) Inverse:

$$v_i = \frac{1}{n} \sum_{j=0}^{n-1} \omega^{-ij} V_j \quad i = 0, \ldots, n-1,$$

where, in an arbitrary field, n is defined as $1 + 1 + 1 + \cdots + 1$ (n terms).

(3) Modulation:

$$[v_i \omega^{i\ell}] \leftrightarrow [V_{((j+\ell))}],$$

where the use of double parentheses $((\cdot))$ denotes modulo n.

(4) Translation:

$$[v_{((i-\ell))}] \leftrightarrow [V_j \omega^{\ell j}].$$

(5) Convolution property:

$$e_i = \sum_{\ell=0}^{n-1} f_{((i-\ell))} g_\ell \quad \leftrightarrow \quad E_j = F_j G_j \quad \text{(convolution to multiplication)}$$

and

$$e_i = f_i g_i \quad \leftrightarrow \quad E_j = \frac{1}{n} \sum_{\ell=0}^{n-1} F_{((j-\ell))} G_\ell \quad \text{(multiplication to convolution)}.$$

(6) Polynomial zeros: the polynomial $v(x) = \sum_{i=0}^{n-1} v_i x^i$ has a zero at ω^j if and only if $V_j = 0$. The polynomial $V(y) = \sum_{i=0}^{n-1} V_j y^j$ has a zero at ω^{-i} if and only if $v_i = 0$.

(7) Linear complexity: the weight of a vector v is equal to the cyclic complexity of its Fourier transform V. (This is explained in Section 1.5 as the statement $\text{wt} v = \mathcal{L}(V)$.)

(8) Reciprocation: the reciprocal of a vector $[v_i]$ is the vector $[v_{n-i}]$. The Fourier transform of the reciprocal of v is the reciprocal of the Fourier transform V:

$$[v_{((n-i))}] \leftrightarrow [V_{((n-j))}].$$

(9) Cyclic decimation: suppose $n = n'n''$; then

$$[v_{((n''i'))} \mid i' = 0, \ldots, n'-1] \leftrightarrow \left[\frac{1}{n''} \sum_{j''=0}^{n''-1} V_{((j'+n'j''))} \mid j' = 0, \ldots, n'-1 \right],$$

where $\gamma = \omega^{n''}$ is the element of order n' used to form the Fourier transform of blocklength n'. (The folding of the spectrum on the right side is called *aliasing*.)

(10) Poisson summation formula: suppose $n = n'n''$; then

$$\sum_{i'=0}^{n'-1} v_{n''i'} = \frac{1}{n''} \sum_{j''=0}^{n''-1} V_{n'j''}.$$

(11) Cyclic permutation: suppose that the integers b and n are *coprime*, meaning that the greatest common divisor, denoted $\text{GCD}(b,n)$, equals 1; then

$$[v_{((bi))}] \leftrightarrow [V_{((Bj))}],$$

where B is such that $Bb = 1 (\text{mod } n)$.

(12) Decimated cyclic permutation: suppose $\text{GCD}(b,n) = n'' \neq 1$, $n = n'n''$, and $b = b'n''$. Let $\gamma = \omega^{n''}$ be used to form the n'-point Fourier transform of any vector of blocklength n'. Then

$$[v_{((bi'))} \mid i' = 0, \ldots, n'-1] \leftrightarrow [\overline{V}_{B'j'(\text{mod } n')} \mid j' = 0, \ldots, n'-1],$$

where B' is such that $B'b' = 1 \ (\text{mod } n')$ and

$$\overline{V}_{j'} = \frac{1}{n''} \sum_{j''=0}^{n''-1} V_{j'+n'j''} \quad j' = 0, \ldots, n'-1.$$

This completes our list of elementary properties of the Fourier transform.

As an example of property (12), let $n = 8$ and $b = 6$. Then $n' = 4, b' = 3, B' = 3$, and the transform

$$[v_0, v_1, v_2, v_3, v_4, v_5, v_6, v_7] \leftrightarrow [V_0, V_1, V_2, V_3, V_4, V_5, V_6, V_7]$$

under decimation by b becomes

$$[v_0, v_6, v_4, v_2] \leftrightarrow \left[\frac{V_0 + V_4}{2}, \frac{V_3 + V_7}{2}, \frac{V_2 + V_6}{2}, \frac{V_1 + V_5}{2} \right].$$

We shall now outline the derivations of most of the properties that have been stated above.

(1) Linearity:

$$\sum_{i=0}^{n-1} \omega^{ij} (\lambda v_i + \mu v_i') = \lambda \sum_{i=0}^{n-1} \omega^{ij} v_i + \mu \sum_{i=0}^{n-1} \omega^{ij} v_i' = \lambda V_j + \mu V_j'.$$

(2) Inverse:

$$\frac{1}{n} \sum_{j=0}^{n-1} \omega^{-ij} \sum_{\ell=0}^{n-1} \omega^{\ell j} v_\ell = \frac{1}{n} \sum_{\ell=0}^{n-1} v_\ell \sum_{j=0}^{n-1} \omega^{(\ell-i)j}$$

$$= \frac{1}{n} \sum_{\ell=0}^{n-1} v_\ell \begin{cases} n \text{ if } \ell = i \\ \dfrac{1 - \omega^{(\ell-i)n}}{1 - \omega^{(\ell-i)}} = 0 \text{ if } \ell \neq i. \end{cases}$$

$$= v_i$$

(3) Modulation:

$$\sum_{i=0}^{n-1} (v_i \omega^{i\ell}) \omega^{ij} = \sum_{i=0}^{n-1} v_i \omega^{i(j+\ell)} = V_{((j+\ell))}.$$

(4) Translation (dual of modulation):

$$\frac{1}{n} \sum_{j=0}^{n-1} (V_j \omega^{\ell j}) \omega^{-ij} = \frac{1}{n} \sum_{j=0}^{n-1} V_j \omega^{-(i-\ell)j} = v_{((i-\ell))}.$$

(5) Convolution property (cyclic):

$$e_i = \sum_{\ell=0}^{n-1} f_{((i-\ell))} g_\ell = \sum_{\ell=0}^{n-1} f_{((i-\ell))} \frac{1}{n} \sum_{j=0}^{n-1} \omega^{-\ell j} G_j$$

$$= \frac{1}{n} \sum_{j=0}^{n-1} \omega^{-ij} G_j \sum_{\ell=0}^{n-1} \omega^{(i-\ell)j} f_{((i-\ell))} = \frac{1}{n} \sum_{j=0}^{n-1} \omega^{-ij} G_j F_j.$$

(6) Polynomial zeros: follows immediately from the equation

$$v(\omega^j) = \sum_{i=0}^{n-1} v_i (\omega^j) = V_j.$$

(7) Linear complexity property: deferred until after discussion of linear complexity in Section 1.5.

(8) Reciprocation. It follows from $\omega^n = 1$ that

$$\sum_{i=0}^{n-1} v_{((n-i))} \omega^{ij} = \sum_{i=0}^{n-1} v_i \omega^{(n-i)j}$$

$$= \sum_{i=0}^{n-1} v_i \omega^{i(n-j)} = V_{((n-j))}.$$

(9) Cyclic decimation. Write the spectral index j in terms of a vernier index j' and a coarse index j'':

$$j = j' + n'j''; \quad j' = 0, \ldots, n' - 1; \quad j'' = 0, \ldots, n'' - 1.$$

Then

$$v_{n''i'} = \frac{1}{n} \sum_{j'=0}^{n'-1} \sum_{j''=0}^{n''-1} \omega^{-n''i'(j'+n'j'')} V_{j'+n'j''}$$

$$= \frac{1}{n} \sum_{j'=0}^{n'-1} \sum_{j''=0}^{n''-1} \omega^{-n''i'j'} \omega^{-n''n'i'j''} V_{j'+n'j''}.$$

Because $\omega^n = 1$, the second term in ω equals 1. Then

$$v_{n''i'} = \frac{1}{n'} \sum_{j'=0}^{n'-1} \gamma^{-i'j'} \left(\frac{1}{n''} \sum_{j''=0}^{n''-1} V_{j'+n'j''} \right),$$

where $\gamma = \omega^{n''}$ has order n'.

(10) Poisson summation. The left side is the direct computation of the zero component of the Fourier transform of the decimated sequence. The right side is the same zero component given by the right side of the decimation formula in property (9).

(11) Cyclic permutation. By assumption, b and n are coprime, meaning that they have no common integer factor. For coprime integers b and n, elementary number theory states that integers B and N always exist that satisfy

$$Bb + Nn = 1.$$

Then we can write

$$\sum_{i=0}^{n-1} \omega^{ij} v_{((bi))} = \sum_{i=0}^{n-1} \omega^{(Bb+Nn)ij} v_{((bi))} = \sum_{i=0}^{n-1} \omega^{(bi)(Bj)} v_{((bi))}.$$

Let $i' = ((bi))$. Because b and n are coprime, this is a permutation, so the sum is unchanged. Then

$$\sum_{i=0}^{n-1} \omega^{ij} v_{((bi))} = \sum_{i'=0}^{n-1} \omega^{i'Bj} v_{i'} = V_{((Bj))}.$$

(12) Decimated cyclic permutation. This is simply a combination of properties (9) and (10). Because $v_{((bi))} = v_{((b'n''i))} = v_{((b'((n''i))))}$, the shortened cyclic permutation can be obtained in two steps: first decimating by n'', then cyclically permuting with b'.

1.4   Univariate and homogeneous bivariate polynomials

A *monomial* is a term of the form x^i. The *degree* of the monomial x^i is the integer i. A *polynomial* of degree r over the field F is a linear combination of a finite number of distinct monomials of the form $v(x) = \sum_{i=0}^{r} v_i x^i$. The *coefficient* of the *term* $v_i x^i$ is the field element v_i from F. The *index* of the term $v_i x^i$ is the integer i. The *leading term* of any nonzero univariate polynomial is the nonzero term with the largest index. The *leading index* of any nonzero univariate polynomial is the index of the leading term. The *leading monomial* of any nonzero univariate polynomial is the monomial corresponding to the leading term. The *leading coefficient* of any nonzero univariate polynomial is the coefficient of the leading term. If the leading coefficient is the field element one, the polynomial is called a *monic polynomial*. The *degree* of the nonzero polynomial $v(x)$ is the largest degree of any monomial appearing as a term of $v(x)$ with a nonzero coefficient. The degree of the zero polynomial is $-\infty$. The *weight* of a polynomial is the number of its nonzero coefficients. A polynomial $v(x)$ may also be called a *univariate polynomial* $v(x)$ when one wishes to emphasize that there is only a single polynomial indeterminate x. Two polynomials $v(x)$ and $v'(x)$ over the same field can be added by the rule

$$v(x) + v'(x) = \sum_i (v_i + v_i')x^i,$$

and can be multiplied by the rule

$$v(x)v'(x) = \sum_i \sum_j v_j v_{i-j}' x^i.$$

The *division algorithm* for univariate polynomials is the statement that, for any two nonzero univariate polynomials $f(x)$ and $g(x)$, there exist uniquely two polynomials $Q(x)$, called the *quotient polynomial*, and $r(x)$, called the *remainder polynomial*, such that

$$f(x) = Q(x)g(x) + r(x),$$

and $\deg r(x) < \deg g(x)$.

The *reciprocal polynomial* of $v(x)$, a polynomial of degree r, is the polynomial $\tilde{v}(x) = \sum_{i=r}^{0} v_{r-i}x^i$. Sometimes $v(x)$ is regarded as an element of the set of polynomials of degree less than n over the field F if this is the set of polynomials under consideration. Then the reciprocal polynomial may be defined as $\tilde{v}(x) = \sum_{l=n-1}^{n-1-r} v_{n-1-i}x^i$, which is in accord with the definition of a reciprocal vector. Thus the coefficients are written into the reciprocal polynomial in reverse order, starting with either the first nonzero coefficient or with coefficient v_{n-1} even though it may be zero. The context will determine which definition of $\tilde{v}(x)$ should be understood.

A polynomial $v(x)$ of degree r can be converted into a *homogeneous bivariate polynomial*, defined as

$$v(x, y) = \sum_{i=0}^{r} v_i x^i y^{r-i}.$$

The term "homogeneous" means that the sum of the exponents of x and y equals r in every term. The conversion of a univariate polynomial to a homogeneous bivariate polynomial is a technical device that is sometimes useful in formulating the discussion of certain topics in a more convenient way.

The nonzero polynomial $v(x)$ over the field F is *reducible* if $v(x) = a(x)b(x)$ for some polynomials $a(x)$ and $b(x)$, neither of which has degree 0. A polynomial of degree larger than 0 that is not reducible is *irreducible*. (A univariate polynomial that is not reducible in the field F will be reducible when viewed in an appropriate algebraic extension of the field F.) The term $a(x)$, if it exists, is called a *factor* of $v(x)$, and is called an *irreducible factor* if it itself is irreducible. For definiteness, we can require the irreducible factors to be monic polynomials. Any two polynomials with no common polynomial factor are called *coprime polynomials*. Any polynomial can be written as a field element times a product of all its irreducible factors, perhaps repeated. This product, called the *factorization* of $v(x)$ into its irreducible factors, is unique up to the order of the factors. This property is known as the *unique factorization theorem*.

The field element β is called a *zero* of polynomial $v(x)$ if $v(\beta) = 0$. Because β is a field element, all the indicated arithmetic operations are operations in the field F. The division algorithm implies that if β is a zero of $v(x)$, then $x - \beta$ is a factor of $v(x)$. In particular, this means that a polynomial $v(x)$ of degree n can have at most n zeros.

The field F is an *algebraically closed field* if every polynomial $v(x)$ of degree 1 or greater has at least one zero. In an algebraically closed field, only polynomials of degree 1 are irreducible. Every field F is contained in an algebraically closed field. The complex field C is algebraically closed.

A zero β is called a *singular point* of the polynomial $v(x)$ if the formal derivative (defined below) of $v(x)$ is also zero at β. A polynomial $v(x)$ is called a *singular polynomial* if $v(x)$ has at least one singular point. A polynomial that has no singular points is called a *nonsingular polynomial* or a *regular polynomial*. A polynomial in one variable over the field F is singular if and only if it has a zero of *multiplicity* at least 2 in some extension field of F.

The set of polynomials over the field F is closed under addition, subtraction, and multiplication. It is an example of a ring. In general, a *ring* is an algebraic system (satisfying several formal, but evident, axioms) that is closed under addition, subtraction, and multiplication. A ring that has an identity under multiplication is called a *ring with identity*. The identity element, if it exists, is called *one*. A nonzero element of a ring need not have an inverse under multiplication. An element that does have an inverse under multiplication is called a *unit* of the ring. The ring of polynomials over the field F is conventionally denoted $F[x]$. The ring of univariate polynomials modulo $x^n - 1$, denoted $F[x]/\langle x^n - 1\rangle$ or $F^\circ[x]$, is an example of a *quotient ring*. In the quotient ring $F[x]/\langle p(x)\rangle$, which consists of the set of polynomials of degree smaller than the degree of $p(x)$, the result of a polynomial product is found by first computing the polynomial product in $F[x]$, then reducing to a polynomial of degree less than the degree of $p(x)$ by taking the remainder modulo $p(x)$. In $F[x]/\langle x^n - 1\rangle$, this remainder can be computed by repeated applications of $x^n = 1$.

Later, we shall speak frequently of a special kind of subset of the ring $F[x]$, called an "*ideal*." Although at this moment we consider primarily the ring $F[x]$, the definition of an ideal can be stated in any ring R. An *ideal* I in $F[x]$ is a nonempty subset of $F[x]$

that is closed under addition and is closed under multiplication by any polynomial of the parent ring $F[x]$. Thus for I to be an ideal, $f(x) + g(x)$ must be in I if both $f(x)$ and $g(x)$ are in I, and $f(x)p(x)$ must be in I if $p(x)$ is any polynomial in $F[x]$ and $f(x)$ is any polynomial in I. An ideal I of the ring R is a *proper ideal* if I is not equal to R or to $\{0\}$. An ideal I of the ring R is a *principal ideal* if I is the set of all multiples of a single element of R. This element is called a *generator* of the ideal. Every ideal of $F[x]$ is a principal ideal. A ring in which every ideal is a principal ideal is called a *principal ideal ring*.

We need to introduce the notion of a *derivative* of a polynomial. In the real field, the derivative is defined as a limit, which is not an algebraic concept. In an arbitrary field, the notion of a limit does not have a meaning. For this reason, the derivative of a polynomial in an arbitrary field is simply defined as a polynomial with the form expected of a derivative. In a general field, the derivative of a polynomial is called a *formal derivative*. Thus we define the formal derivative of $a(x) = \sum_{i=0}^{n-1} a_i x^i$ as

$$a^{(1)}(x) = \sum_{i=1}^{n-1} i a_i x^{i-1},$$

where $i a_i$ means the sum of i copies of a_i (which implies that $p a_i = 0$ in a field of characteristic p because $p = 0 \pmod{p}$). The rth formal derivative, then, is given by

$$a^{(r)}(x) = \sum_{i=r}^{n-1} \frac{i!}{(i-r)!} a_i x^{i-r}.$$

In a field of characteristic p, all pth and higher derivatives are always equal to zero, and so may not be useful. The *Hasse derivative* is an alternative definition of a derivative in a finite field that need not equal zero for pth and higher derivatives. The rth Hasse derivative of $a(x)$ is defined as

$$a^{[r]}(x) = \sum_{i=r}^{n-1} \binom{i}{r} a_i x^{i-r}.$$

It follows that

$$a^{(r)}(x) = (r!) a^{[r]}(x).$$

In particular, $a^{(1)}(x) = a^{[1]}(x)$. It should also be noted that if $b(x) = a^{[r]}(x)$, then, in general,

$$b^{[k]}(x) \neq a^{[r+k]}(x).$$

Hence this useful and well known property of the formal derivative does not carry over to the Hasse derivative. The following theorem gives a property that does follow over.

Theorem 1.4.1 (Hasse) *If $h(x)$ is an irreducible polynomial of degree at least 1, then $[h(x)]^m$ divides $f(x)$ if and only if $h(x)$ divides $f^{[\ell]}(x)$ for $\ell = 0, \ldots, m-1$.*

Proof: This is given as Problem 1.14. ∎

1.5   Linear complexity of sequences

A *linear recursion* (or *recursion*) over the field F is an expression of the form

$$V_j = -\sum_{k=1}^{L} \Lambda_k V_{j-k} \quad j = L, L+1, \ldots,$$

where the terms V_j and Λ_j are elements of the field F. Given the L *connection coefficients* Λ_j for $j = 1, \ldots, L$, the linear recursion produces the terms V_j for $j = L, L+1, \ldots$ from the terms V_j for $j = 0, \ldots, L-1$. The integer L is called the *length* of the recursion. The L coefficients of the recursion are used conventionally to form a polynomial, $\Lambda(x)$, called the *connection polynomial* and defined as

$$\Lambda(x) = 1 + \sum_{j=1}^{L} \Lambda_j x^j$$

$$= \sum_{j=0}^{L} \Lambda_j x^j,$$

where $\Lambda_0 = 1$. The linear recursion is denoted concisely as $(\Lambda(x), L)$, where $\Lambda(x)$ is a polynomial and L is an integer.

The *linear complexity* of the (finite or infinite) sequence $V = (V_0, V_1, \ldots)$ is the smallest value of L for which such a linear recursion exists for that sequence. This is the shortest linear recursion that will produce,[3] from the first L components of the sequence V, the remaining components of that sequence. The linear complexity of V will be denoted $\mathcal{L}(V)$. If, for a nonzero infinite sequence V, no such recursion exists, then $\mathcal{L}(V) = \infty$. The linear complexity of the all-zero sequence of any length is defined to be zero. For a finite sequence of length r, $\mathcal{L}(V)$ is always defined and is not larger than r. For a periodic sequence of period n, $\mathcal{L}(V)$ is always defined and is not larger than n.

The linear complexity can be restated in the language of shift-register circuits. The linear complexity of V is equal to the length of the shortest linear-feedback shift

---

[3] We avoid the term "generates" here to prevent clashes later with the "generator" of ideals. Thus $\Lambda(x)$ *generates* the ideal $\langle \Lambda(x) \rangle$ and *produces* the sequence $V_0, V_1, \ldots$

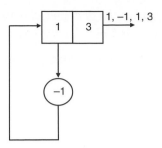

Figure 1.1. Simple linear recursion.

register that will produce all of V when initialized with the beginning of V. The coefficients Λ_k of the recursion are the connection coefficients of the linear-feedback shift register. For example, a shift-register circuit that recursively produces the sequence $(V_0, V_1, V_2, V_3) = (3, 1, -1, 1)$ is shown in Figure 1.1. Because this is the shortest linear-feedback shift register that produces this sequence, the linear complexity of the sequence is two. The linear recursion $(\Lambda(x), L)$ corresponding to this shift-register circuit is $(1 + x, 2)$.

The connection polynomial $\Lambda(x)$ does not completely specify the recursion because (as in the example above) it may be that $\Lambda_L = 0$. This means that we cannot always deduce L from the degree of $\Lambda(x)$. All that we can deduce is the inequality $L \geq \deg \Lambda(x)$. This is why the notation $(\Lambda(x), L)$ mentions both $\Lambda(x)$ and L. Accordingly, one may prefer[4] to work with the reciprocal form of the connection polynomial, denoted $\widetilde{\Lambda}(x)$, and also called the connection polynomial or, better, the *reciprocal connection polynomial*. This monic polynomial is given by

$$\widetilde{\Lambda}(x) = x^L \Lambda(x^{-1})$$

$$= x^L + \sum_{k=1}^{L} \Lambda_{L-k} x^k.$$

Thus $\widetilde{\Lambda}_k = \Lambda_{L-k}$. Now we have, more neatly, the equality $\deg \widetilde{\Lambda}(x) = L$. With this alternative notation, the example of Figure 1.1 is denoted as $(\widetilde{\Lambda}(x), L) = (x^2 + x, 2)$. Possibly, as in the example, $\widetilde{\Lambda}(x)$ is divisible by a power of x because one or more coefficients including $\widetilde{\Lambda}_0$ are zero, but the length L is always equal to the degree of the reciprocal connection polynomial $\widetilde{\Lambda}(x)$.

In the rational field Q, the recursion

$$(\Lambda(x), L) = (-x^2 - x + 1, 2)$$

---

[4] This choice is rather arbitrary here, but in Chapter 7, which studies bivariate recursions, the reciprocal form appears to be unavoidable.

(or $\widetilde{\Lambda}(x) = x^2 - x - 1$) produces the *Fibonacci sequence*

$1, 1, 2, 3, 5, 8, 13, 21, 34, \ldots$

In contrast, the recursion

$(\Lambda(x), L) = (-x^2 - x + 1, 4)$

(or $\widetilde{\Lambda}(x) = x^4 - x^3 - x^2$) produces the modified sequence

$A, B, 2, 3, 5, 8, 13, 21, 34, \ldots$

when initialized with $(A, B, 2, 3)$, where A and B are any two integers. This is true even if A and B both equal 1, but, for that sequence, the recursion is not of minimum length, so then the recursion $(-x^2 - x + 1, 4)$ does not determine the linear complexity.

The linear complexity of the Fibonacci sequence – or any nontrivial segment of it – is 2 because the Fibonacci sequence cannot be produced by a linear recursion of shorter length. If, however, the first n symbols of the Fibonacci sequence are periodically repeated, then the periodic sequence of the form (for $n = 8$)

$1, 1, 2, 3, 5, 8, 13, 21, 1, 1, 2, 3, 5, 8, 13, 21, 1, 1, \ldots$

is obtained. It is immediately obvious that the linear complexity of this periodic sequence is at most 8 because it is produced by the recursion $V_j = V_{j-8}$. It follows from Massey's theorem, which will be given in Section 1.6, that the linear complexity is at least 6. In fact, it is 6.

A linear recursion of length L for a sequence of length n can be written in the form

$$V_j + \sum_{k=1}^{L} \Lambda_k V_{j-k} = 0 \quad j = L, \ldots, n-1$$

(where n may be replaced by infinity). The linear recursion can be expressed concisely as

$$\sum_{k=0}^{L} \Lambda_k V_{j-k} = 0 \quad j = L, \ldots, n-1,$$

where $\Lambda_0 = 1$. The left side is the jth coefficient of the polynomial product $\Lambda(x)V(x)$. Consequently, the jth coefficient of the polynomial product $\Lambda(x)V(x)$ is equal to 0 for $j = L, \ldots, n-1$. To compute a linear recursion of length at most L that produces $V(x)$, one must solve the polynomial equation

$$\Lambda(x)V(x) = p(x) + x^n g(x)$$

for a connection polynomial, $\Lambda(x)$, of degree at most L, such that $\Lambda_0 = 1$, and $p(x)$ and $g(x)$ are any polynomials such that $\deg p(x) < L$. Equivalently, one must find $\Lambda(x)$ and $p(x)$ such that

$$\Lambda(x)V(x) = p(x) \pmod{x^n},$$

where $\Lambda_0 = 1$, $\deg \Lambda(x) \leq L$, and $\deg p(x) < L$.

If the sequence is infinite, then the modulo x^n operation is removed, and the infinite sequence V must be expressed as

$$V(x) = \frac{p(x)}{\Lambda(x)}$$

for $\Lambda(x)$ and $p(x)$ of the stated degrees.

1.6 Massey's theorem for sequences

We will start this section with a useful condition under which two recursions will continue to agree if they agree up to a certain point. The recursion $(\Lambda(x), L)$ produces the finite sequence $V_0, V_1, \ldots, V_{r-1}$ if

$$V_j = -\sum_{k=1}^{L} \Lambda_k V_{j-k} \quad j = L, \ldots, r-1.$$

The recursion $(\Lambda'(x), L')$ produces the same sequence $V_0, V_1, \ldots, V_{r-1}$ if

$$V_j = -\sum_{k=1}^{L'} \Lambda'_k V_{j-k} \quad j = L', \ldots, r-1.$$

Under what condition will the next term, V_r, produced by each of the two recursions be the same?

Theorem 1.6.1 (Agreement theorem) *If $(\Lambda(x), L)$ and $(\Lambda'(x), L')$ both produce the sequence $V_0, V_1, \ldots, V_{r-1}$, and if $r \geq L + L'$, then both produce the sequence $V_0, V_1, \ldots, V_{r-1}, V_r$.*

Proof: We must show that

$$-\sum_{k=1}^{L} \Lambda_k V_{r-k} = -\sum_{k=1}^{L'} \Lambda'_k V_{r-k}.$$

By assumption,

$$V_i = -\sum_{j=1}^{L} \Lambda_j V_{i-j} \quad i = L, \ldots, r-1;$$

$$V_i = -\sum_{j=1}^{L'} \Lambda'_j V_{i-j} \quad i = L', \ldots, r-1.$$

Because $r \geq L + L'$, we can set $i = r - k$ in these two equations, and write

$$V_{r-k} = -\sum_{j=1}^{L} \Lambda_j V_{r-k-j} \quad k = 1, \ldots, L',$$

and

$$V_{r-k} = -\sum_{j=1}^{L'} \Lambda'_j V_{r-k-j} \quad k = 1, \ldots, L,$$

with all terms from the given sequence $V_0, V_1, \ldots, V_{r-1}$. Finally, we have

$$-\sum_{k=1}^{L} \Lambda_k V_{r-k} = \sum_{k=1}^{L} \Lambda_k \sum_{j=1}^{L'} \Lambda'_j V_{r-k-j}$$

$$= \sum_{j=1}^{L'} \Lambda'_j \sum_{k=1}^{L} \Lambda_k V_{r-k-j}$$

$$= -\sum_{j=1}^{L'} \Lambda'_j V_{r-j}.$$

This completes the proof.   ∎

Theorem 1.6.2 (Massey's theorem)   *If $(\Lambda(x), L)$ is a linear recursion that produces the sequence $V_0, V_1, \ldots, V_{r-1}$, but $(\Lambda(x), L)$ does not produce the sequence $V = (V_0, V_1, \ldots, V_{r-1}, V_r)$, then $\mathcal{L}(V) \geq r + 1 - L$.*

Proof: Suppose that the recursion $(\Lambda'(x), L')$ is any linear recursion that produces the longer sequence V. Then $(\Lambda(x), L)$ and $(\Lambda'(x), L')$ both produce the sequence $V_0, V_1, \ldots, V_{r-1}$. If $L' \leq r - L$, then $r \geq L' + L$. By the agreement theorem, both must produce the same value at iteration r, contrary to the assumption of the theorem. Therefore $L' > r - L$.   ∎

If it is further specified that $(\Lambda(x), L)$ is the *minimum-length* linear recursion that produces the sequence $V_0, V_1, \ldots, V_{r-1}$, then Massey's theorem can be strengthened to the statement that $\mathcal{L}(V) \geq \max[L, r + 1 - L]$. Later, we shall show that $\mathcal{L}(V) = \max[L, r + 1 - L]$ by giving an algorithm (the *Berlekamp–Massey algorithm*) that computes such a recursion. Massey's theorem will then allow us to conclude that this algorithm produces a minimum-length recursion.

1.7   Cyclic complexity and locator polynomials

In this section, we shall study first the linear complexity of periodic sequences. For emphasis, the linear complexity of a periodic sequence will also be called the *cyclic complexity*. When we want to highlight the distinction, the linear complexity of a finite, and so nonperiodic, sequence may be called the *acyclic complexity*. The cyclic complexity is the form of the linear complexity that relates most naturally to the Fourier transform and to a polynomial known as the *locator polynomial*, which is the second topic of this chapter.

Thus the cyclic complexity of the vector V, having blocklength n, is defined as the smallest value of L for which a cyclic recursion of the form

$$V_{((j))} = -\sum_{k=1}^{L} \Lambda_k V_{((j-k))} \quad j = L, \ldots, n-1, n, n+1, \ldots, n+L-1,$$

exists, where the double parentheses denote modulo n on the indices. This means that $(\Lambda(x), L)$ will cyclically produce V from its first L components. Equivalently, the linear recursion $(\Lambda(x), L)$ will produce the infinite periodic sequence formed by repeating the n symbols of V in each period. The cyclic complexity of the all-zero sequence is zero.

The distinction between the cyclic complexity and the acyclic complexity is illustrated by the sequence $(V_0, V_1, V_2, V_3) = (3, 1, \ 1, 1)$ of blocklength 4. The linear recursion $(\Lambda(x), L) = (1 + x, 2)$ achieves the acyclic complexity, and the linear recursion $(\Lambda(x), L) = (1 - x + x^2 - x^3, 3)$ achieves the cyclic complexity. These are illustrated in Figure 1.2.

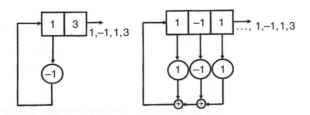

Figure 1.2. Linear-feedback shift registers.

When expressed in the form

$$V_{((j))} = -\sum_{k=1}^{L} \Lambda_k V_{((j-k))} \quad j = 0, \ldots, n-1,$$

it becomes clear that the cyclic recursion can be rewritten as a cyclic convolution,

$$\sum_{k=0}^{L} \Lambda_k V_{((j-k))} = 0 \quad j = 0, \ldots, n-1,$$

where $\Lambda_0 = 1$. The left side of this equation can be interpreted as the set of coefficients of a polynomial product modulo $x^n - 1$. Translated into the language of polynomials, the equation becomes

$$\Lambda(x)V(x) = 0 \quad (\bmod \ x^n - 1),$$

with

$$V(x) = \sum_{j=0}^{n-1} V_j x^j.$$

In the inverse Fourier transform domain, the cyclic convolution becomes $\lambda_i v_i = 0$, where λ_i and v_i are the ith components of the inverse Fourier transform. Thus λ_i must be zero whenever v_i is nonzero. In this way, the connection polynomial $\Lambda(x)$ that achieves the cyclic complexity locates, by its zeros, the nonzeros of the polynomial $V(x)$.

To summarize, the connection polynomial is defined by its role in the linear recursion. If the sequence it produces is periodic, however, then it has another property. Accordingly, we shall now define a polynomial, called a locator polynomial, in terms of this other property. Later, we will find the conditions under which the connection polynomial and the locator polynomial are the same polynomial, so we take the liberty of also calling the locator polynomial $\Lambda(x)$.

A *locator polynomial*, $\Lambda(x)$ or $\Lambda^\circ(x)$, for a finite set of nonzero points of the form β_ℓ or ω^{i_ℓ}, $\ell = 1, \ldots, t$, in the field F, is a polynomial of $F[x]$ or $F[x]/\langle x^n - 1 \rangle$ that has the points of this set among its zeros, where ω is an element of F of order n. The notation $\Lambda^\circ(x)$ is used when it is desired to emphasize that the cyclic complexity is under consideration. Then the polynomial $\Lambda^\circ(x)$ is regarded as an element of $F[x]/\langle x^n - 1 \rangle$. Therefore,

$$\Lambda^\circ(x) = \prod_{\ell=1}^{t} (1 - \omega^{i_\ell x}),$$

where t is the number of points in the set, and the nonzero value ω^{i_ℓ} specifies the ℓth point of the set of points in F. In the context of the locator polynomial, we may refer to the points ω^{i_ℓ} as *locations* in F. With this notation, we may also call i_ℓ (or ω^{i_ℓ}) the index of the ℓth location. If the field F is the finite field $GF(q)$, and $n = q - 1$, then every nonzero element is a power of a primitive element α of the field and, with $\omega = \alpha$,

$$\Lambda^\circ(x) = \prod_{\ell=1}^{t}(1 - \alpha^{i_\ell x}).$$

Because the finite field $GF(q)$ has a primitive element α of order $q - 1$, and $V(x)$ is a polynomial in the ring of polynomials $GF(q)[x]/\langle x^n - 1 \rangle$, we can find the nonzeros of $V(x)$ at nonzero points of $GF(q)$ by computing $V(\alpha^{-i})$ for $i = 0, \ldots, n - 1$. This is the computation of a Fourier transform of blocklength n. The polynomial $V(x)$ has a nonzero at α^{-i} if $V(\alpha^{-i}) \neq 0$. A locator polynomial for the set of nonzeros of $V(x)$ is then a polynomial $\Lambda^\circ(x)$ that satisfies

$$\Lambda^\circ(\alpha^{-i})V(\alpha^{-i}) = 0.$$

This means that a locator polynomial for the nonzeros of $V(x)$ is a polynomial that satisfies

$$\Lambda^\circ(x)V(x) = 0 \quad (\text{mod } x^n - 1).$$

Then, any $\Lambda^\circ(x)$ satisfying this equation "locates" the nonzeros of $V(x)$ by its zeros, which have the form α^{-i}. If V is a vector whose blocklength n is a divisor of $q^m - 1$, then only the nonzeros of $V(x)$ at locations of the form ω^{-i} are of interest, where ω is an element of $GF(q^m)$ of order n. In such a case, the primitive element α can be replaced in the above discussion by an ω of order n. The zeros of $\Lambda(x)$ of the form ω^{-i} locate the indicated nonzeros of $V(x)$.

We have not required that a locator polynomial have the minimal degree, so it need not be unique. The set of all locator polynomials for a given $V(x)$ forms an ideal, called the *locator ideal*. Because $GF(q)[x]/\langle x^n - 1 \rangle$ is a principal ideal ring, meaning that any ideal is generated by a single polynomial of minimum degree, the locator ideal is a principal ideal. All generator polynomials for this ideal have minimum degree and are scalar multiples of any one of them. All elements of the ideal are polynomial multiples of any generator polynomial. It is conventional within the subject of this book to speak of *the* unique locator polynomial by imposing the requirements that it have minimal degree and the constant term Λ_0 is equal to unity. The monic locator polynomial is a more conventional choice of generator within the subject of algebra.

Now we will prove the linear complexity property of the Fourier transform, which was postponed until after the discussion of the cyclic complexity. This property may be stated as follows:

"The weight of a vector $\boldsymbol{v}$ is equal to the cyclic complexity of its Fourier transform $\boldsymbol{V}$."

Let wt v denote the weight of the vector v. Then the linear complexity property can be written as

$$\text{wt } v = \mathcal{L}(V).$$

It is assumed implicitly, of course, that n, the blocklength of v, admits a Fourier transform in the field F (or in an extension of F). Specifically, the field must contain an element of order n, so n must divide $q - 1$, or $q^m - 1$, for some integer m.

The proof of the statement follows. The recursion $(\Lambda(x), L)$ will cyclically produce V if and only if

$$\Lambda(x)V(x) = 0 \quad (\text{mod } x^n - 1).$$

This is the cyclic convolution

$$\Lambda * V = 0.$$

By the convolution theorem, the cyclic convolution transforms into a componentwise product. Then

$$\lambda_i v_i = 0 \quad i = 0, \ldots, n - 1,$$

where λ is the inverse Fourier transform of Λ. Therefore λ_i must be zero everywhere that v_i is not zero. But the polynomial $\Lambda(x)$ cannot have more zeros than its degree, so the degree of $\Lambda(x)$ must be at least as large as the weight of v. In particular, the locator polynomial

$$\Lambda(x) = \prod_{\ell=1}^{t}(1 - x\omega^{-i_\ell})$$

suffices where wt $v = t$ and $(i_1, i_2, i_3, \ldots, i_t)$ are the t values of the index i at which v_i is nonzero and ω is an element of order n, a divisor of $q^m - 1$. Moreover, except for a constant multiplier, this minimum locator polynomial is unique because every locator polynomial must have these same zeros. Clearly, then, any nonzero polynomial multiple of this minimal degree locator polynomial is a locator polynomial, and there are no others. This completes the proof of the linear complexity property.

Later, we shall want to compute the recursion $(\Lambda(x), L)$ that achieves the cyclic complexity of a sequence, whereas the powerful algorithms that are known compute instead the recursion $(\Lambda(x), L)$ that achieves the acyclic complexity. There is a simple condition under which the cyclic complexity and the acyclic complexity are the same. The following theorem gives this condition, usually realized in applications, that allows the algorithm for one problem to be used for the other.

The locator polynomial of $V(x)$ is properly regarded as an element, $\Lambda^\circ(x)$, of the ring $GF(q)[x]/\langle x^n - 1\rangle$. However, we will find it convenient to compute the connection polynomial of $V(x)$ by performing the computations in the ring $GF(q)[x]$.

Given polynomial $V(x)$, a connection polynomial for the sequence of coefficients of $V(x)$ in $GF(q)[x]$ need not be equal to a locator polynomial for $V(x)$ in $GF(q)[x]/\langle x^n - 1\rangle$, and this is why we use different names. However, we shall see that, in cases of interest to us, they are the same polynomial.

Theorem 1.7.1 *The cyclic complexity and the acyclic complexity of a sequence of blocklength n are equal if the cyclic complexity is not larger than $n/2$.*

Proof: This is a simple consequence of the agreement theorem. The acyclic complexity is clearly not larger than the cyclic complexity. Thus, by assumption, the recursions for the two cases are each of length at most $n/2$, and they agree at least until the nth symbol of the sequence. Hence, by the agreement theorem, they continue to agree thereafter. ∎

The linear complexity property can be combined with the cyclic permutation property of the Fourier transform to relate the recursions that produce two periodic sequences that are related by a cyclic permutation. Suppose that the integers b and n are coprime. If the recursion $(\Lambda(x), L)$ produces the periodic sequence $(V_k, k = 0, \dots, n-1)$, where

$$\Lambda(x) = \prod_{\ell=1}^{L}(1 - x\omega^{i_\ell}),$$

then the recursion $(\Lambda_b(x), L)$ produces the periodic sequence $(V_{((bk))}, k = 0, \dots, n-1)$, where

$$\Lambda_b(x) = \prod_{\ell=1}^{L}(1 - x\omega^{hi_\ell}).$$

To prove this, let $V'_k = V_{((bk))}$. In the inverse Fourier transform domain, $v'_i = v_{((b^{-1}i))}$, so $v_i = v'_{((bi))}$. If v_i is nonzero, then $v'_{((bi))}$ is nonzero. Therefore $\Lambda'(x)$ must have its zeros at ω^{-bi_ℓ} for $\ell = 1, \dots, L$.

If b and n are not coprime, a more complicated version of this is true. Then

$$\Lambda'(x) = \prod_{\substack{\text{distinct} \\ \text{terms}}}(1 - x\gamma^{b'i_\ell})$$

is a connection polynomial, not necessarily minimal, for the decimated sequence, where $\gamma = \omega^{n''}$, $\text{GCD}(b, n) = n''$, and $b' = b/n''$.

1.8   Bounds on the weights of vectors

The linear complexity property relates the weight of a vector to the length of the linear recursion that produces the Fourier transform of that vector periodically repeated. By using this property, the Fourier transform of a vector can be constrained to ensure that the vector has a weight at least as large as some desired value d.

The theorems of this section describe how patterns of zeros in the Fourier transform of a vector determine bounds on the weight of that vector. These bounds can also be obtained as consequences of the fundamental theorem of algebra.

Theorem 1.8.1 (BCH bound)   *The only vector of blocklength n of weight $d - 1$ or less that has $d - 1$ (cyclically) consecutive components of its transform equal to zero is the all-zero vector.*

Proof: The linear complexity property says that, because the vector v has weight less than d, its Fourier transform V satisfies the following recursion:

$$V_j = -\sum_{k=1}^{d-1} \Lambda_k V_{((j-k))}.$$

This recursion implies that any $d - 1$ cyclically consecutive components of V equal to zero will be followed by another component of V equal to zero, and so forth. Thus V must be zero everywhere. Therefore v is the all-zero vector.   ∎

Theorem 1.8.2 (BCH bound with cyclic permutation)   *Suppose that b and n are coprime and a is arbitrary. The only vector v of weight $d - 1$ or less, whose Fourier transform satisfies*

$$V_{((a+b\ell))} = 0 \quad \ell = 1, \ldots, d - 1,$$

is the all-zero vector.

Proof: The modulation property of the Fourier transform implies that translation of the spectrum V by a places does not change the weight of v. The cyclic permutation property implies that cyclic permutation of the transform V by $B = b^{-1} \pmod{n}$ places does not change the weight of v. This gives a weight-preserving permutation of v that rearranges the $d - 1$ given zeros of V so that they are consecutive. The BCH bound completes the proof.   ∎

The BCH bound uses the length of the longest string of zero components in the Fourier transform of a vector to bound the weight of that vector. Theorems 1.8.3 and 1.8.4 use other patterns of substrings of zeros. The first of these theorems uses a pattern of evenly spaced substrings of components that are all equal to zero. The second theorem also uses a pattern of evenly spaced substrings of components, most of which are zero, but, in this case, several may be nonzero.

Theorem 1.8.3 (Hartmann–Tzeng bound) *Suppose that b and n are coprime. The only vector* v *of blocklength n of weight* $d - 1$ *or less, whose spectral components satisfy*

$$V_{((a+\ell_1+b\ell_2))} = 0 \qquad \begin{array}{l} \ell_1 = 0, \ldots, d-2-s \\ \ell_2 = 0, \ldots, s, \end{array}$$

is the all-zero vector.

Proof: This bound is a special case of the *Roos bound*, which is given next.   ∎

Notice that the Hartmann–Tzeng bound is based on $s+1$ uniformly spaced substrings of zeros in the spectrum, each substring of length $d-1-s$. The Roos bound, given next, allows the evenly spaced repetition of these $s + 1$ substrings of zeros to be interrupted by some nonzero substrings, as long as there are not too many such nonzero substrings. The Roos bound can be further extended by combining it with the cyclic decimation property.

Theorem 1.8.4 (Roos bound) *Suppose that b and n are coprime. The only vector* v *of blocklength n of weight* $d - 1$ *or less, whose spectral components satisfy*

$$V_{((a+\ell_1+b\ell_2))} = 0 \quad \ell_1 = 0, \ldots, d-2-s,$$

for at least s + 1 values of ℓ_2 *in the range* $0, \ldots, d - 2$, *is the all-zero vector.*

Proof: We only give an outline of a proof. The idea of the proof is to construct a new vector in the transform domain whose cyclic complexity is not smaller and to which the BCH bound can be applied. From $V \leftrightarrow v$, we have the Fourier transform pair

$$\lfloor V_{((j+r))} \rfloor \leftrightarrow \lfloor v_i \omega^{l_i} \rfloor,$$

where $[\cdot]$ denotes a vector with the indicated components. This allows us to write the Fourier transform relationship for a linear combination such as

$$[\beta_0 V_j + \beta_1 V_{((j+r))}] \leftrightarrow [\beta_0 v_i + \beta_1 v_i \omega^{ir}],$$

where β_0 and β_1 are any field elements. The terms v_i and $v_i\omega^{ir}$ on the right are both zero or both nonzero. The linear combination of these two nonzero terms can combine to form a zero, but two zero terms cannot combine to form a nonzero. This means that the weights satisfy

$$\text{wt } v \geq \text{wt } [\beta_0 v_i + \beta_1 v_i \omega^{ir}].$$

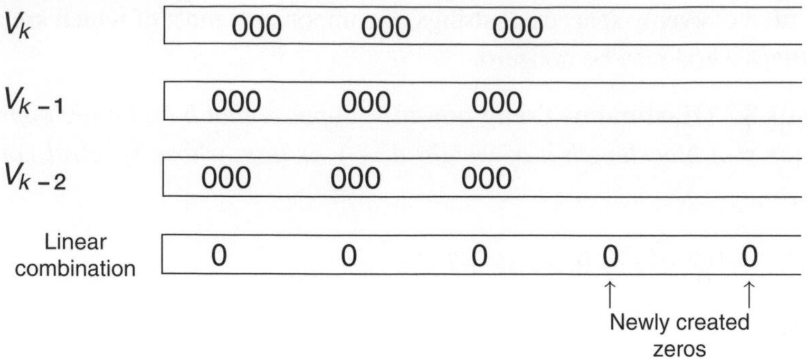

V_k	000	000	000		
V_{k-1}	000	000	000		
V_{k-2}	000	000	000		
Linear combination	0	0	0	0	0

Newly created zeros

Figure 1.3. Construction of new zeros.

The weight on the right side can be bounded by the zero pattern of the vector $[\beta_0 V_j + \beta_1 V_{((j+r))}]$. The bound is made large by the choice of β_0 and β_1 so as to create a favorable pattern of zeros.

In the same way, one can linearly combine multiple translates of V, as suggested in Figure 1.3, to produce multiple new zeros. We then have the following transform pair:

$$\left[\sum_\ell \beta_\ell V_{((j+r_\ell))}\right] \leftrightarrow \left[\sum_\ell \beta_\ell v_i \omega^{ir_\ell}\right].$$

The coefficients of the linear combination are chosen to create new zeros such that the $d-1$ zeros form a regular pattern of zeros spaced by b, as described in Theorem 1.8.2. The new sequence with components $V_{a+\ell b}$ for $\ell = 0, \ldots, d-2$ is zero in at least $s+1$ components, and so is nonzero in at most $d - s - 2$ components. The same is true for the sequence $V_{a+\ell b+\ell_2}$ for each of $d - s - 1$ values of ℓ_2, so all the missing zeros can be created. Theorem 1.8.2 then completes the proof of Theorem 1.8.4 provided the new vector is not identically zero. But it is easy to see that the new vector cannot be identically zero unless the original vector has a string of $d-1$ consecutive zeros in its spectrum, in which case the BCH bound applies. ∎

The final bound of this section subsumes all the other bounds, but the proof is less transparent and the bound is not as easy to use. It uses the notion of a *triangular matrix*, which is a matrix that has only zero elements on one side of its diagonal.

Theorem 1.8.5 (van Lint–Wilson bound) *Given* $V \in GF(q)^m$, *define the components of an n by n matrix* M *by*

$$M_{\ell j} = V_{((j-\ell))} \quad j = 0, \ldots, n-1; \; \ell = 0, \ldots, n-1,$$

and let $\widetilde{M}$ *be any matrix obtained from* M *by row permutations and column permutations. Then* v, *the inverse Fourier transform of* V, *satisfies*

$$\mathrm{wt}\ v \geq \mathrm{rank}\ T,$$

where T *is any submatrix of* $\widetilde{M}$ *that is triangular.*

Proof: The matrix M can be decomposed as

$$M = \Omega \bar{v} \Omega^T,$$

where $\bar{v}$ is an n by n diagonal matrix, whose diagonal elements are equal to the components of the vector v, and Ω is the matrix describing the Fourier transform. The matrix Ω has the elements $\Omega_{ij} = \omega^{ij}$.

Because Ω has full rank,

$$\mathrm{rank}\ M = \mathrm{rank}\ \bar{v}\ =\ \mathrm{wt}\ v.$$

Moreover,

$$\mathrm{rank}\ M = \mathrm{rank}\ \widetilde{M}\ \geq\ \mathrm{rank}\ T,$$

from which the inequality of the theorem follows. ∎

For an application of Theorem 1.8.5, consider a vector of blocklength 7 whose spectrum is given by $(V_0, 0, 0, V_3, 0, V_5, V_6)$. Then

$$M = \begin{bmatrix} V_0 & 0 & 0 & V_3 & 0 & V_5 & V_6 \\ V_6 & V_0 & 0 & 0 & V_3 & 0 & V_5 \\ V_5 & V_6 & V_0 & 0 & 0 & V_3 & 0 \\ 0 & V_5 & V_6 & V_0 & 0 & 0 & V_3 \\ V_3 & 0 & V_5 & V_6 & V_0 & 0 & 0 \\ 0 & V_3 & 0 & V_5 & V_6 & V_0 & 0 \\ 0 & 0 & V_3 & 0 & V_5 & V_6 & V_0 \end{bmatrix}.$$

The bottom left contains the three by three triangular submatrix

$$T = \begin{bmatrix} V_3 & 0 & V_5 \\ 0 & V_3 & 0 \\ 0 & 0 & V_3 \end{bmatrix}.$$

Clearly, this matrix has rank 3 whenever V_3 is nonzero, so, in this case, the weight of the vector is at least 3. (If V_3 is zero, other arguments show that the weight is greater than 3.)

1.9 Subfields, conjugates, and idempotents

The field F has a Fourier transform of blocklength n if F contains an element ω of order n. If F contains no element of order n, then no Fourier transform exists of blocklength n over F. If the extension field E contains an element ω of order n, then there is a Fourier transform of blocklength n in E, which has the same form as before:

$$V_j = \sum_{i=0}^{n-1} \omega^{ij} v_i \quad j = 0, \ldots, n-1.$$

Now, however, the vector V has components in the extension field E even if v has components only in the field F. We wish to describe the nature of any vector V in the vector space E^n that is the Fourier transform of a vector v in the vector space F^n.

Theorem 1.9.1 *The vector V over the complex field C is the Fourier transform of a vector v over the real field R if and only if, for all j,*

$$V_j^* = V_{n-j}.$$

The vector V over the finite field $GF(q^m)$ is the Fourier transform of a vector v over $GF(q)$ if and only if, for all j,

$$V_j^q = V_{((qj))}.$$

Proof: The first statement is well known and straightforward to prove. The second statement is proved by evaluating the following expression:

$$V_j^q = \left(\sum_{i=0}^{n-1} \omega^{ij} v_i \right)^q.$$

In any field of characteristic p, $\binom{p^s}{\ell} = p^s!/((p^s - \ell)!\ell!) = 0 \pmod{p}$ for $0 < \ell < p^s$. This implies that in $GF(q^m)$, $(a+b)^q = a^q + b^q$ if q is a power of p, because all other terms are of the form $\binom{q}{\ell} a^{q-\ell} b^\ell$, and so are equal to zero modulo p because $\binom{q}{\ell}$ is a multiple of p. From this we can write

$$V_j^q = \sum_{i=0}^{n-1} \omega^{qij} v_i^q.$$

Then we use the fact that $a^q = a$ for all a in $GF(q)$ to write

$$V_j^q = \sum_{i=0}^{n-1} \omega^{iqj} v_i = V_{((qj))}.$$

This completes the proof.   ∎

The *conjugacy constraint*, given by

$$V_j^q = V_{((qj))},$$

leads us to a special relationship between an extension field $GF(q^m)$ and a subfield $GF(q)$; this is the relationship of conjugacy. In the finite field $GF(q^m)$, the q^ith powers of an element β, for $i = 1, \ldots, r - 1$, are called the q-ary *conjugates* of β (with r the smallest positive integer for which $\beta^{q^r} = \beta$). The set

$$\{\beta, \beta^q, \beta^{q^2}, \ldots, \beta^{q^{r-1}}\}$$

is called the set of q-ary conjugates of β (or the *Galois orbit* of β). If γ is a conjugate of β, then β is a conjugate of γ. In general, an element has more than one q-ary conjugate. If an element of $GF(q^m)$ has r q-ary conjugates (including itself), it is an element of the subfield $GF(q^r) \subset GF(q^m)$, so r divides m. Thus, under conjugacy, the field decomposes into disjoint subsets called *conjugacy classes*. The term might also be used to refer to the set of exponents on a primitive element of the members of a set of q-ary conjugates.

In the binary field $GF(2^m)$, all binary powers of an element β are called the *binary conjugates* of β. The binary conjugacy classes in the field $GF(16)$, for example, are $\{\alpha^0\}$, $\{\alpha^1, \alpha^2, \alpha^4, \alpha^8\}$, $\{\alpha^3, \alpha^6, \alpha^{12}, \alpha^9\}$, $\{\alpha^5, \alpha^{10}\}$, and $\{\alpha^7, \alpha^{14}, \alpha^{13}, \alpha^{11}\}$. The conjugacy classes might also be identified with the exponents of α as $\{0\}$, $\{1, 2, 4, 8\}$, $\{3, 6, 12, 9\}$, $\{5, 10\}$, and $\{7, 14, 13, 11\}$. These sets can be represented by the four-bit binary representation of the leading term as 0000, 0001, 0011, 0101, and 0111. The cyclic shifts of each of these four-bit numbers then give the binary representation of the other elements of that conjugacy class.

The q-ary conjugacy classes of size 1 form the subfield $GF(q)$ within $GF(q^m)$. To recognize the elements of the subfield, note that every element of $GF(q)$ satisfies $\beta^q = \beta$, and $x^q - x$ can have only q zeros in $GF(q^m)$, so these are the elements in the q-ary conjugacy classes of size 1. For example, the four elements of $GF(64)$ that satisfy $\beta^4 = \beta$ are the four elements of the subfield $GF(4)$.

The sum of all elements of a q-ary conjugacy class of $GF(q^m)$,

$$\beta + \beta^q + \beta^{q^2} + \cdots + \beta^{q^{r-1}},$$

is called the *trace*, or the *q-ary trace*, of β and is denoted $\mathrm{tr}(\beta)$. The *q*-ary trace is an element of $GF(q)$ because

$$(\mathrm{tr}(\beta))^q = (\beta + \beta^q + \beta^{q^2} + \cdots + \beta^{q^{r-1}})^q$$

$$= \beta^q + \beta^{q^2} + \cdots + \beta^{q^{r-1}} + \beta$$

$$= \mathrm{tr}(\beta).$$

In the binary field $GF(2^m)$, the sum of all binary conjugates of β is called the *binary trace* of β. Elements in the same conjugacy class have the same binary trace. In the field $GF(16)$, the binary traces of elements in the conjugacy classes of $\alpha^0, \alpha^1, \alpha^3, \alpha^5$, and α^7 are 1, 0, 1, 1, and 1, respectively.

A *binary idempotent polynomial* (or idempotent) is a polynomial $w(x)$ over $GF(2)$ whose transform has components W_j that only take values 0 and 1. Because $W_j^2 = W_j$, the convolution theorem asserts that an idempotent polynomial satisfies $w(x)^2 = w(x)(\mathrm{mod}\ x^n - 1)$. The conjugacy constraint $W_j^2 = W_{((2j))}$ implies that if $w(x)$ is an idempotent polynomial, then W_j takes the same value, either 0 or 1, on every j for which α^j is in the same conjugacy class.

For example, the binary conjugacy classes of $GF(8)$ are $\{\alpha^0\}$, $\{\alpha^1, \alpha^2, \alpha^4\}$, and $\{\alpha^3, \alpha^6, \alpha^5\}$. Because there are three conjugacy classes and 2^3 ways of taking unions of these, there are 2^3 idempotent polynomials. Of these, two are trivial. The spectra of the nontrivial idempotent polynomials are $\boldsymbol{W} = (1, 0, 0, 0, 0, 0, 0)$, $(0, 1, 1, 0, 1, 0, 0)$, $(0, 0, 0, 1, 0, 1, 1)$, and all pairwise componentwise sums of these three spectra. There are six such nontrivial spectra. These correspond to idempotent polynomials $w(x) = x^6 + x^5 + x^4 + x^3 + x^2 + x + 1$, $x^4 + x^2 + x$, $x^6 + x^5 + x^3$, and all pairwise sums of these polynomials. Each idempotent polynomial satisfies the equation

$$w(x)^2 = w(x) \quad (\mathrm{mod}\ x^7 - 1).$$

There are exactly six nontrivial solutions to this equation, and we have found all of them.

A sequence $V_j, j = 0, \ldots, n-1$, in the field $GF(q^m)$ that arises by evaluating a polynomial $v(x)$ with coefficients in the field $GF(q)$ must obey the conjugacy constraints. What can one say about the connection polynomial of such a sequence? The minimum linear recursion of a sequence always respects conjugacy relationships when they exist. Seemingly mysterious coincidences occur, which are described by Theorems 1.9.2 and 1.9.3.

Theorem 1.9.2 *If for any sequence $V_0, V_1, \ldots, V_{n-1}$ over a field of characteristic 2, satisfying $V_j^2 = V_{((2j))}$, and for any linear recursion $(\Lambda(x), L)$,*

$$V_j = -\sum_{i=1}^{L} \Lambda_i V_{j-i} \quad j = L, \ldots, 2r - 1,$$

then

$$V_{2r} = -\sum_{i=1}^{L} \Lambda_i V_{2r-i}.$$

Proof: By assumption, $V_{2r} = V_r^2$. The proof consists of giving two expressions for the same term. First, using $1 + 1 = 0$ in a field of characteristic 2, we have that

$$V_r^2 = \left(\sum_{i=1}^{L} \Lambda_i V_{r-i} \right)^2 = \sum_{i=1}^{L} \Lambda_i^2 V_{r-i}^2 = \sum_{i=1}^{L} \Lambda_i^2 V_{2r-2i}.$$

Second, we have that

$$V_{2r} = -\sum_{k=1}^{L} \Lambda_k V_{2r-k} = \sum_{k=1}^{L} \sum_{i=1}^{L} \Lambda_k \Lambda_i V_{2r-k-i}.$$

By symmetry, every term with $i \neq k$ appears twice, and, in fields of characteristic 2, these two terms add to 0. Hence only the diagonal terms (with $i = k$) contribute. Thus

$$V_{2r} = -\sum_{k=1}^{L} \Lambda_k V_{2r-k} = \sum_{i=1}^{L} \Lambda_i^2 V_{2r-2i}.$$

Because this agrees with the earlier expression for V_r^2 and $V_r^2 = V_{2r}$, the theorem is proved. ∎

One consequence of the theorem is that if the sequence $V_0, V_1, \ldots, V_{n-1}$ is the Fourier transform of a binary-valued vector, then to test whether $(\Lambda(x), L)$ produces the sequence, only values produced by the recursion for odd values of j need to be verified. For even values of j, the theorem tells us that the recursion is automatically satisfied if it is satisfied for all prior values of j.

Now that we have seen how to prove this theorem for finite fields of characteristic 2, we can understand more readily the proof of the theorem generalized to a finite field of arbitrary characteristic.

Theorem 1.9.3 *For any sequence satisfying $V_j^q = V_{((qj))}$ in the field $GF(q^m)$ of characteristic p, and if the linear recursion*

$$V_j = -\sum_{i=1}^{L} \Lambda_i V_{j-i}$$

holds for $j = L, \ldots, qr - 1$, then it also holds for $j = qr$.

Proof: We shall give two expressions for the same term. By assumption, $V_j^q = V_{((qj))}$. The first expression is given by

$$V_r^q = \left(-\sum_{i=1}^{L} \Lambda_i V_{r-i}\right)^q = -\sum_{i=1}^{L} \Lambda_i^q V_{r-i}^q = -\sum_{i=1}^{L} \Lambda_i^q V_{q(r-i)}.$$

To derive the second expression, embed the linear recursion into itself to obtain

$$V_{qr} = -\sum_{k=1}^{L} \Lambda_k V_{qr-k} = -\sum_{k_1=1}^{L} \Lambda_{k_1}\left[-\sum_{k_2=1}^{L} \Lambda_{k_2} V_{qr-k_1-k_2}\right]$$

$$= (-1)^q \sum_{k_1=1}^{L} \sum_{k_2=1}^{L} \cdots \sum_{k_q=1}^{L} \Lambda_{k_1}\Lambda_{k_2}\cdots\Lambda_{k_q} V_{qr-k_1-k_2-\cdots-k_q}.$$

The final step of the proof is to collapse the sum on the right, because, unless $k_1 = k_2 = k_3 = \cdots = k_q$, each term will recur in multiples of the field characteristic p, and each group of p identical terms adds to zero modulo p. To continue, regard the multiple index $(k_1, k_2, k_3, \ldots, k_q)$ as a q-ary n-tuple. The sum is over all such n-tuples. Two distinct n-tuples that are related by a permutation give the same contribution to the sum. The right side is invariant under permutations of the indices $(k_1, k_2, \ldots, k_q)$. In particular, the right side of the equation is invariant under cyclic shifts. Given any set of indices, consider the set of all of its cyclic shifts, denoted $\{(k_1, k_2, \ldots, k_m)\}$. The number of elements in this set must divide q and so is a power of p, possibly the zero power. If two or more terms are related by a permutation, then there are p such equal terms, and they add to zero modulo p. Therefore the expression collapses to

$$V_{qr} = -\sum_{k=1}^{L} \Lambda_k V_{qr-k} = -\sum_{k=1}^{L} \Lambda_k^q V_{q(r-k)}.$$

Consequently, because two terms equal to the same thing are equal to each other, we have that

$$V_{qr} = V_r^q = -\sum_{k=1}^{L} \Lambda_k V_{qr-k},$$

as required. ∎

1.10   Semifast algorithms based on conjugacy

A *semifast algorithm* for a computation is an algorithm that significantly reduces the number of multiplications compared with the natural form of the computation, but does not reduce the number of additions. A *semifast Fourier transform* in $GF(q)$ is a computational procedure for computing the n-point Fourier transform in $GF(q)$ that uses about $n \log n$ multiplications in $GF(q)$ and about n^2 additions in $GF(q)$. We shall describe a semifast Fourier transform algorithm. It partitions the computation of the Fourier transform into pieces by using the conjugacy classes of $GF(q)$. In contrast to the fast Fourier transform algorithm to be discussed in Section 5.6, the semifast Fourier transform algorithm for the Fourier transform exists even when the blocklength n is a prime.

The vector $\boldsymbol{v}$ of blocklength $n = q - 1$ over the finite field $GF(q)$ has component v_i with index i, which we also associate with element α^i of $GF(q)$. If $q = p^m$, the components of $\boldsymbol{v}$ can be partitioned into sets corresponding to the conjugacy classes of $GF(q)$ over the underlying prime field $GF(p)$. Over the field $GF(2)$, for example, each conjugacy class of $GF(2^m)$ contains at most m elements. For each ℓ, the number m_ℓ of elements in the ℓth conjugacy class divides m, and for most ℓ, m_ℓ equals m. The algorithm that we will describe for the Fourier transform uses not more than m_ℓ^2 multiplications for the ℓth conjugacy class. There are approximately n/m conjugacy classes, each taking at most m^2 multiplications, so the total number of multiplications can be approximated by

$$\sum_\ell m_\ell^2 \approx \frac{n}{m} m^2 = n \log_p q,$$

which is about $n \log_p n$ multiplications.

To formulate the algorithm, we will choose one representative b from each conjugacy class and decompose $\boldsymbol{v}$ as the sum of vectors

$$\boldsymbol{v} = \sum_b \boldsymbol{v}^{(b)},$$

where the vector $\boldsymbol{v}^{(b)}$ has nonzero component $v_i^{(b)}$ only if i is an element of the conjugacy class of b, which is $\mathcal{A}_b = \{b, pb, p^2 b, \ldots, p^{r-1} b\}$, where r is the number of elements in the conjugacy class of b. Thus,

$$v_i^{(b)} = \begin{cases} v_i & i \in \mathcal{A}_b \\ 0 & i \notin \mathcal{A}_b. \end{cases}$$

Then

$$V = \sum_b V^{(b)},$$

where $V^{(b)}$ is the Fourier transform of $v^{(b)}$. Thus it is only necessary to compute the Fourier transform for each conjugacy class. For each representative b, compute

$$V_j^{(b)} = \sum_{i=0}^{n-1} \omega^{ij} v_i^{(b)} \quad j = 0, \ldots, n-1,$$

and then add these vectors together.

Proposition 1.10.1 *For fixed b, let r be the cardinality of conjugacy class $\mathcal{A}_b$. The set of vectors*

$$V_j^{(b)} = \sum_{i=0}^{n-1} \omega^{ij} v_i^{(b)} \quad j = 0, \ldots, n-1,$$

forms a linear subspace of $GF(p^m)^n$ of dimension r spanned by $V_0^{(b)}, V_1^{(b)}, \ldots, V_{r-1}^{(b)}$, the first r components of $V^{(b)}$.

Proof: The set of vectors $[V^{(b)}]$ is a vector space because it is the image of the vector space $[v^{(b)}]$ under a linear map. By restricting the sum in the Fourier transform to only those i where the summands are nonzero, which are those i in the bth conjugacy class $\mathcal{A}_b$, we can write

$$V_j^{(b)} = \sum_{i \in \mathcal{A}_b} \omega^{ij} v_i = \sum_{\ell=0}^{r-1} \omega^{p^\ell bj} v_{p^\ell b} = \sum_{\ell=0}^{r-1} (\omega^{bj})^{p^\ell} v_{p^\ell b}.$$

Recall, however, that x^{p^ℓ} is a linear function over the field $GF(p^m)$. Thus

$$V_j^{(b)} + V_{j'}^{(b)} = \sum_{\ell=0}^{r-1} \left[(\omega^{bj})^{p^\ell} + (\omega^{bj'})^{p^\ell} \right] v_{p^\ell b}$$

$$= \sum_{\ell=0}^{r-1} \left[\omega^{bj} + \omega^{bj'} \right]^{p^\ell} v_{p^\ell b}$$

$$= V_k^{(b)},$$

where k is defined by $\omega^{bj} + \omega^{bj'} = \omega^{bk}$. This relationship provides the necessary structure to compute $V_j^{(b)}$ for values of j from r to $n-1$ from those $V_j^{(b)}$ for values of j from zero to $r-1$. ∎

For example, consider a Fourier transform of blocklength 15 over $GF(16)$. The binary conjugacy classes modulo 15 are $\{\alpha^0\}$, $\{\alpha^1, \alpha^2, \alpha^4, \alpha^8\}$, $\{\alpha^3, \alpha^6, \alpha^{12}, \alpha^9\}$, $\{\alpha^5, \alpha^{10}\}$, and $\{\alpha^7, \alpha^{14}, \alpha^{13}, \alpha^{11}\}$. The vector $\boldsymbol{v}$ is decomposed by conjugacy classes as

$$\boldsymbol{v} = \boldsymbol{v}^{(0)} + \boldsymbol{v}^{(1)} + \boldsymbol{v}^{(3)} + \boldsymbol{v}^{(5)} + \boldsymbol{v}^{(7)},$$

and then the Fourier transform can be written as a sum of Fourier transforms as follows:

$$\boldsymbol{V} = \boldsymbol{V}^{(0)} + \boldsymbol{V}^{(1)} + \boldsymbol{V}^{(3)} + \boldsymbol{V}^{(5)} + \boldsymbol{V}^{(7)}.$$

In the conjugacy class of α, $V_0^{(1)}$, $V_1^{(1)}$, $V_2^{(1)}$, and $V_3^{(1)}$ determine all other $V_j^{(1)}$. Because $\alpha^4 = \alpha + 1$, we find, for example, that $V_4^{(1)} = V_1^{(1)} + V_0^{(1)}$, that $V_5^{(1)} = V_2^{(1)} + V_1^{(1)}$, and so on. Continuing in this way to express all other components, we obtain

$$\boldsymbol{V}^{(1)} = \begin{bmatrix} V_0^{(1)} & V_1^{(1)} & V_2^{(1)} & V_3^{(1)} \end{bmatrix} \begin{bmatrix} 1 & 0 & 0 & 0 & 1 & 0 & 0 & 1 & 1 & 0 & 1 & 0 & 1 & 1 & 1 \\ 0 & 1 & 0 & 0 & 1 & 1 & 0 & 1 & 0 & 1 & 1 & 1 & 1 & 0 & 0 \\ 0 & 0 & 1 & 0 & 0 & 1 & 1 & 0 & 1 & 0 & 1 & 1 & 1 & 1 & 0 \\ 0 & 0 & 0 & 1 & 0 & 0 & 1 & 1 & 0 & 1 & 0 & 1 & 1 & 1 & 1 \end{bmatrix},$$

where

$$\begin{bmatrix} V_0^{(1)} \\ V_1^{(1)} \\ V_2^{(1)} \\ V_3^{(1)} \end{bmatrix} = \begin{bmatrix} 1 & 1 & 1 & 1 \\ \omega & \omega^2 & \omega^4 & \omega^8 \\ \omega^2 & \omega^4 & \omega^8 & \omega \\ \omega^3 & \omega^6 & \omega^{12} & \omega^9 \end{bmatrix} \begin{bmatrix} v_1 \\ v_2 \\ v_4 \\ v_8 \end{bmatrix}.$$

This can be computed with sixteen multiplications. (If the multiplications by one are skipped, it can be computed with only twelve multiplications, but we regard this refinement as a distraction from the main point.)

Similarly, in the conjugacy class of α^3, $V_0^{(3)}$, $V_1^{(3)}$, $V_2^{(3)}$, and $V_3^{(3)}$ determine all others. Because α^3 is a zero of the polynomial $x^4 + x^3 + x^2 + x + 1$, we can write $(\alpha^3)^4 = (\alpha^3)^3 + (\alpha^3)^2 + (\alpha^3)^1 + (\alpha^3)^0$. Then we find, for example, that $V_4^{(3)} = V_3^{(3)} + V_2^{(3)} + V_1^{(3)} + V_0^{(3)}$. Continuing, we obtain

$$\boldsymbol{V}^{(3)} = \begin{bmatrix} V_0^{(3)} & V_1^{(3)} & V_2^{(3)} & V_3^{(3)} \end{bmatrix} \begin{bmatrix} 1 & 0 & 0 & 0 & 1 & 1 & 0 & 0 & 0 & 1 & 1 & 0 & 0 & 0 & 1 \\ 0 & 1 & 0 & 0 & 1 & 0 & 1 & 0 & 0 & 1 & 0 & 1 & 0 & 0 & 1 \\ 0 & 0 & 1 & 0 & 1 & 0 & 0 & 1 & 0 & 1 & 0 & 0 & 1 & 0 & 1 \\ 0 & 0 & 0 & 1 & 1 & 0 & 0 & 0 & 1 & 1 & 0 & 0 & 0 & 1 & 1 \end{bmatrix},$$

where

$$
\begin{bmatrix} V_0^{(3)} \\ V_1^{(3)} \\ V_2^{(3)} \\ V_3^{(3)} \end{bmatrix} = \begin{bmatrix} 1 & 1 & 1 & 1 \\ \omega^3 & \omega^6 & \omega^{12} & \omega^9 \\ \omega^6 & \omega^{12} & \omega^9 & \omega^3 \\ \omega^9 & \omega^3 & \omega^6 & \omega^{12} \end{bmatrix} \begin{bmatrix} v_3 \\ v_6 \\ v_{12} \\ v_9 \end{bmatrix}.
$$

Expressions of the same kind can be written for $V^{(0)}$, $V^{(5)}$, and $V^{(7)}$. The expression for $V^{(7)}$ also involves a four vector. The expression for $V^{(0)}$, corresponding to the conjugacy class of zero, is trivial because every component of $V^{(0)}$ is equal to v_0. The expression for $V^{(5)}$ involves a two vector,

$$
\begin{bmatrix} V_0^{(5)} \\ V_1^{(5)} \end{bmatrix} = \begin{bmatrix} 1 & 1 \\ \omega^5 & \omega^{10} \end{bmatrix} \begin{bmatrix} v_5 \\ v_{10} \end{bmatrix},
$$

and

$$
V^{(5)} = \begin{bmatrix} V_0^{(5)} & V_1^{(5)} \end{bmatrix} \begin{bmatrix} 1 & 0 & 1 & 1 & 0 & 1 & 1 & 0 & 1 & 1 & 0 & 1 & 1 & 0 & 1 \\ 0 & 1 & 1 & 0 & 1 & 1 & 0 & 1 & 1 & 0 & 1 & 1 & 0 & 1 & 1 \end{bmatrix}.
$$

In total, the computation of the Fourier transform requires a total of fifty-two multiplications ($3 \times 4^2 + 2^2$) in the field $GF(16)$. Some of these multiplications are by 1 and can be skipped. In contrast, direct computation of the Fourier transform as defined requires a total of 225 multiplications in the field $GF(16)$. Again, some of these multiplications are by 1 and can be skipped.

1.11   The Gleason–Prange theorem

The final two sections of this chapter present several theorems describing properties, occasionally useful, that are unique to Fourier transforms of prime blocklength. Although these properties are only of secondary interest, we include them to satisfy our goal of presenting a broad compendium of properties of the Fourier transform.

Let p be an odd prime and let F be any field that contains an element ω of order p, or that has an extension field that contains an element ω of order p. This requirement is equivalent to the requirement that the characteristic of F is not p. Let v be a vector of blocklength p. The Fourier transform of blocklength p,

$$
V_j = \sum_{i=0}^{p-1} \omega^{ij} v_i \quad j = 0, \dots, p-1,
$$

has, of course, all the properties that hold in general for a Fourier transform. Moreover, because the blocklength is a prime integer, it has several additional properties worth mentioning. These are the *Gleason–Prange theorem*, which is discussed in this section, and the *Rader algorithm*, which is discussed in Section 1.12.

The indices of v and of V may be regarded as elements of $GF(p)$, and so we call $GF(p)$ the *index field*. The index field, which cannot contain an element of order p, should not be confused with the symbol field F. The elements of $GF(p)$ can be partitioned as

$$GF(p) = \mathcal{Q} \cup \mathcal{N} \cup \{0\},$$

where $\mathcal{Q}$ is the set of (nonzero) squares (called the *quadratic residues*) and $\mathcal{N}$ is the set of (nonzero) nonsquares (called the *quadratic nonresidues*). Not every element of $GF(p)$ can be a square because $\beta^2 = (-\beta)^2$. This means that two elements of $GF(p)$ map into each square. Not more than two elements can map into each square because the polynomial $x^2 - \beta^2$ has only two zeros. Thus there must be $(p-1)/2$ squares. This means that there are $(p-1)/2$ elements in $\mathcal{Q}$ and $(p-1)/2$ elements in $\mathcal{N}$. If π is a primitive element of $GF(p)$, then the squares are the even powers of π and the nonsquares are the odd powers of π. This partitioning of the index set into squares and nonsquares leads to the special properties of the Fourier transform of blocklength p.

The Gleason–Prange theorem holds in any field, but the statement of the general case requires the introduction of Legendre symbols and gaussian sums, which we prefer to postpone briefly. Initially, to simplify the proof, we temporarily restrict the treatment to symbol fields F of the form $GF(2^m)$.

The Gleason–Prange theorem deals with a vector v of blocklength p, with p a prime, augmented by one additional component, denoted v_∞. With this additional component, the vector v has length $p+1$. For the field $GF(2^m)$, the additional component is given by

$$v_\infty - \sum_{i=0}^{p-1} v_i - V_0.$$

The *Gleason–Prange permutation* of the vector

$$v = (v_0, v_1, v_2, \ldots, v_{p-1}, v_\infty)$$

is the vector u with the components $u_i = v_{-i^{-1}}$, and with $u_0 = v_\infty$ and $u_\infty = v_0$. The index $-i^{-1}$ is defined in terms of the operations of the field $GF(p)$. If the Gleason–Prange permutation is applied twice, the original v is restored because $-(-i^{-1})^{-1} = i$ in $GF(p)$.

For example, with $p = 11$, the Gleason–Prange permutation of the vector

$$v = (v_0, v_1, v_2, v_3, v_4, v_5, v_6, v_7, v_8, v_9, v_{10}, v_\infty)$$

is the vector

$$u = (v_\infty, v_{10}, v_5, v_7, v_8, v_2, v_9, v_3, v_4, v_6, v_1, v_0).$$

The Gleason–Prange permutation of the vector u is the vector v.

We shall say that the spectrum V satisfies a *Gleason–Prange condition* if either $V_j = 0$ for every $j \in \mathcal{Q}$, or $V_j = 0$ for every $j \in \mathcal{N}$. For example, for $p = 11$, $\mathcal{Q} = \{1, 4, 9, 5, 3\}$ and $\mathcal{N} = \{2, 6, 7, 8, 10\}$, so both the vector

$$V = (V_0, 0, V_2, 0, 0, 0, V_6, V_7, V_8, 0, V_{10})$$

and the vector

$$V = (V_0, V_1, 0, V_3, V_4, V_5, 0, 0, 0, V_9, 0)$$

satisfy a Gleason–Prange condition.

Theorem 1.11.1 (Gleason–Prange)   *Over $GF(2^m)$, suppose that the extended vectors v and u are related by the Gleason–Prange permutation. If V satisfies a Gleason–Prange condition, then U satisfies the same Gleason–Prange condition.*

Proof: We shall prove the theorem for the case in which $V_j = 0$ for $j \in \mathcal{Q}$. The other case, in which $V_j = 0$ for $j \in \mathcal{N}$, is treated the same way.

Because $V_0 = v_\infty$, the inverse Fourier transform of V can be written as follows:

$$v_i = v_\infty + \sum_{k=1}^{p-1} \omega^{-ik} V_k.$$

Consequently,

$$v_{-i^{-1}} = v_\infty + \sum_{k=1}^{p-1} \omega^{i^{-1}k} V_k \quad i = 1, \ldots, p-1.$$

On the other hand, because $u_0 = v_\infty$, the Fourier transform of u can be written as follows:

$$U_j = u_0 + \sum_{i=1}^{p-1} \omega^{ij} u_i \quad j = 1, \ldots, p-1,$$

$$= v_\infty + \sum_{i=1}^{p-1} \omega^{ij} v_{-i^{-1}},$$

Combining these equations, we obtain

$$U_j = v_\infty \left[1 + \sum_{i=1}^{p-1} \omega^{ij} \right] + \sum_{i=1}^{p-1} \omega^{ij} \sum_{k=1}^{p-1} \omega^{i^{-1}k} V_k \quad j = 1, \ldots, p-1.$$

Because j is not zero, the first term is zero. Therefore, because $V_k = 0$ for $k \in \mathcal{Q}$, we have

$$U_j = \sum_{k=1}^{p-1} V_k \sum_{i=1}^{p-1} \omega^{ij+i^{-1}k} = \sum_{k \in \mathcal{N}} V_k \sum_{i=1}^{p-1} \omega^{ij+i^{-1}k}.$$

We must show that $U_j = 0$ if $j \in \mathcal{Q}$. This will be so if every ω^r that occurs in the sum occurs twice in the sum because then $\omega^r + \omega^r = 0$ in $GF(2^m)$. Given any i, let $\ell = i^{-1}kj^{-1}$. Then $\ell j + \ell^{-1}k = ij + i^{-1}k$. So if $\ell \neq i$, then the exponent of ω occurs twice in the formula for U_j. It only remains to show that $\ell \neq i$. But $j \in \mathcal{Q}$ and $k \in \mathcal{N}$. This means that if $i \in \mathcal{Q}$, then $\ell \in \mathcal{N}$, and if $i \in \mathcal{N}$, then $\ell \in \mathcal{Q}$. Hence ℓ and i are not equal, so every ω^r occurs twice in the formula for U_j, and the proof is complete. ∎

The theorem holds because the array $A_{jk} = \sum_i \omega^{ij+i^{-1}k}$ has an appropriate pattern of zeros. To illustrate an example of this pattern, let $p = 7$, and let ω be an element of $GF(7)$ that satisfies $\omega^3 + \omega + 1 = 0$. Then it is straightforward to calculate the array A as follows:

$$\left[\sum_{i=1}^{6} \omega^{ij+i^{-1}k} \right] = \begin{bmatrix} \omega^3 & \omega^5 & 0 & \omega^6 & 0 & 0 \\ \omega^5 & \omega^6 & 0 & \omega^3 & 0 & 0 \\ 0 & 0 & \omega^5 & 0 & \omega^3 & \omega^6 \\ \omega^6 & \omega^3 & 0 & \omega^5 & 0 & 0 \\ 0 & 0 & \omega^3 & 0 & \omega^6 & \omega^5 \\ 0 & 0 & \omega^6 & 0 & \omega^5 & \omega^3 \end{bmatrix}.$$

By the permutation of its rows and columns, this matrix can be put into other attractive forms. For example, the matrix A can be put into the form of a block diagonal matrix with identical three by three matrices on the diagonal and zeros elsewhere. Alternatively, the matrix can be written as follows:

$$\begin{bmatrix} U_1 \\ U_5 \\ U_2 \\ U_3 \\ U_4 \\ U_6 \end{bmatrix} = \begin{bmatrix} \omega^3 & 0 & \omega^5 & 0 & \omega^6 & 0 \\ 0 & \omega^3 & 0 & \omega^5 & 0 & \omega^6 \\ \omega^6 & 0 & \omega^3 & 0 & \omega^5 & 0 \\ 0 & \omega^6 & 0 & \omega^3 & 0 & \omega^5 \\ \omega^5 & 0 & \omega^6 & 0 & \omega^3 & 0 \\ 0 & \omega^5 & 0 & \omega^6 & 0 & \omega^3 \end{bmatrix} \begin{bmatrix} V_1 \\ V_3 \\ V_2 \\ V_6 \\ V_4 \\ V_5 \end{bmatrix},$$

with each row the cyclic shift of the previous row. Then the Gleason–Prange theorem, stated in the Fourier transform domain, becomes obvious from the above matrix–vector product. A similar arrangement holds for arbitrary p, which will be explained in Section 1.12 as a consequence of the Rader algorithm.

The Gleason–Prange theorem holds more generally for a Fourier transform of block-length p in any field F whose characteristic is not equal to p, provided the definition of the Gleason–Prange permutation is appropriately generalized. For this purpose, let θ denote the *gaussian sum*, which in the field F is defined for any ω of prime order p by

$$\theta = \sum_{i=0}^{p-1} \chi(i)\omega^i,$$

where $\chi(i)$ is the *Legendre symbol*, defined by

$$\chi(i) = \begin{cases} 0 & \text{if } i \text{ is a multiple of } p \\ 1 & \text{if } i \text{ is a nonzero square (mod } p) \\ -1 & \text{if } i \text{ is a nonzero nonsquare (mod } p). \end{cases}$$

An important property of the Legendre symbol for p prime that we shall use is that

$$\sum_{i=0}^{p-1} \chi(i)\omega^{ij} = \chi(j)\theta,$$

which is easy to prove by a change of variables using $\text{GCD}(j, p) = 1$.

Theorem 1.11.2 *For any field F whose characteristic is not p, the gaussian sum satisfies*

$$\theta^2 = p\chi(-1).$$

Proof: To prove this, consider the zero component of the cyclic convolution $(\chi * \chi)_0$. This can be found by computing the inverse Fourier transform of the square of the vector X having the components

$$X_j = \sum_{i=0}^{p-1} \chi(i)\omega^{ij} = \chi(j)\theta \quad j = 0, \ldots, p-1.$$

Hence,

$$(\chi * \chi)_0 = \frac{1}{p}\sum_{j=0}^{p-1} [\chi(j)\theta]^2 = \frac{p-1}{p}\theta^2.$$

But the zero component of the convolution can be computed directly. Thus

$$(\chi * \chi)_0 = \sum_{i=0}^{p-1} \chi(i)\chi(-i)$$

$$= \sum_{i=1}^{p-1} \chi(i^2)\chi(-1)$$

$$= (p-1)\chi(-1),$$

because i^2 is always a square. Hence

$$(p-1)\chi(-1) = \frac{p-1}{p}\theta^2,$$

from which we conclude that

$$\frac{\theta^2}{p} = \chi(-1).$$

This completes the proof of the theorem. ∎

Next, we will generalize the definition of the Gleason–Prange permutation to finite fields of characteristic p. This permutation is defined for any vector $\boldsymbol{v}$ of blocklength $p + 1$, with p a prime and with component v_∞ satisfying

$$v_\infty = -\frac{1}{p}\theta \sum_{i=0}^{p-1} v_i.$$

The Gleason–Prange permutation of the vector $\boldsymbol{v}$ is defined as the vector $\boldsymbol{u}$ with components

$$u_i = \chi(-i^{-1})v_{-i^{-1}} \quad i = 1,\ldots,p-1,$$

and

$$u_0 = \chi(-1)v_\infty,$$

$$u_\infty = v_0.$$

Because $-i^{-1}(\bmod\ p)$ is a permutation, the first line can be written as follows:

$$u_{-i^{-1}} = \chi(i)v_i \quad i = 1,\ldots,p-1.$$

The Gleason–Prange permutation of the vector $\boldsymbol{u}$ returns to the vector $\boldsymbol{v}$.

Theorem 1.11.3 (Gleason–Prange)　*Over the field F, suppose that the vectors v and u are related by the Gleason–Prange permutation. If V satisfies a Gleason–Prange condition, then U satisfies the same Gleason–Prange condition.*

Proof:　The proof proceeds along the same lines as the proof of the earlier theorem for fields of characteristic 2, and it is essentially identical up to the development of the equation

$$U_j = \frac{\chi(-1)}{p} \sum_{k \in \mathcal{N}} V_k \sum_{i=1}^{p-1} \alpha^{ij + i^{-1}k} \chi(i),$$

which differs from the binary case by the term $\chi(-1)/p$ outside the sums and the term $\chi(i)$ inside the sums.

Let $\ell = kj^{-1}i^{-1}$. We must show that $\ell \neq i$ and that the ℓth term of the sum cancels the ith term of the sum. But $k \in \mathcal{N}$ and $j \in \mathcal{Q}$, which implies that i and ℓ have the opposite quadratic character modulo p. This means that $\chi(i) = -\chi(\ell)$, so that $\alpha^{ij+k/i}\chi(i) = -\alpha^{\ell j + k/\ell}\chi(\ell)$. We conclude that terms in the inner sum cancel in pairs. Hence $U_j = 0$.

It remains only to show that $\sum_{i=0}^{p-1} u_i = -(p/\theta)u_\infty$. This proof consists of a string of manipulations, starting with the expression

$$u_\infty = v_0 = \frac{1}{p}\left(V_0 + \sum_{j=1}^{p-1} V_j\right) = \frac{1}{p}\left(-\frac{p}{\theta}v_\infty + \sum_{j=1}^{p-1} V_j\right).$$

Because $\chi(0) = 0$ and $V_j = 0$ unless $\chi(j) = -1$, this can be rewritten as follows:

$$u_\infty = \frac{1}{p}\left(-\frac{p}{\theta}v_\infty - \sum_{j=0}^{p-1} \chi(j)V_j\right) = \frac{1}{p}\left(-\frac{p}{\theta}v_\infty - \sum_{j=0}^{p-1} \chi(j) \sum_{i=0}^{p-1} v_i \omega^{ij}\right)$$

$$= \frac{1}{p}\left(-\frac{p}{\theta}v_\infty - \sum_{i=0}^{p-1} v_i \sum_{j=0}^{p-1} \chi(j)\omega^{ij}\right)$$

$$= -\frac{1}{p}\left(\frac{p}{\theta}v_\infty + \theta \sum_{i=0}^{p-1} v_i \chi(i)\right) = -\frac{1}{p}\left(\frac{p}{\theta}v_\infty + \theta \sum_{i=1}^{p-1} u_{-i^{-1}}\right).$$

The sum at the right is unaffected by a permutation, which means that

$$u_\infty = -\frac{1}{p}\left(\frac{p}{\theta}v_\infty - \theta u_0 + \theta \sum_{i=0}^{p-1} u_i\right).$$

But $u_0 = \chi(-1)v_\infty$ and $p = \theta^2\chi(-1)$, from which we conclude that

$$u_\infty = -\frac{1}{p}\theta\sum_{i=0}^{p-1} u_i,$$

which completes the proof.   ∎

1.12 The Rader algorithm

Each nonzero element of $GF(p)$ can be written as a power of π, where π is a primitive element of the field $GF(p)$. Hence each integer i from 1 to $p-1$ can be written as a power modulo p of π; the power is called the logarithm of i to the base π in $GF(p)$. If $i = \pi^{r(i)}$, then $r(i) = \log_\pi i$. Thus each nonzero index i of the vector v has a logarithm. The nonzero index j in the Fourier transform V_j also has a logarithm. However, it is convenient to treat the index in the transform domain slightly differently. Let $s(j) = -\log_\pi j$ so that $j = \pi^{-s(j)}$.

Now write the Fourier transform as follows:

$$V_j = v_0 + \sum_{i=1}^{p-1} \omega^{ij} v_i \quad j = 1,\ldots,p-1,$$

and

$$V_0 = \sum_{i=0}^{p-1} v_i.$$

The reason that v_0 and V_0 are given special treatment is that zero does not have a base $-\pi$ logarithm. Next, write the indices as powers of π as follows:

$$V_{\pi^{-s(j)}} - v_0 = \sum_{i=1}^{p-1} \omega^{\pi^{r(i)}\pi^{-s(j)}} v_{\pi^{r(i)}} \quad j = 1,\ldots,p-1.$$

But $r(i)$ is a permutation, and it does not change the sum if the terms are reordered, so we can define $V'_s = V_{\pi^{-s}} - v_0$ and $v'_r = v_{\pi^r}$, and write

$$V'_s = \sum_{r=0}^{p-2} \omega^{\pi^{r-s}} v'_r \quad s = 0,\ldots,p-2.$$

This expression is the Rader algorithm for computing V. It is a cyclic convolution because ω^{π^r} is periodic with period $p-1$. The Rader algorithm has replaced the computation of the Fourier transform of blocklength p by the computation of a cyclic convolution of blocklength $p - 1$.

Accordingly, define $V'(x) = \sum_{s=0}^{p-2} V'_s x^s$, and define $v'(x) = \sum_{r=0}^{p-2} v'_r x^r$. Define the *Rader polynomial* as

$$g(x) = \sum_{r=0}^{p-2} \omega^{\pi^r} x^r.$$

The Rader algorithm expresses the Fourier transform of blocklength p as the polynomial product

$$V'(x) = g(x)v'(x) \quad (\text{mod } x^{p-1} - 1),$$

or as a $(p - 1)$-point cyclic convolution

$$V' = g * v'.$$

The components of v' are given as the components of v rearranged. The components of V are easily found as the components of V' rearranged.

For an example of the Rader algorithm, let

$$v = (v_0, v_1, v_2, v_3, v_4, v_5, v_6)$$

be a vector over the field F. Choose the primitive element $\pi = 3$ of $GF(7)$ to write the nonzero indices as $i = 3^r$, so that

$$v = (v_0, v_{\pi^0}, v_{\pi^2}, v_{\pi^1}, v_{\pi^4}, v_{\pi^5}, v_{\pi^3}),$$

from which we obtain

$$v'(x) = v_5 x^5 + v_4 x^4 + v_6 x^3 + v_2 x^2 + v_3 x + v_1.$$

Denote the transform of this vector as

$$V = (V_0, V_1, V_2, V_3, V_4, V_5, V_6)$$
$$= (V_0, V_{\pi^{-0}}, V_{\pi^{-4}}, V_{\pi^{-5}}, V_{\pi^{-2}}, V_{\pi^{-1}}, V_{\pi^{-3}}),$$

from which we obtain

$$V'(x) = V_3 x^5 + V_2 x^4 + V_6 x^3 + V_4 x^2 + V_5 x + V_1.$$

The Rader polynomial is given by

$$g(x) = \sum_{r=0}^{6} \omega^{\pi^r} x^r$$

$$= \omega^3 x^5 + \omega^2 x^4 + \omega^6 x^3 + \omega^4 x^2 + \omega^5 x + \omega.$$

Then, except for the terms v_0 and V_0, the Fourier transforms can be computed as

$$V'(x) = g(x)v'(x) \quad (\mathrm{mod}\ x^6 - 1),$$

as one can verify by direct computation. This is a six-point cyclic convolution. In this way, a p-point Fourier transform has been replaced with a $(p-1)$-point cyclic convolution. This cyclic convolution can be computed in any convenient way, even by using a six-point Fourier transform, if it exists in that field. Although p is a prime, $p-1$ is composite, so the Good–Thomas fast Fourier transform, to be discussed in Section 5.6, can be used to compute the convolution.

Finally, one may combine the Rader algorithm with the Gleason–Prange theorem. This clarifies the example given at the end of Section 1.11. Let V be a vector over $GF(8)$ of blocklength 7 of the form

$$V = (V_0, 0, 0, V_3, 0, V_5, V_6).$$

Let v be the inverse Fourier transform of V and let U be the Fourier transform of u, the Gleason–Prange permutation of v. To form U from V, take the inverse Fourier transform of V to form v, followed by the Gleason–Prange permutation of v to form u, followed by the Fourier transform of u to form U. This is given by

$$U_j = v_\infty \left[1 + \sum_{i=1}^{6} \omega^{ij} \right] + \sum_{i=1}^{6} \omega^{ij} \sum_{k=1}^{6} \omega^{i^{-1}k} V_k \quad j = 1, \ldots, 6.$$

Because j is not zero, the first term is zero. Therefore, because $V_k = 0$ for $k \in Q$, we have

$$U_j = \sum_{i=1}^{6} \omega^{ij} \sum_{k=1}^{6} \omega^{i^{-1}k} V_k.$$

Both sums can be changed into convolutions by using the Rader algorithm. We will rewrite this as

$$U_{j-1} = \sum_{i=1}^{6} \omega^{j^{-1}i} \sum_{k=1}^{6} \omega^{i^{-1}k} V_k$$

by replacing j by j^{-1}. In this way, both summations are changed into identical convolutions, using the same filter $g(x)$. Now one can express the computation as

$$U'(x) = g(x)[g(x)V'(x)]$$
$$= g^2(x)V'(x),$$

where

$$V'(x) = \sum_{s=0}^{p-2} [V_{\pi^{-s}} - v_0] x^s$$

and

$$U'(x) = \sum_{s=0}^{p-2} [U_{\pi^s} - u_0] x^s.$$

Then $V'(x)$, given above, reduces to

$$V'(x) = V_3 x^5 + V_6 x^3 + V_5 x$$

and

$$U'(x) = U_5 x^5 + U_6 x^3 + U_3 x.$$

Modulo $x^6 - 1$, the square of the Rader polynomial $g(x)$ in the field $GF(8)$ is

$$g^2(x) = \omega^5 x^4 + \omega^6 x^2 + \omega^3.$$

Consequently, because $U'(x) = g^2(x)V'(x)$, we can compute

$$U'(x) = (\omega^5 V_5 + \omega^3 V_3 + \omega^6 V_6)x^5 + (\omega^6 V_5 + \omega^5 V_3 + \omega^3 V_6)x^3$$
$$+ (\omega^3 V_5 + \omega^6 V_3 + \omega^5 V_6)x$$
$$= U_5 x^5 + U_6 x^3 + U_3 x,$$

and so U is given by

$$U = \left(V_0, 0, 0, \omega^6 V_3 + \omega^5 V_5 + \omega^3 V_6, 0, \omega^3 V_3 + \omega^6 V_5 + \omega^5 V_6, \omega^5 V_3 \right.$$
$$\left. + \omega^3 V_5 + \omega^6 V_6\right),$$

which satisfies the same Gleason–Prange condition as V.

Problems

1.1 (a) List the properties of the Walsh–Hadamard transform. Is there a convolution property?

(b) The *tensor product* (*outer product*) of matrices A and B is the matrix consisting of blocks of the form $a_{ij}B$. The tensor product can be used to express a multidimensional Fourier transform as a matrix–vector product. Describe a sixteen-point Walsh–Hadamard transform as a matrix–vector product.

1.2 Prove the following properties of the formal derivative:

(a) $[f(x)g(x)]' = f'(x)g(x) + f(x)g'(x)$;

(b) If $(x - a)^m$ divides $f(x)$, then $f^{(m)}(a) = 0$.

1.3 Construct a Fourier transform of blocklength 10 over $GF(11)$. Use the Fourier transform and the convolution property to compute the polynomial product $(x^4 + 9x^3 + 8x^2 + 7x + 6)(x^9 + 2x^3 + 3x^2 + 4x + 6)(\bmod 11)$. Compare the amount of work with the work of computing the polynomial product directly.

1.4 Find an element ω of order 2^{16} in the field $GF(2^{16} + 1)$.

1.5 Prove that the rings $F[x]$ and $F[x]/\langle p(x)\rangle$ are *principal ideal rings*. That is, every ideal can be written as $I = \{g(x)a(x)\}$, where $g(x)$ is a fixed polynomial and $a(x)$ varies over all elements of the ring.

1.6 Generalize the Walsh–Hadamard transform to $Q(i)$ as constructed by using the irreducible polynomial $p(x) = x^2 + 1$ over Q. Express this Fourier transform for blocklength 16 as a matrix with elements ± 1, $\pm i$.

1.7 Use the Fourier transform over $Q^{(8)}$, constructed with $z^8 + 1$, and the convolution theorem to compute the bivariate polynomial product $a(x, y)b(x, y)$, using only monovariate polynomial products, where $a(x, y) = 1 + x - y + x^2 - y^2 + x^3 - y^3$ and $b(x, y) = 1 + x^2y^2 + x^3 + y^3 - x^3y^3$.

1.8 Let b be coprime with n. Suppose that $\Lambda(x)$ is the polynomial of minimal degree that cyclically generates V, a vector of blocklength n. What is the polynomial of minimal degree that cyclically generates the cyclic decimation of V by b?

1.9 Prove the BCH bound using the fact that the number of zeros of the univariate polynomial $p(x)$ is not larger than the degree of $p(x)$.

1.10 Prove the Hartmann–Tzeng bound. Prove the Roos bound.

1.11 Prove that the cyclic complexity of the vector V is not changed by a cyclic decimation with an integer b coprime with the blocklength n.

1.12 (a) Prove that in the field $GF(q)$,

$$(\beta + \gamma)^q = \beta^q + \gamma^q.$$

(b) Prove that in the field $GF(q^m)$, the q-ary trace satisfies

(i) $\mathrm{tr}(\beta + \gamma) = \mathrm{tr}\,\beta + \mathrm{tr}\,\gamma$.

(ii) $\mathrm{tr}(\xi\beta) = \xi(\mathrm{tr}\ \beta)$ if $\xi \in GF(q)$.

(c) Prove that in the field $GF(2^m)$, the binary trace of every element is either 0 or 1.

1.13   What is the linear complexity of the counting sequence ($v_{n+1} = v_n + 1$)?

1.14   Prove the following property of the Hasse derivative (cited earlier without proof): if $h(x)$ is an irreducible polynomial of degree at least 1, then $[h(x)]^m$ divides $f(x)$ if and only if $h(x)$ divides $f^{[\ell]}(x)$ for $\ell = 0, \ldots, m-1$. Does the statement hold if the Hasse derivative $f^{[\ell]}(x)$ is replaced by the formal derivative $f^{(\ell)}(x)$?

1.15   Prove that the gaussian sum has the following property:

$$\sum_{i=0}^{p-1} \chi(i)\omega^{ij} = \chi(j)\theta.$$

Thus the gaussian sum is an eigenvalue of the matrix corresponding to the Fourier transform.

1.16   The *Pascal triangle* in the rational field is the following infinite arrangement of integers:

$$
\begin{array}{ccccccccc}
 & & & & 1 & & & & \\
 & & & 1 & & 1 & & & \\
 & & 1 & & 2 & & 1 & & \\
 & 1 & & 3 & & 3 & & 1 & \\
1 & & 4 & & 6 & & 4 & & 1 \\
\vdots & & \vdots & & & & & &
\end{array}
$$

It is defined recursively, forming the elements of each row by adding the two nearest elements of the previous row.

(a) Describe the Pascal triangle in the field $GF(2^m)$.

(b) Formulate an efficient procedure for computing all Hasse derivatives of a polynomial $p(x)$ over $GF(2^m)$.

1.17   To what does the Poisson summation formula reduce in the two extreme cases $(n', n'') = (1, n)$ and $(n', n'') = (n, 1)$?

1.18   Let v and u be two vectors of length $p+1$ over the field F related by a Gleason–Prange permutation. Prove that for any fixed γ, and if $V_j = \gamma$ whenever j is a quadratic residue in $GF(p)$, then $U_j = \gamma$ whenever j is a quadratic residue.

Notes

The purpose of this chapter is to provide a compendium of properties of the Fourier transform from the engineer's point of view, mingling properties that arise in signal processing with those that arise in algebraic coding theory. For the most part, these properties hold in any algebraic field. This unified presentation, independent of any particular field, is an aid to understanding. In a similar way, placing concrete examples in various algebraic fields side by side can suggest helpful insights.

The classical bounds of coding theory are presented herein simply as relationships between the weight of an individual sequence and the pattern of zeros of its Fourier transform. These bounds are valid in any field. The linear complexity property appeared explicitly in Blahut (1979), though it is implicit in the bounds of coding theory. This was discussed by Massey (1998). The role of the Fourier transform in coding theory, though not under that name, appears in the work of Mattson and Solomon (1961). Schaub's doctoral thesis (Schaub, 1988) reinforced interest in a linear complexity approach to algebraic coding theory. In his thesis, Schaub developed the matrix rank argument for proving the van Lint–Wilson bound. Massey (1969) first introduced his theorem for his formulation of the Berlekamp–Massey algorithm. It suits the purposes of this book to give it an independent identity as a statement concerning linear recurrences.

The Gleason–Prange theorem for finite fields was first published by Mattson and Assmus (1964). The proof for an arbitrary field is due to Blahut (1991) with a later simplification by Huffman (1995). The treatment here also draws on unpublished work of McGuire. The Rader algorithm (Rader, 1968) was introduced for the purpose of simplifying the computation of a Fourier transform of prime blocklength. In addition to the semifast algorithm described here, semifast algorithms for the Fourier transform were given by Goertzel (1968) for the complex field, and by Sarwate (1978) for the finite fields.

2   The Fourier Transform and Cyclic Codes

Error-control codes are now in widespread use in many applications such as communication systems, magnetic recording systems, and optical recording systems. The compact disk and the digital video disk are two familiar examples of such applications.

We shall discuss only block codes for error control. A block code for error control is a set of n-tuples in some finite alphabet, usually the finite field $GF(q)$. The reason for choosing a field as the alphabet is to have a rich arithmetic structure so that practical codes can be constructed and encoders and decoders can be designed as computational algorithms. The most popular block codes are linear. This means that the component-wise sum of two codewords is a codeword, and any scalar multiple of a codeword is a codeword. So that a large number of errors can be corrected, it is desirable that codewords be very dissimilar from each other. This dissimilarity will be measured by the Hamming distance.

The most important class of block codes, the *Reed–Solomon codes*, will be described as an exercise in the complexity of sequences and of Fourier transform theory. Another important class of block codes, the *BCH codes*, will be described as a class of subcodes of the Reed–Solomon codes, all of whose components lie in a subfield. The BCH codes and the Reed–Solomon codes are examples of cyclic codes, which themselves form a subclass of the class of linear block codes.

2.1   Linear codes, weight, and distance

A linear (n, k) block code $\mathcal{C}$ over the field F is a k-dimensional subspace of the vector space of n-tuples over F. We shall be interested primarily in the case where the field F is the finite field $GF(q)$. Then a k-dimensional subspace of $GF(q)^n$ contains q^k vectors of length n. An (n, k) code is used to represent each k symbol *dataword* by an n symbol *codeword*. There is an overhead of $n - k$ symbols that provides redundancy in the codeword so that errors (or other impairments) in the received word, or *senseword*, can be corrected. By an appropriate choice of basis for the subspace, the overhead can be confined to $n - k$ symbols, called *check symbols*. The *rate* of the code is defined as k/n. The *blocklength* of the code is n; the *datalength* or *dimension* of the code is k. (The terms blocklength and datalength also apply to nonlinear codes; the term dimension does not.)

The *Hamming weight* of a vector is defined as the number of components at which the vector is nonzero. The *Hamming distance* between two vectors of the same blocklength is defined as the number of components at which the two vectors are different. The *minimum distance* of a code, denoted d_{min}, is defined as the minimum Hamming distance between any two distinct codewords of the code. For a linear code, the minimum Hamming distance between any pair of codewords is equal to the minimum Hamming weight of any nonzero codeword.

A linear (n, k) code is also described as a linear (n, k, d) code, which properly means that $d = d_{min}$. This notation is sometimes used informally to mean the weaker statement that $d \leq d_{min}$, the context usually indicating the intended meaning. The informal usage arises when the minimum distance is not known or is not evident, but it is known that d is a lower bound on d_{min}.

The *packing radius* of a code, denoted t, is defined as the largest integer smaller than $d_{min}/2$. The packing radius of a code is the largest integer t such that spheres of Hamming radius t about codewords $c(x)$ of $\mathcal{C}$ are disjoint. This should be contrasted with the *covering radius* of a code, which is the smallest integer ρ such that spheres of Hamming radius ρ about codewords cover the whole vector space $GF(q)^n$.

Because the spheres with Hamming radius t centered on codewords are disjoint, the number of elements of $GF(q)^n$ in any such sphere multiplied by the number of such spheres q^k cannot be larger than the total number of elements of $GF(q)^n$. This means that

$$q^k \sum_{\ell=0}^{t} (q-1)^\ell \binom{n}{\ell} \leq q^n,$$

an inequality known as the *Hamming bound*. A linear code that meets the Hamming bound with equality is called a *perfect code*. Perfect codes are very rare.

Because it is a k-dimensional subspace of the vector space $GF(q)^n$, a linear code $\mathcal{C}$ can be specified by any basis of this subspace. Any matrix whose k rows form a basis for the subspace is called a *generator matrix* for the code and is denoted $\boldsymbol{G}$. Any matrix whose $n - k$ rows form a basis for the orthogonal complement is called a *check matrix* for the code and is denoted $\boldsymbol{H}$. Consequently, $\boldsymbol{c}$ is an element of $\mathcal{C}$ if and only if $\boldsymbol{c}$ satisfies either of the two equivalent conditions: $\boldsymbol{c} = \boldsymbol{aG}$ or $\boldsymbol{cH}^\mathrm{T} = \boldsymbol{0}$. The first condition expresses the codeword $\boldsymbol{c}$ in terms of the dataword $\boldsymbol{a}$. The second condition says that $\mathcal{C}$ is in the *null space* of $\boldsymbol{H}^\mathrm{T}$. The code $\mathcal{C}$ has a codeword of weight w if and only if $\boldsymbol{H}$ has w linearly dependent columns.

The *dimension* of the linear code C is equal to the rank of the generator matrix G. Because[1] the number of rows of G plus the number of rows of H is equal to the dimension n of the underlying vector space, the dimension of a code is also equal to n minus the rank of the check matrix H.

Because the rank of a matrix is equal to both its row rank and its column rank, the dimension of a linear code C is also equal to the cardinality of the largest set of linearly independent columns of the generator matrix G. In contrast, the minimum distance of C is equal to the largest integer d such that *every* set of $d - 1$ columns of H is linearly independent.

There is such a strong parallel in these statements that we will coin a term to complement the term "rank."

Definition 2.1.1 *For any matrix M:*

the rank of M is the largest value of r such that some set of r columns of M is linearly independent;

the heft of M is the largest value of r such that every set of r columns of M is linearly independent.

In contrast to the rank, the heft of M need not equal the heft of M^T. We will be interested only in the heft of matrices with at least as many columns as rows.

Clearly, for any matrix M, the inequality

$$\text{heft } M \leq \text{rank } M$$

holds. The rank can also be described as the smallest value of r such that *every* set of $r + 1$ columns is linearly dependent, and the heft can also be described as the smallest value of r such that *some* set of $r + 1$ columns is linearly dependent.

The dimension k of a linear code is equal to the rank of the generator matrix G, and the minimum distance of a linear code is equal to one plus the heft of the check matrix H. Because the rank of a check matrix H equals $n - k$, and the heft of H equals $d_{\min} - 1$, the above inequality relating the heft and the rank of a matrix implies that

$$d_{\min} \leq n - k + 1,$$

an inequality known as the *Singleton bound*. The quest for good linear (n, k) codes over $GF(q)$ can be regarded as the quest for k by n matrices over $GF(q)$ with large heft. For large n, very little is known about finding such matrices.

A linear code that meets the Singleton bound with equality is called a *maximum-distance code*.

---

[1] The *rank-nullity theorem* says the rank of a matrix plus the dimension of its null space is equal to the dimension of the underlying vector space.

Theorem 2.1.2 *In an (n, k) maximum-distance code, any set of k places may be chosen as data places and assigned any arbitrary values from $GF(q)$, thereby specifying a unique codeword.*

Proof: A code of minimum distance $d_{\min}$ can correct any $d_{\min} - 1$ erasures. Because $d_{\min} = n - k + 1$ for maximum-distance codes, the theorem is proved by regarding the $n - k$ unassigned places as erasures. ∎

A linear code $\mathcal{C}$ is a linear subspace of a vector space $GF(q)^n$. As a vector subspace, $\mathcal{C}$ has an *orthogonal complement* $\mathcal{C}^\perp$, consisting of all vectors $\boldsymbol{v}$ of $GF(q)^n$ that are *orthogonal* to every element of $\mathcal{C}$, meaning that the *inner product* $\sum_{i=0}^{n-1} c_i v_i$ equals zero for all $\boldsymbol{c} \in \mathcal{C}$. Thus

$$\mathcal{C}^\perp = \left\{ \boldsymbol{v} \mid \sum_{i=0}^{n-1} c_i v_i = 0 \text{ for all } \boldsymbol{c} \in \mathcal{C} \right\}.$$

The orthogonal complement $\mathcal{C}^\perp$ is itself a linear code, called the *dual code* of $\mathcal{C}$. The matrices $\boldsymbol{H}$ and $\boldsymbol{G}$, respectively, are the generator and check matrices for $\mathcal{C}^\perp$. Over a finite field, $\mathcal{C}$ and $\mathcal{C}^\perp$ may have a nontrivial intersection; the same nonzero vector $\boldsymbol{c}$ may be in both $\mathcal{C}$ and $\mathcal{C}^\perp$. Indeed, it may be true that $\mathcal{C} = \mathcal{C}^\perp$, in which case the code is called a *self-dual code*. Even though $\mathcal{C}$ and $\mathcal{C}^\perp$ may have a nontrivial intersection, it is still true, by the rank-nullity theorem, that $\dim(\mathcal{C}) + \dim(\mathcal{C}^\perp) = n$.

The blocklength n of a linear code can be reduced to form a new code of smaller blocklength n'. Let $\mathcal{B} \subset \{0, \ldots, n-1\}$ be a set of size n' that indexes a fixed set of n' codeword components to be retained in the new code. There are two distinct notions for reducing the blocklength based on $\mathcal{B}$. These are puncturing and shortening.

A *punctured code* $\mathcal{C}(\mathcal{B})$ is obtained from an (n, k) code $\mathcal{C}$ simply by dropping from each codeword all codeword components with indices in the set $\mathcal{B}^c$. This corresponds simply to dropping $n - n'$ columns from the generator matrix $\boldsymbol{G}$ to form a new generator matrix, denoted $\boldsymbol{G}(\mathcal{B})$. If the rows of $\boldsymbol{G}(\mathcal{B})$ are again linearly independent, the dimension of the code is not changed. In this case, the number of data symbols remains the same, the number of check symbols is reduced by $n - n'$, and the minimum distance is reduced by at most $n - n'$.

A *shortened code* $\mathcal{C}'(\mathcal{B})$ is obtained from an (n, k) code $\mathcal{C}$ by first forming the subcode $\mathcal{C}'$, consisting of all codewords whose codeword components with indices in the set $\mathcal{B}^c$ are zero, then dropping all codeword components with indices in the set $\mathcal{B}^c$. This corresponds simply to dropping $n - n'$ columns from the check matrix $\boldsymbol{H}$ to form a new check matrix, denoted $\boldsymbol{H}(\mathcal{B})$. If the rows of $\boldsymbol{H}(\mathcal{B})$ are again linearly independent, the redundancy of the code is not changed. In this case, the number of check symbols remains the same, and the number of data symbols is reduced by $n - n'$.

In summary,

$$\mathcal{C}(\mathcal{B}) = \{(c_{i_1}, c_{i_2}, c_{i_3}, \ldots, c_{i_{n'}}) \mid c \in \mathcal{C}, i_\ell \in \mathcal{B}\};$$
$$\mathcal{C}'(\mathcal{B}) = \{(c_{i_1}, c_{i_2}, c_{i_3}, \ldots, c_{i_{n'}}) \mid c \in \mathcal{C}, i_\ell \in \mathcal{B} \text{ and } c_i = 0 \text{ for } i \in \mathcal{B}^c\}.$$

The notions of a punctured code and a shortened code will play important roles in Chapter 10.

A code $\mathcal{C}$ over $GF(q^m)$ has components in the field $GF(q^m)$. The code $\mathcal{C}$ may include some codewords, all of whose components are in the subfield $GF(q)$. The subset of $\mathcal{C}$ consisting of all such codewords forms a code over the subfield $GF(q)$. This code, which is called a *subfield-subcode*, can be written as

$$\mathcal{C}' = \mathcal{C} \cap GF(q)^n.$$

The minimum distance of $\mathcal{C}'$ is not smaller than the minimum distance of $\mathcal{C}$. A subfield-subcode is a special case of a subcode. In general, a *subcode* is any subset of a code. A linear subcode is a subcode that is linear under the obvious inherited operations.

2.2 Cyclic codes

Cyclic codes, including those codes known as Reed–Solomon codes and BCH codes, which are studied in this chapter, comprise the most important class of block codes for error correction.

We define a *cyclic code* of blocklength n over the field F as the set of all n-vectors c having a specified set of spectral components equal to zero. The set of such vectors is closed under linear combinations, so the definition implies that a cyclic code is a linear code. Spectral components exist only if a Fourier transform exists, so a cyclic code of blocklength n exists in the field F only if a Fourier transform of blocklength n exists in the field F or in an extension field of F. Fix a set of *spectral indices*, $\mathcal{A} = \{j_1, j_2, \ldots, j_r\}$, which is called the *defining set* of the cyclic code. The code $\mathcal{C}$ is the set of all vectors c of blocklength n over the field F whose Fourier transform C satisfies $C_{j_\ell} = 0$ for $\ell = 1, \ldots, r$. Thus

$$\mathcal{C} = \{c \mid C_{j_\ell} = 0 \qquad \ell = 1, \ldots, r\},$$

where

$$C_j = \sum_{i=0}^{n-1} \omega^{ij} c_i$$

and ω is an element of order n in F or an extension field of F. The spectrum $\boldsymbol{C}$ is called the *codeword spectrum*. Moreover, the inverse Fourier transform yields

$$c_i = \frac{1}{n} \sum_{j=0}^{n-1} \omega^{-ij} C_j.$$

If F is the finite field $GF(q)$, then n must divide $q^m - 1$ for some m, and so ω is an element of $GF(q^m)$. If $n = q^m - 1$, then ω is a primitive element of $GF(q^m)$, and the cyclic code is called a *primitive cyclic code*.

To index the q^m-1 codeword components of a primitive cyclic code, each component is assigned to one of the $q^m - 1$ nonzero elements of $GF(q^m)$, which can be described as the $q^m - 1$ powers of a primitive element. Similarly, to index the n codeword components of a cyclic code of blocklength n, each component is assigned to one of the n distinct powers of ω, an element of order n. The components of the codeword can be denoted c_{ω^i} for $i = 0, 1, \ldots, n - 1$. Because this notation is needlessly clumsy, we may also identify i with ω^i; the components are then denoted c_i for $i = 0, \ldots, n - 1$ instead of c_{ω^i}, according to convenience. The field element zero is not used as an index for a cyclic code.

A codeword $\boldsymbol{c}$ of a cyclic code is also represented by a *codeword polynomial*, defined as

$$c(x) = \sum_{i=0}^{n-1} c_i x^i.$$

A codeword spectrum $\boldsymbol{C}$ of a cyclic code is also represented by a *spectrum polynomial*, defined as

$$C(x) = \sum_{j=0}^{n-1} C_j x^j.$$

The Fourier transform and inverse Fourier transform are then given by $C_j = c(\omega^j)$ and $c_i = n^{-1} C(\omega^{-i})$.

If ω is an element of $GF(q)$, then each spectral component C_j is an element of $GF(q)$, and, if j is not in the defining set, C_j can be specified arbitrarily and independently of the other spectral components. If ω is not an element of $GF(q)$, then it is an element of the extension field $GF(q^m)$ for some m, and, by Theorem 1.9.1, the spectral components must satisfy the conjugacy constraint $C_j^q = C_{((qj))}$. This means that qj (modulo n) must be in the defining set $\mathcal{A}$ whenever j is in the defining set. In such a case, the defining set $\mathcal{A}$ may be abbreviated by giving only one member (or several members) of each conjugacy class. In this case, for clarity, the defining set itself may be called the *complete defining set*, then denoted $\mathcal{A}_c$.

A cyclic code always contains the unique codeword polynomial $w(x)$, called the *principal idempotent*, having the property that $w(x)$ is a codeword polynomial and, for any codeword polynomial $c(x)$, $w(x)c(x) = c(x) \pmod{x^n - 1}$. The principal idempotent can be identified by its Fourier transform. Clearly, by the convolution theorem, for any codeword spectrum C, this becomes $W_j C_j = C_j$ for all j. The codeword spectrum W with the required property is given by

$$W_j = \begin{cases} 0 & j \in \mathcal{A}_c \\ 1 & j \notin \mathcal{A}_c, \end{cases}$$

and this spectrum specifies a unique codeword.

A cyclic code always contains the unique codeword polynomial $g(x)$, called the *generator polynomial* of the code, having the property that $g(x)$ is the monic codeword polynomial of minimum degree. Clearly, there is such a monic codeword polynomial of minimum degree. It is unique because if there were two monic codeword polynomials of minimum degree, then their difference would be a codeword polynomial of smaller degree, which could be made monic by multiplication by a scalar. Every codeword polynomial $c(x)$ must have a remainder equal to zero under division by $g(x)$. Otherwise, the remainder would be a codeword polynomial of degree smaller than the degree of $g(x)$. This means that every codeword polynomial must be a polynomial multiple of $g(x)$, written $c(x) = a(x)g(x)$. Thus the dimension of the code is $k = n - \deg g(x)$.

By the translation property of the Fourier transform, if c is cyclically shifted by b places, then C_j is replaced by $C_j \omega^{jb}$, which again is zero whenever C_j is zero. Thus we conclude that the cyclic shift of any codeword of a cyclic code is again a codeword of the same cyclic code, a property known as the *cyclic property*. The cyclic codes take their name from this property, although we do not regard the property, in itself, as important. The cyclic codes are important, not for the cyclic property, but because the Fourier transform properties make it convenient to determine their minimum distances and to develop encoders and decoders. The cyclic property is an example of an *automorphism* of a code, which is defined as any permutation of codeword components that preserves the code. The *automorphism group* of a code is the set of all automorphisms of the code.

Because C is cyclic, if $c(x)$ is in the code, then $xc(x)(\bmod{x^n - 1})$ is in the code as well, as is $a(x)c(x)(\bmod{x^n - 1})$ for any polynomial $a(x)$. By the division algorithm,

$$x^n - 1 = Q(x)g(x) + r(x),$$

where the degree of the remainder polynomial $r(x)$ is smaller than the degree of $g(x)$, so $r(x)$ cannot be a nonzero codeword. But $r(x)$ has the requisite spectral zeros to be a codeword, so it must be the zero codeword. Then $r(x) = 0$, so

$$g(x)h(x) = x^n - 1$$

for some polynomial $h(x)$ called the *check polynomial*.

The central task in the study of a cyclic code is the task of finding the minimum distance of the code. Because a cyclic code is linear, finding the minimum distance of the code is equivalent to finding the smallest Hamming weight of any nonzero codeword of the code. Because the code is completely determined by its defining set, the minimum distance must be a direct consequence of the code's defining set. Thus the relationship between the weight of a vector and the pattern of zeros in its Fourier transform is fundamental to the nature of cyclic codes. This relationship is described in large part, though not completely, by the bounds given in Section 1.8. We consider these bounds as central to the study of cyclic codes – indeed, as a primary reason for introducing the class of cyclic codes.

A polynomial $g(x)$ over $GF(q)$ can also be regarded as a polynomial over $GF(q^m)$. When used as a generator polynomial, $g(x)$ can define a cyclic code over either $GF(q)$ or $GF(q^m)$.

Theorem 2.2.1 *Let $g(x)$, a polynomial over $GF(q)$, divide $x^{q^m-1} - 1$. The cyclic code over $GF(q)$ generated by $g(x)$ and the cyclic code over $GF(q^m)$ generated by $g(x)$ have the same minimum distance.*

Proof: Let C_q and C_{q^m} be the codes over $GF(q)$ and $GF(q^m)$, respectively. Because $C_q \subset C_{q^m}$, it follows that $d_{\min}(C_q) \geq d_{\min}(C_{q^m})$. Let $c(x)$ be a minimum-weight codeword polynomial in C_{q^m}. Then $c(x) = a(x)g(x)$, where the coefficients of $a(x)$ and $c(x)$ are in $GF(q^m)$ and the coefficients of $g(x)$ are in $GF(q)$. The components of $\boldsymbol{c}$ are $c_i = \sum_{j=0}^{k-1} g_{i-j}a_j$. Let $\boldsymbol{c}'$ be the nonzero vector whose ith component is the ith component of the q-ary trace of $\boldsymbol{c}$. We can assume that $\boldsymbol{c}'$ is not the zero vector, because if it were, then we would instead consider the codeword $\gamma\boldsymbol{c}$ for some γ since $\mathrm{tr}(\gamma c_i)$ cannot be zero for all γ unless c_i is zero. Then

$$c_i' = \mathrm{tr}(c_i) = \mathrm{tr}\sum_{j=0}^{k-1} g_{i-j}a_j = \sum_{j=0}^{k-1}\mathrm{tr}(g_{i-j}a_j).$$

Because g_{i-j} is an element of $GF(q)$, it is equal to its own qth power, and so can be factored out of the trace. We can conclude that

$$c_i' = \sum_{j=0}^{k-1} g_{i-j}\mathrm{tr}(a_j) = \sum_{j=0}^{k-1} g_{i-j}a_j'.$$

Thus we see that the polynomial $c'(x)$ is given by $g(x)a'(x)$, and so corresponds to a codeword in C_q. But the trace operation cannot form a nonzero component c_i' from a zero component c_i. Therefore the weight of c' is not larger than the weight of c. Consequently, we have that $d_{\min}(\mathcal{C}_q) \leq d_{\min}(\mathcal{C}_{q^m})$, and the theorem follows. ∎

2.3   Codes on the affine line and the projective line

In general, a primitive cyclic code over $GF(q)$ has blocklength $n = q^m - 1$ for some integer m. When $m = 1$, the code has blocklength $n = q - 1$. A code of larger blocklength may sometimes be desirable. It is possible to extend the length of a cyclic code of blocklength $n = q - 1$ to $n = q$ or to $n = q + 1$ in a natural way. An extended cyclic code is described traditionally in terms of a cyclic code of blocklength $q - 1$ that is extended by one or two extra components. We shall describe these codes more directly, and more elegantly, in terms of the evaluation of polynomials so that codes of blocklength $q - 1$, q, and $q + 1$ are of equal status.

The *affine line* over the finite field $GF(q)$ is the set of all elements of $GF(q)$. The *cyclic line*[2] over the finite field $GF(q)$ is the set denoted $GF(q)^*$ of all nonzero elements of the field. The *projective line* over the finite field $GF(q)$ is the set, which we denote $GF(q)^+$ or $\boldsymbol{P}(GF(q))$, of pairs of elements (β, γ) such that the rightmost nonzero element is 1. The point $(1, 0)$ of the projective line is called the *point at infinity*. The remaining points of the projective line are of the form $(\beta, 1)$ and, because β can take on any value of the field, they may be regarded as forming a copy of the affine line contained within the projective line. The projective line has one point more than the affine line and two points more than the cyclic line, but, for our purposes, it has a more cumbersome structure. The cyclic line has one point less than the affine line and two fewer points than the projective line, but, for our purposes, it has the cleanest structure. Thus, in effect, $GF(q)^* \subset GF(q) \subset GF(q)^+$. The disadvantage of working on the affine line or projective line is that the properties and computational power of the Fourier transform are suppressed.

Let $V(x)$ be any polynomial of degree at most $n - 1$. It can be regarded as the spectrum polynomial of the vector $\boldsymbol{v}$. The coefficients of $V(x)$ are the components V_j, $j = 0, \ldots, n - 1$, of the spectrum $\boldsymbol{V}$. The vector $\boldsymbol{v}$, defined by the inverse Fourier transform

$$v_i = \frac{1}{n} \sum_{j=0}^{n-1} V_j \omega^{-ij},$$

---

[2] For our purposes, this terminology is convenient because it fits the notion of cyclic as used in "cyclic codes," or as in the "cycle" of a primitive element of $GF(q)$, but the similar term "circle" would clash with the point of view that the real affine line $\boldsymbol{R}$, together with the point at infinity, form a topological circle.

is the same as the vector obtained by evaluating the polynomial $V(x)$ on the cyclic line, using the reciprocal powers of ω and writing

$$v_i = \frac{1}{n} V(\omega^{-i}).$$

This vector is given by

$$v = (v_0, v_1, v_2, \ldots, v_{q-2})$$

or

$$v = (v_{\omega^{-0}}, v_{\omega^{-1}}, v_{\omega^{-2}}, \ldots, v_{\omega^{-(q-2)}}).$$

Thus the components of the vector v can be indexed by i or by the reciprocal powers of ω.

Constructing vectors by evaluating polynomials in this way can be made slightly stronger than the Fourier transform because one can also evaluate $V(x)$ at the additional point $x = 0$, which evaluation we call v_-. Thus the element

$$v_- = \frac{1}{n} V(0) = \frac{1}{n} V_0$$

can be used as one more component to lengthen the vector v to

$$v = (v_-, v_0, v_1, \ldots, v_{q-2}).$$

This vector now has blocklength $n = q$. Rather than the subscript "minus," the subscript "infinity" may be preferred. Alternatively, we may write

$$v = (v_-, v_{\omega^{-0}}, v_{\omega^{-1}}, v_{\omega^{-2}}, \ldots, v_{\omega^{-(q-2)}}).$$

In this case, rather than the subscript "minus," the subscript "zero" may be preferred so then all of the q elements of the affine line are used to index the components of the vector.

It is possible to obtain a second additional component if the defining set has the form $\mathcal{A} = \{k, \ldots, n-1\}$. First, replace the spectrum polynomial $V(x)$ by a *homogeneous bivariate polynomial*,

$$V(x, y) = \sum_{j=0}^{k-1} V_j x^j y^{k-1-j},$$

where $k - 1$ is the maximum degree of $V(x)$. Then evaluate $V(x, y)$ at all points of the projective line. This appends one additional component to v because there are $q + 1$

points on the projective line, given by $(0, 1)$, $(\alpha^i, 1)$ for $i = 0, \ldots, q - 2$, and $(1, 0)$. This means that one can evaluate $V(x, y)$ at the point at infinity $(1, 0)$, which evaluation we call v_+. This gives

$$v_+ = \frac{1}{n}V(1, 0) = \frac{1}{n}V_r,$$

which can be used as another component to lengthen the vector v to

$$v = (v_-, v_0, v_1, \ldots, v_{q-2}, v_+).$$

This vector now has blocklength $n = q + 1$. An alternative notation for the vector is

$$v = (v_-, v_{\omega^{-0}}, v_{\omega^{-1}}, v_{\omega^{-2}}, \ldots, v_{\omega^{-(q-2)}}, v_+).$$

Rather than the subscripts "minus" and "plus," the subscripts "zero" and "infinity" may be preferred, so that the components are then indexed by all of the elements of the projective line.

To summarize this discussion, we can extend a cyclic code C by one or two components. For each $c \in C$, let $C(x)$ be the spectrum polynomial. Then with $c_- = (1/n)C(0)$ and $c_+ = (1/n)C(1, 0)$, the *singly extended* cyclic code is given by

$$C' = \{(c_-, c_0, c_1, \ldots, c_{q-2})\}$$

and the *doubly extended* cyclic code is given by

$$C'' = \{(c_-, c_0, c_1, \ldots, c_{q-2}, c_+)\}.$$

In this form, an extended cyclic code is not itself cyclic, although it is linear. There are a few rare examples of doubly extended cyclic codes, however, that do become cyclic under an appropriate permutation of components.

2.4   The wisdom of Solomon and the wizardry of Reed

The BCH bound tells us how to design a linear (n, k, d) cyclic code of minimum weight at least d, where d (or sometimes d^*) is called the *designed distance* of the code. Simply choose $d - 1$ consecutive spectral components as the defining set of the cyclic code so that the BCH bound applies to each codeword.

Definition 2.4.1 *An (n, k, d) cyclic Reed–Solomon code over a field F that contains an element of order n is the set of all vectors over F of blocklength n that have a specified set of $d - 1$ consecutive spectral components in the Fourier transform domain equal to zero.*

A cyclic Reed–Solomon code of blocklength n does not exist over F if a Fourier transform of blocklength n does not exist over F. A *narrow-sense* Reed–Solomon code is a Reed–Solomon code with spectral zeros at $j = n - d + 1, \ldots, n - 1$. A *primitive* Reed–Solomon code over the finite field $GF(q)$ is a cyclic Reed–Solomon code of blocklength $q - 1$.

If codeword c has the spectral component C_j equal to zero for $j = j_0, j_0 + 1, \ldots, j_0 + d - 2$ and codeword c' has the spectral component C'_j equal to zero for $j = j_0, j_0 + 1, \ldots, j_0 + d - 2$, then $c'' = \alpha c + \beta c'$ also has spectral components C''_j equal to zero for these same indices. Hence the Reed–Solomon code is a linear code. The dimension of this linear code is denoted k. Because the dimension is equal to the number of components of the spectrum not constrained to zero, the dimension k satisfies $n - k = d - 1$.

A Reed–Solomon code also can be defined in the language of linear algebra. The Fourier transform is an invertible linear transformation from F^n to F^n. When the result of the Fourier transform map is truncated to any specified set of $d - 1$ consecutive components, then the truncated Fourier transform can be regarded as a linear map from an n-dimensional vector space to an $(n - k)$-dimensional vector space, where $n - k = d - 1$. The Reed–Solomon code is defined as the null space of this map. Likewise, the inverse Fourier transform is an invertible linear transformation from F^n to F^n. When applied to a subspace of F^n of dimension k, consisting of all vectors with a specified set of $d - 1$ consecutive components all equal to zero, the inverse Fourier transform can be regarded as a map from a k-dimensional vector space to an n-dimensional vector space. The Reed–Solomon code is the image of this map. Hence it has dimension k.

The BCH bound says that every nonzero codeword of the Reed–Solomon code has weight at least d and the code is linear, so the minimum distance of the Reed–Solomon code is at least $d = n - k + 1$. Consequently, the minimum distance is exactly $n - k + 1$ because, as asserted by the Singleton bound, no linear code can have a minimum distance larger than $n - k + 1$. This means that an (n, k, d) Reed–Solomon code is a maximum-distance code, and that the packing radius t of a Reed–Solomon code is $(n - k)/2$ if $n - k$ is even and $(n - k - 1)/2$ if $n - k$ is odd.

A simple nontrivial example of a Reed–Solomon code is a $(7, 5, 3)$ Reed–Solomon code over $GF(8)$. Choose $\mathcal{A} = \{1, 2\}$ as the defining set of the code. Every codeword c has $C_1 = C_2 = 0$, while C_0, C_3, C_4, C_5, and C_6 are arbitrary. We may visualize a list of these codewords c where codeword components are elements of $GF(8)$ given in an octal notation, as shown in Table 2.1. Even though this is a rather small Reed–Solomon code, it would be unreasonable to write out this list in full because the full list contains $8^5 = 32\,768$ codewords. Because this code was constructed to satisfy the BCH bound with $d_{\min} = 3$, every two codewords on the list must differ in at least three places.

Although the definition of a Reed–Solomon code holds in any field F, it appears that practical applications of Reed–Solomon codes have always used codes over a finite

Table 2.1. *The (7,5) Reed–Solomon code*

0	0	0	0	0	0	0
0	0	0	0	1	6	3
0	0	0	0	2	7	6
0	0	0	0	3	1	5
			$\vdots$			
0	0	0	1	0	1	1
0	0	0	1	1	7	2
0	0	0	1	2	6	7
0	0	0	1	3	0	4
			$\vdots$			
0	0	0	7	0	7	7
0	0	0	7	1	1	4
0	0	0	7	2	0	1
0	0	0	7	3	6	2
			$\vdots$			
0	0	1	0	0	7	3
0	0	1	0	1	1	0
0	0	1	0	2	0	5
0	0	1	0	3	6	6
			$\vdots$			

field $GF(q)$. Then n must be a divisor of $q - 1$. A primitive cyclic Reed–Solomon code over $GF(q)$ has blocklength $q - 1$. For those values of n that do not divide $q - 1$, an element ω of order n does not exist in $GF(q)$, so a Reed–Solomon code on the cyclic line does not exist for such an n. However, shortened Reed–Solomon codes do exist. Longer Reed–Solomon codes – those of blocklength q on the affine line and of blocklength $q + 1$ on the projective line – also exist.

A Reed–Solomon code of blocklength q or $q+1$ can be defined by extending a Reed–Solomon code of blocklength $q - 1$, or by evaluating polynomials on the affine line or on the projective line. To define a Reed–Solomon code in the language of polynomial evaluation, let

$$\mathcal{S} = \{C(x) \mid \deg C(x) \leq k - 1\},$$

and let the defining set be $\mathcal{A} = \{k, k + 1, \ldots, n - 1\}$.

The Reed–Solomon code on the cyclic line is given by

$$\mathcal{C} = \left\{ c \mid c_i = \frac{1}{n} C(\omega^{-i}), C(x) \in \mathcal{S} \right\}.$$

The Reed–Solomon code on the affine line is given by

$$\mathcal{C} = \left\{ \boldsymbol{c} \mid c_i = \frac{1}{n} C(\beta_i), \beta_i \in GF(q), C(x) \in \mathcal{S} \right\}.$$

The Reed–Solomon code on the projective line is given by

$$\mathcal{C} = \left\{ \boldsymbol{c} \mid c_i = \frac{1}{n} C(\beta, \gamma), C(x, 1) \in \mathcal{S} \right\},$$

where $C(x, y)$ is a homogeneous polynomial and (β, γ) ranges over the points of the projective line. That is, $\beta, \gamma \in GF(q)$, and either $\gamma = 1$ or $(\beta, \gamma) = (1, 0)$.

These three versions of the Reed–Solomon code have blocklengths $n = q - 1$, q, and $q + 1$. Accordingly, the latter two are sometimes called *singly extended* and *doubly extended Reed–Solomon codes*. We shall prefer to use the term Reed–Solomon code inclusively to refer to any of the three cases. When it is necessary to be precise, we shall refer to Reed–Solomon codes of blocklength $q - 1$, q, or $q + 1$, respectively, as cyclic, affine, or projective Reed–Solomon codes.

The extra one or two components that are appended to the cyclic Reed–Solomon codewords increase the minimum distance of the code by 1 or by 2. This can be seen by noting that the polynomials $C(x)$ have coefficients C_j equal to zero for $j = k, \ldots, n - 1$. There are $n - k$ consecutive zeros, so the BCH bound says that each codeword of the cyclic code has minimum weight at least $n - k + 1$. But the extended symbols are C_0 and C_{k-1} divided by n. If either or both are zero for any codeword, then the number of consecutive zeros in the spectrum increases by one or two, so the BCH bound says that the weight is larger accordingly. If, instead, either or both of C_0 and C_{k-1} are nonzero, then either or both of c_- or c_+ are nonzero, and again the weight is larger accordingly. Finally, because the code is linear, the minimum distance is equal to the minimum weight of the code.

The dual of a $(q - 1, k, q - k)$ cyclic Reed–Solomon code $\mathcal{C}$ over $GF(q)$ with defining set $\mathcal{A}$ is the $(q - 1, q - 1 - k, k + 1)$ cyclic Reed–Solomon code $\mathcal{C}^\perp$ over $GF(q)$ with defining set $\mathcal{A}^c$, the complement of $\mathcal{A}$. To see this, let $\boldsymbol{c} \in \mathcal{C}$ and $\boldsymbol{c}^\perp \in \mathcal{C}^\perp$ be represented by codeword polynomials $c(x)$ and $c^\perp(x)$, respectively, and observe that the codeword polynomials satisfy $c(\omega^j) c^\perp(\omega^j) = 0$ for all j, from which the convolution property implies orthogonality of $\boldsymbol{c}$ and $\boldsymbol{c}^\perp$ (as well as orthogonality of $\boldsymbol{c}$ and cyclic shifts of $\boldsymbol{c}^\perp$).

The dual of a $(q, k, q - k + 1)$ affine Reed–Solomon code over $GF(q)$, with defining set $\mathcal{A} = \{k, \ldots, q - 2\}$ is a $(q, n - k + 1)$ affine Reed–Solomon code over $GF(q)$ with defining set $\mathcal{A}^\perp = \{q - 1 - k, \ldots, q - 2\}$, but defined with α^{-1} in place of α.

2.5 Encoders for Reed–Solomon codes

An *encoder* is a rule for mapping a k symbol dataword into an n symbol codeword, or it is a device for performing that mapping. A code will have many possible encoders. Any encoder that satisfies the taste or the requirements of the designer can be used.

One encoding rule is simply to insert the k data symbols a_ℓ for $\ell = 0, \ldots, k-1$ into the k unconstrained components of the spectrum

$$C_j = \begin{cases} a_{j-j_0-n+k} & j = j_0 + n - k, \ldots, j_0 + n - 1 \\ 0 & j = j_0, j_0 + 1, \ldots, j_0 - 1 + n - k, \end{cases}$$

as illustrated in Figure 2.1. An inverse Fourier transform completes the encoding. The codeword is given by

$$c_i = \frac{1}{n} \sum_{i=0}^{n-1} \omega^{-ij} C_j \quad i = 0, \ldots, n-1.$$

Note that the k data symbols are not immediately visible among the n components of c. To recover the dataword, one must compute the Fourier transform of c. We refer to this decoder as a *transform-domain encoder*.

A more popular encoder, which we call a *code-domain encoder*,[3] is as follows. Simply define the *generator polynomial* as follows:

$$g(x) = (x - \omega^{j_0})(x - \omega^{j_0+1}) \cdots (x - \omega^{j_0+n-k-1})$$

$$= x^{n-k} + g_{n-k-1} x^{n-k-1} + g_{n-k-2} x^{n-k-2} + \cdots + g_1 x + g_0.$$

The coefficients of $g(x)$ provide the components of a vector g whose Fourier transform G is zero in the required components $j_0, j_0 + 1, \ldots, j_0 + n - k - 1$. Thus $g(x)$ itself is a codeword of weight not larger than $n - k + 1$. (The BCH bound says that it has a minimum weight not smaller than $n - k + 1$, so this is an alternative demonstration that the Reed–Solomon code has minimum weight equal to $n - k + 1$.)

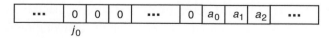
j_0

Figure 2.1. Placement of spectral zeros.

---

[3] In the language of signal processing, these encoders would be called *time-domain* and *frequency-domain* encoders, respectively.

An encoder is as follows. The k data symbols are used to define the data polynomial

$$a(x) = \sum_{i=0}^{k-1} a_i x^i.$$

Then

$$c(x) = g(x)a(x).$$

The degree of $c(x)$ is given by

$$\deg c(x) = \deg g(x) + \deg a(x).$$

If $a(x)$ has its maximum degree of $k - 1$, then $c(x)$ has its maximum degree of

$$\deg c(x) = (n - k) + (k - 1) = n - 1.$$

Thus multiplication of $a(x)$ by $g(x)$ precisely fills out the n components of the codeword.

Again, the k data symbols are not immediately visible in $c(x)$. They are easily recovered, however, by polynomial division:

$$a(x) = \frac{c(x)}{g(x)}.$$

Note that this code-domain encoder gives exactly the same set of codewords as the transform-domain encoder described earlier. The correspondence between datawords and codewords, however, is different.

A code-domain encoder is immediately suitable for a shortened Reed–Solomon code, just as it is suitable for a primitive Reed–Solomon code. To choose n smaller than $q - 1$, simply reduce the dimension and the blocklength by the same amount.

The encoder we shall describe next is useful because the data symbols are explicitly visible in the codeword. An encoder with this property is called a *systematic encoder*. First, observe that the polynomial $x^{n-k} a(x)$ has the same coefficients as $a(x)$, except they are shifted in the polynomial by $n - k$ places. The $n - k$ coefficients of $x^{n-k} a(x)$ with the indices $n - k - 1, n - k - 2, \ldots, 0$ are all zero. We will insert the check symbols into these $n - k$ positions to produce a codeword of the Reed–Solomon code. Specifically, let

$$c(x) = x^{n-k} a(x) - R_{g(x)}[x^{n-k} a(x)],$$

where $R_{g(x)}[x^{n-k}a(x)]$ denotes the remainder polynomial, obtained when $x^{n-k}a(x)$ is divided by $g(x)$. The coefficients of the remainder polynomial occur exactly where $x^{n-k}a(x)$ itself has all coefficients equal to zero. Thus the two pieces making up $c(x)$ do not overlap. The coefficients of the data polynomial $a(x)$ are immediately visible in $c(x)$.

To see that $c(x)$ is indeed a valid codeword polynomial, compute the remainder polynomial of $c(x)$ divided by $g(x)$ by using the facts that remaindering can be distributed across addition (or subtraction) and that the remainder of the remainder is the remainder. Thus,

$$\begin{aligned} R_{g(x)}[c(x)] &= R_{g(x)}[x^{n-k}a(x) - R_{g(x)}[x^{n-k}a(x)]] \\ &= R_{g(x)}[x^{n-k}a(x)] - R_{g(x)}[x^{n-k}a(x)] \\ &= 0. \end{aligned}$$

Therefore $c(x)$, so defined, is a multiple of $g(x)$. This means that it must have the correct spectral zeros, and so it is a codeword of the Reed–Solomon code. The systematic encoder produces the same set of codewords as the other two encoders, but the correspondence between datawords and codewords is different.

2.6   BCH codes

The 8^5 codewords of the $(7, 5, 3)$ Reed–Solomon code over $GF(8)$ are partially listed in Table 2.2. If this list is examined, one finds several codewords that are binary-valued: each component is either 0 or 1. The full list of codewords of the $(7, 5, 3)$ Reed–Solomon code contains exactly sixteen such binary codewords. These sixteen codewords form a linear cyclic code over $GF(2)$. This code is called a $(7, 4, 3)$ BCH code over $GF(2)$; it is also called a $(7, 4, 3)$ Hamming code.

Any subfield-subcode over $GF(q)$ of a Reed–Solomon code over $GF(q^m)$ is called a *BCH code*. A *primitive* BCH code is a subfield-subcode of a primitive Reed–Solomon code. A *narrow-sense* BCH code is a subfield-subcode of a narrow-sense Reed–Solomon code.

The BCH codes satisfy the need for codes of blocklength n over $GF(q)$ when $GF(q)$ contains no element ω of order n, but the extension field $GF(q^m)$ does contain such an element. Then the codeword components are elements of the subfield $GF(q)$, but the codeword spectral components are elements of $GF(q^m)$. Simply choose an (n, k, d) Reed–Solomon code $\mathcal{C}$ over $GF(q^m)$ and define the BCH code to be

$$\mathcal{C}' = \mathcal{C} \cap GF(q)^n.$$

Table 2.2. *Extracting a subfield-subcode from a* $(7,5)$ *code*

Reed–Solomon code	Subfield-subcode
0 0 0 0 0 0 0	0 0 0 0 0 0 0
0 0 0 0 1 6 3	
0 0 0 0 2 7 6	
0 0 0 0 3 1 5	
$\vdots$	
0 0 0 1 0 1 1	0 0 0 1 0 1 1
0 0 0 1 1 7 2	
0 0 0 1 2 6 7	
0 0 0 1 3 0 4	
$\vdots$	
0 0 0 7 0 7 7	
0 0 0 7 1 1 4	
0 0 0 7 2 0 1	
0 0 0 7 3 6 2	
$\vdots$	
0 0 1 0 0 7 3	
0 0 1 0 1 1 0	0 0 1 0 1 1 0
0 0 1 0 2 0 5	
0 0 1 0 3 6 6	
$\vdots$	
0 0 1 1 0 6 2	
0 0 1 1 1 0 1	0 0 1 1 1 0 1
0 0 1 1 2 1 4	
0 0 1 1 3 7 7	
$\vdots$	

The minimum distance of the BCH code is at least as large as the designed distance d. Whereas the packing radius of the Reed–Solomon code $\mathcal{C}$ is always $t = \lfloor (d-1)/2 \rfloor$, the packing radius of the BCH code $\mathcal{C}$ may be larger than $\lfloor (d-1)/2 \rfloor$ because $d_{\min}$ may be larger than d. When this point needs to be emphasized, the designed distance may be denoted d^*, and t may be called the *BCH radius*. The BCH bound guarantees that the packing radius is at least as large as the BCH radius t.

The conjugacy constraint $C_j^q = C_{((qj))}$ gives a complete prescription for defining a spectrum whose inverse Fourier transform is in $GF(q)$. Every codeword of the BCH code must satisfy this constraint on its spectrum.

The BCH bound gives a condition for designing a code whose minimum distance is at least as large as the specified designed distance. A code over $GF(q)$, constructed by imposing both the BCH bound and the conjugacy constraint on the spectrum components in $GF(q^m)$, is a BCH code.

To obtain a binary double-error-correcting code of blocklength 15, choose four consecutive components of the spectrum as the defining set. We shall choose $j = 1, 2, 3, 4$ as the defining set so that $C_1 = C_2 = C_3 = C_4 = 0$. The other components of the spectrum are elements of $GF(16)$, interrelated by the conjugacy condition

$$C_j^2 = C_{((2j))}.$$

This means that a spectral component with index j, corresponding to conjugacy class $\{\alpha^j, \alpha^{2j}, \ldots\}$, determines all other spectral components with an index corresponding to this same conjugacy class. The conjugacy classes modulo 15 lead to the following partition of the set of spectral components:[4]

$$\{C_0\}, \{C_1, C_2, C_4, C_8\}, \{C_3, C_6, C_{12}, C_9\}, \{C_5, C_{10}\}, \{C_7, C_{14}, C_{13}, C_{11}\}.$$

To satisfy the BCH bound for a distance-5 code, we choose the defining set $\mathcal{A} = \{1, 2, 3, 4\}$. By appending all elements of all conjugacy classes of these elements, one obtains the complete defining set, which is $\{1, 2, 3, 4, 6, 8, 9, 12\}$. Because

$$C_0^2 = C_0,$$

component C_0 is an element of $GF(2)$; it can be specified by one bit. Because

$$C_5^4 = C_{10}^2 = C_5,$$

component C_5 is an element of $GF(4)$; it can be specified by two bits. Component C_7 is an arbitrary element of $GF(16)$; it can be specified by four bits. In total, it takes seven bits to specify the codeword spectrum. Thus the code is a $(15, 7, 5)$ BCH code over $GF(2)$.

The $(15, 7, 5)$ binary BCH code can also be described in the code domain in terms of its generator polynomial. This polynomial must have one zero, α^j, corresponding to each element of the complete defining set. Therefore,

$$\begin{aligned} g(x) &= (x - \alpha^1)(x - \alpha^2)(x - \alpha^4)(x - \alpha^8)(x - \alpha^3)(x - \alpha^6)(x - \alpha^{12})(x - \alpha^9) \\ &= x^8 + x^7 + x^6 + x^4 + 1, \end{aligned}$$

which, as required, has coefficients only in $GF(2)$. Because $\deg g(x) = 8$, $n - k = 8$ and $k = 7$.

To obtain a binary triple-error-correcting code of blocklength 63, choose six consecutive spectral indices as the defining set. We shall choose the defining set

---

[4] The term *chord* provides a picturesque depiction of the set of frequencies in the same conjugacy class.

$\mathcal{A} = \{1, 2, 3, 4, 5, 6\}$, which indirectly constrains all components in the respective conjugacy classes. The appropriate conjugacy classes are $\{1, 2, 4, 8, 16, 32\}$, $\{3, 6, 12, 24, 48, 33\}$, and $\{5, 10, 20, 40, 17, 34\}$. The complete defining set is the union of these three sets, so the generator polynomial has degree 18. This cyclic code is a $(63, 45, 7)$ BCH code.

A BCH code may have minimum distance larger than its designed distance. To obtain a binary double-error-correcting code of blocklength 23, choose four consecutive spectral indices as the defining set. We shall choose $j = 1, 2, 3, 4$ as the defining set so that $C_1 = C_2 = C_3 = C_4 = 0$. Because $23 \cdot 89 = 2^{11} - 1$, the spectral components are in the field $GF(2^{11})$. The other components of the spectrum are elements of $GF(2^{11})$, interrelated by the conjugacy condition. There are three conjugacy classes modulo 23, and these partition the Fourier spectrum into the following "chords":

$\{C_0\}$,

$\{C_1, C_2, C_4, C_8, C_{16}, C_9, C_{18}, C_{13}, C_3, C_6, C_{12}\}$,

$\{C_{22}, C_{21}, C_{19}, C_{15}, C_7, C_{14}, C_5, C_{10}, C_{20}, C_{17}, C_{11}\}$.

Only the elements in the set containing C_1 are constrained to zero. Therefore the code is a $(23, 12, 5)$ BCH code. However, this code actually has a minimum distance larger than its designed distance. We shall see later that the true minimum distance of this code is 7. This nonprimitive BCH code is more commonly known as the $(23, 12, 7)$ binary Golay code.

2.7   Melas codes and Zetterberg codes

Let m be odd and $n = 2^m - 1$. Using the Hartmann–Tzeng bound, it is not hard to show that the binary cyclic code of blocklength $n = 2^m - 1$, with spectral zeros at $j = \pm 1$, has minimum distance equal to at least 5. Such a code is called a *Melas double-error-correcting code*. The dimension of the code is $k = 2^m - 1 - 2m$. Thus the parameters of the Melas codes are $(31, 21, 5)$, $(127, 113, 5)$, $(511, 493, 5)$, and so on.

Let m be even and $n = 2^m + 1$. A Fourier transform over $GF(2)$ of blocklength n lies in $GF(2^{2m})$. Using the Hartmann–Tzeng bound, it is not hard to show that the binary cyclic code of blocklength $n = 2^m + 1$, with a spectral zero at $j = 1$, has minimum distance equal to at least 5. Such a code is called a *Zetterberg double-error-correcting code*. The dimension of the Zetterberg code is $k = 2^m + 1 - 2m$. Thus the parameters of the Zetterberg codes are $(33, 23, 5)$, $(129, 115, 5)$, $(513, 495, 5)$, and so on.

2.8   Roos codes

There are some binary cyclic codes whose minimum distance is given by the Roos bound. We call these *Roos codes*. The Roos codes are interesting cyclic codes, but they do not have any special importance. We shall describe two codes of blocklength 127. These are the first members of a family of codes of blocklength $2^m - 1$ for $m = 7, 9, 11, \ldots$ In each case, we will outline the argument used to construct the Roos bound to show that the weight of any nonzero codeword is at least 5.

The $(127, 113)$ binary cyclic code, with defining set $\mathcal{A} = \{5, 9\}$, has its spectral zeros at components with indices in the complete defining set $\mathcal{A}_c = \{5, 10, 20, 40, 80, 33, 66\} \cup \{9, 18, 36, 72, 17, 34, 68\}$. This set of spectral indices contains the subset $\{9, 10, 17, 18, 33, 34\}$. The Roos construction is based on the fact that if c' is defined as

$$c'_i = c_i + A c_i \omega^i,$$

then $C'_j = C_j + A C_{j+1}$. By the modulation property, the vector $[c_i \omega^i]$ has spectral zeros in the set $\{8, 9, 16, 17, 32, 33\}$. Therefore c' has spectral zeros in the set $\{9, 17, 33\}$ and, unless it is identically zero, has weight not larger than the weight of c. But c' cannot be identically zero unless C'_8 and C'_{11} are both zero, which means that either c is zero or has weight at least 5. Furthermore, if C_{25} is not zero, the constant A can be chosen so that $C_{25} + A C_{26}$ is equal to zero. Then c' has spectral zeros at $\{9, 17, 25, 33\}$. Thus there are four spectral zeros regularly spaced with a spacing coprime to n. The BCH bound, suitably generalized, implies that c' has weight at least five. The weight of c cannot be less. Therefore the code is a $(127, 113, 5)$ binary code. It is not a BCH code, but it has the same minimum distance as the $(127, 113, 5)$ binary BCH code.

A second example of a Roos code is the $(127, 106)$ binary cyclic code with defining set $\mathcal{A} = \{1, 5, 9\}$. The complete defining set $\mathcal{A}_c$ contains the subset $\{(8, 9, 10), (16, 17, 18), (32, 33, 34)\}$. By the construction leading to the Roos bound, the two vectors $[c_i \omega^i]$ and $[c_i \omega^{2i}]$ have spectra that are two translates of C. Thus $[c_i \omega^i]$ has spectral zeros at $\{(7, 8, 9), (15, 16, 17), (31, 32, 33)\}$, and $[c_i \omega^{2i}]$ has spectral zeros at $\{(6, 7, 8), (14, 15, 16), (30, 31, 32)\}$. Then the vector c, with components $c_i + A c_i \omega^i + A' c_i \omega^{2i}$, has spectral zeros at indices in the set $\{8, 16, 32\}$. Moreover, for some choice of the constants A and A', spectral zeros can be obtained at $\{8, 16, 24, 32, 40\}$. Thus unless c' is all zero, the weight of codeword c is not smaller than a word with spectral zeros in the set $\{8, 16, 24, 32, 40\}$. By the BCH bound, the weight of that word is at least 6.

Furthermore, if the weight of c is even, the complete defining set contains the set

$$\{(0, 1, 2), (8, 9, 10), (16, 17, 18), (32, 33, 34)\}.$$

Hence by the Roos bound, the weight of an even-weight codeword c is at least as large as a word with defining set $\{0, 8, 16, 24, 32, 40\}$. Because 8 is coprime to 127, we can conclude that the weight of an even-weight codeword is at least 8. Hence the code is a $(127, 106, 7)$ binary cyclic code. It is not a BCH code, but it has the same minimum distance as the $(127, 106, 7)$ BCH code.

2.9   Quadratic residue codes

The binary quadratic residue codes are cyclic codes of blocklength p, with p a prime, and dimension $k = (p+1)/2$. When extended by one bit, a quadratic residue code has blocklength $p + 1$ and rate $1/2$. The family of quadratic residue codes contains some very good codes of small blocklength. No compelling reason is known why this should be so, and it is not known whether any quadratic residue code of large blocklength is good.

Binary quadratic residue codes only exist for blocklengths of the form $p = 8\kappa \pm 1$ for some integer κ. In general, the minimum distance of a quadratic residue code is not known. The main facts known about the minimum distance of quadratic residue codes, which will be proved in this section, are the following. If $p = 8\kappa - 1$,

(i) $d_{\min} = 3 \pmod 4$;
(ii) $d_{\min}(d_{\min} - 1) > p$.

If $p = 8\kappa + 1$,

(i) $d_{\min}$ is odd;
(ii) $d_{\min}^2 > p$.

The Hamming bound, together with the above facts, provide upper and lower bounds on the minimum distance of quadratic residue codes. These bounds are useful for quadratic residue codes of small blocklength, but are too weak to be useful for quadratic residue codes of large blocklength.

Table 2.3 gives a list of the parameters of some binary quadratic residue codes for which the minimum distance is known. Most codes on this list have the largest known minimum distance of any binary code with the same n and k, and this is what makes quadratic residue codes attractive. However, not all quadratic residue codes are this good. Also, satisfactory decoding algorithms are not known for quadratic residue codes of large blocklength, nor are their minimum distances known.

The most notable entry in Table 2.3 is the $(23, 12, 7)$ Golay code, which will be studied in Section 2.10. When extended by one bit, it becomes the $(24, 12, 8)$ code that is known as the *extended Golay code*. Among the quadratic residue codes, extended by

Table 2.3. *Parameters of some binary quadratic residue codes*

n	k	d_{min}
7	4	3[a]
17	9	5[a]
23	12	7[a]
31	16	7[a]
41	21	9[a]
47	24	11[a]
71	36	11
73	37	13
79	40	15[a]
89	45	17[a]
97	49	15
103	52	19[a]
113	57	15
127	64	19
151	76	19

[a]As good as the best code known of this n and k.

one bit, are some very good codes, including binary codes with parameters $(24, 12, 8)$, $(48, 24, 12)$, and $(104, 52, 20)$. All of these codes are self-dual codes.

Quadratic residue codes take their name from their relationship to those elements of a prime field $GF(p)$ that have a square root. We have already stated in Section 1.11 that in the prime field $GF(p)$, $p \neq 2$, exactly half of the nonzero field elements are squares in $GF(p)$ – those $(p-1)/2$ elements that are an even power of a primitive element; these field elements are the elements that have a square root. The set of quadratic residues, denoted $\mathcal{Q}$, is the set of nonzero squares of $GF(p)$, and the set of quadratic nonresidues of $GF(p)$, denoted $\mathcal{N}$, is the set of nonzero elements that are not squares.

Let π be a primitive element of $GF(p)$. Then every element of $GF(p)$, including -1, can be written as a power of π, so the nonzero elements of $GF(p)$ can be written as the sequence

$$\pi^1, \pi^2, \pi^3, \ldots, -1, -\pi^1, -\pi^2, -\pi^3, \ldots, -\pi^{p-2}, 1,$$

in which nonsquares and squares alternate and π is a nonsquare (otherwise, every power of π would also be a square, which cannot be). Because $\pi^{p-1} = 1$ and $(-1)^2 = 1$, it is clear that $\pi^{(p-1)/2} = -1$. If $(p-1)/2$ is even (which means that $p-1$ is a multiple of 4), then -1 appears in the above sequence in the position of a square; otherwise, it appears in the position of a nonsquare.

Theorem 2.9.1 *In the prime field $GF(p)$, the element -1 is a square if and only if $p = 4\kappa + 1$ for some integer κ; the element 2 is a square if and only if $p = 8\kappa \pm 1$ for some integer κ.*

Proof: The first statement of the theorem follows immediately from the remarks prior to the theorem. To prove the second statement, let β be an element of order 8, possibly in an extension field. This element must exist because 8 and p are coprime. Then $\beta^8 = 1$ and $\beta^4 = -1$, so $\beta^2 = -\beta^{-2}$. Let $\gamma = \beta + \beta^{-1}$, and note that

$$\gamma^2 = (\beta + \beta^{-1})^2 = \beta^2 + 2 + \beta^{-2} = 2.$$

Thus 2 is a square. It remains only to show that γ is in the field $GF(p)$ if and only if $p = 8\kappa \pm 1$. But in a field of characteristic p,

$$(\beta + \beta^{-1})^p = \beta^p + \beta^{-p}.$$

If $p = 8\kappa \pm 1$, then because (by the definition of β) $\beta^8 = 1$,

$$(\beta + \beta^{-1})^p = \beta^{8\kappa \pm 1} + \frac{1}{\beta^{8\kappa \pm 1}}$$

$$= \beta^{\pm 1} + \frac{1}{\beta^{\pm 1}}$$

$$= \beta + \beta^{-1}.$$

Thus $\gamma^p = \gamma$, so γ is an element of $GF(p)$. On the other hand, if $p \neq 8\kappa \pm 1$, then $p = 8\kappa \pm 3$. Because $\beta^8 = 1$,

$$(\beta + \beta^{-1})^p = \beta^{8\kappa \pm 3} + \frac{1}{\beta^{8\kappa \pm 3}}$$

$$= \beta^{\pm 3} + \frac{1}{\beta^{\pm 3}}$$

$$= \beta^3 + \beta^{-3}.$$

But $\beta^2 = -\beta^{-2}$, so $\beta^3 = -\beta^{-1}$ and $\beta^{-3} = -\beta$. We conclude that $\gamma^p = -\gamma$. But every element of $GF(p)$ satisfies $\gamma^p = \gamma$, which means that γ is not an element of $GF(p)$. This completes the proof of the theorem. $\blacksquare$

When the field $GF(p)$ is written in the natural way as

$$GF(p) = \{0, 1, 2, \ldots, p - 1\},$$

the squares and nonsquares appear in an irregular pattern. When studying quadratic residue codes, it may be preferable to write the nonzero elements in the order of powers of a primitive element π of $GF(p)$. Thus we will think of $GF(p)$ in the alternative order

$$GF(p) = \{0, \pi^0, \pi^1, \pi^2, \ldots, \pi^{p-2}\}.$$

We may list the coordinates of the codeword c in permuted order to give the equivalent codeword:

$$c' = (c_0, c_{\pi^0}, c_{\pi^1}, c_{\pi^2}, \ldots, c_{\pi^{p-2}}).$$

Definition 2.9.2 *A binary quadratic residue code of blocklength p, with p a prime, is a binary cyclic code whose complete defining set $\mathcal{A} \subset GF(p)$ is the set of nonzero squares in $GF(p)$.*

The locator field $GF(2^m)$ should not be confused with the index field $GF(p)$, which is not a subfield of the locator field. The locator field $GF(2^m)$ of the binary quadratic residue code is the smallest extension of the symbol field $GF(2)$ that contains an element of order p. Although the binary quadratic residue codes are over the symbol field $GF(2)$, the quadratic residues used in the definition are in the field $GF(p)$, $p \neq 2$.

The generator polynomial $g(x)$ of the cyclic binary quadratic residue code has a zero in $GF(2^m)$ at ω^j whenever j is a quadratic residue, where $GF(2^m)$ is the smallest extension field of $GF(2)$ that contains an element ω of order p. It follows from this definition that

$$g(x) = \prod_{j \in Q} (x - \omega^j).$$

The binary quadratic residue code of blocklength p exists only if the generator polynomial $g(x)$ has all its coefficients in the symbol field $GF(2)$. This means that every conjugate of a square in $GF(p)$ must also be a square in $GF(p)$. Consequently, the complete defining set of the code must be equal to Q. The following theorem will allow us to specify when this is so.

Theorem 2.9.3 *A binary quadratic residue code has a blocklength of the form $p = 8\kappa \pm 1$.*

Proof: A binary cyclic code must have $2j$ in the defining set whenever j is in the defining set. But if j is a quadratic residue, then $2j$ is a quadratic residue only if 2 is also a quadratic residue. We have already seen that 2 is a quadratic residue only if $p = 8\kappa \pm 1$, so the proof is complete. ∎

Definition 2.9.4 *An extended binary quadratic residue code of blocklength $p+1$, with p a prime, is the set of vectors of the form*

$$c = (c_0, c_1, \ldots, c_{p-1}, c_\infty),$$

where $(c_0, c_1, \ldots, c_{p-1})$ *is a codeword of the binary quadratic residue code of blocklength p and*

$$c_\infty = \sum_{i=0}^{p-1} c_i.$$

In a moment, we will show that the minimum weight of a quadratic residue code is always odd. But when any codeword of odd weight is extended, $c_\infty = 1$. Therefore the minimum weight of the extended quadratic residue code is always even.

Our general understanding of the minimum distance of quadratic residue codes comes mostly from the following two theorems.

Theorem 2.9.5   *The minimum weight of a binary quadratic residue code is odd.*

Proof: A quadratic residue code is defined so that its spectrum satisfies the quadratic residue condition given in Section 1.11 in connection with the Gleason–Prange theorem. This quadratic residue code can be extended by one component by appending the symbol

$$c_\infty = \sum_{i=0}^{p-1} c_i.$$

Choose any nonzero codeword c of the cyclic code of even weight. The symbol c_∞ of the extended code must be zero for a cyclic codeword of even weight. Because the code is cyclic, we may choose the nonzero codeword so that $c_0 = 1$. The Gleason–Prange theorem tells us that the Gleason–Prange permutation is an automorphism of every extended quadratic residue code, so there is a permutation that interchanges c_0 and c_∞. This permutation produces another codeword of the extended code of the same weight that has a 1 in the extended position. Dropping this position gives a nonzero codeword of the cyclic code with weight smaller by 1. Therefore for any nonzero codeword of even weight in the cyclic code, there is another codeword of weight smaller by 1. Hence the minimum weight of the cyclic code must be odd.   ■

Theorem 2.9.6   *Let c be a codeword of odd weight w from a quadratic residue code of blocklength p. There exists a nonnegative integer r such that*

$$w^2 = p + 2r.$$

Moreover, if $p = -1$ (mod 4), then w satisfies the stronger condition

$$w^2 - w + 1 = p + 4r.$$

Proof: Every nonsquare can be expressed as an odd power of the primitive element π. Let s be any nonsquare element of $GF(p)$. Then js is an even power of π and hence is a square if and only if j is not a square.

Let $c(x)$ be any codeword polynomial of odd weight, and let $\tilde{c}(x) = c(x^s)$ (mod $x^p - 1$), where s is any fixed nonsquare. The coefficients of $\tilde{c}(x)$ are a permutation of the coefficients of $c(x)$. We will show that $c(x)\tilde{c}(x)$ (mod $x^p - 1$) is the all-ones polynomial. It must have odd weight because $c(x)$ and $\tilde{c}(x)$ both have odd weight.

By assumption, $C_j = c(\omega^j) = 0$ for all nonzero j that are squares modulo p, and so $\tilde{C}_j = \tilde{c}(\omega^j) = 0$ for all j that are nonsquares modulo p. Thus $C_j\tilde{C}_j = 0$ for all nonzero j. Further, because $c(x)$ and $\tilde{c}(x)$ each has an odd weight, $C_0 = c(\omega^0) = 1$ and $\tilde{C}_0 = \tilde{c}(\omega^0) = 1$. Therefore

$$C_j\tilde{C}_j = \begin{cases} 1 & \text{if } j = 0 \\ 0 & \text{otherwise.} \end{cases}$$

Then, by the convolution theorem, the inverse Fourier transform of both sides leads to

$$c(x)\tilde{c}(x) = x^{p-1} + x^{p-2} + \cdots + x + 1.$$

Therefore $c(x)\tilde{c}(x)$ (mod $x^p - 1$) has weight p, as can be seen from the right side of this equation.

To prove the first statement of the theorem, we calculate the weight of $c(x)\tilde{c}(x)$ (mod $x^p - 1$) by an alternative method. Consider the computation of $c(x)\tilde{c}(x)$. There are w^2 terms in the raw polynomial product $c(x)\tilde{c}(x)$, and these terms cancel in pairs to produce a polynomial with p ones. Thus $p = w^2 - 2r$, where r is a nonnegative integer. The proof of the first statement of the theorem is complete.

To prove the second statement of the theorem, recall that if $p = -1$ (mod 4), then -1 is a nonsquare modulo p. A codeword polynomial $c(x)$ of odd weight w can be written as follows:

$$c(x) = \sum_{\ell=1}^{w} x^{i_\ell}.$$

Choose $s = -1$ so that $\tilde{c}(x) = c(x^{-1})$, which can be written as follows:

$$\tilde{c}(x) = \sum_{\ell=1}^{w} x^{-i_\ell}.$$

The raw polynomial product $c(x)\tilde{c}(x)$ has w^2 distinct terms before modulo 2 cancellation. Of these w^2 terms, there are w terms of the form $x^{i_\ell}x^{-i_\ell}$, all of which are equal to 1. Because w is odd, $w - 1$ of these terms cancel modulo 2, leaving $w^2 - w + 1$ terms. The remaining terms cancel four at a time, because if

$$x^{i_\ell}x^{-i_{\ell'}} = x^{i_k}x^{-i_{k'}},$$

and so cancel modulo 2, then

$$x^{i\ell'}x^{-i\ell} = x^{ik'}x^{-ik},$$

and these two terms also cancel modulo 2. Thus such terms drop out four at a time. We conclude that, altogether, $w - 1 + 4r$ terms cancel for some, as yet undetermined, nonnegative integer r. Hence, for some r, the weight of the product $c(x)\tilde{c}(x)$ is given by

$$\text{wt}[c(x)\tilde{c}(x)] = w^2 - (w - 1 + 4r),$$

which completes the proof of the theorem. ∎

Corollary 2.9.7 (Square-root bound) *The minimum distance of a quadratic residue code of blocklength p satisfies*

$$d_{\min} \geq \sqrt{p}.$$

Proof: The code is linear so the minimum distance is equal to the weight of the minimum-weight codeword. The minimum-weight codeword of a binary quadratic residue code has odd weight. Thus Theorem 2.9.6 applies. ∎

Corollary 2.9.8 *Every codeword of a binary cyclic quadratic residue code of blocklength p of the form $p = 4\kappa - 1$ has weight either 3 or 0 modulo 4.*

Proof: If the codeword c of the cyclic quadratic residue code has odd weight w, then the theorem allows us to conclude that

$$w^2 - w + 1 = p + 4r$$
$$= 4\kappa - 1 + 4r.$$

Hence,

$$w^2 - w = -2 \quad (\text{mod } 4).$$

This is satisfied for odd w only if $w = 3$ modulo 4.

An argument similar to the one used in the proof of Theorem 2.9.5 allows us to conclude that for every codeword of even weight, there is a codeword of weight smaller by 1. Thus a nonzero codeword of the cyclic quadratic residue code of even weight w can be cyclically shifted into a codeword of the same weight with $c_0 = 1$. Because the weight is even, the codeword can be extended to a codeword with $c_\infty = 0$. The Gleason–Prange permutation produces a new extended codeword c' with $c'_\infty = 1$ and $c'_0 = 0$. When this symbol c'_∞ is purged to obtain a codeword of the cyclic quadratic residue code, that codeword has odd weight which must equal 3 modulo 4. Hence the original codeword has weight equal to 0 modulo 4. ∎

Corollary 2.9.9 *The weight of every codeword of an extended binary quadratic residue code of blocklength p is a multiple of 4 if $p = -1$ modulo 4.*

Proof: This follows immediately from the theorem. ∎

An extended quadratic residue code of blocklength $p + 1$ has a rather rich automorphism group. There are three permutations that suffice to generate the automorphism group for most quadratic residue codes, though, for some p, the automorphism group may be even larger.

Theorem 2.9.10 *The automorphism group of the extended binary quadratic residue code of blocklength $p + 1$ contains the group of permutations generated by the three following permutations:*

(i) $i \rightarrow i + 1 \pmod{p}$, $\infty \rightarrow \infty$;
(ii) $i \rightarrow \pi^2 i \pmod{p}$, $\infty \rightarrow \infty$;
(iii) $i \rightarrow -i^{-1} \pmod{p}$, $i \neq 0$, $0 \rightarrow \infty$, $\infty \rightarrow 0$.

Proof: The first permutation is the cyclic shift that takes index i into index $i + 1$ modulo p (and ∞ into ∞). It is an automorphism because the underlying quadratic residue code is a cyclic code.

The second permutation takes codeword index i into index $\pi^2 i$ (and ∞ into ∞), where π is a primitive element of $GF(p)$. Let d be the permuted sequence. Let C and D be the Fourier transforms of c and d. The cyclic permutation property of the Fourier transform (applied twice) says that if d_i is equal to $c_{\pi^2 i}$, then D_j is equal to $C_{\sigma^2 j}$, where $\pi\sigma = 1 \pmod{p}$. Because π is primitive in $GF(p)$, σ is also primitive in $GF(p)$. But $C_{\sigma^2 j}$ is an automorphism because every nonzero j can be written as σ^r for some r, so $j = \sigma^r$ goes to $\sigma^2 j = \sigma^{r+2}$. If σ^r is a square, then so is σ^{r+2}.

The third permutation uses the structure of the field $GF(p)$ within which the indices of the quadratic residue code lie. Because each index i is an element of $GF(p)$, both the inverse i^{-1} and its negative $-i^{-1}$ are defined. Thus the permutation

$$i \rightarrow -i^{-1}$$

is defined on the extended quadratic residue code, with the understanding that $-(1/0) = \infty$ and $-(1/\infty) = 0$.

Let $d_i = c_{-1/i}$, for $i = 0, \ldots, p - 1$, be the permuted sequence. The sequence d is the Gleason–Prange permutation of the sequence c, and c satisfies the Gleason–Prange condition. Hence by the Gleason–Prange theorem, the sequence d is a codeword of the quadratic residue code, as was to be proved. ∎

Recall that every cyclic code contains a unique idempotent $w(x)$ that satisfies

$$w(x)c(x) = c(x) \pmod{x^n - 1}$$

for all codeword polynomials $c(x)$. In a binary quadratic residue code, the principal idempotent has an attractive form. There are two possibilities, depending on the choice of ω.

Theorem 2.9.11 *Let $w(x)$ be the principal idempotent of the binary cyclic quadratic residue code of blocklength p. If $p = 4\kappa - 1$, then, depending on the choice of ω, either*

$$w(x) = \sum_{i \in \mathcal{Q}} x^i \quad or \quad w(x) = \sum_{i \in \mathcal{N}} x^i.$$

If $p = 4\kappa + 1$, then, depending on the choice of ω, either

$$w(x) = 1 + \sum_{i \in \mathcal{Q}} x^i \quad or \quad w(x) = 1 + \sum_{i \in \mathcal{N}} x^i.$$

Proof: The spectrum of the principal idempotent satisfies

$$W_j = \begin{cases} 0 & \text{if } j \text{ is a nonzero square} \\ 1 & \text{otherwise.} \end{cases}$$

It is only necessary to evaluate the inverse Fourier transform:

$$w_i = \sum_{j=0}^{p-1} \omega^{-ij} W_j.$$

If $p - 1 = 4\kappa - 2$, the equation

$$w_0 = \sum_{j=0}^{p-1} W_j$$

sums an even number of ones, and so $w_0 = 0$. If $p - 1 = 4\kappa$, the first equation sums an odd number of ones, and so $w_0 = 1$.

We use the Rader algorithm to express w_i in the form

$$w'_r = W_0 + \sum_{s=0}^{p-2} g_{r-s} W'_s,$$

where $w'_r = w_{\pi^r}$ and $W'_s = W_{\pi^{-s}}$. This equation can be rewritten as follows:

$$w'_{r+2} = W_0 + \sum_{s=0}^{p-2} g_{r-s} W'_{s+2}.$$

But $W'_{s+2} = W'_s$, so $w'_{r+2} = w'_r$. Because $\boldsymbol{w}$ has only zeros and ones as components, we conclude that $\boldsymbol{w}'$ is an alternating sequence of zeros and ones. It is easy to exclude the possibility that $\boldsymbol{w}'$ is all zeros or is all ones. That is, either w_i is zero when i is a square and one otherwise, or w_i is zero when i is a nonsquare and one otherwise. Hence the theorem is proved.   ∎

Both cases in the theorem are possible, and the theorem cannot be tightened. To show this, we evaluate the single component w_1:

$$w_1 = \sum_{j=0}^{p-1} \omega^{-j} W_j = 1 + \sum_{j \in \mathcal{N}} \omega^{-j} = 1 + \sum_{j \in \mathcal{Q}} \omega^{-\pi j},$$

where π is a primitive element of $GF(p)$. But $\nu = \omega^{\pi}$ also has order p, and therefore could have been used instead of ω in the definition of $g(x)$. With this choice, w_1 would be as follows:

$$w_1 = \sum_{j=0}^{p-1} \nu^{-j} W_j = 1 + \sum_{j \in \mathcal{N}} \omega^{-\pi j} = 1 + \sum_{j \in \mathcal{Q}} \omega^{-j}.$$

So, if ω is chosen as the element of order n, then $w_1 = 1 + \sum_{\mathcal{N}} \omega^{-j}$, while if ν is chosen as the element of order n, then $w_1 = 1 + \sum_{\mathcal{Q}} \omega^{-j}$. But

$$0 = \sum_{j=0}^{p-1} \omega^{-j} = 1 + \sum_{j \in \mathcal{N}} \omega^{-j} + \sum_{j \in \mathcal{Q}} \omega^{-j},$$

so w_1 cannot be invariant under the choice of ω.

2.10   The binary Golay code

The $(23, 12, 7)$ binary quadratic residue code is a remarkable code that deserves special attention. This code was discovered earlier than the other quadratic residue codes. For this reason, and because of its special importance, the $(23, 12, 7)$ binary quadratic residue code is also called the *binary Golay code*. The binary Golay code is unique up to the permutation of components. It is the only $(23, 12, 7)$ binary code. When extended by one additional check bit, the Golay code becomes the $(24, 12, 8)$ extended Golay code. The $(24, 12, 8)$ extended Golay code is also unique up to the permutation of components.

According to Theorem 2.9.6, the minimum distance of this code satisfies

$$d_{\min}^2 - d_{\min} = p - 1 + 4r$$

for some unspecified nonnegative integer value of r. The possibilities for the right side are

$$p - 1 + 4r = 22, 26, 30, 34, 38, 42, 46, \ldots$$

Because $d_{\min}$ is known to be odd for a binary quadratic residue code, the possibilities for the left side are

$$3^2 - 3 = 6,$$
$$5^2 - 5 = 20,$$
$$7^2 - 7 = 42,$$
$$9^2 - 9 = 72,$$
$$11^2 - 11 = 110,$$

and so forth.

The integer 42 occurs on both sides of the equation. Thus the value $d_{\min} = 7$ satisfies the square-root bound. Larger integers, such as 11, also solve the square-root bound, but these can be excluded by the Hamming bound, which is given by the following counting argument. Because $d_{\min}$ is at least 7, spheres of radius 3 about codewords do not intersect. There are $\sum_{\ell=0}^{3} \binom{23}{\ell} = 2^{11}$ points within distance 3 from a codeword, and there are 2^{12} codewords. Because there are $2^{23} = 2^{11}2^{12}$ points in $GF(2)^{23}$, every point of the space is not more than distance 3 from a codeword. Hence spheres of radius 4 around codewords cannot be disjoint, so the minimum distance between codewords is at most 7.

Because of its importance, we will summarize what has been proved as a theorem.

Theorem 2.10.1 *The binary Golay code is a perfect triple-error-correcting code.*

Proof: By the square root bound, the minimum distance is at least 7. By the Hamming bound, the minimum distance is at most 7. Moreover, because

$$2^{12}\left[\binom{23}{0} + \binom{23}{1} + \binom{23}{2} + \binom{23}{3}\right] = 2^{23},$$

the Golay code is a perfect code. ∎

The number of codewords of each weight of the Golay code is tabulated in Table 2.4. This table is easy to compute by examining all codewords. It agrees with the assertion of Corollary 2.9.8, which says that every codeword in the Golay code has weight, modulo 4, equal to 3 or 4.

The next task is to find the generator polynomial of the Golay code. Because it is a quadratic residue code, we know that the Golay code is a cyclic code and so has a

Table 2.4. *Weight distribution of Golay codes*

Weight	(23, 12) code	Extended (24, 12) code
0	1	1
7	253	0
8	506	759
11	1288	0
12	1288	2576
15	506	0
16	253	759
23	1	0
24	—	1
	4096	4096

generator polynomial. Let $g(x)$ and $\tilde{g}(x)$ be the following two reciprocal polynomials in the ring $GF(2)[x]$:

$$g(x) = x^{11} + x^{10} + x^6 + x^5 + x^4 + x^2 + 1,$$
$$\tilde{g}(x) = x^{11} + x^9 + x^7 + x^6 + x^5 + x + 1.$$

By direct multiplication, it is easy to verify that

$$(x - 1)g(x)\tilde{g}(x) = x^{23} - 1.$$

Hence either $g(x)$ or $\tilde{g}(x)$ can be used as the generator polynomial of a $(23, 12)$ cyclic code. To show that these codes are the only $(23, 12)$ cyclic codes, it is enough to show that these polynomials are irreducible, because this means that there could be no other factors of $x^{23} - 1$ of degree 11.

Because $2047 = 23 \times 89$, we know that if α is a primitive element in the field $GF(2048)$, then $\omega = \alpha^{89}$ has order 23, as does ω^{-1}. Let $f(x)$ and $\tilde{f}(x)$ denote the minimal polynomials of ω and ω^{-1}, respectively. The conjugates of ω are the elements of the set

$$\mathcal{B} = \{\omega, \omega^2, \omega^4, \omega^8, \omega^{16}, \omega^9, \omega^{18}, \omega^{13}, \omega^3, \omega^6, \omega^{12}\},$$

which has eleven members. The conjugates of ω^{-1} are the inverses of the conjugates of ω. Because the conjugates of ω and their inverses altogether total 22 field elements, and the 23rd power of each element equals 1, we conclude that both $f(x)$ and $\tilde{f}(x)$ have degree 11. Hence,

$$(x - 1)f(x)\tilde{f}(x) = x^{23} - 1,$$

which, by the unique factorization theorem, is unique. But we have already seen that this is satisfied by the $g(x)$ and $\tilde{g}(x)$ given earlier. Hence the generator polynomials $g(x)$ and $\tilde{g}(x)$ are the minimal polynomials of α^{89} and α^{-89} in the extension field $GF(2048)$. These polynomials must generate the Golay code and the reciprocal Golay code.

2.11   A nonlinear code with the cyclic property

Not all codes that satisfy the cyclic property are cyclic codes. A code may satisfy the cyclic property and yet not be linear. We shall construct a nonlinear binary code that satisfies the cyclic property. We shall still refer to this as a $(15, 8, 5)$ code even though the code is not linear. The datalength is 8, which means that there are 2^8 codewords. Because the cyclic code is nonlinear, it does not have a dimension. The comparable linear cyclic code is the $(15, 7, 5)$ binary BCH code, which is inferior because it contains only 2^7 codewords. The nonlinear $(15, 8, 5)$ cyclic code may be compared with the $(15, 8, 5)$ *Preparata code*, which is a noncyclic, nonlinear binary code that will be studied in other ways in Section 2.16.

Let ω be any element of $GF(16)$ of order 15 (thus a primitive element) used to define a fifteen-point Fourier transform. Define the code $\mathcal{C}$ as the set of binary words of blocklength 15 whose spectra satisfy the constraints $C_1 = 0$, $C_3 = A$, and $C_5 = B$, where either

(1) $B = 0$ and $A \in \{1, \omega^3, \omega^6, \omega^9, \omega^{12}\}$

or

(2) $A = 0$ and $B \in \{1, \omega^5, \omega^{10}\}$,

and all other spectral components are arbitrary insofar as the conjugacy constraints allow. Clearly, this code is contained in the $(15, 11, 3)$ Hamming code and contains the $(15, 5, 7)$ BCH code.

By the modulation property of the Fourier transform, a cyclic shift of b places replaces C_3 by $\omega^{3b}C_3$ and replaces C_5 by $\omega^{5b}C_5$. This means that the cyclic shift of every codeword is another codeword. However, this code is not a linear code. In particular, the all-zero word is not a codeword.

Because C_0 is an arbitrary element of $GF(2)$, and C_7 is an arbitrary element of $GF(16)$, C_0 and C_7 together represent five bits. An additional three bits describe the eight choices for A and B. Altogether it takes eight bits to specify a codeword. Hence there are 256 codewords; the code is a nonlinear $(15, 8)$ code.

The minimum distance of the code is 5, as will be shown directly by an investigation of the linear complexity of codewords. Because the code is a subcode of the BCH

$(15, 11, 3)$ code, the distance between every pair of codewords is at least 3, so we need only to prove that it is not 3 or 4. Thus we must prove that if codeword c has spectral components $C_1 = 0$, $C_3 = A$, and $C_5 = B$, and c' has spectral components $C_1' = 0$, $C_3' = A'$, and $C_5' = B'$, as described above, then the difference $v = c - c'$ does not have weight 3 or 4. This is so by the BCH bound if $A = A' = 0$. We need only consider the cases where either A or A' is nonzero or both A and A' are nonzero.

The method of proof is to show that, for both of these cases, there is no linear recursion of the form $V_j = \sum_{k=1}^{4} \Lambda_k V_{j-k}$ that is satisfied by the spectrum V. Starting with $V_j^2 = V_{((2j))}$ and repeating the squaring operation twice more, we have

$$V_3^8 = V_{((24))} = V_9.$$

The recursion provides the following equations:

$$V_4 = \Lambda_1 V_3 + \Lambda_2 V_2 + \Lambda_3 V_1 + \Lambda_4 V_0,$$
$$V_5 = \Lambda_1 V_4 + \Lambda_2 V_3 + \Lambda_3 V_2 + \Lambda_4 V_1,$$
$$V_6 = \Lambda_1 V_5 + \Lambda_2 V_4 + \Lambda_3 V_3 + \Lambda_4 V_2,$$
$$V_7 = \Lambda_1 V_6 + \Lambda_2 V_5 + \Lambda_3 V_4 + \Lambda_4 V_3,$$
$$V_8 = \Lambda_1 V_7 + \Lambda_2 V_6 + \Lambda_3 V_5 + \Lambda_4 V_4,$$
$$V_9 = \Lambda_1 V_8 + \Lambda_2 V_7 + \Lambda_3 V_6 + \Lambda_4 V_5,$$
$$V_{10} = \dots$$

If the weight of v is less than 5, then the weight is either 3 or 4. If the weight of v is 4, then $V_0 = 0$. If the weight of v is 3, then $\Lambda_4 = 0$. In either case, $\Lambda_4 V_0 = 0$. Because $V_1 = V_2 = V_4 = V_8 = 0$, and $V_6 = V_3^2$, these equations reduce to the following:

$$0 = \Lambda_1 V_3,$$
$$V_5 = \Lambda_2 V_3,$$
$$V_3^2 = \Lambda_1 V_5 + \Lambda_3 V_3,$$
$$V_7 = \Lambda_1 V_3^2 + \Lambda_2 V_5 + \Lambda_4 V_3,$$
$$0 = \Lambda_2 V_3^2 + \Lambda_3 V_5,$$
$$V_9 = \Lambda_2 V_7 + \Lambda_3 V_3^2 + \Lambda_4 V_5.$$

We have already remarked that if $V_3 = 0$, then the BCH bound asserts that the weight of v is at least 5. We need only consider the case in which V_3 is nonzero. Then the first equation requires that $\Lambda_1 = 0$, so the recursion reduces to the following simplified equations:

$$V_5 = \Lambda_2 V_3,$$

$$V_3 = \Lambda_3,$$

$$V_7 = \Lambda_2 V_5 + \Lambda_4 V_3,$$

$$V_3^8 = \Lambda_2 V_7 + \Lambda_3 V_3^2 + \Lambda_4 V_5.$$

If A is nonzero and A' is zero, then B is zero and B' is nonzero. By the definition of A and B', V_3 is a nonzero cube and V_5 is a nonzero fifth power. Because 3 and 5 are coprime integers, there is a cyclic shift of b places such that $V_3\omega^{3b} = 1$ and $V_5\omega^{5b} = 1$. Then we may take $V_3 = V_5 = 1$ without changing the weight of v. We need only show that no vector v with $V_1 = 0$, $V_3 = 1$, and $V_5 = 1$ has weight 3 or 4. In this case, the equations from the recursion become

$$1 = \Lambda_2,$$

$$1 = \Lambda_3,$$

$$V_7 = \Lambda_2 + \Lambda_4,$$

$$1 = \Lambda_2 V_7 + \Lambda_3 + \Lambda_4,$$

which reduce to

$$V_7 = 1 + \Lambda_4$$

and

$$1 = V_7 + 1 + \Lambda_4.$$

However, these two equations are not consistent. The contradiction implies that under the stated assumptions, the linear complexity cannot be 3 or 4. This means that the weight of v is at least 5 if both A and B' are nonzero.

Finally, if A and A' are both nonzero, then B and B' are both zero. Moreover, $A - A'$ must be a noncube because the sum of two cubes in $GF(16)$ is never a cube, which is easy to see from a table of $GF(16)$ (see Table 3.1). Thus, we must show that no vector v, with $V_1 = 0$, V_3 a noncube, and $V_5 = 0$, has weight 3 or 4.

If $V_3 = A - A'$ is a nonzero noncube and $V_5 = B - B' = 0$, then the equations of the recursion require that $\Lambda_2 = 0$, $\Lambda_3 = V_3$, and $V_3^5 = 1$. But if $V_3^5 = 1$ in $GF(16)$, then V_3 is a cube, while $V_3 = A - A'$ must be a noncube as the sum of two nonzero cubes. This contradiction implies that the weight of v cannot be 3 or 4 under the stated assumptions, which means that the weight of v is at least 5.

We conclude that the minimum distance of the code is at least 5, so the code is a nonlinear $(15, 8, 5)$ code.

2.12 Alternant codes

A BCH code over $GF(q)$ of blocklength $n = q^m - 1$ is a subfield-subcode of a Reed–Solomon code over $GF(q^m)$, and so it has at least as large a minimum distance as the Reed–Solomon code. Unfortunately, even though the original Reed–Solomon code has a great many codewords, the subfield-subcode uses very few of them. BCH codes of large blocklength and large minimum distance have dimensions that are small and quite disappointing. In this section, we shall study a method to form better codes by reducing the Reed–Solomon code to a subfield-subcode in another way. This construction produces a large class of codes known as *alternant codes* and a subclass of alternant codes known as *Goppa codes*. The alternant codes are studied in this section, and the Goppa codes are studied in Section 2.13.

Let $\mathcal{C}_{RS}$ be an (n, K, D) Reed–Solomon code over $GF(q^m)$. Let $\boldsymbol{g}$ be a fixed vector[5] of length n, called a *template*, all of whose components are nonzero elements of $GF(q^m)$. A *generalized Reed–Solomon code*, $\mathcal{C}_{GRS}(\boldsymbol{g})$, is a code formed by componentwise multiplication of $\boldsymbol{g}$ with each of the Reed–Solomon codewords. That is,

$$\mathcal{C}_{GRS}(\boldsymbol{g}) = \left\{ \boldsymbol{c} \mid \boldsymbol{c} = \boldsymbol{g}\boldsymbol{c}', \boldsymbol{c}' \in \mathcal{C}_{RS} \right\},$$

where $\boldsymbol{g}\boldsymbol{c}'$ denotes the vector whose ith component is $g_i c_i'$ for $i = 0, \ldots, n-1$. The code $\mathcal{C}_{GRS}(\boldsymbol{g})$ is a linear code. This code contains $(q^m)^K$ vectors, as does the code $\mathcal{C}_{RS}$, and the minimum distance of $\mathcal{C}_{GRS}(\boldsymbol{g})$ is the same as the minimum distance of $\mathcal{C}_{RS}$. Both are equal to D. Thus the generalized Reed–Solomon code is also an (n, K, D) code.

A few of the vectors of $\mathcal{C}_{GRS}(\boldsymbol{g})$ may have all of their components in the smaller field $GF(q)$, and the set of such vectors forms a linear code over $GF(q)$. This subfield-subcode of $\mathcal{C}_{GRS}(\boldsymbol{g})$ is known as an *alternant code*. Specifically, the alternant code $\mathcal{C}_A(\boldsymbol{g})$ is defined as follows:

$$\begin{aligned}\mathcal{C}_A(\boldsymbol{g}) &= \mathcal{C}_{GRS}(\boldsymbol{g}) \cap GF(q)^n \\ &= \left\{ \boldsymbol{c} \mid c_i \in GF(q); \ \boldsymbol{c} = \boldsymbol{g}\boldsymbol{c}', \boldsymbol{c}' \in \mathcal{C}_{RS} \right\}.\end{aligned}$$

Because all g_i are nonzero, we may also write this statement in terms of an inverse template denoted $\boldsymbol{g}^{-1}$ with components g_i^{-1}. Then

$$\mathcal{C}_A(\boldsymbol{g}) = \{ \boldsymbol{c} \mid c_i \in GF(q); \boldsymbol{g}^{-1}\boldsymbol{c} = \boldsymbol{c}'; \ \boldsymbol{c}' \in \mathcal{C}_{RS} \}.$$

---

[5] The use of the notation $\boldsymbol{g}$ and, in Section 2.13, $\boldsymbol{h}$ for the template and inverse template is not to be confused with the use of the notation $\boldsymbol{G}$ and $\boldsymbol{H}$ for the generator matrix and check matrix.

Table 2.5. *Extracting binary codes from a* $(7, 5, 3)$ *Reed–Solomon code*

BCH code							Reed–Solomon code							$g = (5, 6, 1, 4, 1, 1, 7)$ alternant code						
0	0	0	0	0	0	0	0	0	0	0	0	0	0	0	0	0	0	0	0	0
							0	0	0	0	1	6	3							
							0	0	0	0	2	7	6							
							0	0	0	0	3	1	5							
							$\vdots$													
0	0	0	1	0	1	1	0	0	0	1	0	1	1							
							0	0	0	1	1	7	2							
							0	0	0	1	3	0	4							
							$\vdots$													
							0	0	0	7	0	7	7							
							0	0	0	7	1	1	4	0	0	0	1	1	1	1
							0	0	0	7	2	0	1							
							0	0	0	7	3	6	2							
							$\vdots$													
							0	0	1	0	0	7	3							
0	0	1	0	1	1	0	0	0	1	0	1	1	0	0	0	1	0	1	1	0
							0	0	1	0	2	0	5							
							0	0	1	0	3	6	6							
							$\vdots$													

An alternant code, in general, is not a cyclic code. It is easy to see that an alternant code is a linear code and that the minimum distance is at least as large as the minimum distance of the underlying Reed–Solomon code, though it may be larger.

The way that this construction extracts a binary code from a Reed–Solomon code over $GF(2^m)$ is illustrated in Table 2.5. This small example is based on the $(7, 5, 3)$ Reed–Solomon code over $GF(8)$, which expressed in octal notation with template $g = (5, 6, 1, 4, 1, 1, 7)$. Each component g_i of g is a nonzero element of $GF(8)$, which is expressed in octal notation. Of course, one cannot hope to find a binary code better than the $(7, 4, 3)$ Hamming code, so the alternant code constructed in Figure 2.5 cannot contain more than sixteen codewords. This example is too small to give an interesting code. For larger examples, however, it may be that codes better than BCH codes can be found in this way. Indeed, it can be proved that for large values of n and k, by choosing an appropriate template g, one will obtain an (n, k, d) alternant code whose dimension k is large – much larger than that of a BCH code of comparable n and d. Unfortunately, no constructive procedure for choosing the template g is known.

For a more complete example, let $\mathcal{C}_{RS}$ be the extended $(8, 6, 3)$ Reed–Solomon code with defining set $\{6, 0\}$. Choosing the template $\boldsymbol{g} = (1\ 1\ \alpha^5\ \alpha^3\ \alpha^5\ \alpha^6\ \alpha^6\ \alpha^3)$ gives the alternant code with check matrix given by

$$
\boldsymbol{H} = \begin{bmatrix} 1 & 1 & \alpha^2 & \alpha^4 & \alpha^2 & \alpha & \alpha & \alpha^4 \\ 0 & 1 & \alpha^3 & \alpha^6 & \alpha^5 & \alpha^5 & \alpha^6 & \alpha^3 \end{bmatrix}.
$$

The first column corresponds to the symbol appended to give the extended code. Replacing each element of $GF(8)$ by its three-bit representation yields

$$
\boldsymbol{H} = \begin{bmatrix} 1 & 1 & 0 & 0 & 0 & 0 & 0 & 0 \\ 0 & 0 & 0 & 1 & 0 & 1 & 1 & 1 \\ 0 & 0 & 1 & 1 & 1 & 0 & 0 & 1 \\ 0 & 1 & 1 & 1 & 1 & 1 & 1 & 1 \\ 0 & 0 & 1 & 0 & 1 & 1 & 0 & 1 \\ 0 & 0 & 0 & 1 & 1 & 1 & 1 & 0 \end{bmatrix}.
$$

The six rows of $\boldsymbol{H}$ are linearly independent, and hence this check matrix specifies an $(8, 2, 5)$ alternant code. It is easy to verify that a generator matrix for this code is given by

$$
\boldsymbol{G} = \begin{bmatrix} 1 & 1 & 1 & 1 & 0 & 1 & 0 & 0 \\ 1 & 1 & 0 & 0 & 1 & 0 & 1 & 1 \end{bmatrix}.
$$

We shall see in Section 2.13 that this particular alternant code is actually a Goppa code.

To appreciate why one cannot hope to find the template $\boldsymbol{g}$ by unstructured search methods for large codes, note that over $GF(q^m)$ there are $(q^m - 1)^n$ templates with nonzero coefficients, and each of these templates produces a generalized Reed–Solomon code with $(q^m)^K$ codewords. To find a binary code of blocklength 255, one would have to search over 255^{255} templates, approximately 10^{600}, and each template would produce a generalized Reed–Solomon code over $GF(256)$ with 256^K codewords, from which the binary codewords would be extracted to form the binary code. Many of the codes constructed in this way would be worthless, and others would be worthwhile. We do not know how to find the templates that produce good binary codes – we will show only that they do exist. And, of course, even if a good template were known, it would not be practical, in general, simply to list all the codewords; there would be too many. One would need a practical encoding algorithm that would produce the appropriate codeword when it was needed.

Because of the way in which an alternant code is related to the Reed–Solomon code, it is apparent that the minimum distance is at least as large as the designed distance of the Reed–Solomon code. The following theorem says, further, that the dimension satisfies $k \geq n - (d-1)m$.

Theorem 2.12.1 *Let $\mathcal{C}_{GRS}$ be an (n, K, D) generalized Reed–Solomon code over $GF(q^m)$, and let $\mathcal{C}_A$ be an (n, k, d) subfield-subcode of $\mathcal{C}_{GRS}$ over $GF(q)$. Then $D \leq d$ and $n - (d-1)m \leq k \leq K$.*

Proof: The inequality $D \leq d$ is apparent. This inequality leads immediately to the inequality $D + K \leq d + K$, whereas the inequality $d + k \leq D + K$ holds because the Reed–Solomon code satisfies the Singleton bound with equality and the subfield-subcode need not. Together these two inequalities lead to the inequality $k \leq K$.

The only inequality still requiring proof is $n - (d-1)m \leq k$. The generalized Reed–Solomon code is a linear code determined by $n - K$ check equations over $GF(q^m)$. Each check equation is a linear combination of elements of $GF(q)$ with coefficients in $GF(q^m)$. Each such linear combination can be viewed as m check equations with coefficients in $GF(q)$ that the subfield-subcode must satisfy. These $m(n - K)$ check equations over $GF(q)$ need not be linearly independent. The inequality $(n - k) \leq m(n - K)$ follows. To complete the proof, set $n - K$ for the Reed–Solomon code equal to $D - 1$, so $n \leq k + m(D - 1) \leq k + m(d - 1)$. ∎

Because a BCH code is actually a special case of an alternant code in which the template is all ones, the theorem holds for the class of BCH codes. With reference to Theorem 2.12.1, one wishes to choose the template of an alternant code such that the inequality bound $n - (d - 1)m \leq k$ is satisfied as loosely as possible, and, more to the point, that the code is better than the corresponding BCH code. This may occur either because $d \geq D$ or because $k \geq n - (D - 1)m$, or both.

For example, let $\mathcal{C}_{RS}$ be a primitive cyclic Reed–Solomon code over $GF(2^m)$, with defining set $\{0, 1\}$. If the template is all ones, then, because $C_0 = 0$, all codewords of the binary alternant code have even weight, and, because $C_1 = 0$, all codewords of that code are binary Hamming codewords. Thus $d_{\min} = 4$ and $k = n - (m + 1)$. If, instead, the template is $g_i = \alpha^i$ for $i = 0, \ldots, n - 1$, then the generalized Reed–Solomon code is actually a Reed–Solomon code with defining set $\{1, 2\}$. Hence the alternant code is a Hamming code with $d_{\min} = 3$ and $k = n - m$. Both of these examples are actually BCH codes: one has a larger dimension and one has a larger minimum distance.

Alternant codes are attractive because, as we shall see, there are templates that give much better alternant codes than the BCH code. For blocklength $n = 2^m - 1$, there are n^n templates. Some of these give good codes. In particular, there are sequences of alternant codes of increasing blocklength such that the rate k/n and relative minimum distance $d_{\min}/n$ both remain bounded away from zero as n goes to infinity. This is a consequence of the following theorem.

Theorem 2.12.2   *For any prime power q and integer m, let $n = q^m - 1$, and let d and r be any integers that satisfy*

$$\sum_{j=1}^{d-1} \binom{n}{j} (q-1)^j < (q^m - 1)^r.$$

Then there exists an alternant code over $GF(q)$ of blocklength n, dimension $k \geq n - mr$, and minimum distance $d_{\min} \geq d$.

Proof: The method of proof is to fix an (n, k) Reed–Solomon code over $GF(q^m)$ and an arbitrary vector v over $GF(q)$ of weight j. Then count the number of templates for which v belongs to the alternant code formed by that template from the fixed Reed–Solomon code. We conclude that there are not enough v of weight less than d to allow every template to produce at least one such v. Thus at least one of the templates gives an alternant code that has no v of weight less than d. This alternant code must have minimum distance at least as large as d.

Step (1) Let $\mathcal{C}_{\text{RS}}$ be a fixed Reed–Solomon code over $GF(q^m)$ of blocklength n and dimension $K = n - r$. For each template g, let $\mathcal{C}_A(g)$ be the alternant code over $GF(q)$ generated from $\mathcal{C}_{\text{RS}}$ by g. Then

$$\mathcal{C}_A(g) = \{c \in GF(q)^n \mid g^{-1} c \in \mathcal{C}_{\text{RS}}\},$$

and $g^{-1}c$ denotes the vector $\{g_i^{-1} c_i \mid i = 0, \ldots, n-1\}$. Because $g_i \neq 0$ for all i, there are $(q^m - 1)^n$ such templates that can be used with the Reed–Solomon code $\mathcal{C}_{\text{RS}}$ to form an alternant code, possibly not all of the alternant codes are different. Each alternant code is a subfield-subcode of the generalized Reed–Solomon code $\{c \in GF(q^m)^n \mid g^{-1}c \in \mathcal{C}_{\text{RS}}\}$. The generalized Reed–Solomon code is linear and has r check equations over $GF(q^m)$ that become at most mr check equations over $GF(q)$. For each such code, it follows from Theorem 2.12.1 that

$$k \geq n - mr.$$

Step (2) Choose any vector v over $GF(q)$ of nonzero weight $j < d$. This vector v may appear as a codeword in one or more of the alternant codes defined in Step (1). There are $\sum_{j=1}^{d-1} \binom{n}{j} (q-1)^j$ such vectors of nonzero weight less than d.

Step (3) A vector v of weight j appears $(q^m - 1)^{n-r}$ times in the collection of alternant codes defined in Step (1). This is because, as asserted by Theorem 2.1.2, any $n - r$ places in a Reed–Solomon codeword specify the codeword. If we fix v, there are exactly $n - r$ places in g that can be independently specified such that $g^{-1}v$ is in $\mathcal{C}_{\text{RS}}$.

Step (4) The number of templates that give rise to an alternant code containing a codeword of weight less than d is not larger than the product of the number of vectors of weight less than d and the number of templates for which a given vector could be a codeword in the alternant code produced by that template. From Steps (2) and (3), this product is given by $(q^m - 1)^{n-r} \sum_{j=1}^{d-1} \binom{n}{j}$ $(q - 1)^j$. From Step (1), the number of templates is $(q^m - 1)^n$. Suppose

$$(q^m - 1)^n > (q^m - 1)^{n-r} \sum_{j=1}^{d-1} \binom{n}{j} (q - 1)^j.$$

Then some code of dimension at least $n - mr$ does not contain any codeword of weight smaller than d, and so has minimum distance at least as large as d. This is equivalent to the statement of the theorem. ∎

Corollary 2.12.3   *An (n, k) binary alternant code exists that satisfies*

$$\sum_{j=1}^{d-1} \binom{n}{j} < 2^{n-k}.$$

Proof: With $q = 2$, the theorem states that if

$$\sum_{j=1}^{d-1} \binom{n}{j} < (2^m - 1)^r,$$

then there exists a binary alternant code with minimum distance at least as large as d and with $k \geq n - mr$, so such a code exists with $k = n - mr$. The corollary then follows because $2^m - 1 < 2^m$. ∎

The class of alternant codes is very large because the number of templates over $GF(q)$ of blocklength $q^m - 1$ is $(q^m - 1)^{q^m-1}$. Theorem 2.12.12 and Corollary 2.12.13 only tell us that some of these templates give good alternant codes, but they do not indicate how to find them. In fact, little is known about how to find the good alternant codes.

The following corollary is a restatement of the previous corollary in a somewhat more convenient form, using the function

$$H_2(x) = -x \log_2 x - (1 - x) \log_2(1 - x) \quad 0 \leq x \leq 1,$$

which is known as the (*binary*) *entropy*.

Corollary 2.12.4 (Varshamov–Gilbert bound)   *A binary code of rate R and relative minimum distance d/n exists for sufficiently large n, provided that*

$$H_2\left(\frac{d}{n}\right) < 1 - R.$$

Proof: The weak form of Stirling's approximation is given by

$$n! = 2^{n\log_2 n + o(1)},$$

where $o(1)$ is a term that goes to zero as n goes to infinity. Using the weak form of Stirling's approximation, we can form the following bound:

$$\sum_{j=1}^{d-1}\binom{n}{j} > \binom{n}{d-1} = \frac{n!}{(d-1)!(n-d+1)!} = 2^{n[H_2(p)+o'(1)]},$$

where $p = d/n$ and $o'(1)$ is a term that goes to zero as n goes to infinity. The difference between $p = d/n$ and $(d-1)/n$ is absorbed into $o'(1)$. Therefore Corollary 2.12.3 can be written $2^{n[H_2(p)+o'(1)]} < 2^{n(1-R)}$, where $R = k/n$. Under the statement of the corollary, the condition of Corollary 2.12.3 will be satisfied for sufficiently large n. The corollary follows. ∎

The Varshamov–Gilbert bound can also be proved for other classes of codes. At present, it is not known whether a class of binary codes exists that is asymptotically better than the Varshamov–Gilbert bound. The alternant codes form a very large class, however, and without some constructive methods for isolating the good codes, the performance statement of Corollary 2.12.4 is only an unfulfilled promise.

Because an alternant code is closely related to a Reed–Solomon code, any procedure for decoding the Reed–Solomon code can be used to decode the alternant code out to the designed distance. The only change that is needed is a new initial step to modify the senseword, using the inverse of the template to reconstruct a noisy Reed–Solomon codeword. This observation, however, misses the point. The appeal of an alternant code is that its minimum distance can be much larger than its designed distance. A binary alternant code used with a Reed–Solomon decoder has little advantage over a binary BCH code used with that decoder. The only advantage is that, although the Reed–Solomon decoder can only correct to the designed distance, it can detect error patterns up to the minimum distance. This might be a minor reason to use an alternant code in preference to a BCH code, but it does not fulfil the real purpose of a code.

Finally, we remark that, though we do not have decoders for alternant codes that decode to their minimum distances, this lack remains of little importance because we cannot even find the good codes.

2.13   Goppa codes

A special subclass of alternant codes, the subclass of *Goppa codes*, was discovered earlier than the general class and remains worthy of individual attention. We know the subclass of Goppa codes retains the property that it contains many good codes of large blocklength, but we do not yet know how to find the good Goppa codes because this subclass is still of such a large size. However, the Goppa codes of small blocklength can be constructed. These small codes are interesting because there are some codes with combinations of blocklength, dimension, and minimum distance that cannot be achieved with BCH codes or other known codes.

Recall that an alternant code of blocklength $n = q^m - 1$ is associated with a template g with nonzero components g_i. Define an inverse template h with nonzero components h_i such that $(1/n)g_i h_i = 1$. Thus $h_i = ng_i^{-1}$. The template and the inverse template have Fourier transforms G and H, which can be represented as polynomials $G(x)$ and $H(x)$. The convolution property of the Fourier transform converts the expression $(1/n)g_i h_i = 1$ to the expression $(1/n)G(x)H(x) = n$, so that

$$g_i h_i = \frac{1}{n}G(\omega^{-i})\frac{1}{n}H(\omega^{-i}).$$

To develop this same statement in a roundabout way, note that $G(x)$ has no zeros in $GF(q^m)$ because $G(\omega^{-i}) = ng_i \neq 0$. Hence $G(x)$ is coprime to $x^n - 1$, and, by the extended euclidean algorithm for polynomials, the polynomials $F(x)$ and $E(x)$ over $GF(q)$ exist such that

$$G(x)F(x) + (x^n - 1)E(x) = 1.$$

That is, over $GF(q)$,

$$G(x)F(x) = 1 \quad (\mathrm{mod}\ x^n - 1),$$

so the asserted $H(x)$ does indeed exist as the polynomial $n^2 F(x)$.

The definition of the alternant codes is easily restated in the transform domain, and this is the setting in which the Goppa codes will be defined. Let ω be an element of $GF(q^m)$ of order n. Let $H(x)$ be a fixed polynomial such that $H(\omega^{-i}) \neq 0$ for $i = 0, \ldots, n - 1$, and let j_0 and t be fixed integers. The alternant code $\mathcal{C}_A$ is the set containing every vector c whose transform C satisfies two conditions:

$$C_j^q = C_{((qj))}$$

and

$$\sum_{k=0}^{n-1} H_{((j-k))} C_k = 0 \quad j = j_0, \dots, j_0 + 2t - 1.$$

The first of these two conditions ensures that the code-domain codewords are $GF(q)$-valued. The second condition is a convolution, corresponding to the componentwise product $g_i^{-1} c_i$ of the code-domain definition of the alternant code given in Section 2.12. The vector

$$C_j' = \sum_{k=0}^{n-1} H_{((j-k))} C_k \quad j = 0, \dots, n-1$$

might be called the *filtered spectrum* of the alternant codeword. The second condition states that the filtered spectrum C' must be the spectrum of a Reed–Solomon codeword. In the language of polynomials, this becomes $C'(x) = H(x)C(x)$, where $C'(x)$ is the spectrum polynomial of a Reed–Solomon code.

An equivalent statement of this formulation can be written in terms of G as follows:

$$C_j = \sum_{k=0}^{n-1} G_{((j-k))} C_k' \quad j = 0, \dots, n-1.$$

In the language of polynomials, this becomes $C(x) = G(x)C'(x)$.

All of the preceding remarks hold for any alternant code. For an alternant code to be a Goppa code, $G(x)$ is required to satisfy the additional condition given in the following definition.

Definition 2.13.1 *A Goppa code of designed distance d is an alternant code of designed distance d with nonzero template components of the form $g_i = (1/n)G(\omega^{-i})$, where $G(x)$ is a polynomial of degree $d - 1$.*

The new condition is that the polynomial $G(x)$ is now required to have degree $d - 1$. This polynomial is called the *Goppa polynomial*. If $G(x)$ is an irreducible polynomial, then the Goppa code is called an *irreducible Goppa code*. Because of the restriction that $\deg G(x) = d - 1$, the Goppa code is a special case of an alternant code. As for the general case of alternant codes, $G(x)$ can have no zeros in the field $GF(q^m)$, so all g_i are nonzero, unless the code is a shortened code.

A *narrow-sense Goppa code* is a Goppa code for which the underlying Reed–Solomon code is a narrow-sense Reed–Solomon code. Thus if $C(x)$ is the spectrum polynomial of the narrow-sense Goppa code, then the polynomial $C(x) = G(x)C'(x)$ has degree at most $n - 1$, so a modulo $x^n - 1$ reduction would be superfluous. This is because $C'(x)$ has degree at most $n - d - 2$, and $G(x)$ has degree $d - 1$, so the polynomial $C(x)$ has degree at most $n - 1$, even without the modulo $x^n - 1$ reduction.

Theorem 2.13.2   *In a Goppa code with Goppa polynomial $G(x)$ and defining set $j_0, \ldots, j_0 + d - 2$, c is a codeword if and only if*

$$\sum_{i=0}^{n-1} c_i \frac{\omega^{ij}}{G(\omega^{-i})} = 0 \quad j = j_0, \ldots, j_0 + d - 2.$$

Proof: The proof follows directly from the convolution property of the Fourier transform. ∎

Theorem 2.13.3   *A Goppa code with Goppa polynomial of degree $d - 1$ has minimum distance $d_{\min}$ and dimension k satisfying*

$$d_{\min} \geq d,$$
$$k \geq n - (d - 1)m.$$

Proof: The proof follows immediately from Theorem 2.12.1. ∎

As a subclass of the class of alternant codes, the class of Goppa codes retains the property that it includes many codes whose minimum distance is much larger than d. Just as for the general case of an alternant code, however, not much is known about finding the good Goppa codes. Similarly, no good encoding algorithms for general Goppa codes are known, and no algorithms are known for decoding Goppa codes up to the minimum distance.

It is possible to define the Goppa codes in a more direct way without mentioning the underlying Reed–Solomon codes. This alternative description of the Goppa codes is the content of the following theorem. The theorem can be proved as an immediate consequence of the $GF(q)$ identity,

$$\prod_{i' \neq i} (1 - x\omega^{i'}) = \sum_{j=0}^{n-1} \omega^{ij} x^j,$$

which can be verified by multiplying both sides by $(1 - x\omega^i)$. Instead we will give a proof using the convolution property of the Fourier transform.

Theorem 2.13.4   *The narrow-sense Goppa code over $GF(q)$, with blocklength $n = q^m - 1$ and with Goppa polynomial $G(x)$, is given by the set of all vectors $c = (c_0, \ldots, c_{n-1})$ over $GF(q)$ satisfying*

$$\sum_{i=0}^{n-1} c_i \prod_{i' \neq i} (1 - x\omega^{i'}) = 0 \quad (\mathrm{mod}\ G(x)).$$

Proof: The condition of the theorem can be written

$$\sum_{i=0}^{n-1} c_i \prod_{i' \neq i}(1 - x\omega^{i'}) = C'(x)G(x),$$

where $C'(x)$ is a polynomial of degree at most $n - d$ because $G(x)$ is a polynomial of degree $d - 1$, and the left side is a polynomial of degree at most $n - 1$. That is, $C'(x)$ is the spectrum polynomial of a narrow-sense Reed–Solomon code of dimension $k = n - d - 1$. We only need to show that the polynomial on the left side, denoted

$$C(x) = \sum_{i=0}^{n-1} c_i \prod_{i' \neq i}(1 - x\omega^{i'}),$$

is the spectrum polynomial of the Goppa codeword. Consequently, we shall write

$$C(\omega^{-i}) = c_i \prod_{i' \neq i}(1 - \omega^{-i}\omega^{i'}),$$

$$= c_i \prod_{k=1}^{n}(1 - \omega^{k}).$$

Recall the identity

$$\prod_{k=1}^{n}(x - \omega^{k}) = \sum_{\ell=1}^{n} x^{\ell},$$

which is equal to n when $x = 1$. Therefore $C(\omega^{-i}) = nc_i$, so $C(x)$ is indeed the spectrum polynomial of codeword c. Thus the condition of the theorem is equivalent to the condition defining the narrow-sense Goppa code:

$$C(x) = C'(x)G(x),$$

which completes the proof of the theorem.   ■

The representation given in Theorem 2.13.4 makes it easy to extend the Goppa code by one symbol to get a code with blocklength q^m. Simply append the field element zero as another location number. Then we have the following definition.

Definition 2.13.5 *The Goppa code over $GF(q)$ of blocklength $n = q^m$ and with Goppa polynomial $G(x)$ is given by the set of all vectors $\mathbf{c} = (c_0, \ldots, c_{n-1})$ over $GF(q)$ satisfying*

$$\sum_{i=0}^{n-1} c_i \prod_{i' \neq i} (1 - \beta_{i'} x) = 0 \quad (\mathrm{mod}\ G(x)),$$

where β_i ranges over all q^m elements of $GF(q^m)$.

We now turn to the special case of *binary* Goppa codes, restricting attention to those binary codes whose Goppa polynomial has no repeated zeros in any extension field. Such a code is called a *separable binary Goppa code*. For separable binary Goppa codes, we shall see that the minimum distance is at least $2d - 1$, where $d - 1$ is the degree of $G(x)$. This is more striking than the general bound for any Goppa code, $d_{\min} \geq d$, although of less significance than it might seem.

Theorem 2.13.6 *Suppose that $G(x)$, an irreducible polynomial of degree $d - 1$, is a Goppa polynomial of a narrow-sense binary Goppa code. Then $d_{\min} \geq 2d - 1$.*

Proof: The polynomial $G(x)$ has no zeros in $GF(2^m)$. For a binary code, c_i is either 0 or 1. Let $\mathcal{A}$ be the set of integers that index the components in which c_i is 1. Then Theorem 2.13.4 can be rewritten as follows:

$$\sum_{i \in \mathcal{A}} \prod_{i' \neq i} (1 - x\omega^{i'}) = 0 \quad (\mathrm{mod}\ G(x)).$$

Many of the factors in the product (those for which $i' \in \mathcal{A}$) are in every one of the terms of the sum and can be brought outside the sum as

$$\prod_{i' \notin \mathcal{A}} (1 - x\omega^{i'}) \left[\sum_{i \in \mathcal{A}} \prod_{\substack{i' \in \mathcal{A} \\ i' \neq i}} (1 - x\omega^{i'}) \right] = 0 \quad (\mathrm{mod}\ G(x)).$$

Because $G(x)$ has no zeros in $GF(2^m)$, it must divide the second term on the left. Now write those i in the set $\mathcal{A}$ as i_ℓ for $\ell = 1, \ldots, \nu$. Then the second term on the left can be written as $\sum_{\ell=1}^{\nu} \prod_{\ell' \neq \ell} (1 - x\beta_{\ell'})$, where $\beta_\ell = \omega^{i_\ell}$.

To interpret this term, consider the reciprocal form of the locator polynomial of codeword $\mathbf{c}$, which is given by

$$\Lambda_c(x) = \prod_{\ell=1}^{\nu} (x - \beta_\ell),$$

where β_ℓ is the field element corresponding to the ℓth one of codeword $\mathbf{c}$, and ν is the weight of codeword $\mathbf{c}$. The degree of $\Lambda_c(x)$ is ν. The formal derivative of $\Lambda_c(x)$ is given by

$$\Lambda'_c(x) = \sum_{\ell=1}^{\nu} \prod_{\ell' \neq \ell} (x - \beta_{\ell'}).$$

The right side is the term that we have highlighted earlier. Because that term is divided by $G(x)$, we can conclude that $\Lambda'_c(x)$ is divided by $G(x)$.

Moreover, $\Lambda'_c(x)$ itself can be zero only if $\Lambda_c(x)$ is a square, which is not the case, so $\Lambda'_c(x)$ is a nonzero polynomial and all coefficients of $\Lambda'_c(x)$ of odd powers of x are equal to zero because it is the formal derivative of a polynomial over a finite field of characteristic 2. Thus, it can be written

$$\Lambda'_c(x) = \sum_{\ell=0}^{L} a_\ell x^{2\ell}$$

$$= \left(\sum_{\ell=0}^{L} a_\ell^{1/2} x^\ell \right)^2$$

because in a field of characteristic 2, every element has a square root.

Suppose we have a separable Goppa code with Goppa polynomial $G(x)$. Then not only does $G(x)$ divide $\Lambda'_c(x)$, but because $\Lambda'_c(x)$ is a nonzero square, $G(x)^2$ must also divide $\Lambda'_c(x)$. This shows that $\nu - 1 \geq \deg \Lambda'_c(x) \geq \deg G(x)^2$. Because $G(x)^2$ has a degree of $2(d - 1)$, we conclude that $d_{\min} \geq 2d - 1$.   ∎

The theorem says that for any designed distance $d = r + 1$, a binary Goppa code exists with $d_{\min} \geq 2r + 1$ and $k \geq 2^m - mr$. This code can be compared with an extended binary BCH code with designed distance $r + 1$ for which the extended code satisfies $d_{\min} \geq r + 2$ and $k \geq 2^m - (1/2)mr - 1$. To facilitate the comparison, replace r by $2r'$ for the BCH code. Then $d_{\min} \geq 2r' + 2$ and $k \geq 2^m - mr' - 1$. Thus the significance of Theorem 2.13.6 is that, whereas an extended binary BCH code is larger by 1 in minimum distance, an extended binary Goppa code is larger by 1 in dimension. The theorem promises nothing more than this.

Although the theorem appears to make a separable Goppa code rather attractive because it has minimum distance of at least $2r + 1$, we should point out that the definition produces only r syndromes rather than $2r$, and the usual locator decoding techniques of Reed–Solomon codes do not apply directly. One would need to design a decoding algorithm for these codes that uses only r syndromes.

This concludes our discussion of the theory of Goppa codes. We have presented all the known facts of significance about Goppa codes except for the statement that Goppa codes achieve the Varshamov–Gilbert bound, which proof we omit.

Good examples of large Goppa codes remain undiscovered. The smallest interesting example of a Goppa code is an $(8, 2, 5)$ binary Goppa code, which was used as an example of an alternant code in Section 2.12. Take $G(x) = x^2 + x + 1$. The zeros of this polynomial are distinct and are in $GF(4)$ and in all extensions of $GF(4)$. Thus none are in $GF(8)$. Hence $G(x)$ can be used to obtain a Goppa code with blocklength 8, minimum distance at least 5, and dimension at least 2. We shall see that the dimension is 2 and the minimum distance is 5.

The definition of the Goppa code in Theorem 2.13.2 is not suitable for encoding because it defines the Goppa code in terms of a check matrix over the extension field $GF(2^m)$. To find a generator matrix for the $(8, 2, 5)$ code of our example, using this theorem, one must write out a check matrix over $GF(8)$, convert it to a check matrix over $GF(2)$, extract a set of linearly independent rows, and then manipulate the resulting matrix into a systematic form. For our example of the $(8, 2, 5)$ code, this is straightforward. The check matrix for the $(8, 2, 5)$ code, with the Goppa polynomial $x^2 + x + 1$, is given by

$$
H = \begin{bmatrix} 1 & 1 & \alpha^2 & \alpha^4 & \alpha^2 & \alpha & \alpha & \alpha^4 \\ 0 & 1 & \alpha^3 & \alpha^6 & \alpha^5 & \alpha^5 & \alpha^6 & \alpha^3 \end{bmatrix}.
$$

Replacing each field element by its three-bit representation yields

$$
H = \begin{bmatrix} 1 & 1 & 0 & 0 & 0 & 0 & 0 & 0 \\ 0 & 0 & 0 & 1 & 0 & 1 & 1 & 1 \\ 0 & 0 & 1 & 1 & 1 & 0 & 0 & 1 \\ 0 & 1 & 1 & 1 & 1 & 1 & 1 & 1 \\ 0 & 0 & 1 & 0 & 1 & 1 & 0 & 1 \\ 0 & 0 & 0 & 1 & 1 & 1 & 1 & 0 \end{bmatrix}.
$$

These six rows are linearly independent, so H is a nonsystematic check matrix for the $(8, 2, 5)$ binary code. It can be used to form a generator matrix, G, by elementary methods. This process for finding G would be elaborate for large Goppa codes and gives an encoder in the form of an n by k binary generator matrix. Accordingly, we will describe an alternative encoder for the code by a process in the transform domain.

The Goppa polynomial $G(x) = x^2 + x + 1$ leads to the inverse Goppa polynomial $H(x) = x^6 + x^5 + x^3 + x^2 + 1$, because $H(x)G(x) + x(x^7 - 1) = 1$. The underlying $(7, 5, 3)$ cyclic Reed–Solomon code C' of blocklength 7 has spectral zeros at α^{-1} and α^{-2}, and $C'(x) = H(x)C(x)$. Thus $C'_5 = C'_6 = 0$, and C'_k satisfies the equation

$$
C'_k = \sum_{j=0}^{n-1} H_{k-j} C_j,
$$

from which we get the check equations

$$
0 = C_0 + C_1 + C_3 + C_4 + C_6,
$$
$$
0 = C_6 + C_0 + C_2 + C_3 + C_5.
$$

Any spectrum that satisfies these check equations and the conjugacy constraints $C_j^2 = C_{((2j))}$ is a codeword spectrum. Clearly, $C_0 \in GF(2)$ because $C_0^2 = C_0$. Using the conjugacy constraints to eliminate C_2, C_4, C_5, and C_6 from the above equations yields

$$0 = C_0 + (C_1 + C_1^4) + (C_3 + C_3^2),$$
$$0 = C_0 + C_1^2 + (C_3 + C_3^2 + C_3^4).$$

These can be manipulated to give

$$C_0 = C_1 + C_1^2,$$
$$C_3 = C_1 + C_1^2 + C_1^4.$$

The first equation can be solved in $GF(8)$ only if $C_0 = 0$. Then $C_1 \in \{0, 1\}$, and C_3 is determined by C_1. Hence there are only two codewords, as determined by the value of C_1. However, this is not the end of the story. The Reed–Solomon code can be extended by an additional component, c_+, which is then also used to extend the Goppa code. The above equations now become

$$c_+ = C_0 + (C_1 + C_1^4) + (C_3 + C_3^2),$$
$$0 = C_0 + C_1^2 + (C_3 + C_3^2 + C_3^4),$$

which can be manipulated into

$$C_0 = C_1 + C_1^2 + c_+,$$
$$C_3 = C_1 + C_1^2 + C_1^4.$$

These are satisfied if we take $c_+ = C_0$, with the encoding rule

$$C_0 \in \{0, 1\},$$
$$C_1 \in \{0, 1\},$$

and C_3 is equal to C_1. Thus we have two binary data symbols encoded by the values of C_0 and C_1.

Thus in summary, to encode two data bits, a_0 and a_1, set $C_0 = c_+ = a_0$, set $C_1 = C_3 = a_1$, and set all conjugates to satisfy $C_{2j} = C_j^2$. An inverse Fourier transform then produces the codeword.

The $(8, 2, 5)$ binary extended Goppa code, given in this example, might be compared with an $(8, 4, 4)$ binary extended Hamming code. Both are binary alternant codes constructed from the same $(8, 5, 4)$ extended Reed–Solomon code.

For an example of a larger binary Goppa code we will choose the Goppa polynomial $G(x) = x^3 + x + 1$, which has three distinct zeros in $GF(8)$ or in any extension of

$GF(8)$, and hence has no zeros in $GF(32)$. Then by Theorem 2.13.6, $G(x)$ can be used as the Goppa polynomial for a $(31, 16, 7)$ Goppa code, or a $(32, 17, 7)$ extended Goppa code. The $(31, 16, 7)$ binary Goppa code is not better than a $(31, 16, 7)$ binary BCH code. However, the $(32, 17, 7)$ extended Goppa code has a larger dimension, whereas the $(32, 16, 8)$ extended BCH code has a larger minimum distance.

This Goppa code can be described explicitly by writing out a 32 by 17 binary generator matrix or a 32 by 15 binary check matrix. Instead, we will work out an encoder in the transform domain. The Goppa polynomial $G(x) = x^3 + x + 1$ leads to the inverse Goppa polynomial:

$$H(x) = x^{30} + x^{27} + x^{24} + x^{23} + x^{20} + x^{18} + x^{17} + x^{16} + x^{13}$$
$$+ x^{11} + x^{10} + x^9 + x^6 + x^4 + x^3 + x^2,$$

because $H(x)G(x) = (x^2 + 1)(x^{31} - 1) + 1$.

The underlying cyclic Reed–Solomon code C' for the Goppa code C has the defining set $\{28, 29, 30\}$. To examine the structure of this Goppa code, recall that

$$C'_k = \sum_{j=0}^{n-1} H_{k-j} C_j,$$

from which we have the three check equations

$$0 = C_0 + C_3 + C_6 + C_7 + C_{10} + C_{12} + C_{13} + C_{14} + C_{17}$$
$$+ C_{19} + C_{20} + C_{21} + C_{24} + C_{26} + C_{27} + C_{28},$$
$$0 = C_{30} + C_2 + C_5 + C_6 + C_9 + C_{11} + C_{12} + C_{13} + C_{16}$$
$$+ C_{18} + C_{19} + C_{20} + C_{23} + C_{25} + C_{26} + C_{27},$$
$$0 = C_{29} + C_1 + C_4 + C_5 + C_8 + C_{10} + C_{11} + C_{12} + C_{15}$$
$$+ C_{17} + C_{18} + C_{19} + C_{22} + C_{24} + C_{25} + C_{26},$$

and the conjugacy constraints

$$C_j^2 = C_{2j}.$$

Straightforward algebraic manipulation reduces these to three encoding constraint equations:

$$C_3 = \left[C_1^2 + C_1^4 + C_1^8 + C_1^{16}\right] + \left[C_7^4 + C_7^{16}\right] + \left[C_{11} + C_{11}^2 + C_{11}^8 + C_{11}^{16}\right],$$
$$C_5 = c_+ + C_0 + \left[C_1 + C_1^8 + C_1^{16}\right] + \left[C_{11} + C_{11}^2 + C_{11}^4 + C_{11}^8\right],$$
$$C_{15} = \left[C_1^2 + C_1^{16}\right] + \left[C_7^2 + C_7^4 + C_7^8 + C_7^{16}\right] + \left[C_{11} + C_{11}^2 + C_{11}^4 + C_{11}^8 + C_{11}^{16}\right].$$

To encode seventeen data bits, set c_+ and C_0 each to one data bit, set C_1, C_7, and C_{11} each to five data bits, and set C_3, C_5, and C_{15} by the above constraint equations. An inverse Fourier transform completes the encoding. This Goppa code can correct three errors, but, because the defining set of the underlying Reed–Solomon code has only three consecutive elements, the methods of locator decoding, discussed in Chapter 3, cannot be used as such. However, an alternant code exists with the same performance and with six consecutive elements of the defining set, so locator decoding applies as such to that code. The polynomial $x^6 + x^2 + 1$, which is the square of the Goppa polynomial $x^3 + x + 1$, can be used with a similar construction and with a defining set of size 6 to produce a code with the same performance as the Goppa code with the Goppa polynomial $x^3 + x + 1$.

Our two examples of Goppa codes, the $(8, 2, 5)$ binary Goppa code and the $(32, 17, 7)$ binary Goppa code, are the best linear binary codes known of their respective block-lengths and dimensions. Their performance is described by the theorems of this section. However, the main attraction of the class of Goppa codes is their good asymptotic performance, and the examples given do not illustrate this asymptotic behavior. Specific classes of Goppa codes that illustrate the asymptotic behavior have never been found.

2.14   Codes for the Lee metric

The ring of integers modulo q is denoted, as is standard, by $\mathbf{Z}_q$. The *Lee weight* of an element β of $\mathbf{Z}_4$, with the elements written $\{0, 1, 2, 3\}$, is defined as

$$w_L(\beta) = \begin{cases} 0 & \text{if} \quad \beta = 0 \\ 1 & \text{if} \quad \beta = 1 \text{ or } 3 \\ 2 & \text{if} \quad \beta = 2. \end{cases}$$

The Lee weight can be written as $w_L(\beta) = \min[\beta, 4 - \beta]$. The *Lee distance* between two elements of $\mathbf{Z}_4$ is the Lee weight of their difference magnitude: $d_L(\beta, \gamma) = w_L(|\beta - \gamma|)$.

Similarly, the Lee weight of an element β of $\mathbf{Z}_q$, with the elements of $\mathbf{Z}_q$ written $\{0, \ldots, q - 1\}$, is the integer

$$w_L(\beta) = \min[\beta, q - \beta].$$

If the elements of $\mathbf{Z}_q$ are regarded as a cyclic group, then the Lee weight of β is the length of the shortest path on the cycle from β to the zero element. The Lee distance between two elements of $\mathbf{Z}_q$ is the Lee weight of their difference magnitude, $d_L(\beta, \gamma) = w_L(|\beta - \gamma|)$. The Lee distance is the length of the shortest path on the cycle from β to γ.

The Lee weight of sequence $\mathbf{c} \in \mathbf{Z}_q^n$ is the sum of the Lee weights of the n components of $\mathbf{c}$. Thus $w_L(\mathbf{c}) = \sum_{i=0}^{n-1} w_L(c_i)$. The Lee distance, denoted $d_L(\mathbf{c}, \mathbf{c}')$, between two sequences, c and c', of equal length, is defined as the sum of the Lee weights of the componentwise difference magnitudes $\sum_{i=0}^{n-1} w_L |c_i - c'_i|$.

An alternative to the Hamming weight of a sequence on an alphabet of size q is the Lee weight of that sequence. An alternative to the Hamming distance between two finite sequences of the same length on an alphabet of size q is the Lee distance between the two sequences. The Lee weight and the Lee distance are closely related to the modulo-q addition operation. Thus it is natural to introduce the ring $\mathbf{Z}_q$ into the discussion. Indeed, the full arithmetic structure of $\mathbf{Z}_q$ will be used later to design codes based on Lee distance.

A *code* $\mathcal{C}$ of blocklength n and size M over the ring $\mathbf{Z}_q$ is a set of M sequences of blocklength n over the ring $\mathbf{Z}_q$. The code $\mathcal{C}$ is a subset of $\mathbf{Z}_q^n$. A code over the ring $\mathbf{Z}_q$ may be judged either by its minimum Hamming distance or by its minimum Lee distance. In the latter case, we may refer to these codes as Lee-distance codes, thereby implying that Lee distance is the standard of performance. Codes over $\mathbf{Z}_4$ might also be called *quadary codes* to distinguish them from codes over $GF(4)$, often called *quaternary codes*.

Only the addition operation in $\mathbf{Z}_q$ is needed to determine the Lee distance between two codewords. The multiplication operation in $\mathbf{Z}_q$ comes into play only if the code is a linear code. A *linear code* over $\mathbf{Z}_q$ is a code such that the $\mathbf{Z}_q$ componentwise sum of two codewords is a codeword, and the componentwise product of any codeword with any element of $\mathbf{Z}_q$ is a codeword. Even though $\mathbf{Z}_q^n$ is not a vector space, the notions of generator matrix and check matrix of a code do apply.

For example, over $\mathbf{Z}_4$ let

$$G = \begin{bmatrix} 1 & 1 & 1 & 3 \\ 0 & 2 & 0 & 2 \\ 0 & 0 & 2 & 2 \end{bmatrix}.$$

Let $a = [a_0 \ a_1 \ a_2]$ be a dataword over $\mathbf{Z}_4$. Then the codeword $c = [c_0 \ c_1 \ c_2 \ c_3]$ over $\mathbf{Z}_4$ is given by $c = aG$. Although this representation of the code in terms of a generator matrix appears very familiar, the usual operations that exist for a generator matrix over a field need not apply. For example, it is not possible to make the leading nonzero element of the second row of G equal to 1 by rescaling because the inverse of 2 does not exist in $\mathbf{Z}_4$.

A *cyclic code* over the ring $\mathbf{Z}_q$ is a linear code over $\mathbf{Z}_q$ with the property that the cyclic shift of any codeword is another codeword. The codewords of a cyclic code can be represented as polynomials. Then the codewords of a cyclic code can be regarded as elements of $\mathbf{Z}_q[x]$ or, better, of $\mathbf{Z}_q[x]/\langle x^n - 1 \rangle$. One way to form a cyclic code is as the set of polynomial multiples of a polynomial, $g(x)$, called the generator polynomial. Of course, because $\mathbf{Z}_q$ is not a field, the familiar properties of cyclic codes over a field need not apply.

Our first example of a Lee-distance cyclic code over $\mathbf{Z}_4$ is a $(7, 4, 5)$ cyclic code, which can be extended to an $(8, 4, 6)$ code over $\mathbf{Z}_4$ in the usual way by appending an overall check sum. The generator polynomial for the cyclic code is given by

$$g(x) = x^3 + 2x^2 + x + 3.$$

The check polynomial is given by

$$h(x) = x^4 + 2x^3 + 3x^2 + x + 1.$$

This $(7, 4, 5)$ cyclic code has the generator matrix

$$G = \begin{bmatrix} 3 & 1 & 2 & 1 & 0 & 0 & 0 \\ 0 & 3 & 1 & 2 & 1 & 0 & 0 \\ 0 & 0 & 3 & 1 & 2 & 1 & 0 \\ 0 & 0 & 0 & 3 & 1 & 2 & 1 \end{bmatrix}.$$

When extended by one additional check symbol, this code is an $(8, 4, 6)$ code over $\mathbf{Z}_4$, known as the *octacode*. The octacode has the following generator matrix:

$$G = \begin{bmatrix} 1 & 3 & 1 & 2 & 1 & 0 & 0 & 0 \\ 1 & 0 & 3 & 1 & 2 & 1 & 0 & 0 \\ 1 & 0 & 0 & 3 & 1 & 2 & 1 & 0 \\ 1 & 0 & 0 & 0 & 3 & 1 & 2 & 1 \end{bmatrix}$$

as a matrix over $\mathbf{Z}_4$.

The subject of cyclic codes over $\mathbf{Z}_q$ has many similarities to the subject of cyclic codes over a field, but there are also considerable differences. Various properties that hold for cyclic codes over a field do not hold for cyclic codes over a ring. One difference is that the degree of the product $a(x)b(x)$ can be smaller than the sum of the degrees of $a(x)$ and $b(x)$. Indeed, it may be that $a(x)b(x) = 1$, even though both $a(x)$ and $b(x)$ have degrees larger than 0. Thus such an $a(x)$ has an *inverse* under multiplication. Any such $a(x)$ is a *unit* of the ring $\mathbf{Z}_q[x]$. For example, the square of $2x^2 + 2x + 1$ over $\mathbf{Z}_4$ is equal to 1, which means that $2x^2 + 2x + 1$ is a unit of $\mathbf{Z}_4[x]$. Moreover, there is no unique factorization theorem in the ring of polynomials over a ring. For example, observe that

$$\begin{aligned} x^4 - 1 &= (x - 1)(x + 1)(x^2 + 1) \\ &= (x + 1)^2(x^2 + 2x - 1), \end{aligned}$$

so there are (at least) two distinct factorizations over $\mathbf{Z}_4$ of the polynomial $x^4 - 1$. This behavior is typical. Many polynomials over $\mathbf{Z}_q$ have multiple distinct factorizations.

To eliminate this ambiguity, we will choose to define a preferred factorization by using a preferred kind of irreducible polynomial known as a *basic irreducible polynomial*. A basic irreducible polynomial $f(x)$ over $\mathbf{Z}_4$ is a polynomial such that $f(x) \pmod 2$ is an irreducible polynomial over $GF(2)$. Thus the polynomial $f(x)$ over $\mathbf{Z}_4$ is mapped into a polynomial over $GF(2)$ by mapping coefficients 0 and 2 into 0, and mapping coefficients 1 and 3 into 1. The polynomial $f(x)$ is a basic irreducible polynomial over $\mathbf{Z}_4$ if the resulting polynomial over $GF(2)$ is irreducible. The polynomial $f(x)$ is a *primitive* basic irreducible polynomial over $\mathbf{Z}_4$ (or *primitive polynomial*) if the resulting polynomial over $GF(2)$ is primitive.

For example, the irreducible factorization

$$x^7 - 1 = (x - 1)(x^3 + 2x^2 + x - 1)(x^3 - x^2 + 2x - 1)$$

is a factorization over $\mathbf{Z}_4$ into basic irreducible polynomials because, modulo 2, it becomes

$$x^7 - 1 = (x + 1)(x^3 + x + 1)(x^3 + x^2 + 1),$$

which is an irreducible factorization over $GF(2)$. The polynomial $x^3 + 2x^2 + x - 1$ is called the *Hensel lift* of polynomial $x^3 + x + 1$. The Hensel lift to $\mathbf{Z}_4$ of a polynomial over $GF(2)$ can be computed by a procedure called the *Graeffe method*.

Starting with the irreducible polynomial $f(x)$ over $GF(2)$, the Graeffe method first sets

$$f(x) = f_e(x) + f_o(x),$$

where $f_e(x)$ and $f_o(x)$ are made up of the terms of $f(x)$ with even and odd indices, respectively. Then

$$\tilde{f}(x^2) = \pm \left[f_e(x)^2 - f_o(x)^2 \right]$$

determines the Hensel lift $\tilde{f}(x)$. The sign is chosen to make the leading coefficient positive, given that $\mathbf{Z}_4$ is written $\{0, \pm 1, 2\}$.

For example, starting with $f(x) = x^3 + x^2 + 1$ over $GF(2)$, we have $f_e(x) = x^2 + 1$ and $f_o(x) = x^3$. Then

$$\tilde{f}(x^2) = \pm \left[(x^2 + 1)^2 - x^6 \right]$$
$$= x^6 - x^4 + 2x^2 - 1,$$

because $-2 = 2 \pmod 4$. Therefore

$$\tilde{f}(x) = x^3 - x^2 + 2x - 1$$

is the corresponding basic irreducible polynomial over $\mathbf{Z}_4$.

Using the Graeffe method, the factorization

$$x^7 - 1 = (x - 1)(x^3 + x^2 + 1)(x^3 + x + 1)$$

over $GF(2)$ is easily "lifted" to the basic factorization

$$x^7 - 1 = (x - 1)(x^3 + 2x^2 + x - 1)(x^3 - x^2 + 2x - 1)$$

over $\mathbf{Z}_4$. All the factors are primitive basic irreducible polynomials. The expression over $\mathbf{Z}_4$ is easily "dropped" to the original expression over $GF(2)$ by setting -1 equal to $+1$ and setting 2 equal to 0.

Not every polynomial over $\mathbf{Z}_4$ is suitable as a generator polynomial for a cyclic code over $\mathbf{Z}_4$. For the code to be a proper cyclic code, one must respect the algebraic structure of $\mathbf{Z}_4[x]$. Just as one can form cyclic codes of blocklength n over $GF(2)$ by using the irreducible factors of $x^n - 1$ over $GF(2)$ and their products, one can also form cyclic codes of blocklength n over $\mathbf{Z}_4$ by using the basic irreducible factors of $x^n - 1$ and their products. However, the possibilities are more extensive. Let $g(x)$ be any basic irreducible factor of $x^n - 1$ over $\mathbf{Z}_4$. Then $g(x)$ can be used as the generator polynomial of a cyclic code over $\mathbf{Z}_4$ of blocklength n. Moreover, $2g(x)$ can also be used as the generator polynomial of a different cyclic code, also of blocklength n. Besides these, there are other possibilities. One can take two basic irreducible factors, $g_1(x)$ and $g_2(x)$, of $x^n - 1$ as generator polynomials and form the code whose codeword polynomials are of the form

$$c(x) = a_1(x)g_1(x) + 2a_2(x)g_2(x),$$

where the degrees of $a_1(x)$ and $a_2(x)$ are restricted so that each of the two terms on the right has a degree not larger than $n - 1$. An instance of a cyclic code with this form is based on a factorization of the polynomial $g(x)$ as $g(x) = g'(x)g''(x)$, where $g(x)$ is a basic irreducible factor of $x^n - 1$. The code as

$$\mathcal{C} = \{a_1(x)g'(x)g''(x) + 2a_2(x)g''(x)\},$$

with the understanding that the degrees of $a_1(x)$ and $a_2(x)$ are restricted so that each of the two terms in the sum has a degree not larger than $n - 1$. However, this code may be unsatisfactory unless $g_1(x)$ and $g_2(x)$ are appropriately paired. This is because the same codeword may arise in two different ways. For example, if the degree conditions allow, set $a_1(x) = 2g_2(x)$ and $a_2(x) = g_1(x)$. Then $c(x)$ is the zero codeword polynomial even though $a_1(x)$ and $a_2(x)$ are nonzero.

The way to obtain a cyclic code with this form without this disadvantage is to begin with

$$x^n - 1 = h(x)g(x)f(x),$$

where $f(x)$, $g(x)$, and $h(x)$ are monic polynomials over $\mathbf{Z}_4$. Then define the code

$$\mathcal{C} = \{a_1(x)g(x)f(x) + 2a_2(x)h(x)f(x)\},$$

with the understanding that the degrees of $a_1(x)$ and $a_2(x)$ are restricted so that each of the two terms in the sum has a degree not larger than $n - 1$. Thus $g_1(x) = g(x)f(x)$ and $g_2(x) = h(x)f(x)$. In this way, the polynomial $f(x)$ has filled out the degrees of $g_1(x)$ and $g_2(x)$ so that the choice $a_1(x) = 2g_2(x)$ violates the degree condition on $a_1(x)$.

To see that the code $\mathcal{C}$ is a cyclic code over $\mathbf{Z}_4$, let codeword $c(x)$ have the leading coefficient $c_{n-1} = a + 2b$, where $a \in \{0, 1\}$ and $b \in \{0, 1\}$. Then

$$xc(x) = xc(x) - a(x^n - 1) - 2b(x^n - 1)$$
$$= xc(x) - ah(x)g(x)f(x) - 2bh(x)g(x)f(x)$$
$$= [xa_1(x) - ah(x)]g(x)f(x) + 2(xa_2(x) - bg(x))h(x)f(x),$$

which, modulo $x^n - 1$, is an element of the code $\mathcal{C}$. Of course, one need not restrict the code in this way. One could use the more general form by restricting the encoder so that the same codeword does not represent two different datawords.

The theory of cyclic codes over rings has not been developed in depth for the general case. It has been developed primarily for codes of the form $\{a(x)f(x)\}$ and, even then, not in great depth.

2.15   Galois rings

A cyclic code over $\mathbf{Z}_4$ can be studied entirely in terms of the polynomials of the ring $\mathbf{Z}_4[x]$. However, just as it is productive to study codes over the field $GF(q)$ in the larger algebraic field $GF(q^m)$, so, too, it is productive to study codes over $\mathbf{Z}_4$ in a larger algebraic system called a *Galois ring*. A Galois ring over $\mathbf{Z}_4$ is defined in a way analogous to the definition of an extension field of $GF(q)$. Let $h(x)$ be a primitive basic irreducible polynomial (a primitive polynomial) of degree m over $\mathbf{Z}_4$. Then, with the natural definitions of addition and multiplication, $\mathbf{Z}_4[x]$ is a ring of polynomials over $\mathbf{Z}_4$, and the *Galois ring* $\mathbf{Z}_4[x]/\langle h(x)\rangle$ is the ring of polynomials modulo $h(x)$. This Galois ring has 4^m elements, and is denoted $GR(4^m)$. Although some properties of

Table 2.6. *The cycle of a primitive element in $GR(4^m)$*

$$\xi^1 = x$$
$$\xi^2 = x^2$$
$$\xi^3 = 2x^2 - x + 1$$
$$\xi^4 = -x^2 - x + 2$$
$$\xi^5 = x^2 - x - 1$$
$$\xi^6 = x^2 + 2x + 1$$
$$\xi^7 = 1 = \xi^0.$$

Galois fields carry over to Galois rings, other properties do not. In particular, the Galois ring $GR(4^m)$ cannot be generated by a single element. However, there will always be an element with order 2^m, which we will call ξ. It is a zero of a primitive polynomial over $\mathbf{Z}_4$, and hence may be called a primitive element, though it does not generate the Galois ring in the manner of a primitive element of a Galois field. If ξ is a primitive element of $GR(4^m)$, then every element of $GR(4^m)$ can be written as $a + 2b$, where a and b are elements of the set $\{0, 1, \xi, \xi^2, \ldots, \xi^{2^m-2}\}$. Because $2^m \cdot 2^m = 4^m$, this representation accounts for all 4^m elements of the ring $GR(4^m)$. With the convention that $\xi^{-\infty} = 0$, every element of $GR(4^m)$ can be written in the *biadic representation* as $\xi^i + 2\xi^j$.

For example, to construct the Galois ring $GR(4^3)$, choose the primitive polynomial $x^3 + 2x^2 + x - 1$ over $\mathbf{Z}_4$. Then let $\xi = x$, and write the cycle of ξ, as shown in Table 2.6. The 64 elements of $GR(64)$, then, are those of the form $a + 2b$, where $a, b \in \{0, 1, \xi, \xi^2, \ldots, \xi^6\}$. Of course, the biadic representation is not the only representation. Each element of $GR(64)$ can also be written as a polynomial over $\mathbf{Z}_4$ in x of degree at most 6, with multiplication modulo $h(x)$.

It is now an easy calculation in this Galois ring to verify the following factorizations:

$$x^3 + 2x^2 + x - 1 = (x - \xi)(x - \xi^2)(x - \xi^4),$$
$$x^3 - x^2 + 2x - 1 = (x - \xi^3)(x - \xi^6)(x - \xi^5),$$
$$x - 1 = (x - \xi^0).$$

Each such factorization can be regarded as a kind of lift to $GR(4^3)$ of a like factorization over $GF(2^3)$. The primitive element ξ of $GR(4^m)$ becomes the primitive element α of $GF(2^m)$ when $GR(4^m)$ is mapped into $GF(2^m)$. This means that the cyclic orbit of ξ, taken modulo 2, becomes the cyclic orbit of α.

In general, the elements of the Galois ring $GR(4^m)$ may be represented in a variety of ways. One, of course, is the definition as $\sum_i a_i x^i$, a polynomial in x of degree at most

Table 2.7. *Galois orbits in $GR(4^m)$ and $GF(2^m)$*

$\xi^1 = x$		$\alpha^1 = x$
$\xi^2 = x^2$		$\alpha^2 = x^2$
$\xi^3 = x + 1$	$+2(x^2 + x)$	$\alpha^3 = x + 1$
$\xi^4 = x^2 + x$	$+2(x^2 + x + 1)$	$\alpha^4 = x^2 + x$
$\xi^5 = x^2 + x + 1$	$+2(x + 1)$	$\alpha^5 = x^2 + x + 1$
$\xi^6 = x^2 + 1$	$+2x$	$\alpha^6 = x^2 + 1$
$\xi^7 = 1 = \xi^0$		$\alpha^7 = 1 = \alpha^0$

$m - 1$. We have also seen that we may write an arbitrary ring element, β, in the biadic representation

$$\beta = \xi^i + 2\xi^j$$
$$= a + 2b,$$

where a and b, or $a(\beta)$ and $b(\beta)$, denote the left part and right part of β, respectively. Each part is a power of ξ. This representation is convenient for some calculations. As a third representation, it may be helpful to see the elements of $GR(4^m)$ lying above the elements of $GF(2^m)$. For this purpose, regard the element β of $GR(4^m)$ to be written as $\beta_o + 2\beta_e$, where β_o and β_e, called the odd part and the even part of the ring element β, are both polynomials in x with all coefficients from $\{0, 1\}$.

To find the representation $\beta = \beta_o + 2\beta_e$, write the odd part as $\beta_o = \beta$ modulo 2, then the even part β_e is determined as the difference between β and β_o. With due care, both β_o and β_e can be informally regarded as elements of the extension field $GF(2^m)$, though operations in $GR(4^m)$ are actually modulo 4, not modulo 2.

To see the relationship between ξ and α, the comparison of the cycles of ξ and α, given in Table 2.7, is useful: The cycle of ξ is the same as in Table 2.6, but expressed to show the role of the two. We may summarize this relationship by writing $\xi^j = \alpha^j + 2\gamma_j$, where $2\gamma_j$ is defined as $\xi^j - \alpha^j$. Thus α^j is the odd part of ξ^j and γ_j is the even part of ξ^j.

The following proposition tells how to calculate the representation $\xi^i + 2\xi^j$ from any other representation of β.

Proposition 2.15.1 *Let $\beta = a + 2b$ denote the biadic representation of $\beta \in GR(4^m)$. Then*

$$a = \beta^{2^m}$$

and

$$2b = \beta - a.$$

Proof: To prove the first expression, observe that

$$\beta^2 = (a + 2b)^2$$
$$= a^2 + 4ab + 4b^2 \ (\text{mod } 4)$$
$$= a^2.$$

Because a is a power of ξ, and so has order dividing $2^m - 1$, repeated squaring now gives $\beta^{2^m} = a^{2^m} = a$, which is the first expression of the proposition. The second expression is then immediate.   ∎

Proposition 2.15.2   *Let $a(\beta) + 2b(\beta)$ be the biadic representation of any $\beta \in GR(4^m)$. Then*

$$a(\beta + \gamma) = a(\beta) + a(\gamma) + 2(\beta\gamma)^{2^{m-1}},$$
$$a(\beta\gamma) = a(\beta)a(\gamma).$$

Proof: Using Proposition 2.15.1, the statement to be proved can be restated as

$$(\beta + \gamma)^{2^m} = \beta^{2^m} + \gamma^{2^m} + 2(\beta\gamma)^{2^{m-1}}.$$

For $m = 1$, this is elementary:

$$(\beta + \gamma)^2 = \beta^2 + \gamma^2 + 2\beta\gamma.$$

Because $4 = 0$ in this ring, it is now clear that

$$(\beta + \gamma)^4 = (\beta^2 + \gamma^2)^2 + 4(\beta\gamma)(\beta^2 + \gamma^2) + 4(\beta^2\gamma^2)$$
$$= (\beta^2 + \gamma^2)^2$$
$$= \beta^4 + \gamma^4 + 2\beta^2\gamma^2.$$

The recursion is now clear, so the proof of the first identity is complete. The proof of the second identity follows from $\beta\gamma = (a + 2b)(a' + 2b') = aa' + 2(ab' + a'b)$.   ∎

The statement of Proposition 2.15.2 will now be extended to the generalization in which there are n terms in the sum.

Proposition 2.15.3   *Let $a(\beta_\ell) + 2b(\beta_\ell)$ denote the biadic representation of $\beta_\ell \in GR(4^m)$. Then*

$$a\left(\sum_{\ell=1}^{n} \beta_\ell\right) = \sum_{\ell=1}^{n} a(\beta_\ell) + 2\sum_{\ell=1}^{n}\sum_{\ell' \neq \ell}(\beta_\ell\beta_{\ell'})^{2^{m-1}}.$$

Proof: If there are two terms in the sum, the statement is true by Proposition 2.15.2. Suppose that the expression is true if there are $n - 1$ terms in the sum. Then

$$a\left(\sum_{\ell=1}^{n} \beta_\ell\right) = a\left(\sum_{\ell=1}^{n-1} \beta_\ell + \beta_n\right)$$

$$= a\left(\sum_{\ell=1}^{n-1} \beta_\ell\right) + a(\beta_n) + 2(\beta_n^{2^{m-1}})\left(\sum_{\ell=1}^{n-1} \beta_\ell\right)^{2^{m-1}}$$

$$= \sum_{\ell=1}^{n-1} a(\beta_\ell) + 2\sum_{\ell=1}^{n-1}\sum_{\ell' \neq \ell}(\beta_\ell\beta_{\ell'})^{2^{m-1}} + a(\beta_n) + 2(\beta_n^{2^{m-1}})\left(\sum_{\ell=1}^{n-1} \beta_\ell\right)^{2^{m-1}}$$

$$= \sum_{\ell=1}^{n} a(\beta_\ell) + 2\sum_{\ell=1}^{n}\sum_{\ell' \neq \ell}(\beta_\ell\beta_{\ell'})^{2^{m-1}},$$

as was to be proved.   ∎

In $GR(4^m)$, the square of the ring element $c = a + 2b$ is always $c^2 = a^2$, independent of b because $4 = 0$ in this ring. In this sense, squaring is a lossy operation. A useful variant of the squaring function is the *frobenius function*, defined in the Galois ring $GR(4^m)$ as $c^{\mathrm{f}} = a^2 + 2b^2$. Now the *trace* in $GR(4^m)$ can be defined as $\mathrm{tr}(c) = c + c^{\mathrm{f}} + \cdots + c^{\mathrm{f}^{m-1}}$.

There is also a Fourier transform in the Galois ring $GR(4^m)$. A "vector" c of blocklength $n = 2^m - 1$ over the ring $GR(4^m)$ has a Fourier transform, defined as

$$C_j = \sum_{i=0}^{n-1} \xi^{ij} c_i,$$

where ξ is a primitive element of $GR(4^m)$ and $n = 2^m - 1$ is the order of ξ. The Fourier transform C is also a vector of blocklength n over the ring $GR(4^m)$. Because $\mathbf{Z}_4$ is contained in $GR(4^m)$, a vector c over $\mathbf{Z}_4$ of blocklength n is mapped into a vector C over $GR(4^m)$ by the Fourier transform. Moreover, by setting $2 = 0$, the Fourier transform in the ring $GR(4^m)$ can be dropped to a Fourier transform in the field $GF(2^m)$, with components $C_j = \sum_{i=0}^{n-1} \alpha^{ij} c_i$.

Many elementary properties of the Fourier transform hold for the Galois ring Fourier transform. The inverse Fourier transform can be verified in the usual way by using the relationship

$$\sum_{i=0}^{n-1} \xi^i = \frac{1 - \xi^n}{1 - \xi} = 0$$

unless $\xi = 1$. Therefore, because an inverse Fourier transform exists, each c corresponds to a unique spectrum C.

There is even a kind of conjugacy relationship in the transform domain. Let c be a vector over $\mathbf{Z}_4$, written $c = a + 2b$, with components displayed in the biadic representation as $c_i = a_i + 2b_i$. Because c is a vector over $\mathbf{Z}_4$, the components satisfy $a_i \in \{0, 1\}$ and $b_i \in \{0, 1\}$. Then the spectral components are given by

$$C_j = \sum_{i=0}^{n-1} (a_i + 2b_i)\xi^{ij}.$$

Although C_j is not itself in the biadic representation, each term within the sum is in the biadic representation, because a_i and b_i can only be zero or one.

We now express the spectral component C_j in the biadic representation as $C_j = A_j + 2B_j$. By Proposition 2.15.3, the left term of the biadic representation of $C_j = \sum_i c_i \xi^{ij}$ is given by

$$A_j = \sum_i a_i \xi^{ij} + 2 \sum_i \sum_{i' \neq i} ((a_i + 2b_i)\xi^{ij}(a_{i'} + 2b_{i'})\xi^{i'j})^{2^{m-1}}.$$

Because $4 = 0$ in this ring, the second term can be simplified so that

$$A_j = \sum_i a_i \xi^{ij} + 2 \sum_i \sum_{i' \neq i} (a_i \xi^{ij})^{2^{m-1}} (a_{i'} \xi^{i'j})^{2^{m-1}}$$

and $2B_j = C_j - A_j$. We conclude that

$$C_j = A_j + 2B_j$$

$$= \left(\sum_i a_i \xi^{ij} + 2 \sum_i \sum_{i' \neq i} (a_i a_{i'} \xi^{ij} \xi^{i'j})^{2^{m-1}} \right)$$

$$+ 2 \left(\sum_i b_i \xi^{ij} + \sum_i \sum_{i' \neq i} (a_i a_{i'} \xi^{ij} \xi^{i'j})^{2^{m-1}} \right)$$

is the biadic representation of C_j.

Although this representation for C_j seems rather complicated, it is the starting point for proving the following useful theorem. This theorem characterizes the spectral components of a vector over $\mathbf{Z}_4$. In particular, the theorem says that component C_{2j}, which is given by

$$C_{2j} = \sum_{i=0}^{n-1} a_i \xi^{2ij} + 2 \sum_{i=0}^{n-1} b_i \xi^{2ij},$$

is related to C_j by a conjugacy constraint. The theorem also implies that if $C_j = 0$, then $C_{2j} = 0$ as well.

Theorem 2.15.4 *Let c be a vector of blocklength $n = 2^m - 1$ over Z_4. Then the components of the Fourier transform C satisfy $C_{2j} = C_j^f$, where C_j^f denotes the frobenius function of C_j.*

Proof: We will give an explicit computation using the formula derived prior to the statement of the theorem. Write

$$
C_j^f = \left(\sum_i a_i \xi^{ij} + 2 \sum_i \sum_{i' \neq i} (a_i a_{i'} \xi^{ij} \xi^{i'j})^{2^{m-1}} \right)^2
$$

$$
+ 2 \left(\sum_i b_i \xi^{ij} + \sum_i \sum_{i' \neq i} (a_i a_{i'} \xi^{ij} \xi^{i'j})^{2^{m-1}} \right)^2 .
$$

The first term has the form $(x + 2y)^2$, which expands to $x^2 + 4xy + 4y^2 = x^2 \pmod 4$. The second term has the form $2(x + y)^2$, which expands to $2(x^2 + 2xy + y^2) = 2(x^2 + y^2) \pmod 4$. Therefore

$$
C_j^f = \left(\sum_i a_i \xi^{ij} \right)^2 + 2 \left[\left(\sum_i b_i \xi^{ij} \right)^2 + \left(\sum_i \sum_{i' \neq i} (a_i a_{i'} \xi^{ij} \xi^{i'j})^{2^{m-1}} \right)^2 \right].
$$

Now rewrite each of these three squares. The first square is expanded as

$$
\left(\sum_i a_i \xi^{ij} \right)^2 = \sum_i (a_i \xi^{ij})^2 + 2 \sum_i \sum_{i' \neq i} (a_i a_{i'} \xi^{ij} \xi^{i'j}).
$$

Each of the second two squares can be expanded in this way as well, but the cross terms drop out because $4 = 0$ in the ring Z_4. The summands in these latter two terms then become $(b_i \xi^{ij})^2$ and $((a_i a_{i'} \xi^{ij} \xi^{i'j})^{2^{m-1}})^2$. Therefore because each a_i or b_i can only be a zero or a one,

$$
C_j^f = \sum_i a_i \xi^{2ij} + 2 \sum_i \sum_{i' \neq i} a_i a_{i'} \xi^{ij} \xi^{i'j} + 2 \sum_i b_i \xi^{2ij} + 2 \sum_i \sum_{i' \neq i} a_i a_{i'} \xi^{ij} \xi^{i'j}
$$

$$
= \sum_{i=0}^{n-1} a_i \xi^{2ij} + 2 \sum_{i=0}^{n-1} b_i \xi^{2ij}
$$

$$
= C_{2j},
$$

as was to be proved. ∎

This theorem allows us to conclude that, as in the case of a Galois field, if $g(x)$ is a polynomial over Z_4 with a zero at the element ξ^i of the Galois ring $GR(4^m)$, then it also has a zero at the element ξ^{2i}. In particular, a basic irreducible polynomial over $\mathbf{Z}_4$, with a zero at β, has the form

$$p(x) = (x - \beta)(x - \beta^2) \dots (x - \beta^{2^{r-1}}),$$

where r is the number of conjugates of β in $GR(4^m)$.

A cyclic code over $GR(4^m)$ that is defined in terms of the single generator polynomial $g(x)$ consists of all polynomial multiples of $g(x)$ of degree at most $n - 1$. Every codeword has the form $c(x) = a(x)g(x)$. Although $g(x)$ is the Hensel lift of a polynomial over $GF(2^m)$, $a(x)g(x)$ need not be the Hensel lift of a polynomial over $GF(2^m)$. In particular, not every $a(x)$ over $GR(4^m)$ is the Hensel lift of a polynomial over $GF(2^m)$.

One way to define a cyclic code over $\mathbf{Z}_4$ – but not every cyclic code over $\mathbf{Z}_4$ – is as the set of polynomials in $\mathbf{Z}_4[x]/\langle x^n - 1 \rangle$ with zeros at certain fixed elements of $GR(4^m)$. This is similar to the theory of cyclic codes over a field. For example, the cyclic code with the primitive polynomial $x^3 + 2x^2 + x - 1$ as the generator polynomial $g(x)$ can be defined alternatively as the set of polynomials over $\mathbf{Z}_4$ of degree at most 7 with a zero at the primitive element ξ, a zero of $g(x)$. Thus $c(x)$ is a codeword polynomial if $c(\xi) = 0$. Then Theorem 2.15.4 tells us that $c(\xi^2) = 0$ as well, and so forth.

In the case of a cyclic code over a Galois field, the generator polynomial $g(x)$ can be specified by its spectral zeros. Similarly, a single generator polynomial for a cyclic code over $\mathbf{Z}_4$ can be specified by its spectral zeros. Because the spectral zeros define a simple cyclic code over $\mathbf{Z}_4$, the minimum distance of that code is somehow implicit in the specification of the spectral zeros of the single generator polynomial. Thus, we might hope for a direct statement of this relationship analogous to the BCH bound. However, a statement with the simplicity of the BCH bound for a Lee-distance code over $\mathbf{Z}_4$ is not known. For this reason, it is cumbersome to find the minimum distance of a cyclic code over $\mathbf{Z}_4$ that is defined in this way.

A cyclic code over $\mathbf{Z}_4$ can be dropped to the underlying code over $GF(2)$, where the BCH bound does give useful, though partial, information about the given Lee-distance code. If codeword $\bar{c}$ over $\mathbf{Z}_4$ is dropped to codeword $\underline{c}$ over $GF(2)$, then the codeword $\underline{c}$ will have a 1 at every component where the codeword $\bar{c}$ has either a 1 or a 3. Hence the minimum Lee distance of the $\mathbf{Z}_4$ code is at least as large as the minimum Hamming distance of that binary code, and that minimum distance satisfies the BCH bound.

Our two examples of cyclic codes over $\mathbf{Z}_4$ that will conclude this section are known as *Calderbank–McGuire codes*. These codes over $\mathbf{Z}_4$ are defined by reference to the Galois ring $GR(4^5)$. They are related to the binary $(32, 16, 8)$ self-dual code based on the binary $(31, 16, 7)$ cyclic BCH code, in the sense that the Calderbank–McGuire codes can be dropped to these binary codes. The cyclic versions of the two Calderbank–McGuire codes are a $(31, 18.5, 11)$ cyclic Lee-distance code over $\mathbf{Z}_4$ and a $(31, 16, 13)$ cyclic Lee-distance code over $\mathbf{Z}_4$. When extended by a single check symbol, these cyclic codes over $\mathbf{Z}_4$ are, respectively, a $(32, 18.5, 12)$ Lee-distance code over $\mathbf{Z}_4$ and a $(32, 16, 14)$

Lee-distance code over $\mathbf{Z}_4$. When the symbols of $\mathbf{Z}_4$ are represented by pairs of bits by using the Gray map (described in Section 2.16), these codes become *nonlinear* $(64, 37, 12)$ and $(64, 32, 14)$ binary Hamming-distance codes, with *datalengths* 37 and 32, respectively. Their performance is better than the best linear codes known. The comparable known linear codes are the $(64, 36, 12)$ and $(64, 30, 14)$ BCH codes, with the dimensions 36 and 30.

The first Calderbank–McGuire cyclic code is the set of polynomials $c(x)$ of block-length 31 over $\mathbf{Z}_4$ that satisfy the conditions $c(\xi) = c(\xi^3) = 2c(\xi^5) = 0$, where ξ is a primitive element of $GR(4^5)$. The condition $2c(\xi^5) = 0$ means that $c(\xi^5)$ must be even, but not necessarily zero, which accounts for the unusual datalength of this $(31, 18.5, 11)$ cyclic Calderbank–McGuire code over $\mathbf{Z}_4$. Accordingly, the check matrix of this cyclic code is given by

$$ H = \begin{bmatrix} 1 & \xi^1 & \xi^2 & \cdots & \xi^{30} \\ 1 & \xi^3 & \xi^6 & \cdots & \xi^{90} \\ 2 & 2\xi^5 & 2\xi^{10} & \cdots & 2\xi^{150} \end{bmatrix}. $$

In the Galois ring $GR(4^5)$, the elements ξ, ξ^3, and ξ^5 each have five elements in their conjugacy classes. This means that the first two rows of H each reduce the datalength by 5. The third row only eliminates half of the words controlled by the conjugacy class of ξ^5. Thus $n - k = 12.5$ and $n = 31$, so $k = 18.5$.

The cyclic $(31, 18.5, 11)$ Calderbank–McGuire code over $\mathbf{Z}_4$ can be lengthened by a simple check symbol to form the $(32, 18.5, 12)$ extended Calderbank–McGuire code over $\mathbf{Z}_4$. The lengthened code has the check matrix

$$ H = \begin{bmatrix} 1 & 1 & \xi^1 & \xi^2 & \cdots & \xi^{30} \\ 0 & 1 & \xi^3 & \xi^6 & \cdots & \xi^{90} \\ 0 & 2 & 2\xi^5 & 2\xi^{10} & \cdots & 2\xi^{150} \end{bmatrix}. $$

There are two noteworthy binary codes that are closely related to this code. A linear code of blocklength 32 is obtained by simply dropping the codewords into $GF(2)$, which reduces every symbol of $\mathbf{Z}_4$ to one bit – a zero or a one according to whether the Lee weight of the $\mathbf{Z}_4$ symbol is even or odd. This map takes the $\mathbf{Z}_4$ code into a linear binary $(32, 22, 5)$ code. It is an extended BCH code. The other binary code is obtained by using the Gray map to represent each symbol of $\mathbf{Z}_4$ by two bits. The Gray map takes the $\mathbf{Z}_4$ code into a nonlinear binary $(64, 37, 12)$ code. The performance of this code is better than any known linear binary code.

If the 2 is struck from the last row of H of the cyclic code, then we have the second Calderbank–McGuire cyclic code, which has $c(\xi) = c(\xi^3) = c(\xi^5) = 0$ in $GR(4^5)$. This gives a cyclic $(31, 16, 13)$ Lee-distance code over $\mathbf{Z}_4$ with datalength 16. It can be lengthened by a simple check symbol to form a $(32, 16, 14)$ Lee-distance code over $\mathbf{Z}_4$.

The lengthened code has the check matrix

$$
H = \begin{bmatrix} 1 & 1 & \xi^1 & \xi^2 & \cdots & \xi^{30} \\ 0 & 1 & \xi^3 & \xi^6 & \cdots & \xi^{90} \\ 0 & 1 & \xi^5 & \xi^{10} & \cdots & \xi^{150} \end{bmatrix}.
$$

The Gray map takes the Z_4 code into a nonlinear $(64, 32, 14)$ binary Hamming-distance code.

Inspection of the check matrices makes it clear that the two cyclic Calderbank–McGuire codes over Z_4, of blocklength 31, are contained in the cyclic Preparata code over Z_4 of blocklength 31, which is defined in Section 2.16 and has the check matrix

$$
H = \begin{bmatrix} 1 & \xi^1 & \xi^2 & \cdots & \xi^{30} \end{bmatrix}.
$$

Likewise, the extended Calderbank–McGuire codes over Z_4, of blocklength 32, are contained in the extended Preparata code over Z_4 of blocklength 32.

We do not provide detailed proofs of the minimum distances of the Calderbank–McGuire codes here. Instead, we leave this as an exercise. Some methods of finding the minimum Lee distance of a code over Z_4 are given in Section 2.16. There we state that every codeword can be written as $c(x) = c_1(x) + 2c_2(x)$, where $c_1(x)$ and $c_2(x)$ have all coefficients equal to zero or one. Thus by reduction modulo 2, the Z_4 polynomial $c(x)$ can be dropped to the binary codeword $c_1(x)$. As a binary codeword, $c_1(x)$ has zeros at α^1, α^3, and α^5, and so has minimum Hamming weight at least equal to 7. If $c_1(x)$ is zero, then $c_2(x)$ drops to a binary codeword with spectral zeros at α^1 and α^3. This means that $c_2(x)$ has Hamming weight at least 5, so the Z_4 codeword $2c_2(x)$ has Lee weight at least 10. This codeword extends to a codeword with Lee weight at least equal to 12. For the second Calderbank–McGuire code, the codeword $c(x) = 2c_2(x)$ has Lee distance at least 14 and this codeword extends to a codeword with Lee weight at least 14. Other codewords of the Calderbank–McGuire code – those for which both $c_1(x)$ and $c_2(x)$ are nonzero – are much harder to analyze.

2.16   The Preparata, Kerdock, and Goethals codes

A nonlinear binary code is interesting whenever the code has more codewords than any comparable linear code that is now known or, in some cases, better than any linear code that exists. Some well known families of such nonlinear binary codes are the *Preparata codes*, the *Kerdock codes*, and the *Goethals codes*. Other notable examples are the Calderbank–McGuire codes that were mentioned in the previous section. The exemplar code of this kind is the binary $(15, 8, 5)$ *Nordstrom–Robinson nonlinear code* that can be extended to a binary $(16, 8, 6)$ nonlinear code. The Nordstrom–Robinson code is both the simplest of the Preparata codes and the simplest of the Kerdock codes. Because the Nordstrom–Robinson code is a nonlinear code, the notion of a dimension does not apply. Because there are 2^8 codewords, we may still refer to this code as a $(15, 8, 5)$ code. Now the second term of the notation (n, k, d) is the *datalength* of the code, referring to the base-2 logarithm of the number of codewords. The datalength of the Nordstrom–Robinson code is 8.

The Nordstrom–Robinson code can be generalized to binary codes of longer blocklength of the form $2^{m+1} - 1$, m odd, either with the minimum distance fixed or with the redundancy fixed. The first case gives a family of $(2^{m+1} - 1, 2^{m+1} - 2(m + 1), 5)$ nonlinear binary codes, known as Preparata codes, and the second case gives a family of $(2^{m+1} - 1, 2(m + 1), 2^m - 2^{(m-1)/2} - 1)$ nonlinear binary codes, known as Kerdock codes. A binary Preparata code is a double-error-correcting code, and a binary Kerdock code is a multiple-error-correcting code. As binary codes, the Preparata codes and the Kerdock codes can be extended by a single check bit that increases the minimum Hamming distance by 1.

We also briefly mention another family of binary nonlinear triple-error-correcting codes known as the family of Goethals codes. The binary Goethals codes have minimum Hamming distance 7 and can be extended by a single check bit that increases the minimum Hamming distance by 1.

The nonlinear codes of this section can be constructed by converting linear codes over $\mathbf{Z}_4$ into nonlinear codes over $GF(2)$, using a map known as the *Gray map*. The resulting nonlinear codes are the best binary codes known of their blocklength and datalength. The Gray map is the following association of elements of $\mathbf{Z}_4$ with pairs of elements of $GF(2)$:

$$0 \rightarrow 00,$$

$$1 \rightarrow 01,$$

$$2 \rightarrow 11,$$

$$3 \rightarrow 10.$$

The Gray map is intrinsically nonlinear if the binary image is to be added componentwise. Thus, for example, in $\mathbf{Z}_4$, consider $3 + 1 = 0$. Adding the Gray map of both terms componentwise on the left side gives $10 + 01$, which equals 11, whereas the Gray map of the right side is 00, which is not the same. Addition is not preserved by the Gray map. The Gray map, when applied componentwise to a sequence $\mathbf{c}$ in $\mathbf{Z}_4^n$, produces a sequence $\tilde{\mathbf{c}}$ in $GF(2)^{2n}$. The sequence $\tilde{\mathbf{c}}$ has twice as many components as the sequence $\mathbf{c}$.

The Lee weight and the Lee distance are defined so that, under the Gray map, the Hamming weight satisfies $w_H(\tilde{\mathbf{c}}) = w_L(\mathbf{c})$ and the Hamming distance satisfies

$d_{\mathrm{H}}(\tilde{c}, \tilde{c}') = d_{\mathrm{L}}(c, c')$. Thus the Lee distance between two sequences in Z_4^n is equal to the Hamming distance between their Gray images in $GF(2)^{2n}$.

Recall that the linear code C over the ring Z_q is a code over Z_q, such that $ac + bc'$ is a codeword of C whenever c and c' are codewords of C. A code $C \in Z_4^n$ is converted to a code $\tilde{C} \in GF(2)^{2n}$ by the Gray map. In general, even though the code C is a linear code in Z_4^n, the code $\tilde{C}$ will be a nonlinear code in $GF(2)^{2n}$.

By applying the Gray map to every codeword, the linear code C in Z_4^n is converted into a nonlinear binary code, called the *Gray image* of C. From a concrete point of view, the Gray map relates two codes, one over Z_4 and one over $GF(2)$. From an abstract point of view, there is only one code, but with two different notations and two different notions of distance.

This method of constructing codes in $GF(2)^{2n}$ yields some noteworthy binary codes. For example, let C be the Lee-distance code over Z_4 that is produced by the generator matrix

$$G = \begin{bmatrix} 1 & 0 & 1 \\ 0 & 1 & 3 \end{bmatrix}.$$

Table 2.8 gives the sixteen codewords of this Lee-distance code and the 16 binary codewords of its Gray image. By inspection of the table, it is easy to see that the binary code is nonlinear. What is harder to see from the table is that for any d, the number of codewords at distance d from any codeword is the same for every codeword. A code with this property is known as a *distance-invariant code*. Because the original

Table 2.8. *A code over Z_4 and its Gray image*

Codewords of C			Codewords of C'					
0	0	0	0	0	0	0	0	0
1	0	1	0	1	0	0	0	1
2	0	2	1	1	0	0	1	1
3	0	3	1	0	0	0	1	0
0	1	3	0	0	0	1	1	0
1	1	0	0	1	0	1	0	0
2	1	1	1	1	0	1	0	1
3	1	2	1	0	0	1	1	1
0	2	2	0	0	1	1	1	1
1	2	3	0	1	1	1	1	0
2	2	0	1	1	1	1	0	0
3	2	1	1	0	1	1	0	1
0	3	1	0	0	1	0	0	1
1	3	2	0	1	1	0	1	1
2	3	3	1	1	1	0	1	0
3	3	0	1	0	1	0	0	0

code in Z_4^n is linear, it is obviously a distance-invariant code. The Gray image of this Lee-distance code must also be distance-invariant under Hamming distance because the Gray map preserves the distance structure.

For a more interesting example, recall that the octacode is an $(8, 4, 6)$ extended cyclic code over Z_4, corresponding to the generator polynomial $g(x) = x^3 + 2x^2 + x + 3$ and the check polynomial $h(x) = x^4 + 2x^3 + 3x^2 + x + 1$. Therefore its Gray image is a $(16, 8, 6)$ nonlinear binary code. The Gray image of the octacode, in fact, may be taken to be the definition of the (extended) Nordstrom–Robinson code.

We want to generalize the Nordstrom–Robinson code to longer blocklengths. Toward this end, first recall that the factorization of $x^7 - 1$ into basic irreducible polynomials is given by

$$x^7 - 1 = (x - 1)(x^3 + 2x^2 + x - 1)(x^3 - x^2 + 2x - 1).$$

Either of the two factors of degree 3 can be used to form a cyclic code over Z_4. These two codes are equivalent codes, one polynomial being the reciprocal of the other. The Gray image of either of these codes is the $(16, 8, 6)$ Nordstrom–Robinson code of blocklength 16 and datalength 8.

The generalization of the code is based on the factorization

$$x^{2^m - 1} - 1 = (x - 1)g(x)h(x),$$

in Z_4, where one of the two nontrivial factors is a primitive basic irreducible polynomial over Z_4, and the other is a product of the remaining basic irreducible polynomials of the factorization of $x^{2^m - 1} - 1$. Each of these polynomials, $h(x)$ and $g(x)$, can be used to form a cyclic code over Z_4 of blocklength $2^m - 1$ or an extended cyclic code over Z_4 of blocklength 2^m.

The cyclic codes over Z_4, with generator polynomials $h(x)$ and (the reciprocal of) $g(x)$, are dual codes using the natural definition of an inner product over Z_4. One is the Preparata code over Z_4 and one is the Kerdock code over Z_4. First, we will discuss the Preparata code over Z_4 and its binary image over $GF(2)$.

Definition 2.16.1 *A cyclic Preparata code over Z_4 of blocklength $n = 2^m - 1$, m odd, is a cyclic code over Z_4 whose generator polynomial $g(x)$ is a primitive basic irreducible factor of $x^{2^m - 1} - 1$ in the Galois ring $GR(4^m)$. An extended Preparata code over Z_4 of blocklength 2^m is a cyclic Preparata code over Z_4 augmented by a simple Z_4 check sum.*

Because the degree of $g(x)$ is m, a cyclic Preparata code over Z_4 has dimension $2^m - 1 - m$. Thus it has $4^{2^m - 1 - m}$ codewords, as does the extended Preparata code. We shall see in Theorem 2.16.2 that these cyclic and extended codes are $(2^m - 1, 2^m - 1 - m, 4)$ and $(2^m, 2^m - 1 - m, 6)$ codes over Z_4, respectively. A *binary Preparata code* of blocklength 2^{m+1} is the Gray image of the extended Preparata code of blocklength

2^m over $\mathbf{Z}_4$. An *original Preparata code* of blocklength $2^{m+1} - 1$ is a binary Preparata code of blocklength 2^{m+1} that is punctured by one bit, and so has blocklength $2^{m+1} - 1$. Because each of the $2^m - 1 - m$ data symbols of the Preparata code over $\mathbf{Z}_4$ is represented by two bits in a binary Preparata code, the datalength of the binary code is $2^{m+1} - 2 - 2m$. These binary Preparata codes, then, are $(2^{m+1}, 2^{m+1} - 2 - 2m, 6)$ and $(2^{m+1} - 1, 2^{m+1} - 2 - 2m, 5)$ nonlinear codes over $GF(2)$, respectively.

The cyclic Preparata code over $\mathbf{Z}_4$ of blocklength $2^m - 1$ can be described as the set of polynomials over $\mathbf{Z}_4$ of degree at most $2^m - 2$ with a zero at ξ, where ξ is a primitive element of $\mathbf{Z}_4$. This means that a cyclic Preparata code over $\mathbf{Z}_4$ has a check matrix of the form

$$H = [1 \quad \xi^1 \quad \xi^2 \quad \cdots \quad \xi^{n-1}],$$

and an extended Preparata code has a check matrix of the form

$$H = \begin{bmatrix} 1 & 1 & 1 & 1 & \cdots & 1 \\ 0 & 1 & \xi^1 & \xi^2 & \cdots & \xi^{n-1} \end{bmatrix}.$$

For the cyclic code, by Theorem 2.15.4, the codeword polynomials have a second zero at ξ^2, and so forth. Therefore, because a codeword polynomial $c(x)$ satisfies $c(\xi^1) = c(\xi^2) = 0$ over $\mathbf{Z}_4$, there are two consecutive spectral zeros. Writing the "vectors" of $\mathbf{Z}_4^n$ as $c = a + 2b$, where $a_i \in \{0, 1\}$ and $b_i \in \{0, 1\}$, the codewords of a cyclic Preparata code with b set to the zero vector are the codewords of the binary Hamming code of the same blocklength. Thus the codeword c can be dropped to the binary Hamming codeword a with spectral zeros at α^1 and α^2. By the BCH bound, if the Hamming codeword a is nonzero, it has Hamming weight at least 3. If, instead a is the zero vector but codeword c is nonzero, then $c = 2b$ where b is a nonzero Hamming codeword, so c has Lee weight at least 6. This much is easy to infer from the BCH bound. Before treating the general case, we examine some examples.

The cyclic Preparata code of blocklength 7 with the generator polynomial

$$g(x) = x^3 + 2x^2 + x - 1$$

has spectral zeros at ξ^1, ξ^2, and ξ^4. One codeword polynomial of this code is given by

$$\begin{aligned} c(x) &= g(x) \\ &= x^3 + 2x^2 + x - 1 \\ &= (x^3 + x + 1) + 2(x^2 + 1). \end{aligned}$$

The BCH bound applied to the odd part of $c(x)$ says that the odd part must have Hamming weight at least 3. This codeword has an odd part with Hamming weight 3 and an even part with Hamming weight 2. The Lee weight of the combined Z_4 codeword is 5. When extended by a simple check sum, the extended code has Lee weight equal to 6. The BCH bound does not apply to the even part of this codeword because the odd part is nonzero. However, a different codeword polynomial of this code is given by

$$c(x) = 2g(x)$$
$$= 2(x^3 + x + 1).$$

The odd part of this codeword is zero, so the BCH bound applies to the even part. It says that the even part of this codeword has Hamming weight at least 3, and so the codeword has Lee weight at least 6. Yet another codeword polynomial of this code is given by

$$c(x) = (x + 1)g(x)$$
$$= x^4 - x^3 - x^2 - 1$$
$$= (x^4 + x^3 + x^2 + 1) + 2(x^3 + x^2 + 1)$$
$$= (x + 1)(x^3 + x + 1) + 2(x^2 + x) + 2(x^3 + x + 1).$$

The last line decomposes this codeword by exhibiting separately three terms: the underlying binary Hamming codeword; the even term in the middle formed by the Hensel lifting of the underlying binary Hamming codeword; and the last term added as an even multiple of another Hamming codeword. The Z_4 codeword has Lee weight equal to 4, and viewed as integers the components of the codeword sum to -2. Thus when extended by a simple check sum, the Lee weight of the codeword is 6 because the Z_4 check sum is 2. This example shows how a cyclic codeword of Lee weight 4 may extend to a codeword of Lee weight 6 In fact, Theorem 2.16.2 states that *every* codeword of a cyclic Preparata code over Z_4 has Lee weight at least 4 and has Lee weight at least 6 when extended by a simple Z_4 check sum.

We will give an elementary proof of the following theorem. In Chapter 3, we will give a decoding algorithm for a Preparata code that serves as an alternative proof of the theorem.

Theorem 2.16.2 *A cyclic Preparata code over Z_4 has minimum Lee distance equal to 4. An extended Preparata code over Z_4 has minimum Lee distance equal to 6.*

Proof: Let $\mathcal{C}$ be the cyclic Preparata code over Z_4 of blocklength n. We will prove that every nonzero codeword of $\mathcal{C}$ has Lee weight at least 4, and that every codeword of the cyclic code with Lee weight 4 must have components summing to 2. This implies that for every such codeword, $c_\infty = 2$, and so the extended codeword has Lee weight 6. Furthermore, every codeword with Lee weight 5 must have components summing to ± 1. For every such codeword $c_\infty = \pm 1$, and the extended codeword again has Lee weight 6.

If a nonzero codeword c has no components of Lee weight 1, it must drop to the all-zero codeword. Then c has the form of a Hamming codeword multiplied by 2 ($c = 2b$), and so has Lee weight at least 6. Furthermore, if a nonzero codeword has any components of Lee weight 1, then there must be at least three components of Lee weight 1, because every such codeword can be dropped to a binary Hamming codeword. In such a codeword, if there is also at least one component of Lee weight 2, then the codeword c has Lee weight at least 5 and extends to a codeword of Lee weight at least 6. Therefore we only need consider codewords in which all nonzero components have Lee weight 1. We will first show that there are no such codewords of Lee weight 3.

A codeword with three components of Lee weight 1 could only have the form

$$c(x) = x^i + x^j \pm x^k$$

(possibly after multiplying $c(x)$ by -1), in which the coefficient of x^k may be either $+1$ or -1. Evaluating $c(x)$ at ξ and ξ^2 gives

$$c(\xi) = \xi^i + \xi^j \pm \xi^k = 0,$$
$$c(\xi^2) = \xi^{2i} + \xi^{2j} \pm \xi^{2k} = 0.$$

These equations can be rewritten as

$$(\xi^i + \xi^j)^2 = (\mp\xi^k)^2,$$
$$\xi^{2i} + \xi^{2j} = \mp\xi^{2k}.$$

If the coefficient of x^k in $c(x)$ is negative, these combine to give $2\xi^i\xi^j = 0$, which is a contradiction because ξ^i and ξ^j are both nonzero. If, instead, the coefficient of x^k is positive, the two equations combine to give $2\xi^i\xi^j = 2\xi^{2k}$, which means that $\xi^{i-k}\xi^{j-k} = 1$. We also know that $\xi^{i-k} + \xi^{j-k} = 1$, which means that $(x - \xi^{i-k})(x - \xi^{j-k}) = x^2 + x + 1$. But the polynomial $x^2 + x + 1$ has no zeros in $GF(2^m)$ if m is odd, so $x^i + x^j + x^k$ cannot be a codeword. Thus there are no codewords of Lee weight 3.

To show that a cyclic codeword with four components of Lee weight 1 must have an extension symbol with Lee weight 2, we will show that no such cyclic codeword whose components sum to 0 or 4 can exist. Such a codeword with four components, each of Lee weight 1, would have the form

$$c(x) = (x^i + x^j) \pm (x^k + x^\ell)$$

(possibly after multiplying $c(x)$ by -1), in which the coefficients of x^k and x^ℓ are both 1 or both -1. Evaluating $c(x)$ at ξ and ξ^2 gives

$$c(\xi) = (\xi^i + \xi^j) \pm (\xi^k + \xi^\ell) = 0,$$
$$c(\xi^2) = (\xi^{2i} + \xi^{2j}) \pm (\xi^{2k} + \xi^{2\ell}) = 0.$$

These equations can be rewritten as

$$(\xi^i + \xi^j)^2 = (\xi^k + \xi^\ell)^2,$$
$$(\xi^{2i} + \xi^{2j}) = \mp(\xi^{2k} + \xi^{2\ell}).$$

If the coefficients of x^k and x^ℓ in $c(x)$ are both negative, these combine to give $2\xi^i\xi^j = 2\xi^k\xi^\ell$, and we already know that $\xi^i + \xi^j = \xi^k + \xi^\ell = 0$. Next, dropping these equations to the underlying field $GF(2^m)$ gives $\alpha^i\alpha^j = \alpha^k\alpha^\ell$ and $\alpha^i + \alpha^j + \alpha^k + \alpha^\ell = 0$. These combine to give $\alpha^{i-\ell} + 1 = \alpha^{j-\ell}(\alpha^{i-\ell} + 1)$, which means that $\alpha^{j-\ell} = 1$. This contradiction, that $x^j = x^\ell$, proves there is no codeword of the form $x^i + x^j - x^k - x^\ell$.

To show a contradiction for a codeword of the form $x^i + x^j + x^k + x^\ell$, combine the two equations

$$(\xi^i + \xi^j + \xi^k + \xi^\ell)^2 = 0$$

and

$$\xi^{2i} + \xi^{2j} + \xi^{2k} + \xi^{2\ell} = 0$$

to give

$$2(\xi^i\xi^j + \xi^i\xi^k + \xi^i\xi^\ell + \xi^j\xi^k + \xi^j\xi^\ell + \xi^k\xi^\ell) = 0.$$

We already know that $\xi^i + \xi^j + \xi^k + \xi^\ell = 0$. Drop these equations to the underlying field to write

$$\alpha^i + \alpha^j + \alpha^k + \alpha^\ell = 0$$

and

$$\alpha^i\alpha^j + \alpha^i\alpha^k + \alpha^i\alpha^\ell + \alpha^j\alpha^k + \alpha^j\alpha^\ell + \alpha^k\alpha^\ell = 0.$$

Then

$$(x - \alpha^i)(x - \alpha^j)(x - \alpha^k)(x - \alpha^\ell) = x^4 + (\alpha^i + \alpha^j + \alpha^k + \alpha^\ell)x^3 + (\alpha^i\alpha^j + \alpha^i\alpha^k$$
$$+ \alpha^i\alpha^\ell + \alpha^j\alpha^k + \alpha^j\alpha^\ell + \alpha^k\alpha^\ell)x^2 + (\alpha^i\alpha^j\alpha^k$$
$$+ \alpha^i\alpha^j\alpha^\ell + \alpha^i\alpha^k\alpha^\ell + \alpha^j\alpha^k\alpha^\ell)x + \alpha^i\alpha^j\alpha^k\alpha^\ell.$$

The coefficients of x^3 and x^4 are zero, so we have

$$(x - \alpha^i)(x - \alpha^j)(x - \alpha^k)(x - \alpha^\ell) = x^4 + Ax + B$$

for some constants A and B. But if any polynomial of degree 4 has four zeros, it can be written as the product of two quadratics, each with two of the zeros. Then

$$x^4 + Ax + B = (x^2 + ax + b)(x^2 + cx + d)$$
$$= x^4 + (a + c)x^3 + (ac + b + d)x^2 + (ad + bc)x + bd.$$

This means that $a + c = 0$, $ac + b + d = 0$, and $ad + bc = A$. Then $a = c$, $b + d = a^2$, and so $a^3 = A$. Such an a exists only if A has a cube root. But if A has a cube root, then by the substitution $y = A^{1/3}x$, the original equation becomes $y^4 + y + B/A^{4/3} = 0$, which, as before, has quadratic factors only if $a^3 = 1$. Such an a does not exist in $GF(2^m)$ if m is odd, so such a polynomial with four distinct zeros does not exist.

Thus every codeword of Lee weight 4 has an odd number of components equal to -1. For such a codeword the extension symbol is 2, so the extended codeword has Lee weight 6. This completes the proof of the theorem. ∎

This concludes the discussion of the Preparata codes. Next we discuss the Kerdock codes which over $\mathbf{Z}_4$ are the duals of the Preparata codes.

Definition 2.16.3 *A cyclic Kerdock code over $\mathbf{Z}_4$ of blocklength $2^m - 1$, m odd, is a cyclic code over $\mathbf{Z}_4$ whose generator polynomial $g(x)$ is $(x^{2^m-1} - 1)/(x-1)h(x)$, where $h(x)$ is a primitive basic irreducible factor of $x^{2^m-1} - 1$ in the Galois ring $GR(4^m)$. An extended Kerdock code over $\mathbf{Z}_4$ of blocklength 2^m is a cyclic Kerdock code augmented by a simple $\mathbf{Z}_4$ check sum.*

Because the degree of $g(x)$ is $n - (m + 1)$, where $n = 2^m - 1$, a cyclic Kerdock code over $\mathbf{Z}_4$ has dimension $m + 1$. Thus it has 4^{m+1} codewords, as does the extended Kerdock code. A *binary Kerdock code* of blocklength 2^{m+1} is the Gray image of the extended Kerdock code over $\mathbf{Z}_4$. An *original Kerdock code* of blocklength $2^{m+1} - 1$ is a binary Kerdock code of blocklength 2^{m+1} punctured by one bit. The binary code is nonlinear as a consequence of the nonlinearity of the Gray map. It inherits the distance-invariance property from the underlying cyclic code over $\mathbf{Z}_4$. Because the binary codes have $4^{m+1} = 2^{2(m+1)}$ codewords, these codes have datalength $2(m + 1)$.

Theorem 2.16.4 *A cyclic Kerdock code over $\mathbf{Z}_4$ of blocklength $2^m - 1$ has minimum Lee distance equal to $2^m - 2^{(m-1)/2} - 2$. An extended Kerdock code over $\mathbf{Z}_4$ of blocklength 2^m has minimum Lee distance equal to $2^m - 2^{(m-1)/2}$.*

Proof: The proof of this theorem is not given. ∎

The theorem allows us to conclude that an original binary Kerdock code of blocklength $2^{m+1} - 1$ has minimum distance

$$d_{\min} = 2^m - 2^{(m-1)/2} - 1$$

because distance is preserved under the Gray map, and puncturing a binary code by one place can reduce the distance between two codewords by at most 1.

Because

$$x^7 - 1 = (x - 1)(x^3 + 2x^2 + x - 1)(x^3 - x^2 + 2x - 1)$$

and the latter two factors are reciprocals (but for sign), the Kerdock code over $\mathbf{Z}_4$ of blocklength 7 is the same as the Preparata code over $\mathbf{Z}_4$ of blocklength 7. Furthermore, it is clear from their definitions that the Preparata code and the (reciprocal) Kerdock code of the same blocklength over $\mathbf{Z}_4$ are duals. However, because the binary Preparata codes and the binary Kerdock codes are nonlinear, the notion of a dual code does not properly apply. Nevertheless, the binary codes do inherit some residual properties of this kind from the fact that the overlying Kerdock and Preparata codes over $\mathbf{Z}_4$ are duals. For these reasons, the binary Kerdock code and the binary Preparata code of blocklength 2^m are sometimes called *formal duals*.

There is one other class of codes over $\mathbf{Z}_4$ that will be mentioned. This is the class of *Goethals codes* over $\mathbf{Z}_4$, which codes have minimum distance 7. We will end the section with a brief description of these codes. The *cyclic Goethals code* over the ring $\mathbf{Z}_4$ of blocklength $n = 2^m - 1$, for m odd and at least 5, is defined by the check matrix

$$H = \begin{bmatrix} 1 & \xi^1 & \xi^2 & \xi^3 & \cdots & \xi^{(n-1)} \\ 2 & 2\xi^3 & 2\xi^6 & 2\xi^9 & \cdots & 2\xi^{3(n-1)}. \end{bmatrix}$$

where ξ is an element of $GR(4^m)$ of order n. This check matrix specifies that $c(x)$ is a codeword if and only if $c(\xi)$ is zero and $c(\xi^3)$ is even. It is not required that $c(\xi^3)$ be zero. Indeed, the Goethals code over $\mathbf{Z}_4$ of blocklength $2^m - 1$ is the set of codewords of the Preparata code over $\mathbf{Z}_4$ of blocklength $2^m - 1$ for which $c(\xi^3)$ is even. The *extended Goethals code* of blocklength 2^m is the cyclic Goethals code of blocklength $2^m - 1$ augmented by a simple $\mathbf{Z}_4$ check sum. The extended Goethals code has the check matrix

$$H = \begin{bmatrix} 1 & 1 & 1 & 1 & 1 & \cdots & 1 \\ 0 & 1 & \xi^1 & \xi^2 & \xi^3 & \cdots & \xi^{(n-1)} \\ 0 & 2 & 2\xi^3 & 2\xi^6 & 2\xi^9 & \cdots & 2\xi^{3(n-1)} \end{bmatrix}.$$

This means that $c(x)$ is a codeword of the extended Goethals code if and only if $c_\infty + c(1)$ is zero, $c(\xi)$ is zero, and $c(\xi^3)$ is even.

In the Galois ring $GR(4^m)$, the element ξ^0 has only itself in its conjugancy class, and both ξ^1 and ξ^3 have m elements in their conjugancy classes. The third row of H, however, only eliminates half of the words controlled by the conjugacy class of ξ^3. Hence the redundancy satisfies $n - k = 1 + m + m/2$ so the dimension of a Goethals code over $\mathbf{Z}_4$ is $k = 2^m - 3m/2 - 1$.

The extended Goethals code can be shortened by taking all codewords for which the extension symbol is zero, then dropping that extension symbol. The shortened code has the check matrix

$$
H = \begin{bmatrix}
1 & 1 & 1 & 1 & \cdots & 1 \\
1 & \xi^1 & \xi^2 & \xi^3 & \cdots & \xi^{(n-1)} \\
2 & 2\xi^3 & 2\xi^6 & 2\xi^9 & \cdots & 2\xi^{3(n-1)}
\end{bmatrix}.
$$

This is the check matrix of another cyclic code over Z_4 contained within the cyclic Goethals code.

The binary Goethals code is the image under the Gray map of the extended Goethals code over the ring Z_4. The binary Goethals code is a nonlinear $(2^{m+1}, 2^{m+1} - 3m - 2, 8)$ binary code. The datalength of the nonlinear binary Goethals code is $2^m - 3m - 2$. It may be presented in a punctured form as a nonlinear $(2^{m+1} - 1, 2^{m+1} - 3m - 2, 7)$ binary triple-error-correcting code.

For example, for $m = 5$, the cyclic Goethals code over Z_4 is a $(31, 23.5)$ code, and the extended Goethals code is a $(32, 23.5, 8)$ code. The Gray map yields a $(64, 47, 8)$ binary code that can be punctured to obtain a nonlinear $(63, 47, 7)$ binary code. The datalength of these codes is 47. For $m = 7$, the cyclic Goethals code is a $(127, 116.5)$ code over Z_4, and the extended Goethal code is a $(128, 116.5, 8)$ code over Z_4. The Gray map yields a $(256, 233, 8)$ code that can be punctured to obtain a $(255, 233, 7)$ code. The datalength of this code is 233. The comparable BCH code is a linear $(255, 231, 7)$ binary code. The $(63, 47, 7)$ and the $(255, 233, 7)$ binary Goethals codes are the best distance-7 binary codes known of their respective blocklengths. No linear binary codes are known with parameters as good or better than these.

Problems

2.1   Prove that the Hamming distance is a metric. (A metric is nonnegative, symmetric, and satisfies the triangle inequality.)

2.2   Prove that a generator polynomial of a cyclic code, defined as a monic codeword polynomial of minimum degree, is unique.

2.3   (a) Prove that the dual of a cyclic Reed–Solomon code is a cyclic Reed–Solomon code.
      (b) What is the dual of an affine Reed–Solomon code?
      (c) What is the dual of a projective Reed–Solomon code?

2.4   Prove that a BCH code of blocklength 17 over $GF(16)$ is a maximum-distance code. Prove that it is equivalent to a doubly extended Reed–Solomon code. Can these remarks be generalized to other blocklengths?

2.5 Prove or disprove the following generalization of the BCH bound. The only vector in $GF(q)^m$ of weight $d-1$ or less that has $d-1$ sequential components of its filtered spectrum $T = H*C$ equal to zero ($T_k = 0$, for $k = k_0, \ldots, k_0+d-2$), where H is an invertible filter, is the all-zero vector.

2.6 Suppose that A is any invertible matrix.
 (a) Prove that if $\tilde{H} = HA$, then heft $\tilde{H} = $ heft H.
 (b) Let H be a check matrix for the cyclic code C. Let $\tilde{H}$ be the row-wise Fourier transform of H. That is, $\tilde{H} = H\Omega$, where

$$\Omega = \begin{bmatrix} 1 & 1 & 1 & \ldots & 1 \\ 1 & \omega & \omega^2 & \ldots & \omega^{n-1} \\ 1 & \omega^2 & \omega^4 & \ldots & \omega^{2(n-1)} \end{bmatrix}.$$

 What can be said relating heft $\tilde{H}$ to heft H?
 (c) Prove the BCH bound from this property.

2.7 Let $c'(x)$ and $c''(x)$ be two minimum-weight codewords in the cyclic code C. Must it always be true that $c''(x) = Ax^\ell c'(x)$ for some field element A and integer ℓ?

2.8 Is the dual of an $(8, 4, 4)$ extended binary Hamming code equivalent to the extension of the dual of the $(7, 4, 3)$ cyclic binary Hamming code?

2.9 A Vandermonde matrix is a square matrix in which the elements in the ℓth row are the ℓth powers of corresponding elements of the first row.
 (a) Show that a Vandermonde matrix is full rank if the elements in the first row are nonzero and distinct.
 (b) Find the minimum distance of the Reed–Solomon code as a consequence of this property of the Vandermonde matrix.

2.10 Prove that the sum of two cubes in $GF(16)$ is never a cube.

2.11 Find the generator polynomial for an $(11, 6, d)$ code over $GF(3)$. Is this code a quadratic residue code? Is it a perfect code? What is d? (This code is known as the *ternary Golay code*).

2.12 Using $\{1, \alpha, \alpha^6\}$ as a basis for $GF(8)$, show that the binary expansion of a $(7, 5, 3)$ Reed–Solomon code, obtained by replacing each symbol of $GF(8)$ by three symbols of $GF(2)$, is equivalent to a $(21, 15, 3)$ BCH code.

2.13 Use the van Lint–Wilson bound to show that the $(23, 12)$ binary Golay code has minimum distance 7. (Row permutation 0, 22, 19, 3, 7, 20, 18, and column permutation 5, 6, 9, 11, 21, 8, 10 will be helpful.)

2.14 Prove that the Singleton bound also holds for nonlinear codes.

2.15 Prove that the $(127, 113)$ Roos code has minimum distance 5. Prove that the $(127, 106)$ Roos code has minimum distance 7.

2.15 Suppose that $g(x) = (x+1)g'(x)$ generates a binary cyclic code with minimum distance d. Show that $g'(x)$ need not generate a binary code with minimum distance at least $d - 1$. (**Hint**: Choose $g'(x)$ to have zeros at α and α^{-1}.) Is the following statement true? "Puncturing a code by eliminating one check symbol reduces the minimum distance by at most one."

2.16   Verify that

$$\binom{90}{0} + \binom{90}{1} + \binom{90}{2} = 2^{12}.$$

Despite this suggestive formula, a $(90, 78, 5)$ linear code does not exist, so there is no linear perfect code with these parameters. Is there a simple proof?

2.17   Prove that the binary Golay code is not the Gray image of a linear code in $\mathbf{Z}_4$.

2.18   Let $G(x) = x^2 + x + 1$ be the Goppa polynomial for a $(32, 22, 5)$ Goppa code. Derive an encoding rule and give a decoding procedure.

2.19   Let $G(x) = x^2 + x + \alpha^3$ be the Goppa polynomial for a $(16, 8, 5)$ Goppa code, where α is a primitive element in $GF(16)$. Find a check matrix for this code.

2.20   The *tetracode* is a $(4, 2, 3)$ linear code over $GF(3)$ with the generator matrix given by

$$G = \begin{bmatrix} 1 & 0 & 1 & 1 \\ 0 & 1 & 1 & 2 \end{bmatrix}.$$

The *hexacode* is a $(6, 3, 4)$ over $GF(4)$ with generator matrix given by

$$G = \begin{bmatrix} 1 & 0 & 0 & 1 & \alpha & \alpha \\ 0 & 1 & 0 & \alpha & 1 & \alpha \\ 0 & 0 & 1 & \alpha & \alpha & 1 \end{bmatrix}.$$

In each case, prove that the code is a unique self-dual code with the stated parameters.

2.21   Factor the polynomial $x^4 - 1$ over $\mathbf{Z}_4$ in two distinct ways into irreducible polynomials over $\mathbf{Z}_4[x]$.

2.22   The *trace code* of a linear code $\mathcal{C}$ is obtained by replacing each component of each codeword by its trace. Prove that the dual of the subfield-subcode of $\mathcal{C}$ is equal to the trace code of the dual code of $\mathcal{C}$.

2.23   Prove that the class of Goppa codes satisfies the Varshamov–Gilbert bound.

2.24   (a) Is the Hensel lift to $\mathbf{Z}_4$ of the product of two polynomials over $GF(2)$ equal to the product of the Hensel lifts?

(b) Is the Hensel lift to $\mathbf{Z}_4$ of the sum of two polynomials over $GF(2)$ equal to the sum of the Hensel lifts?

(c) Let $\bar{g}(x)$ be the Hensel lift of $g(x)$, a primitive binary polynomial dividing $x^{2^m-1} - 1$. Is every codeword polynomial of the Preparata code over $\mathbf{Z}_4$ generated by $\bar{g}(x)$ the Hensel lift of a binary codeword of the code generated by $g(x)$?

2.25 Find the basic irreducible factors over $\mathbf{Z}_4$ of $x^{15} - 1$.

2.26 Does the Singleton bound hold for Lee-distance codes?

2.27 Prove that $\xi^i + \xi^j \neq \xi^k$ for all i, j, and k, where ξ is a primitive element of the Galois ring $GR(4^m)$.

2.28 Let $g'(x)$ and $g''(x)$ be products of distinct irreducible factors of $x^n - 1$ over $\mathbf{Z}_4$.

    (a) Define the code $\mathcal{C}$ over $\mathbf{Z}_4$ as

$$\mathcal{C} = \{a_1(x)g'(x)g''(x) + 2a_2(x)g''(x)\},$$

with the understanding that the degrees of $a_1(x)$ and $a_2(x)$ are restricted so that each of the two terms in the sum has a degree not larger than $n - 1$. Prove that $\mathcal{C}$ is a cyclic code over $\mathbf{Z}_4$.

    (b) Express the two Calderbank–McGuire codes in this form.

2.29 Prove that the two extended Calderbank–McGuire codes over $\mathbf{Z}_4$ have minimum distance 12 and 14, respectively.

Notes

Cyclic codes have long occupied a central position in the subject of algebraic coding. We have slightly de-emphasized the cyclic property in order to give equal importance to codes of blocklengths $q-1$, q, and $q+1$. The terms "codes on the cyclic line," "codes on the affine line," and "codes on the projective line" were chosen because they have an appealing symmetry and give the desired starting point for similar classifications that we want to make for codes on the plane in Chapter 6 and for codes on curves in Chapter 10. Moreover, with these terms, the Reed–Solomon codes of blocklengths $q-1$, q, and $q+1$ are more nearly on an equal footing. There is some merit in the term "code on the cyclic line" in preference to "cyclic code" because it de-emphasizes the cyclic property, which really, in itself, is not an important property of a code. The cyclic form of the Reed–Solomon codes was discovered by Reed and Solomon (1960), independently by Arimoto (1961), and was interpreted only later as a construction on the projective line. The doubly extended form was discovered by Wolf (1969). The BCH codes were discovered independently of the Reed–Solomon codes by Bose and Ray–Chaudhuri (1960), and also by Hocquenghem (1959).

    The class of cyclic codes was introduced by Prange (1957). The Golay code (Golay, 1949) is special; it can be viewed in many ways. The quadratic residue codes were introduced by Prange (1958) as examples of cyclic codes. The binary Golay code, which is one example of a quadratic residue code, had been discovered earlier than the general class of quadratic residue codes. The binary Golay code was shown to be the only $(23, 12, 7)$ binary code by Pless (1968). The quadratic residue codes were first studied by Prange (1957) and others, Assmus and Mattson (1974) includes a compendium of this work. No satisfactory statement describing the minimum distance of quadratic residue codes of large blocklength is known.

The alternant codes were introduced by Helgert (1974), under this name because the check matrix can be put in the form of an alternant matrix. The Goppa codes (Goppa, 1970), are now seen as a subclass of the alternant codes.

The nonlinear $(15, 8, 5)$ cyclic code, discussed in Section 2.10, was discussed by Blahut (1983). It can be compared with the $(15, 8, 5)$ Preparata code. Preparata codes exist for blocklengths n of the form $2^{2m} - 1$. The Preparata codes are examples of a family of nonlinear codes, also including Goethals codes and Kerdock codes, which can be constructed as a representation in $GF(2)$ of a linear code over the ring $\mathbf{Z}_4$.

The Preparata codes have an interesting history. Preparata (1968) discovered that class based on studying the properties of the smallest code in the class, the $(15, 8, 5)$ code, which was already known under the name of the Nordstrom–Robinson code (Nordstrom and Robinson, 1967). Using a computer, Nordstrom and Robinson had constructed the $(15, 8, 5)$ nonlinear code as an extension of both the still earlier $(12, 5, 5)$ nonlinear Nadler code (Nadler, 1962) and the $(13, 6, 5)$ nonlinear Green code (Green, 1966). In turn, the class of Preparata codes has stimulated the discovery of other nonlinear codes: the Kerdock low-rate codes (Kerdock, 1972) and the triple-error-correcting Goethals codes (Goethals, 1976). The recognition that these nonlinear codes (slightly altered) are images of linear codes over $\mathbf{Z}_4$ came simultaneously to several people and was jointly published by Hammons *et al.* (1994). We take the liberty of using the original names for the modern version of the codes over $\mathbf{Z}_4$, regarding the Gray map as simply a way of representing the elements of $\mathbf{Z}_4$ by pairs of bits. The structure of cyclic codes over $\mathbf{Z}_4$ was studied by Calderbank *et al.* (1996) and by Pless and Qian (1996). Calderbank and McGuire (1997) discovered their nonlinear $(64, 37, 12)$ binary code that led them, with Kumar and Helleseth, directly to the discovery of the nonlinear $(64, 32, 14)$ binary code. The octacode, which is a notable code over $\mathbf{Z}_4$, was described by Conway and Sloane (1992). The role of the basic irreducible polynomial was recognized by Solé (1989). The relationship between the octacode and the Nordstrom–Robinson code was observed by Forney, Sloane, and Trott (1993).

3 The Many Decoding Algorithms for Reed–Solomon Codes

Decoding large linear codes, in general, is a formidable task. For this reason, the existence of a practical decoding algorithm for a code can be a significant factor in selecting a code. Reed–Solomon codes – and other cyclic codes – have a distance structure that is closely related to the properties of the Fourier transform. Accordingly, many good decoding algorithms for Reed–Solomon codes are based on the Fourier transform.

The algorithms described in this chapter form the class of decoding algorithms known as "locator decoding algorithms." This is the richest, the most interesting, and the most important class of algebraic decoding algorithms. The algorithms for locator decoding are quite sophisticated and mathematically interesting. The appeal of locator decoding is that a certain seemingly formidable nonlinear problem is decomposed into a linear problem and a well structured and straightforward nonlinear problem. Within the general class of locator decoding algorithms, there are many options, and a variety of algorithms exist.

Locator decoding can be used whenever the defining set of a cyclic code is a set of consecutive zeros. It uses this set of consecutive zeros to decode, and so the behavior of locator decoding is closely related to the BCII bound rather than to the actual minimum distance. Locator decoding, by itself, reaches the BCH radius, which is the largest integer smaller than half of the BCII bound, but reaches the packing radius of the code only if the packing radius is equal to the BCH radius. For a Reed–Solomon code (and most BCH codes), the minimum distance is equal to the BCH bound, so, for these codes, locator decoding does reach the packing radius. Locator decoding is the usual choice for the Reed–Solomon codes.

Locator decoding algorithms are based on the use of the polynomial $\Lambda(x)$, known as the *locator polynomial*. Because locator decoding exploits much of the algebraic structure of the underlying field, it forms a powerful family of decoding algorithms that are especially suitable for large codes. The choice of an algorithm may depend both on the specific needs of an application and on the taste of the designer. The most important decoding algorithms depend on the properties of the Fourier transform. For this reason, the topic of decoding Reed–Solomon codes may be considered a branch of the subject of signal processing. Here, however, the methods of signal processing are used in a

Galois field instead of the real or complex field. Another instance of these methods, now in a Galois ring, is also briefly discussed in this case for decoding a Preparata code. Except for the decoding of Preparata codes, the methods of locator decoding are not yet worked out for codes on rings.

3.1   Syndromes and error patterns

A codeword c is transmitted and the channel makes errors. If there are errors in not more than t places, where $t = \lfloor (d_{\min} - 1)/2 \rfloor$ is the packing radius of the code, then the decoder should recover the codeword (or the data symbols contained in the codeword).

The vector v, which will be called the *senseword*, is the received word in a data communication system and is the read word in a data storage system. The senseword v is the codeword c corrupted by an error vector e. The ith component of the senseword is given by

$$v_i = c_i + e_i \quad i = 0, \ldots, n-1,$$

and e_i is nonzero for at most t values of i. If not more than t components are in error, then a *bounded-distance decoder* is one that must recover the unique codeword (or the data symbols contained in the codeword) from the senseword v. In contrast, a *complete decoder* must recover a codeword that is nearest to the senseword regardless of how many components are in error. For large codes, a complete decoder is neither tractable nor desirable.

We only consider codes whose alphabet is a field, so it is meaningful to define the error in the ith component of the codeword to be $e_i = v_i - c_i$. Consequently, the senseword v can be regarded as the codeword c corrupted by an additive error vector e, and the error vector e is nonzero in at most t components.

A linear code over the field F, usually the finite field $GF(q)$, is associated with a check matrix, H, such that $c H^{\mathrm{T}} = 0$ for every codeword c. Therefore

$$v H^{\mathrm{T}} = (c + e) H^{\mathrm{T}} = e H^{\mathrm{T}}.$$

For a general linear code, the *syndrome vector* s, with components called *syndromes*, is defined as

$$s = v H^{\mathrm{T}} = e H^{\mathrm{T}}.$$

For a linear code, the task of decoding can be decomposed into a preliminary task and a primary task. The preliminary task is to compute the syndrome vector $s = v H^{\mathrm{T}}$, which is a linear operation taking the n vector e to an $(n-k)$ vector s. The primary task is to solve the equation

$$s = eH^{\mathrm{T}}$$

for that vector e with weight not larger than t. This is the task of solving $n - k$ equations for the n-vector e of minimum weight. The set of such n vectors of weight at most t is not a linear subspace of $GF(q)^n$, which means that the map from the set of syndrome vectors s back to the set of error vectors e is not a linear map. To invert requires a nonlinear operation from the space of $(n - k)$ vectors to the space of n vectors. Because every correctable error pattern must have a unique syndrome, the number of vectors in the space of syndromes that have such a solution is $\sum_{\ell=0}^{t}(q-1)^{\ell}\binom{n}{\ell}$. This number, which is not larger than q^{n-k}, is the number of elements in the space of $(n - k)$ vectors that have correctable error patterns as inverse images, under a bounded-distance decoder, in the n-dimensional space of error patterns.

For small binary codes, one can indeed form a table of the correctable error patterns and the corresponding syndromes. A fast decoder, which we call a *boolean-logic decoder*, consists of a logic tree that implements the look-up relationship between syndromes and error patterns. A boolean-logic decoder can be extremely fast, but can be used only for very simple binary codes.

For cyclic Reed–Solomon codes and other cyclic codes, it is much more convenient to use an alternative definition of syndromes in terms of the Fourier transform. The senseword v has the following Fourier transform:

$$V_j = \sum_{i=0}^{n-1} \omega^{ij} v_i \quad j = 0, \ldots, n-1,$$

which is easily computed. By the linearity of the Fourier transform,

$$V_j = C_j + E_j \quad i = 0, \ldots, n-1,$$

Furthermore, by the construction of the Reed–Solomon code,

$$C_j = 0 \quad j = 0, \ldots, n-k-1.$$

Hence

$$V_j = E_j \quad j = 0, \ldots, n-k-1.$$

To emphasize that these are the $n - k$ components of the error spectrum E that are immediately known from V, they are frequently denoted by the letter S and called *(spectral) syndromes*, though they are not the same as the syndromes introduced earlier. To distinguish the two definitions, one might call the former *code-domain syndromes* and the latter *transform-domain syndromes*. Thus the transform-domain syndrome is given by

$$S_j = E_j = V_j \quad j = 0, \ldots, n - k - 1.$$

Here we are treating the special case where $j_0 = 0$. There is nothing lost here because the modulation property of the Fourier transform tells us what happens to c when C is cyclically translated. By using the modulation property, the entire discussion holds for any value of j_0.

Represented as polynomials, the *error-spectrum polynomial* is given by

$$E(x) = \sum_{j=0}^{n-1} E_j x^j,$$

and the *syndrome polynomial* is given by

$$S(x) = \sum_{j=0}^{n-k-1} S_j x^j.$$

The error-spectrum polynomial has degree at most $n - 1$, and the syndrome polynomial has degree at most $n - k - 1$.

Because $d_{\min} = n - k + 1$ for a Reed–Solomon code, the code can correct t errors, where $t = \lfloor (n - k)/2 \rfloor$. Our task, then, is to solve the equation

$$S_j = \sum_{i=0}^{n-1} \omega^{ij} e_i \quad j = 0, \ldots, n - k - 1$$

for the error vector e of smallest weight, given that this weight is at most t. An alternative task is to find the error transform E of blocklength n, given that E_j is equal to the known S_j for $j = 0, \ldots, n - k - 1$, and e has weight at most $t = \lfloor (n - k)/2 \rfloor$. Any decoder that uses such syndromes in the Fourier transform domain is called a *transform-domain decoder*.

The first decoding step introduces, as an intermediate, an auxiliary polynomial $\Lambda(x)$ known as the *locator polynomial* or the *error-locator polynomial*. We shall see that the nonlinear relationship between the set of syndromes and the error spectrum E is replaced by a linear relationship between the syndromes S_j and the coefficients of the error-locator polynomial $\Lambda(x)$. The nonlinear operations that must show up somewhere in the decoder are confined to the relationship between $\Lambda(x)$ and the remaining components of E, and that nonlinear relationship has the simple form of a linear recursion. The obvious linear procedure of finding the coefficients of the linear recursion from the syndromes by direct matrix inversion is known as the *Peterson algorithm*.

Given the error vector e of (unknown) weight v, at most t, consider the polynomial given by

$$\Lambda(x) = \prod_{\ell=1}^{v}(1 - x\omega^{i_\ell}),$$

where the indices i_ℓ for $\ell = 1, \ldots, v$ point to the $v \leq t$ positions that are in error. These positions correspond to the nonzero components of e. Then $\lambda_i = (1/n)\Lambda(\omega^{-i})$ is equal to zero if and only if an error e_i occurred at the component with index i, and this cannot hold for any $\Lambda(x)$ of smaller degree. Therefore $e_i\lambda_i = 0$ for all i. By the convolution property of the Fourier transform, this implies that

$$\Lambda(x)E(x) = 0 \pmod{x^n - 1},$$

which confirms that $\Lambda(x)$ is indeed the error-locator polynomial. Written in terms of its coefficients, this polynomial equation becomes

$$\sum_{k=0}^{v} \Lambda_k E_{((j-k))} = 0 \quad j = 0, \ldots, n-1,$$

where the double parentheses on the indices denote modulo n. Because $\Lambda_0 = 1$, this equation can be rewritten as follows:

$$E_j = -\sum_{k=1}^{v} \Lambda_k E_{((j-k))} \quad j = 0, \ldots, n-1,$$

which is a simple linear recursion that the components of the error spectrum must satisfy.

The statement that the length v of the recursion is equal to the weight of e follows from the previous discussion. This is an instance of the linear complexity property, which was discussed in Section 1.5. The error vector e has weight v at most t, so the linear complexity property says that all components of E can be cyclically produced by a linear recursion of length at most t,

$$E_j = -\sum_{k=1}^{t} \Lambda_k E_{((j-k))} \quad j = 0, \ldots, n-1,$$

where $\Lambda(x)$ is the locator polynomial of the error vector e. The important reason for developing this cyclic recursion is that it is a set of linear equations relating the unknown coefficients Λ_k and the components of E. Of the n equations contained in the above recursion, there are t equations that involve only the $2t$ known components of E and the t unknown components of $\mathbf{\Lambda}$. These are as follows:

$$E_t = -(\Lambda_1 E_{t-1} + \Lambda_2 E_{t-2} + \cdots + \Lambda_t E_0),$$

$$E_{t+1} = -(\Lambda_1 E_t + \Lambda_2 E_{t-1} + \cdots + \Lambda_t E_1),$$

$$\vdots$$

$$E_{2t-1} = -(\Lambda_1 E_{2t-2} + \Lambda_2 E_{2t-3} + \cdots + \Lambda_t E_{t-1}).$$

These t equations, expressed in matrix form, are given by

$$
\begin{bmatrix}
E_{t-1} & E_{t-2} & \cdots & E_0 \\
E_t & E_{t-1} & & E_1 \\
\vdots & & & \vdots \\
E_{2t-2} & E_{2t-3} & \cdots & E_{t-1}
\end{bmatrix}
\begin{bmatrix}
\Lambda_1 \\
\Lambda_2 \\
\vdots \\
\Lambda_t
\end{bmatrix}
= -
\begin{bmatrix}
E_t \\
E_{t+1} \\
\vdots \\
E_{2t-1}
\end{bmatrix}.
$$

This matrix equation can be solved for the connection coefficients Λ_j by any convenient computational procedure for solving matrix equations. One such procedure is the method of gaussian elimination. Because it is assumed that the error vector e has weight at most t, the matrix equation must have a solution. If the determinant of the matrix is zero, then there are fewer than t errors. This means that the leading coefficient Λ_t is zero. If the determinant is zero, simply replace t by $t-1$ in the matrix equation and solve the smaller problem in the same way. In this way, the matrix is eventually reduced to a ν by ν matrix with a nonzero determinant.

Once $\mathbf{\Lambda}$ is known, the other components of the error spectrum E can be computed, one by one, by using the following recursion:

$$E_j = -\sum_{k=1}^{t} \Lambda_k E_{((j-k))} \quad j = 2t, \ldots, n-1.$$

This recursion provides the unavoidable nonlinear function that must be part of the decoding algorithm. An inverse Fourier transform then gives the error vector e. Next, componentwise subtraction yields

$$c_i = v_i - e_i \quad i = 0, \ldots, n-1.$$

Finally, the data symbols are recovered from the code symbols by inverting the operation used by the encoder, normally an easy calculation.

This completes the development of an elementary decoding algorithm for bounded-distance decoding of Reed–Solomon and other BCH codes. However, this is only the start of a line of development that goes much further. Locator decoding has now grown far more sophisticated, driven by a need to simplify the computations of the decoding

Table 3.1. *A representation of GF*(16)

$$
\begin{aligned}
\alpha^0 &= && && && 1 \\
\alpha^1 &= && && z && \\
\alpha^2 &= && z^2 && && \\
\alpha^3 &= z^3 && && && \\
\alpha^4 &= && && z &+& 1 \\
\alpha^5 &= && z^2 &+& z && \\
\alpha^6 &= z^3 &+& z^2 && && \\
\alpha^7 &= z^3 && &+& z &+& 1 \\
\alpha^8 &= && z^2 && &+& 1 \\
\alpha^9 &= z^3 && &+& z && \\
\alpha^{10} &= && z^2 &+& z &+& 1 \\
\alpha^{11} &= z^3 &+& z^2 &+& z && \\
\alpha^{12} &= z^3 &+& z^2 &+& z &+& 1 \\
\alpha^{13} &= z^3 &+& z^2 && &+& 1 \\
\alpha^{14} &= z^3 && && &+& 1
\end{aligned}
$$

algorithm. There are many different ways to organize the computations, using ideas from signal processing to reduce the computational burden. We shall discuss the various enhancements of the Peterson algorithm, beginning in Section 3.4.

As an example of the Peterson algorithm, we shall work through the decoding of a $(15, 9, 7)$ Reed–Solomon code over $GF(16)$. Because $n = 15$ is a primitive blocklength, we can choose $\omega = \alpha$, where α is a primitive element of $GF(16)$. We will choose α such that $\alpha^4 + \alpha + 1 = 0$. The field representation is as shown in Table 3.1. We will choose the particular $(15, 9, 7)$ Reed–Solomon code with the defining set $\{1, 2, 3, 4, 5, 6\}$. For this example, note that we have chosen a defining set that starts at $j_0 = 1$ rather than at $j_0 = 0$, as was the case chosen earlier.

Suppose that the dataword, the codeword, and the senseword are, respectively, given by

$$\boldsymbol{d} = 0, 0, 0, 0, 0, 0, 0, 0, 0,$$

$$\boldsymbol{c} = 0, 0, 0, 0, 0, 0, 0, 0, 0, 0, 0, 0, 0, 0, 0,$$

$$\boldsymbol{v} = 0, 0, 0, 0, 0, 0, 0, \alpha, 0, \alpha^5, 0, 0, \alpha^{11}, 0, 0,$$

with indices running from high to low.

The first step of decoding is to compute the Fourier transform of $\boldsymbol{v}$; only six components are needed. This computation yields

$$V = -, \alpha^{12}, 1, \alpha^{14}, \alpha^{13}, 1, \alpha^{11}, -, -, -, -, -, -, -, -.$$

These six components are equal to the corresponding six components of E. Next, solve for Λ from the equation

$$
\begin{bmatrix} E_3 & E_2 & E_1 \\ E_4 & E_3 & E_2 \\ E_5 & E_4 & E_3 \end{bmatrix} \begin{bmatrix} \Lambda_1 \\ \Lambda_2 \\ \Lambda_3 \end{bmatrix} = - \begin{bmatrix} E_4 \\ E_5 \\ E_6 \end{bmatrix}.
$$

Thus

$$
\begin{bmatrix} \Lambda_1 \\ \Lambda_2 \\ \Lambda_3 \end{bmatrix} = \begin{bmatrix} \alpha^{14} & 1 & \alpha^{12} \\ \alpha^{13} & \alpha^{14} & 1 \\ 1 & \alpha^{13} & \alpha^{14} \end{bmatrix}^{-1} \begin{bmatrix} \alpha^{13} \\ 1 \\ \alpha^{11} \end{bmatrix} = \begin{bmatrix} \alpha^{14} \\ \alpha^{11} \\ \alpha^{14} \end{bmatrix},
$$

and one can conclude that the error-locator polynomial is given by

$$
\Lambda(x) = 1 + \alpha^{14}x + \alpha^{11}x^2 + \alpha^{14}x^3.
$$

This polynomial can be factored as follows:

$$
\Lambda(x) = (1 + \alpha^2 x)(1 + \alpha^5 x)(1 + \alpha^7 x).
$$

This, in turn, means that the errors are at locations $i = 2, 5$, and 7.

3.2   Computation of the error values

The Peterson algorithm, described in the previous section, decomposes the problem of error correction into the task of finding the error-locator polynomial and the task of computing the error values. Once the locator polynomial is known, it only remains to compute the error values from the locator polynomial and the syndromes. The computation can proceed in any of several ways. We shall describe three approaches.

The first method of computing the error values, called *recursive extension*, is to use the recursion

$$
E_j = - \sum_{k=1}^{t} \Lambda_k E_{((j-k))}
$$

to produce the complete error spectrum. Thus, in our running example,

$$
\begin{aligned}
E_7 &= \Lambda_1 E_6 + \Lambda_2 E_5 + \Lambda_3 E_4 \\
&= \alpha^{14} \cdot \alpha^{11} + \alpha^{11} \cdot 1 + \alpha^{14} \cdot \alpha^{13} \\
&= \alpha^5.
\end{aligned}
$$

Similarly,

$$E_8 = \alpha^{14} \cdot \alpha^5 + \alpha^{11} \cdot \alpha^{11} + \alpha^{14} \cdot 1$$
$$= 1$$

and

$$E_9 = \alpha^{14} \cdot 1 + \alpha^{11} \cdot \alpha^5 + \alpha^{14} \cdot \alpha^{11}$$
$$= \alpha^6.$$

The process continues in this way until all components of E are known. This yields

$$E = (\alpha^9, \alpha^{12}, 1, \alpha^{14}, \alpha^{13}, 1, \alpha^{11}, \alpha^5, 1, \alpha^6, \alpha^7, 1, \alpha^{10}, \alpha^3, 1).$$

An inverse Fourier transform of E yields

$$e = 0, 0, 0, 0, 0, 0, 0, 0, \alpha, 0, \alpha^5, 0, 0, \alpha^{11}, 0, 0$$

as the error pattern.

The second method of computing the error values is called the *Gorenstein–Zierler algorithm*. Because $\Lambda(x)$ can be factored and written as

$$\Lambda(x) = (1 + \alpha^2 x)(1 + \alpha^5 x)(1 + \alpha^7 x),$$

we know that the errors are at locations $i = 2, 5,$ and 7. Then the Fourier transform relationship $E_j = \sum_{i=0}^{n-1} e_i \omega^{ij}$ can be truncated to write the following matrix equation:

$$
\begin{bmatrix}
\alpha^2 & \alpha^5 & \alpha^7 \\
\alpha^4 & \alpha^{10} & \alpha^{14} \\
\alpha^6 & 1 & \alpha^6
\end{bmatrix}
\begin{bmatrix}
e_2 \\
e_5 \\
e_7
\end{bmatrix}
=
\begin{bmatrix}
E_1 \\
E_2 \\
E_3
\end{bmatrix},
$$

which can be inverted to find the three error values e_2, e_5, and e_7.

The third method of computing the error values is called the *Forney algorithm*. The Forney algorithm computes the error vector e with the aid of a new polynomial, $\Gamma(x)$, called the *error-evaluator polynomial*, and the formal derivative of $\Lambda(x)$, given by

$$\Lambda'(x) = \sum_{j=1}^{t} j \Lambda_j x^{j-1}.$$

To derive the Forney algorithm, recall that

$$\Lambda(x)E(x) = 0 \quad (\text{mod } x^n - 1).$$

This can be written

$$\Lambda(x)E(x) = -\Gamma(x)(x^n - 1)$$

for some polynomial $\Gamma(x)$. Because $\deg \Lambda(x) \le t$ and $\deg E(x) \le n - 1$, the degree of the product $\Lambda(x)E(x)$ is at most $t + n - 1$. From this we conclude that $\deg \Gamma(x) < t$, and the jth coefficient of $\Lambda(x)E(x)$ is zero for $j = t, \dots, n - 1$. Consequently, we can write

$$\Gamma(x) = \Lambda(x)E(x) \pmod{x^\ell}$$

for any ℓ satisfying $t \le \ell \le n - 1$. If ℓ is chosen larger than $2t$, however, the expression would involve unknown components of E, because E_j is known only for $j < 2t$.

We will choose $\ell = 2t$ and write

$$\Gamma(x) = \Lambda(x)E(x) \pmod{x^{2t}}.$$

This choice allows the equation to be expressed in matrix form:

$$
\begin{bmatrix}
E_0 & 0 & 0 & \cdots & 0 & 0 & 0 \\
E_1 & E_0 & 0 & \cdots & 0 & 0 & 0 \\
\vdots & & & & & & \vdots \\
E_{t-1} & E_{t-2} & E_{t-3} & \cdots & E_1 & E_0 & 0 \\
E_t & E_{t-1} & E_{t-2} & \cdots & E_2 & E_1 & E_0 \\
E_{t+1} & E_{t-2} & E_{t-1} & \cdots & E_3 & E_2 & E_1 \\
\vdots & & & & & & \vdots \\
E_{2t-1} & E_{2t-2} & E_{2t-3} & \cdots & E_{t+1} & E_t & E_{t-1}
\end{bmatrix}
\begin{bmatrix}
1 \\
\Lambda_1 \\
\Lambda_2 \\
\vdots \\
\Lambda_\nu \\
0 \\
\vdots \\
0
\end{bmatrix}
=
\begin{bmatrix}
\Gamma_0 \\
\Gamma_1 \\
\Gamma_2 \\
\vdots \\
\Gamma_{\nu-1} \\
0 \\
0 \\
\vdots \\
0
\end{bmatrix},
$$

where $\nu \le t$ is the actual degree of $\Lambda(x)$. This matrix equation could also be written in terms of the monic form of the reciprocal of $\Lambda(x)$, denoted $\widetilde{\Lambda}(x)$. This alternative is given by $\widetilde{\Lambda}(x) = \Lambda_\nu^{-1}x^\nu\Lambda(x^{-1})$. Then

$$
\begin{bmatrix}
0 & 0 & 0 & \cdots & 0 & 0 & E_0 \\
0 & 0 & 0 & \cdots & 0 & E_0 & E_1 \\
\vdots & & & & & & \vdots \\
0 & E_0 & E_1 & \cdots & E_{t-3} & E_{t-2} & E_{t-1} \\
E_0 & E_1 & E_2 & \cdots & E_{t-2} & E_{t-1} & E_t \\
E_1 & E_2 & E_3 & \cdots & E_{t-1} & E_t & E_{t+1} \\
\vdots & & & & & & \vdots \\
E_{t-1} & E_t & E_{t+1} & \cdots & E_{2t-3} & E_{2t-2} & E_{2t-1}
\end{bmatrix}
\begin{bmatrix}
1 \\
\sigma_1 \\
\sigma_2 \\
\sigma_3 \\
\vdots \\
\sigma_\nu \\
0 \\
\vdots \\
0
\end{bmatrix}
=
\begin{bmatrix}
\Gamma_0 \\
\Gamma_1 \\
\Gamma_2 \\
\vdots \\
\Gamma_{\nu-1} \\
0 \\
0 \\
\vdots \\
0
\end{bmatrix}.
$$

As written, the matrix and the vector on the right have been partitioned into two parts. The bottom part of the matrix is a *Hankel matrix*.

Theorem 3.2.1 (Forney)   *Given the locator polynomial $\Lambda(x)$, the nonzero components of the error vector e occur where $\Lambda(\omega^{-i})$ equals zero and are given by*

$$e_i = -\frac{\Gamma(\omega^{-i})}{\omega^{-i}\Lambda'(\omega^{-i})},$$

where

$$\Gamma(x) = \Lambda(x)E(x) \quad (\text{mod } x^{2t}).$$

Proof: The formal derivative of the equation

$$\Lambda(x)E(x) = -\Gamma(x)(x^n - 1)$$

is

$$\Lambda'(x)E(x) + \Lambda(x)E'(x) = \Gamma'(x)(1 - x^n) - nx^{n-1}\Gamma(x).$$

Set $x = \omega^{-i}$, noting that $\omega^{-n} = 1$. This yields

$$\Lambda'(\omega^{-i})E(\omega^{-i}) + \Lambda(\omega^{-i})E'(\omega^{-i}) = -n\omega^i\Gamma(\omega^{-i}).$$

But e_i is nonzero only if $\Lambda(\omega^{-i}) = 0$, and, in that case,

$$e_i = \frac{1}{n}\sum_{j=0}^{n-1} E_j\omega^{-ij} = \frac{1}{n}E(\omega^{-i}),$$

from which the equation of the theorem can be obtained.   ■

The Forney formula was derived for $j_0 = 0$, but the properties of the Fourier transform show how to modify this for any j_0. Simply cyclically shift the spectrum of zeros by j_0 places so that the pattern of zeros begins at $j = 0$. By the modulation property of the Fourier transform, this multiplies the error spectrum by ω^{ij_0}. Thus the computed error pattern must be corrected by this factor. Therefore

$$e_i = -\frac{\Gamma(\omega^{-i})}{\omega^{-i(j_0-1)}\Lambda'(\omega^{-i})}$$

is the general formula.

Our running example of decoding is completed by using the Forney algorithm to compute the error magnitudes as follows. Because $j_0 = 1$, the matrix equation is given by

$$
\begin{bmatrix}
E_1 & 0 & 0 & 0 \\
E_2 & E_1 & 0 & 0 \\
E_3 & E_2 & E_1 & 0 \\
E_4 & E_3 & E_2 & E_1 \\
E_5 & E_4 & E_3 & E_2 \\
E_6 & E_5 & E_4 & E_3
\end{bmatrix}
\begin{bmatrix}
1 \\
\Lambda_1 \\
\Lambda_2 \\
\Lambda_3
\end{bmatrix}
=
\begin{bmatrix}
\Gamma_0 \\
\Gamma_1 \\
\Gamma_2 \\
0 \\
0 \\
0
\end{bmatrix},
$$

where $\Lambda(x) = 1 + \alpha^{14}x + \alpha^{11}x^2 + \alpha^{14}x^3$. Because the left side is known, the right side can be computed by using Table 3.1 as follows:

$$
\begin{bmatrix}
\alpha^{12} & 0 & 0 & 0 \\
1 & \alpha^{12} & 0 & 0 \\
\alpha^{14} & 1 & \alpha^{12} & 0 \\
\alpha^{13} & \alpha^{14} & 1 & \alpha^{12} \\
1 & \alpha^{13} & \alpha^{14} & 1 \\
\alpha^{11} & 1 & \alpha^{13} & \alpha^{14}
\end{bmatrix}
\begin{bmatrix}
1 \\
\alpha^{14} \\
\alpha^{11} \\
\alpha^{14}
\end{bmatrix}
=
\begin{bmatrix}
\alpha^{12} \\
\alpha^{12} \\
\alpha^{8} \\
0 \\
0 \\
0
\end{bmatrix}.
$$

This gives $\Gamma(x) = \alpha^{12} + \alpha^{12}x + \alpha^{8}x^2$. Furthermore, $\Lambda'(x) = \alpha^{14} + \alpha^{14}x^2$. Finally, because $j_0 = 1$,

$$
e_i = \frac{\Gamma(\omega^{-i})}{\Lambda'(\omega^{-i})}.
$$

The errors are known to be at locations $i = 8, 10, 13$, so ω^{-i} takes the values α^7, α^5, and α^2. Therefore $e_8 = \alpha$, $e_{10} = \alpha^5$, and $e_{13} = \alpha^{11}$.

3.3   Correction of errors of weight 2

A binary extended BCH code over $GF(2)$ with minimum distance 5 can be decoded in a simple way by factoring a quadratic equation over $GF(2)$. An extended Preparata code over Z_4 can be decoded in a similar simple way by dropping combinations of the Z_4 syndromes into $GF(2)$, then factoring a quadratic equation over $GF(2)$. We shall develop these decoding procedures for distance-5 codes both over $GF(2)$ and over Z_4 in this section. In each case, because the extended code must have even weight, the existence of a two-error-correcting decoder implies further that the extended code has weight 6 so it can detect triple errors. This means that the decoders that we describe can be easily extended – though we will not do so – to detect triple errors as well as correct double errors.

A BCH code of minimum distance 5 over $GF(2)$ is the set of polynomials $c(x)$ such that $c(\alpha) = c(\alpha^3) = 0$. The syndromes are $S_1 = v(\alpha) = e(\alpha)$ and $S_3 = v(\alpha^3) = e(\alpha^3)$. For a single-error pattern, $e(x) = x^i$, so $S_1 = \alpha^i$ and $S_3 = \alpha^{3i}$. This case can be recognized by noting that $S_1^3 + S_3 = 0$. Then the exponent of α, in S_1, points to the location of the error. For a double-error pattern, $e(x) = x^i + x^{i'}$. The syndromes are $S_1 = X_1 + X_2$, which is nonzero, and $S_3 = X_1^3 + X_2^3$, where $X_1 = \alpha^i$ and $X_2 = \alpha^{i'}$. This case can be recognized by noting that $S_1^3 + S_3 \neq 0$. Then

$$(x - X_1)(x - X_2) = x^2 + S_1 x + (S_1^3 + S_3)/S_1.$$

The polynomial on the right side depends only on the syndromes. By factoring the right side, as by trial and error, one obtains X_1 and X_2. The exponents of α in X_1 and X_2 point to the two error locations. Thus this process forms a decoding algorithm for a distance-5 BCH code.

We will now give a comparable decoding algorithm for an extended Preparata code over the ring $\mathbf{Z}_4$ that corrects all error patterns of Lee weight 2 or less. This means that the minimum Lee distance of the code is at least 5, and since the minimum distance of the extended code must be even, it is at least 6. This decoding algorithm provides an alternative proof that the minimum distance of the extended Preparata code is (at least) 6. The decoding algorithm also applies to an extended Preparata code over $GF(2^m)$ simply by regarding each pair of bits as a symbol of $\mathbf{Z}_4$.

Let $v(x)$ be any senseword polynomial with Lee distance at most 2 from a codeword $c(x)$ of the Preparata code. The senseword polynomial $v(x)$ can be regarded as a codeword polynomial $c(x)$ corrupted by an error polynomial as $v(x) = c(x) + e(x)$. The correctible sensewords consist of two cases: single errors, either with $e(x) = 2x^i$ or with $e(x) = \pm x^i$, and double errors, with $e(x) = \pm x^i \pm x^{i'}$. (Detectable error patterns, which we do not consider here, have Lee weight 3 and consist of double errors with $e(x) = 2x^i \pm x^{i'}$ and triple errors with $e(x) = \pm x^i \pm x^{i'} \pm x^{i''}$.) Because the codewords are defined by $c_\infty + c(1) = 0$ and $c(\xi) = c(\xi^2) = 0$, the senseword $v(x)$ can be reduced to three pieces of data. These are the three syndromes $S_0 = v_\infty + v(1) = e_\infty + e(1)$, $S_1 = v(\xi) = e(\xi)$, and $S_2 = v(\xi^2) = e(\xi^2)$. We will give an algorithm to compute $e(x)$ from S_0, S_1, and S_2. (Although syndrome S_2 contains no information not already in S_1, it is more convenient to use both S_2 and S_1 as the input to the decoder.)

The syndrome S_0 is an element of $\mathbf{Z}_4$, corresponding to one of three situations. If $S_0 = 0$, either there are no errors or there are two errors with values $+1$ and -1. If $S_0 = \pm 1$, then there is one error with value ± 1. If $S_0 = 2$, then either there are two errors, both with the same value $+1$ or -1, or there is a single error with value 2.

Suppose there is a single error, so $e(x) = e_i x^i$, and

$$S_0 = e_i,$$
$$S_1 = e(\xi) = e_i \xi^i,$$
$$S_2 = e(\xi^2) = e_i \xi^{2i},$$

where $e_i = \pm 1$ or 2. The case of a single error of Lee weight 1 can be recognized by noting that S_0 is ± 1, or by computing $S_1^{2^m}$ and finding that $S_1^{2^m} = S_1$. The case of a single error of Lee weight 2 can be recognized by noting that $S_0 = 2$ and $S_1^2 = 0$. If either instance of a single error is observed, then $\xi^i = S_1/S_0$. The exponent of ξ uniquely points to the single term of $e(x)$ with a nonzero coefficient. Syndrome S_0 is the value of the error.

For the case of a correctable pattern with two errors, $S_0 = 0$ or 2 and $S_1^{2^m} \neq S_1$. Because the Lee weight is 2, $e_i = \pm 1$ and $e_{i'} = \pm 1$. Then

$$S_0 = e(1) = \pm 1 \pm 1,$$
$$S_1 = e(\xi) = \pm \xi^i \pm \xi^{i'},$$
$$S_2 = e(\xi^2) = \pm \xi^{2i} \pm \xi^{2i'}.$$

Syndrome $S_0 = 0$ if the two errors have opposite signs. In this case, without loss of generality, we write $e(x) = x^i - x^{i'}$. If, instead, the two errors have the same sign, syndrome $S_2 = 2$. Thus for any correctable pattern with two errors, we have

$$S_1^2 - S_2 = \xi^{2i} + 2\xi^i \xi^{i'} + \xi^{2i'} \mp \xi^{2i} \mp \xi^{2i'}$$

$$= \begin{cases} 2(\xi^i \xi^{i'} + \xi^{2i'}) & \text{if } S_0 = 0 \\ 2(\xi^i \xi^{i'}) & \text{if } S_0 = 2 \text{ and both errors are } +1 \\ 2(\xi^{2i} + \xi^i \xi^{i'} + \xi^{2i'}) & \text{if } S_0 = 2 \text{ and both errors are } -1. \end{cases}$$

In any case, this is an element of Z_4 that has only an even part, which is twice the odd part of the term in parentheses. Every even element of $GR(4^m)$ is a unique element of $GF(2^m)$ multiplied by 2, so the terms in parentheses can be dropped into $GF(2^m)$. Let B be the term $S_1^2 - S_2$, reduced by the factor of 2. Then

$$B = \begin{cases} \alpha^i \alpha^{i'} + \alpha^{2i} & \text{if } S_0 = 0 \\ \alpha^i \alpha^{i'} & \text{if } S_0 = 2 \text{ and both errors are } +1 \\ \alpha^{2i} + \alpha^i \alpha^{i'} + \alpha^{2i'} & \text{if } S_0 = 2 \text{ and both errors are } -1. \end{cases}$$

Let A denote S_1 modulo 2 and let X_1 and X_2 denote ξ^i and $\xi^{i'}$ modulo 2. Then the equations, now in $GF(2)$, become

$$A = X_1 + X_2$$

and

$$B + X_1^2 = X_1 X_2 \quad \text{if } S_0 = 0,$$
$$B = X_1 X_2 \quad \text{if } S_0 = 2 \text{ and both errors are } +1,$$
$$B + A^2 = X_1 X_2 \quad \text{if } S_0 = 2 \text{ and both errors are } -1.$$

It is trivial to solve the first case by substituting $X_2 = A + X_1$ into $B + X_1^2 = X_1 X_2$. This yields $X_1 = B/A$ and $X_2 = A + B/A$. The latter two cases, those with $S_0 = 2$, though not distinguished by the value of S_0, can be distinguished because only one of them has a solution. To see this, reduce the equations to observe that, in the two cases, X_1 and X_2 are the solutions of either

$$z^2 + Az + B = 0$$

or

$$z^2 + Az + A^2 + B = 0.$$

The first of these can be solved only if $\text{trace}(B/A^2) = 0$. The second can be solved only if $\text{trace}(B/A^2) = 1$. Only one of these can be true, so only one of the two polynomials has its zeros in the locator field.

3.4   The Sugiyama algorithm

The euclidian algorithm is a well known algorithm for computing the greatest common divisor of two polynomials over a field (up to a scalar multiple if the GCD is required to be a monic polynomial). The euclidean algorithm consists of an iterative application of the division algorithm for polynomials. The division algorithm for polynomials says that any $a(x)$ and $b(x)$ with $\deg b(x) \leq \deg a(x)$ can be written as follows:

$$a(x) = Q(x)b(x) + r(x),$$

where $Q(x)$ is called the *quotient polynomial* and $r(x)$ is called the *remainder polynomial*.

Theorem 3.4.1 (euclidean algorithm for polynomials)   *Given two polynomials $a(x)$ and $b(x)$ over the field F with $\deg a(x) \geq \deg b(x)$, their greatest common divisor is the last nonzero remainder of the recursion*

$$a^{(r-1)}(x) = Q^{(r)}(x)b^{(r-1)}(x) + b^{(r)}(x),$$
$$a^{(r)}(x) = b^{(r-1)}(x),$$

for $r = 0, 1, \ldots$, *with* $a^{(0)}(x) = a(x)$ *and* $b^{(0)}(x) = b(x)$, *halting at that* r *for which the remainder is zero.*

Proof: At iteration r, the division algorithm can be used to write

$$a^{(r-1)}(x) = Q^{(r)}(x)b^{(r-1)}(x) + b^{(r)}(x),$$

where the remainder polynomial $b^{(r)}(x)$ has a degree smaller than that of $b^{(r-1)}(x)$. The quotient polynomial will be written as follows:

$$Q^{(r)}(x) = \left\lfloor \frac{a^{(r-1)}(x)}{b^{(r-1)}(x)} \right\rfloor.$$

In matrix form, the iteration is given by

$$\begin{bmatrix} a^{(r)}(x) \\ b^{(r)}(x) \end{bmatrix} = \begin{bmatrix} 0 & 1 \\ 1 & -Q^{(r)}(x) \end{bmatrix} \begin{bmatrix} a^{(r-1)}(x) \\ b^{(r-1)}(x) \end{bmatrix}.$$

Also, define the two by two matrix $A^{(r)}(x)$ by

$$A^{(r)}(x) = \begin{bmatrix} 0 & 1 \\ 1 & -Q^{(r)}(x) \end{bmatrix} A^{(r-1)}(x),$$

with the initial value

$$A^{(0)}(0) = \begin{bmatrix} 1 & 0 \\ 0 & 1 \end{bmatrix}.$$

This halts at the iteration R at which $b^{(R)}(x) = 0$. Thus

$$\begin{bmatrix} a^{(R)}(x) \\ 0 \end{bmatrix} = A^{(R)}(x) \begin{bmatrix} a(x) \\ b(x) \end{bmatrix}.$$

Any polynomial that divides both $a(x)$ and $b(x)$ must divide $a^{(R)}(x)$. On the other hand, any polynomial that divides both $a^{(r)}(x)$ and $b^{(r)}(x)$ divides $a^{(r-1)}(x)$ and $b^{(r-1)}(x)$, and in turn, $a^{(r-2)}(x)$ and $b^{(r-2)}(x)$ as well. Continuing, we can conclude that any polynomial that divides $a^{(R)}(x)$ divides both $a(x)$ and $b(x)$. Hence $a^{(R)}(x) = \text{GCD}[a(x), b(x)]$, as was to be proved.   ■

Corollary 3.4.2   *The greatest common divisor of $a(x)$ and $b(x)$ can be expressed as a polynomial combination of $a(x)$ and $b(x)$.*

Proof: This follows from the expression

$$\begin{bmatrix} a^{(R)}(x) \\ 0 \end{bmatrix} = A^{(R)}(x) \begin{bmatrix} a(x) \\ b(x) \end{bmatrix}$$

by observing that $A^{(r)}(x)$ is a matrix of polynomials.   ■

Corollary 3.4.2 is known as the *extended euclidean algorithm for polynomials*. In particular, if $a(x)$ and $b(x)$ are coprime, then polynomials $A(x)$ and $B(x)$ exist, sometimes called *Bézout coefficients*, such that

$$a(x)A(x) + b(x)B(x) = 1.$$

In this section, our task is the decoding of Reed–Solomon codes, for which we want to invert a system of equations over the field F of the form

$$
\begin{bmatrix}
E_{t-1} & E_{t-2} & \cdots & E_0 \\
E_t & E_{t-1} & \cdots & E_1 \\
\vdots & & \vdots & \\
E_{2t-2} & E_{2t-3} & \cdots & E_{t-1}
\end{bmatrix}
\begin{bmatrix}
\Lambda_1 \\
\Lambda_2 \\
\vdots \\
\Lambda_t
\end{bmatrix}
= -
\begin{bmatrix}
E_t \\
E_{t+1} \\
\vdots \\
E_{2t-1}
\end{bmatrix}.
$$

A matrix of the form appearing here is known as a *Toeplitz matrix*, and the system of equations is called a *Toeplitz system of equations*. This system of equations is a description of the recursion

$$E_j = -\sum_{k=1}^{t} \Lambda_k E_{((j-k))} \quad j = t, \ldots, 2t - 1.$$

We saw in Section 3.2 that this can be expressed as the polynomial equation

$$\Lambda(x)E(x) = \Gamma(x)(1 - x^n),$$

where $\deg \Lambda(x) \leq t$ and $\deg \Gamma(x) \leq t - 1$ because all coefficients on the right side for $j = t, \ldots, 2t - 1$ are equal to zero and involve only known coefficients of $E(x)$. Solving the original matrix equation is equivalent to solving this polynomial equation for $\Lambda(x)$.

The *Sugiyama algorithm*, which is the topic of this section, interprets this computation as a problem in polynomial algebra and solves the equivalent polynomial equation

$$\Lambda(x)E(x) = \Gamma(x) \quad (\bmod\ x^{2t})$$

for a $\Lambda(x)$ of degree less than t, with $E(x)$ given as the input to the computation.

The internal iteration step of the Sugiyama algorithm is the same as the internal iteration step of the euclidean algorithm for polynomials. In this sense, a substep of the euclidean algorithm for polynomials is used as a substep of the Sugiyama algorithm. For this reason, the Sugiyama algorithm is sometimes referred to as the euclidean algorithm. We regard the Sugiyama algorithm as similar to, but different from, the euclidean algorithm because the halting condition is different.

Let $a^{(0)}(x) = x^{2t}$ and $b^{(0)}(x) = E(x)$. Then the rth iteration of the euclidean algorithm can be written as follows:

$$
\begin{bmatrix} a^{(r)}(x) \\ b^{(r)}(x) \end{bmatrix} = \begin{bmatrix} A_{11}^{(r)}(x) & A_{12}^{(r)}(x) \\ A_{21}^{(r)}(x) & A_{22}^{(r)}(x) \end{bmatrix} \begin{bmatrix} x^{2t} \\ E(x) \end{bmatrix},
$$

and

$$
b^{(r)}(x) = A_{22}^{(r)}(x)E(x) \quad (\bmod \, x^{2t}).
$$

Such an equation holds for each r. This has the form of the required decoding computation. For some r, if the degrees satisfy $\deg A_{22}^{(r)}(x) \le t$ and $\deg b^{(r)}(x) \le t - 1$, then this equation provides the solution to the problem. The degree of $b^{(r)}(x)$ decreases at every iteration, so we can stop when $\deg b^{(r)}(x) < t$. Therefore define the stopping index $\bar{r}$ by

$$
\deg b^{(\bar{r}-1)}(x) \ge t,
$$
$$
\deg b^{(\bar{r})}(x) < t.
$$

It remains to show that the inequality

$$
\deg A_{22}^{(\bar{r})}(x) \le t
$$

is satisfied, thereby proving that the problem is solved with $\Lambda(x) = A_{22}^{(\bar{r})}(x)$. Toward this end, observe that because

$$
\det \begin{bmatrix} 0 & 1 \\ 1 & -Q^{(\ell)}(x) \end{bmatrix} = -1,
$$

it is clear that

$$
\det \prod_{\ell=1}^{r} \begin{bmatrix} 0 & 1 \\ 1 & -Q^{(\ell)}(x) \end{bmatrix} = (-1)^r
$$

and

$$
\begin{bmatrix} A_{11}^{(r)}(x) & A_{12}^{(r)}(x) \\ A_{21}^{(r)}(x) & A_{22}^{(r)}(x) \end{bmatrix}^{-1} = (-1)^r \begin{bmatrix} A_{22}^{(r)}(x) & -A_{12}^{(r)}(x) \\ -A_{21}^{(r)}(x) & A_{11}^{(r)}(x) \end{bmatrix}.
$$

Therefore

$$
\begin{bmatrix} x^{2t} \\ E(x) \end{bmatrix} = (-1)^r \begin{bmatrix} A_{22}^{(r)}(x) & -A_{12}^{(r)}(x) \\ -A_{21}^{(r)}(x) & A_{11}^{(r)}(x) \end{bmatrix} \begin{bmatrix} a^{(r)}(x) \\ b^{(r)}(x) \end{bmatrix}.
$$

Finally, we conclude that

$$\deg x^{2t} = \deg A_{22}^{(r)}(x) + \deg a^{(r)}(x)$$

because $\deg A_{22}^{(r)}(x) > \deg A_{12}^{(r)}(x)$ and $\deg a^{(r)}(x) \geq \deg b^{(r)}(x)$. Then, for $r \leq \bar{r}$,

$$\deg A_{22}^{(r)}(x) = \deg x^{2t} - \deg a^{(r)}(x)$$
$$\leq 2t - t$$
$$= t,$$

which proves that the algorithm solves the given problem.

Note that the Sugiyama algorithm is initialized with two polynomials, one of degree $2t$ and one of degree $2t - 1$, and that during each of its iterations the algorithm repeatedly reduces the degrees of these two polynomials eventually to form the polynomial $\Lambda(x)$ having degree t or less. Thus the computational work is proportional to t^2. In the next section, we shall give an alternative algorithm, called the *Berlekamp–Massey algorithm*, that starts with two polynomials of degree 0 and increases their degrees during its iterations to form the same polynomial $\Lambda(x)$ of degree t or less. It, too, requires computational work proportional to t^2, but with a smaller constant of proportionality. The Berlekamp–Massey algorithm and the Sugiyama algorithm both solve the same system of equations, so one may inquire whether the two algorithms have a common structural relationship. In Section 3.10, we shall consider the similarity of the two algorithms, which share similar structural elements but are essentially different algorithms.

3.5   The Berlekamp Massey algorithm

The Berlekamp–Massey algorithm inverts a Toeplitz system of equations, in any field F, of the form

$$\begin{bmatrix} E_{t-1} & E_{t-2} & \cdots & E_0 \\ E_t & E_{t-1} & \cdots & E_1 \\ \vdots & & & \vdots \\ E_{2t-2} & E_{2t-3} & \cdots & E_{t-1} \end{bmatrix} \begin{bmatrix} \Lambda_1 \\ \Lambda_2 \\ \vdots \\ \Lambda_t \end{bmatrix} = - \begin{bmatrix} E_t \\ E_{t+1} \\ \vdots \\ E_{2t-1} \end{bmatrix}.$$

The Berlekamp–Massey algorithm is formally valid in any field, but it may suffer from problems of numerical precision in the real field or the complex field. The computational problem it solves is the same problem solved by the Sugiyama algorithm.

The Berlekamp–Massey algorithm can be described as a fast algorithm for finding a linear recursion, of shortest length, of the form

$$E_j = -\sum_{k=1}^{\nu} \Lambda_k E_{((j-k))} \quad j = \nu, \ldots, 2t - 1.$$

This is the shortest linear recursion that produces $E_\nu, \ldots, E_{2t-1}$ from $E_0, \ldots, E_{\nu-1}$. This formulation of the problem statement is actually stronger than the problem of solving the matrix equation because the matrix equation may have no solution. If the matrix equation has a solution, then a minimum-length linear recursion of this form exists, and the Berlekamp–Massey will find it. If the matrix equation has no solution, then the Berlekamp–Massey algorithm finds a linear recursion of minimum length, but with $\nu > t$, that produces the sequence. Thus the Berlekamp–Massey algorithm actually provides more than was initially asked for. It always computes the linear complexity $L = \mathcal{L}(E_0, \ldots, E_{2t-1})$ and a corresponding shortest linear recursion that will produce the given sequence.

The Berlekamp–Massey algorithm, shown in Figure 3.1, is an iterative procedure for finding a shortest cyclic recursion for producing the initial r terms, $E_0, E_1, \ldots, E_{r-1}$, of the sequence E. At the rth step, the algorithm has already computed the linear recursions $(\Lambda^{(i)}(x), L_i)$ for all i smaller than r. These are the linear recursions that, for each i, produce the first i terms of the sequence E. Thus for each $i = 0, \ldots, r - 1$, we have already found the linear recursion

$$E_j = -\sum_{k=1}^{L_i} \Lambda_k^{(i)} E_{j-k} \quad j = L_i, \ldots, i$$

for each i smaller than r. Then for $i = r$, the algorithm finds a shortest linear recursion that produces all the terms of the sequence E. That is, it finds the linear recursion $(\Lambda^{(r)}(x), L_r)$ such that

$$E_j = -\sum_{k=1}^{L_r} \Lambda_k^{(r)} E_{j-k} \quad j = L_r, \ldots, r.$$

The rth step of the algorithm begins with a shortest linear recursion, $(\Lambda^{(r-1)}(x), L_{r-1})$, that produces the truncated sequence $E^{r-1} = E_0, E_1, \ldots, E_{r-1}$. Define

$$\delta_r = E_r - \left(-\sum_{k=1}^{L_{r-1}} \Lambda_k^{(r-1)} E_{r-k} \right)$$

$$= \sum_{k=0}^{L_{r-1}} \Lambda_k^{(r-1)} E_{r-k}$$

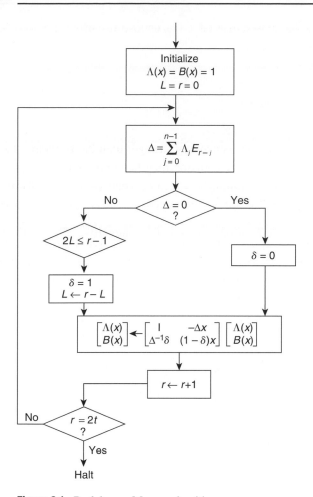

Figure 3.1. Berlekamp–Massey algorithm.

as the discrepancy in the output of the recursion at the rth iteration. The discrepancy δ_r need not be zero. If δ_r is zero, the output of the recursion is the desired field element. It is then trivial to specify the next linear recursion. It is the same linear recursion as found in the previous iteration. In this case, set

$$(\Lambda^{(r)}(x), L_r) = (\Lambda^{(r-1)}(x), L_{r-1})$$

as a shortest linear recursion that produces the truncated sequence E^r, and the iteration is complete. In general, however, δ_r will be nonzero. Then

$$(\Lambda^{(r)}(x), L_r) \neq (\Lambda^{(r-1)}(x), L_{r-1}).$$

To see how $\Lambda^{(r-1)}(x)$ must be revised to get $\Lambda^{(r)}(x)$, choose an earlier iteration count m, smaller than r, such that

$$\sum_{k=0}^{L_{m-1}} \Lambda_k^{(m-1)} E_{j-k} = \begin{cases} 0 & j < m-1 \\ \delta_m & j = m-1. \end{cases}$$

By translating indices so that $j + m - r$ replaces j and then scaling, this becomes

$$\frac{\delta_r}{\delta_m} \sum_{k=0}^{L_{m-1}} \Lambda_k^{(m-1)} E_{j-(r-m)-k} = \begin{cases} 0 & j < r-1 \\ \delta_r & j = r-1, \end{cases}$$

where δ_m is nonzero and E_j is regarded as zero for j negative. This suggests the polynomial update

$$\Lambda^{(r)}(x) = \Lambda^{(r-1)}(x) - \frac{\delta_r}{\delta_m} x^{r-m} \Lambda^{(m-1)}(x),$$

and $\deg \Lambda^{(r)}(x) = L_r \leq \max[L_{r-1}, r - m + L_{m-1}]$. To verify that this works, write

$$\sum_{k=0}^{L_r} \Lambda_k^{(r)} E_{j-k} = \sum_{k=0}^{L_r} \Lambda_k^{(r-1)} E_{j-k} - \frac{\delta_r}{\delta_m} \sum_{k=0}^{L_r} \Lambda_k^{(m-1)} E_{j-(r-m)-k}.$$

If $j < r$, the first sum is zero, and because $j - (r-m) < m$, the second sum is defined and is also zero. If $j = r$, the first sum equals δ_r, and because $r - (r-m) = m$, the second sum equals δ_m. Therefore

$$\sum_{k=0}^{L_r} \Lambda_k^{(r)} E_{j-k} = \begin{cases} 0 & j < r \\ \delta_r - (\delta_r/\delta_m)\delta_m = 0 & j = r. \end{cases}$$

Consequently,

$$E_j = -\sum_{k=1}^{L_r} \Lambda_k^{(r)} E_{j-k} \quad j = L_r, \ldots, r,$$

and the new polynomial $\Lambda^{(r)}(x)$ provides a new linear recursion that produces one more symbol than the previous linear recursion.

To ensure that the recursion is a *minimum-length* recursion, we need to place an additional condition on the choice of $\Lambda^{(m-1)}(x)$. Until now, we only required that m be chosen so that $\delta_m \neq 0$. Now we will further require that m be the most recent index such that $L_m > L_{m-1}$. This requirement implies the earlier requirement that $\delta_m \neq 0$, so that condition need not be checked. The following theorem shows that this last condition ensures that the new recursion will be of minimum length. By continuing this process for $2t$ iterations, the desired recursion is found.

Theorem 3.5.1 (Berlekamp–Massey) *Suppose that* $\mathcal{L}(E_0, E_1, \ldots, E_{r-2}) = L$. *If the recursion* $(\Lambda(x), L)$ *produces* $E_0, E_1, \ldots, E_{r-2}$, *but* $(\Lambda(x), L)$ *does not produce* E_0, $E_1, \ldots, E_{r-1}$, *then* $\mathcal{L}(E_0, E_1, \ldots, E_{r-1}) = \max[L, r - L]$.

Proof: Let $E^{(r)} = E_0, E_1, \ldots, E_{r-1}$. Massey's theorem states that

$$\mathcal{L}(E^{(r)}) \geq \max[L, r - L].$$

Thus it suffices to prove that

$$\mathcal{L}(E^{(r)}) \leq \max[L, r - L].$$

Case (1) $E^{(r)} = (0, 0, \ldots, 0, E_{r-1} \neq 0)$. The theorem is immediate in this case because a linear shift register of length zero produces $E^{(r-1)} = (0, 0, \ldots, 0)$, while a linear shift register of length r is needed to produce $E^{(r)} = (0, 0, \ldots, 0, E_{r-1})$.

Case (2) $E^{(r-1)} \neq (0, 0, \ldots, 0)$. The proof is by induction. Let m be such that $\mathcal{L}(E^{(m-1)}) < \mathcal{L}(E^{(m)}) = \mathcal{L}(E^{(r-1)})$. The induction hypothesis is that $\mathcal{L}(E^{(m)}) = \max[L_{m-1}, m - L_{m-1}]$. By the construction described prior to the theorem,

$$\mathcal{L}(E^{(r)}) \leq \max[L, L_{m-1} + r - m].$$

Consequently,

$$\mathcal{L}(E^{(r)}) \leq \max[\mathcal{L}(E^{(r-1)}), r - \mathcal{L}(E^{(r-1)})]$$
$$= \max[L, r - L],$$

which proves the theorem. ∎

Corollary 3.5.2 (Berlekamp–Massey algorithm) *In any field, let* $S_1, \ldots, S_{2t}$ *be given. Under the initial conditions* $\Lambda^{(0)}(x) = 1$, $B^{(0)}(x) = 1$, *and* $L_0 = 0$, *let the following set of equations be used iteratively to compute* $\Lambda^{(2t)}(x)$:

$$\delta_r = \sum_{j=0}^{L_{r-1}} \Lambda_j^{(r-1)} S_{r-j},$$

$$L_r = \epsilon_r(r - L_{r-1}) + (1 - \epsilon_r)L_{r-1},$$

$$\begin{bmatrix} \Lambda^{(r)}(x) \\ B^{(r)}(x) \end{bmatrix} = \begin{bmatrix} 1 & -\delta_r x \\ \delta_r^{-1} \epsilon_r & (1 - \epsilon_r)x \end{bmatrix} \begin{bmatrix} \Lambda^{(r-1)}(x) \\ B^{(r-1)}(x) \end{bmatrix},$$

$r = 1, \ldots, 2t$, *where* $\epsilon_r = 1$ *if both* $\delta_r \neq 0$ *and* $2L_{r-1} \leq r - 1$, *and otherwise* $\epsilon_r = 0$. *Then* $\Lambda^{(2t)}(x)$ *is the polynomial of smallest degree with the properties that* $\Lambda_0^{(2t)} = 1$, *and*

$$S_r + \sum_{j=1}^{L_{r-1}} \Lambda_j^{(2t)} S_{r-j} = 0 \quad r = L_{2t}, \ldots, 2t - 1.$$

The compact matrix formulation given in the corollary includes the term $\delta_r^{-1}\epsilon_r$. Because δ_r can be zero only when ϵ_r is zero, the term $\delta_r^{-1}\epsilon_r$ is then understood to be zero. The Berlekamp–Massey algorithm, as shown in Figure 3.1, saves the polynomial $\Lambda(x)$ whenever there is a length change as the "interior polynomial" $B(x)$. This $B(x)$ will play the role of $\Lambda^{(m-1)}(x)$ when it is needed in a later iteration. In Corollary 3.5.2, the interior polynomial $B(x)$ is equal to $\delta_m^{-1}x^{r-m}\Lambda^{(m)}(x)$. When $\epsilon_r = 1, B(x)$ is replaced by $\Lambda(x)$, appropriately scaled, and when $\epsilon_r = 0$ it is multiplied by x to account for the increase in r.

Note that the matrix update requires at most $2t$ multiplications per iteration, and the calculation of δ_r requires no more than t multiplications per iteration. There are $2t$ iterations and hence at most $6t^2$ multiplications. Thus using the algorithm will usually be much better than using a matrix inversion, which requires on the order of t^3 multiplications.

The Berlekamp–Massey algorithm is formally valid in any field. However, the decision to branch is based on whether or not δ_r equals zero, so in the real field the algorithm is sensitive to problems of computational precision.

A simple example of the iterations of the Berlekamp–Massey algorithm in the rational field is shown in Table 3.2. In this example, the algorithm computes the shortest recursion that will produce the sequence 1, 1, 0, 1, 0, 0 in the rational field.

A second example of the iterations of the Berlekamp–Massey algorithm in the field $GF(16)$ is shown in Table 3.3. In this example, the algorithm computes the shortest recursion that will compute the sequence $\alpha^{12}, 1, \alpha^{14}, \alpha^{13}, 1, \alpha^{11}$ in the field $GF(16)$. This is the sequence of syndromes for the example of the $(15, 9, 7)$ Reed–Solomon code, using the same error pattern that was studied earlier in Section 3.1. As before, the senseword is the all-zero codeword, and α is the primitive element of $GF(16)$ that satisfies $\alpha^4 = \alpha + 1$.

Now we turn to the final task of this section, which is to exploit the structure of the Berlekamp–Massey algorithm to improve the Forney formula by eliminating the need to compute $\Gamma(x)$.

Corollary 3.5.3 (Horiguchi–Koetter) *Suppose $\Lambda(x)$ has degree v. The components of the error vector $\mathbf{e}$ satisfy*

$$e_i = \begin{cases} 0 & \text{if } \Lambda(\omega^{-i}) \neq 0 \\ \dfrac{\omega^{-i(v-1)}}{\omega^{-i}B(\omega^{-i})\Lambda'(\omega^{-i})} & \text{if } \Lambda(\omega^{-i}) = 0, \end{cases}$$

Table 3.2. *Example of Berlekamp–Massey algorithm for a sequence of rationals*

$S_0 = 1$
$S_1 = 1$
$S_2 = 0$
$S_3 = 1$
$S_4 = 0$
$S_5 = 0$

r	δ_r	$B(x)$	$\Lambda(x)$	L
0		1	1	0
1	1	1	$1-x$	1
2	0	x	$1-x$	1
3	-1	$-1+x$	$1-x+x^2$	2
4	2	$-x+x^2$	$1+x-x^2$	2
5	1	$1+x-x^2$	$1+x-x^3$	3
6	0	$x+x^2-x^3$	$1+x-x^3$	3

$$\Lambda(x) = 1 + x - x^3$$

Table 3.3. *Example of Berlekamp–Massey algorithm for a Reed–Solomon (15, 9, 7) code*

$g(x) = x^6 + \alpha^{10}x^5 + \alpha^{14}x^4 + \alpha^4 x^3 + \alpha^6 x^2 + \alpha^9 x + \alpha^6$
$v(x) = \alpha x^7 + \alpha^5 x^5 + \alpha^{11} x^2 = e(x)$
$S_1 = \alpha\alpha^7 + \alpha^5\alpha^5 + \alpha^{11}\alpha^2 = \alpha^{12}$
$S_2 = \alpha\alpha^{14} + \alpha^5\alpha^{10} + \alpha^{11}\alpha^4 = 1$
$S_3 = \alpha\alpha^{21} + \alpha^5\alpha^{15} + \alpha^{11}\alpha^6 = \alpha^{14}$
$S_4 = \alpha\alpha^{28} + \alpha^5\alpha^{20} + \alpha^{11}\alpha^8 = \alpha^{13}$
$S_5 = \alpha\alpha^{35} + \alpha^5\alpha^{25} + \alpha^{11}\alpha^{10} = 1$
$S_6 = \alpha\alpha^{42} + \alpha^5\alpha^{30} + \alpha^{11}\alpha^{12} = \alpha^{11}$

r	δ_r	$B(x)$	$\Lambda(x)$	L
0		1	1	0
1	α^{12}	α^3	$1+\alpha^{12}x$	1
2	α^7	$\alpha^3 x$	$1+\alpha^3 x$	1
3	1	$1+\alpha^3 x$	$1+\alpha^3 x+\alpha^3 x^2$	2
4	1	$x+\alpha^3 x^2$	$1+\alpha^{14}x$	2
5	α^{11}	$\alpha^4+\alpha^3 x$	$1+\alpha^{14}x+\alpha^{11}x^2+\alpha^{14}x^3$	3
6	0	$\alpha^4 x+\alpha^3 x^2$	$1+\alpha^{14}x+\alpha^{11}x^2+\alpha^{14}x^3$	3

$$\Lambda(x) = 1 + \alpha^{14}x + \alpha^{11}x^2 + \alpha^{14}x^3$$
$$= (1+\alpha^7 x)(1+\alpha^5 x)(1+\alpha^2 x)$$

where $B(x)$ is the interior polynomial computed by the Berlekamp–Massey algorithm.

Proof: The actual number of errors is ν, the degree of $\Lambda(x)$. Define the modified error vector $\tilde{e}$ by the components $\tilde{e}_i = e_i B(\omega^{-i})$. To prove the corollary, we will first show that $B(\omega^{-i})$ is nonzero everywhere that e_i is nonzero. Then we will apply the Forney formula to the modified error vector $\tilde{e}$, and finally divide out $B(\omega^{-i})$.

The iteration equation of the Berlekamp–Massey algorithm can be inverted as follows:

$$
\begin{bmatrix} (1 - \epsilon_r)x & \delta_r x \\ -\delta_r^{-1}\epsilon_r & 1 \end{bmatrix} \begin{bmatrix} \Lambda^{(r)}(x) \\ B^{(r)}(x) \end{bmatrix} = x \begin{bmatrix} \Lambda^{(r-1)}(x) \\ B^{(r-1)}(x) \end{bmatrix}.
$$

If $\Lambda^{(r)}(x)$ and $B^{(r)}(x)$ have a common factor other than x, then $\Lambda^{(r-1)}(x)$ and $B^{(r-1)}(x)$ have that factor also. Hence by induction, $\Lambda^{(0)}(x)$ and $B^{(0)}(x)$ also have that same factor. Because $\Lambda^{(r)}(x)$ does not have x as a factor, and because $\Lambda^{(0)}(x) = B^{(0)}(x) = 1$, there is no common factor. Therefore

$$
\text{GCD}[\Lambda(x), B(x)] = 1.
$$

Because $\Lambda(x)$ and $B(x)$ are coprime, they can have no common zero. This means that the modified error component $\tilde{e}_i$ is nonzero if and only if error component e_i is nonzero. Consequently, $\Lambda(x)$ is also the error-locator polynomial for the modified error vector $\tilde{e}$. For the modified error vector, the syndromes are

$$
\tilde{S}_j = \sum_{i=0}^{n-1} e_i B(\omega^{-i})\omega^{ij} \quad j = 0, \ldots, 2t - 1
$$

$$
= \sum_{k=0}^{n-1} B_k S_{j-k}
$$

$$
= \begin{cases} 0 & j < \nu - 1 \\ 1 & j = \nu - 1 \end{cases}
$$

where the second line is a consequence of the convolution theorem, and the third line is a consequence of the structure of the Berlekamp–Massey algorithm. Thus $\tilde{S}(x) = x^{\nu-1}$.

The modified error-evaluator polynomial for the modified error vector is given by

$$
\tilde{\Gamma}(x) = \Lambda(x)\tilde{S}(x) \quad (\text{mod } x^\nu)
$$

$$
= x^{\nu-1}.
$$

The Forney algorithm, now applied to the modified error vector, yields

$$\tilde{e}_i = -\frac{\tilde{\Gamma}(\omega^{-i})}{\omega^{-i}\Lambda'(\omega^{-i})},$$

from which the conclusion of the corollary follows.   ∎

3.6   Decoding of binary BCH codes

The decoding algorithms for BCH codes hold for BCH codes over any finite field. When the field is $GF(2)$, however, it is only necessary to find the error location; the error magnitude is always equal to 1. Table 3.4 shows the computations of the Berlekamp–Massey algorithm used to decode a noisy senseword of the $(15, 5, 7)$ triple-error-correcting binary BCH code. The calculations can be traced by passing six times around the main loop of Figure 3.1. An examination of Table 3.4 suggests the possibility of a further simplification. Notice that δ_r is always zero on even-numbered iterations, because the trial recursion produces the correct syndrome. We shall see that this is always the case for binary codes, so even-numbered iterations can be skipped. For example, tracing through the algorithm of Figure 3.1, and using the fact that $S_4 = S_2^2 = S_1^4$ for all binary codes, gives

$$\begin{aligned}
\delta_1 &= S_1 & \Lambda^{(1)}(x) &= S_1 x + 1 \\
\delta_2 &= S_2 + S_1^2 = 0 & \Lambda^{(2)}(x) &= S_1 x + 1 \\
\delta_3 &= S_3 + S_1 S_2 & \Lambda^{(3)}(x) &= (S_1^{-1} S_3 + S_2)x^2 + S_1 x + 1 \\
\delta_4 &= S_4 + S_1 S_3 + S_1^{-1} S_2 S_3 + S_2^2 = 0.
\end{aligned}$$

This calculation shows that δ_2 and δ_4 will always be zero for any binary BCH code. Indeed, Theorem 1.9.2 leads to the more general statement that $\delta_r = 0$ for all even r for any binary BCH code. Specifically, if any syndrome sequence $S_1, S_2, \ldots, S_{2\nu-1}$ that satisfies $S_j^2 = S_{2j}$ and the recursion

$$S_j = -\sum_{i=1}^{\nu} \Lambda_i S_{j-i} \quad j = \nu, \ldots, 2\nu - 1,$$

then that recursion will next produce the term

$$S_{2\nu} = S_\nu^2.$$

Thus there is no need to test the recursion for even values of j; the term δ_j is then always zero.

Table 3.4. *Sample Berlekamp–Massey computation for a BCH* $(15, 5, 7)$ *code*

$$g(x) = x^{10} + x^8 + x^5 + x^4 + x^2 + x + 1$$
$$v(x) = x^7 + x^5 + x^2 = e(x)$$
$$S_1 = \alpha^7 + \alpha^5 + \alpha^2 = \alpha^{14}$$
$$S_2 = \alpha^{14} + \alpha^{10} + \alpha^4 = \alpha^{13}$$
$$S_3 = \alpha^{21} + \alpha^{15} + \alpha^6 = 1$$
$$S_4 = \alpha^{28} + \alpha^{20} + \alpha^8 = \alpha^{11}$$
$$S_5 = \alpha^{35} + \alpha^{25} + \alpha^{10} = \alpha^5$$
$$S_6 = \alpha^{42} + \alpha^{30} + \alpha^{12} = 1$$

r	δ_r	$B(x)$	$\Lambda(x)$	L
0		1	1	0
1	α^{14}	α	$1 + \alpha^{14}x$	1
2	0	αx	$1 + \alpha^{14}x$	1
3	α^{11}	$\alpha^4 + \alpha^3 x$	$1 + \alpha^{14}x + \alpha^{12}x^2$	2
4	0	$\alpha^4 x + \alpha^3 x^2$	$1 + \alpha^{14}x + \alpha^{12}x^2$	2
5	α^{11}	$\alpha^4 + \alpha^3 x + \alpha x^2$	$1 + \alpha^{14}x + \alpha^{11}x^2 + \alpha^{14}x^3$	3
6	0	$\alpha^4 x + \alpha^3 x^2$	$1 + \alpha^{14}x + \alpha^{11}x^2 + \alpha^{14}x^3$	3

$$\Lambda(x) = 1 + \alpha^{14}x + \alpha^{11}x^2 + \alpha^{14}x^3$$
$$= (1 + \alpha^7 x)(1 + \alpha^5 x)(1 + \alpha^2 x)$$

Because δ_r is zero for even r, we can analytically combine two iterations to give, for odd r, the following:

$$\Lambda^{(r)}(x) = \Lambda^{(r-2)}(x) - \delta_r x^2 B^{(r-2)}(x),$$

$$B^{(r)}(x) = \epsilon_r \delta_r^{-1} \Lambda^{(r-2)}(x) + (1 - \epsilon_r) x^2 B^{(r-2)}(x).$$

Using these formulas, iterations with even r can be skipped, thereby resulting in a faster decoder for binary codes.

3.7   Putting it all together

A complete decoding algorithm starts with the senseword v and from it computes first the codeword c, then the user dataword. Locator decoding treats this task by breaking it down into three main parts: computation of the syndromes, computation of the locator polynomial, and computation of the errors. We have described several options for these various parts of the decoding algorithm. Now we will discuss putting some of these options together. Alternatives are algorithms such as the code-domain algorithm and the

Welch–Berlekamp algorithm, both of which are described later in the chapter, which suppress the transform-domain syndrome calculations.

Computation of the transform-domain syndromes has the structure of a Fourier transform and can be computed by the *Good–Thomas algorithm*, the *Cooley-Tukey algorithm*, or by other methods. Because, in a finite field, the blocklength of a Fourier transform is not a power of a prime, and because all components of the Fourier transform will not be needed, the advantages of a decimation algorithm may not be fully realized.

Figure 3.2 shows one possible flow diagram for a complete decoding algorithm based on the Berlekamp–Massey algorithm, describing how all n components of the spectrum vector E are computed, starting only with the $2t$ syndromes. The $2t$ iterations of the Berlekamp–Massey algorithm are on the path to the left. First, the Berlekamp–Massey algorithm computes the error-locator polynomial; then the remaining $n - 2t$ unknown components of E are computed. After the $2t$ iterations of the Berlekamp–Massey algorithm, the path to the right is taken. The purpose of the path to the right is to change the other $n - 2t$ components of the computed error spectrum, one by one, into the corresponding $n - 2t$ components of the actual error spectrum E.

The most natural test for deciding that the original $2t$ iterations are finished is the test "$r > 2t$." We will provide an alternative test that suggests methods that will be used for the decoders for two-dimensional codes (which will be given in Chapter 12). The alternative test is $r - L > t$. Once this test is passed, it will be passed for all subsequent r; otherwise, if L were updated, Massey's theorem would require the shift register to have a length larger than t.

The most natural form of the recursive extension of the known syndromes to the remaining syndromes, for $r > 2t$ (or for $r > t + L$), is as follows:

$$E_r = -\sum_{j=1}^{l} \Lambda_j E_{r-j} \quad r = 2t, \ldots, n - 1.$$

To derive an alternative form of the recursive extension, as is shown in Figure 3.2, recall that the Berlekamp–Massey algorithm uses the following equation:

$$\delta_r = V_r - \left[-\sum_{j=1}^{L} \Lambda_j V_{r-j} \right].$$

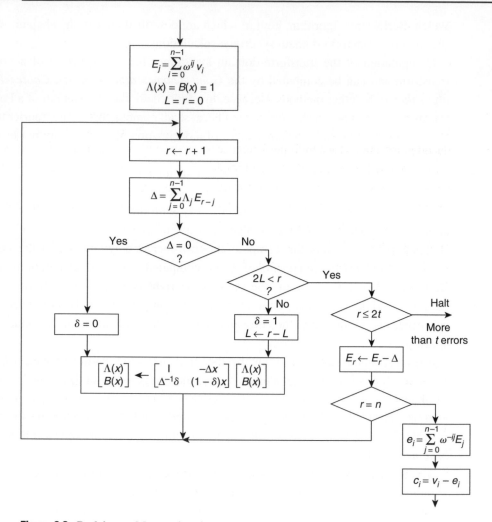

Figure 3.2. Berlekamp–Massey decoder.

If $V_j = E_j$ for $j < r$, this becomes

$$\delta_r = V_r - \left[-\sum_{j=1}^{L} \Lambda_j E_{r-j} \right]$$

$$= V_r - E_r.$$

This equation is nearly the same as the equation for recursive extension and can be easily adjusted to provide that computation. Therefore, instead of using the natural equation for recursive extension, one can use the equation of the Berlekamp–Massey algorithm, followed by the adjustment

$$E_r = V_r - \delta_r.$$

This slightly indirect way of computing E_r (V_r is added into δ_r, then subtracted out) has the minor advantage that the equation for δ_r is used in the flow diagram with no change. It may seem that such tricks are pointless, but they can be significant in structuring the decoder operations for a high-performance implementation. Moreover, this trick seems to be unavoidable in the code-domain implementation in the following section because the syndromes needed by the linear recursion are never computed.

Finally, we recall that recursive extension is not the only way to compute the error pattern from the locator polynomial. In Section 3.2, we studied the Gorenstein–Zierler algorithm and the Forney formula. These procedures may be faster computationally, but with a more complicated structure.

3.8 Decoding in the code domain

The Berlekamp–Massey algorithm takes as its input the sequence of syndromes S_j, which is obtained as a sequence of $2t$ components of the Fourier transform

$$S_j = E_j = \sum_{i=0}^{n-1} v_i \omega^{ij} \quad j = 0, \dots, 2t - 1,$$

and computes the locator polynomial $\Lambda(x)$. The syndromes are computed from the senseword v, then the error-locator polynomial is computed from the syndromes. Finally, all components of the error spectrum E_j for $j = 0, \dots, n - 1$ are computed from the error-locator polynomial. After all components of E are found, an inverse Fourier transform computes the error vector e. There are several alternatives to this procedure after the locator polynomial is computed. These all use some form of an inverse Fourier transform. Thus the decoder has the general structure of a Fourier transform, followed by the computational procedure of the Berlekamp–Massey algorithm, followed by an inverse Fourier transform.

It is possible to eliminate the Fourier transform at the input and the inverse Fourier transform at the output of the computation by analytically taking the inverse Fourier transform of the equations of the Berlekamp–Massey algorithm. This is illustrated in Figure 3.3. Now the senseword v itself plays the role of the syndrome. With this approach, rather than push the senseword into the transform domain to obtain the syndromes, push the equations of the Berlekamp–Massey algorithm into the code domain by means of the inverse Fourier transform on the equations. Replace the locator polynomial $\Lambda(x)$ and the interior polynomial $B(x)$ by their inverse Fourier transforms

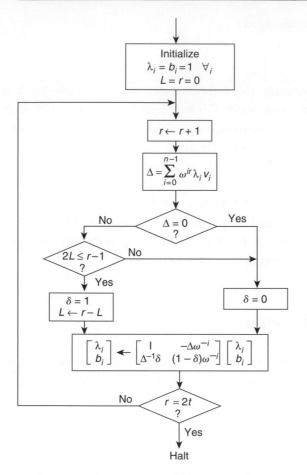

Figure 3.3. Code-domain Berlekamp–Massey algorithm.

λ and b, respectively:

$$\lambda_i = \frac{1}{n}\sum_{j=0}^{n-1}\Lambda_j\omega^{-ij}; \quad b_i = \frac{1}{n}\sum_{j=0}^{n-1}B_j\omega^{-ij}.$$

In the Berlekamp–Massey equations, simply replace the transform-domain variables Λ_k and B_k with the code-domain variables λ_i and b_i; replace the delay operator x with multiplication by ω^{-i}; and replace componentwise products with convolutions. Replacement of the delay operator with a multiplication by ω^{-i} is justified by the translation property of the Fourier transform. Replacement of a componentwise product with a convolution is justified by the convolution theorem. Now the raw senseword v, unmodified, plays the role of the syndrome. The code-domain algorithm, in the form of the following set of recursive equations, is used to compute $\lambda_i^{(2t)}$ for $i = 0, \ldots, n-1$

and $r = 1, \ldots, 2t$:

$$\delta_r = \sum_{i=0}^{n-1} \omega^{i(r-1)} \left[\lambda_i^{(r-1)} v_i \right],$$

$$L_r = \epsilon_r(r - L_{r-1}) + (1 - \epsilon_r)L_{r-1},$$

$$\begin{bmatrix} \lambda_i^{(r)} \\ b_i^{(r)} \end{bmatrix} = \begin{bmatrix} 1 & -\delta_r \omega^{-i} \\ \delta_r^{-1}\epsilon_r & (1 - \epsilon_r)\omega^{-i} \end{bmatrix} \begin{bmatrix} \lambda_i^{(r-1)} \\ b_i^{(r-1)} \end{bmatrix}.$$

The initial conditions are $\lambda_i^{(0)} = 1$ for all i, $b_i^{(0)} = 1$ for all i, $L_0 = 0$, and $\epsilon_r = 1$ if both $\delta_r \neq 0$ and $2L_{r-1} \leq r - 1$, and, otherwise, $\epsilon_r = 0$. Then $\lambda_i^{(2t)} = 0$ if and only if $e_i \neq 0$.

For nonbinary codes, it is not enough to compute only the error locations; we must also compute the error magnitudes. After the $2t$ iterations of the Berlekamp–Massey algorithm are completed, an additional $n - 2t$ iterations may be executed to change the vector v to the vector e. If the computations were in the transform domain, these would be computed by the following recursion:

$$E_k = -\sum_{j=1}^{t} \Lambda_j E_{k-j} \quad k = 2t, \ldots, n - 1.$$

It is not possible just to write the Fourier transform of this equation – some restructuring is necessary. Write the equation as

$$\delta_r = V_r - \left[-\sum_{j=1}^{L} \Lambda_j E_{k-j} \right]$$

$$= \sum_{j=0}^{L} \Lambda_j V_{k-j}^{(r-1)},$$

which is valid if $V_j^{(r-1)} = E_j$ for $j < r$. This is so if $r = 2t$, and we will set up the equation so that it continues to be true.

The following equivalent set of recursive equations for $r = 2t, \ldots, n - 1$ is suitably restructured:

$$\delta_r = \sum_{i=0}^{n-1} \omega^{ir} v_i^{(r-1)} \lambda_i,$$

$$v_i^{(r)} = v_i^{(r-1)} - \frac{1}{n}\delta_r \omega^{-ri}.$$

Starting with $v_i^{(2t)} = v_i$ and $\lambda_i = \lambda_i^{(2t)}$ for $i = 0, \ldots, n - 1$, the last iteration results in

$$v_i^{(n)} = e_i \quad i = 0, \ldots, n - 1.$$

This works because $E_k = V_k$ for $k = 0, \ldots, 2t - 1$, and the new equations, although written in the code domain, in effect are sequentially changing V_k to E_k for $k = 2t, \ldots, n - 1$.

The code-domain decoder deals with vectors of length n rather than with vectors of length t used by the transform-domain decoder. The decoder has no Fourier transforms, but has the complexity n^2. Its advantage is that it has only one major computational module, which is easily designed into digital logic or a software module. For high-rate codes, the time complexity of the code-domain algorithm may be acceptable instead of the space complexity of the transform-domain algorithm.

3.9　The Berlekamp algorithm

If the Forney formula is to be used to compute error values, then the error-evaluator polynomial must be computed first. The expression

$$\Gamma(x) = \Lambda(x)E(x) \quad (\bmod\ x^{2t})$$

presented in Section 3.2 has a simple form but cannot be computed until after $\Lambda(x)$ is computed. An alternative approach is to compute iteratively $\Lambda(x)$ and $\Gamma(x)$ simultaneously in lockstep. The method of simultaneous iterative computation of $\Lambda(x)$ and $\Gamma(x)$ is called the *Berlekamp algorithm*. Figure 3.4 shows how the Berlekamp algorithm can be used with the Forney formula. However, the Horiguchi–Koetter formula is an alternative to the Forney formula that does not use $\Gamma(x)$, so it may be preferred to the Berlekamp algorithm.

Algorithm 3.9.1 (Berlekamp algorithm)　*If*

$$\left[\begin{array}{c} \Gamma^{(0)}(x) \\ A^{(0)}(x) \end{array} \right] = \left[\begin{array}{c} 0 \\ -x^{-1} \end{array} \right]$$

and, for $r = 1, \ldots, 2t$,

$$\left[\begin{array}{c} \Gamma^{(r)}(x) \\ A^{(r)}(x) \end{array} \right] = \left[\begin{array}{cc} 1 & -\delta_r x \\ \delta_r^{-1}\epsilon_r & (1 - \epsilon_r)x \end{array} \right] \left[\begin{array}{c} \Gamma^{(r-1)}(x) \\ A^{(r-1)}(x) \end{array} \right],$$

with δ_r and ϵ_r as in the Berlekamp–Massey algorithm, then $\Gamma^{(2t)}(x) = \Gamma(x)$.

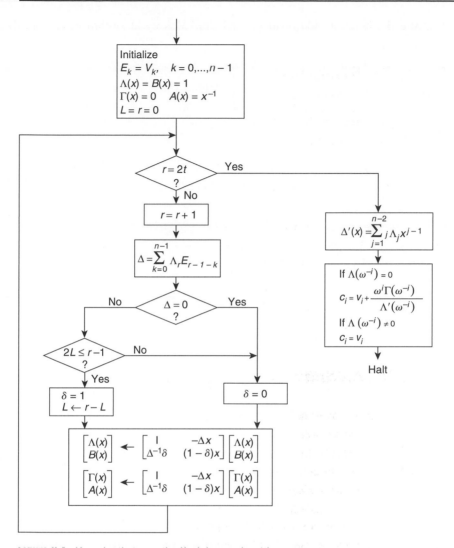

Figure 3.4. Decoder that uses the Berlekamp algorithm.

Proof: For $r = 1, \ldots, 2t$, define the polynomials

$$\Gamma^{(r)}(x) = E(x)\Lambda^{(r)}(x) \pmod{x^r},$$
$$A^{(r)}(x) = E(x)B^{(r)}(x) - x^{r-1} \pmod{x^r},$$

for $r = 1, \ldots, 2t$, with $\Lambda^{(r)}(x)$ and $B^{(r)}(x)$ as in the Berlekamp–Massey algorithm. Clearly, $\Gamma(x)$ is equal to $\Gamma^{(2t)}(x)$.

Using the iteration rule of the Berlekamp–Massey algorithm to expand the right side of

$$
\begin{bmatrix} \Gamma^{(r)}(x) \\ A^{(r)}(x) \end{bmatrix} = \begin{bmatrix} E(x)\Lambda^{(r)}(x) \\ E(x)B^{(r)}(x) - x^{r-1} \end{bmatrix} \quad (\mathrm{mod}\ x^r)
$$

leads to

$$
\begin{bmatrix} \Gamma^{(r)}(x) \\ A^{(r)}(x) \end{bmatrix} = \begin{bmatrix} 1 & -\delta_r \\ \delta_r^{-1}\epsilon_r & (1-\epsilon_r) \end{bmatrix} \begin{bmatrix} E(x)\Lambda^{(r-1)}(x) \\ xE(x)B^{(r-1)}(x) \end{bmatrix} - \begin{bmatrix} 0 \\ x^{r-1} \end{bmatrix} \quad (\mathrm{mod}\ x^r).
$$

But

$$
\delta_r = \sum_{j=0}^{n-1} \Lambda_j^{(r-1)} E_{r-1-j},
$$

so

$$
\Gamma^{(r-1)}(x) + \delta_r x^{r-1} = E(x)\Lambda^{(r-1)}(x) \quad (\mathrm{mod}\ x^r).
$$

Then

$$
\begin{bmatrix} \Gamma^{(r)}(x) \\ A^{(r)}(x) \end{bmatrix} = \begin{bmatrix} 1 & -\delta_r \\ \delta_r^{-1}\epsilon_r & (1-\epsilon_r) \end{bmatrix} \begin{bmatrix} \Gamma^{(r-1)}(x) + \delta_r x^{r-1} \\ xA^{(r-1)}(x) + x^{r-1} \end{bmatrix} - \begin{bmatrix} 0 \\ x^{r-1} \end{bmatrix}
$$

$$
= \begin{bmatrix} 1 & -\delta_r x \\ \delta_r^{-1}\epsilon_r & (1-\epsilon_r)x \end{bmatrix} \begin{bmatrix} \Gamma^{(r-1)}(x) \\ A^{(r-1)}(x) \end{bmatrix}.
$$

This is the iteration asserted in the statement of the algorithm.

To verify the initialization of the algorithm, we will verify that the first iteration yields the correct result. This would be

$$
\Gamma^{(1)}(x) = E(x)(1 - \delta_1 x) \quad (\mathrm{mod}\ x^1),
$$

$$
= E_0,
$$

and, if $E_0 = 0$,

$$
A^{(1)}(x) = E(x)x - x^0 = -1 \quad (\mathrm{mod}\ x^1),
$$

or, if $E_0 \neq 0$,

$$
A^{(1)}(x) = E(x)E_0^{-1} - x^0 = 0 \quad (\mathrm{mod}\ x^1).
$$

Therefore the first iteration yields

$$
\begin{bmatrix} \Gamma^{(1)}(x) \\ A^{(1)}(x) \end{bmatrix} = \begin{bmatrix} 1 & -\delta_1 x \\ \epsilon\delta_1^{-1} & (1-\epsilon)x \end{bmatrix} \begin{bmatrix} 0 \\ -x^{-1} \end{bmatrix}.
$$

Because $E_0 = \delta_1$, this reduces to $\Gamma^{(1)}(x) = E_0$, and

$$
A^{(1)}(x) = \begin{cases} -1 & \text{if } E_0 = 0 \\ 0 & \text{if } E_0 \neq 0, \end{cases}
$$

as required for the first iteration. Thus the first iteration is correct, and iteration r is correct if iteration $r-1$ is correct.   ∎

3.10   Systolic and pipelined algorithms

The performance of a fast algorithm for decoding is measured by its computational complexity, which can be defined in a variety of ways. The most evident way to define the computational complexity of an algorithm is its total number of elementary arithmetic operations. These are the four operations of addition, multiplication, subtraction, and division. In a large problem, however, these elementary operations may be less significant sources of complexity than is the pattern of movement of data flow as it passes between operations. The complexity of such movement of data, however, is hard to quantify.

A *systolic algorithm* is one in which the computations can be partitioned into small repetitive pieces, which will be called *cells*. The cells are arranged in a regular, usually square, array. If, further, the cells can be arranged as a one-dimensional array with data transferred in only one direction along the array, the algorithm would instead be called a *pipelined algorithm*. During one iteration of a systolic algorithm, each cell is allowed to exchange data with neighboring cells, but a cell is not normally allowed to exchange any (or much) data with distant cells. The complexity of a cell is considered to be less important than the interaction between cells. In this sense, a computational algorithm has a structure that may be regarded as something like a topology. In such a situation, the topology of the computation may be of primary importance, while the number of multiplications and additions may be of secondary importance.

We shall examine the structure of the Berlekamp–Massey algorithm and the Sugiyama algorithm from this point of view. The Berlekamp–Massey algorithm and the Sugiyama algorithm solve the same system of equations, so one may inquire whether the two algorithms have a common structure. We shall see in this section that the two algorithms can be arranged to have a common computational element, but the way that this element is used by the two algorithms is somewhat different. Indeed, there must be a difference because the polynomial iterates of the Berlekamp–Massey algorithm have increasing degree, whereas the polynomial iterates of the Sugiyama algorithm have decreasing degree.

The Berlekamp–Massey algorithm begins with two polynomials of degree 0, $\Lambda(x)$ and $B(x)$, and, at each iteration, may increase the degree of either or both polynomial iterates. The central computation of the rth iteration of the algorithm has the form

$$
\begin{bmatrix} \Lambda(x) \\ B(x) \end{bmatrix} \longleftarrow \begin{bmatrix} 1 & -\delta_r x \\ \epsilon_r \delta_r^{-1} & \bar{\epsilon}_r x \end{bmatrix} \begin{bmatrix} \Lambda(x) \\ B(x) \end{bmatrix},
$$

where δ_r is the discrepancy computed during the rth iteration, $\bar{\epsilon}_r = 1 - \epsilon_r$ is either zero or one, and δ_r can be zero only when ϵ_r is zero. Depending on the values of the parameters δ_r and ϵ_r, the update matrix takes one of the following three forms:

$$
A^{(r)} = \begin{bmatrix} 1 & 0 \\ 0 & x \end{bmatrix}, \begin{bmatrix} 1 & -\delta_r x \\ 0 & x \end{bmatrix}, \text{ or } \begin{bmatrix} 1 & -\delta_r x \\ \delta_r^{-1} & 0 \end{bmatrix}.
$$

Each of the $2t$ iterations involves multiplication of the current two-vector of polynomial iterates by one of the three matrices on the right. The Berlekamp–Massey algorithm terminates with a locator polynomial, $\Lambda(x)$, of degree ν at most equal to t.

In analyzing the structure of the Berlekamp–Massey algorithm, it is important to note that the iterate δ_r is a global variable because it is computed from all coefficients of $\Lambda(x)$ and $B(x)$. (It is interesting that a similar iterate with this global attribute does not occur in the Sugiyama algorithm.) An obvious implementation of a straightforward decoder using the Berlekamp–Massey algorithm might be used in a computer program, but the deeper structure of the algorithm is revealed by formulating high-speed hardware implementations.

A systolic implementation of the Berlekamp–Massey algorithm might be designed by assigning one cell to each coefficient of the locator polynomial. This means that, during iteration r, the jth cell is required to perform the following computation:

$$
\Lambda_j = \Lambda_j - \delta_r B_{j-1},
$$
$$
B_j = \epsilon_r \delta_r^{-1} + \bar{\epsilon}_r B_{j-1},
$$
$$
\delta_{r+1,j} = \Lambda_j S_{r-j}.
$$

The computations within a single cell require that B_j and S_j be passed from neighbor to neighbor at each iteration, as shown in Figure 3.5, with cells appropriately initialed to zero. In addition to the computations in the cells, there is one global computation for the discrepancy, given by $\delta_{r+1} = \sum_j \delta_{r+1,j}$, in which data from all cells must be combined into the sum δ_{r+1}, and the sum δ_{r+1} returned to all cells. During the rth iteration, the jth cell computes Λ_j, the jth coefficient of the current polynomial iterate

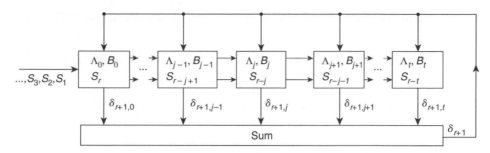

Figure 3.5. Structure of systolic Berlekamp–Massey algorithm.

$\Lambda^{(r)}(x)$. After $2t$ iterations, the computation of $\Lambda(x)$ is complete, with one polynomial coefficient in each cell.

An alternative version of the Berlekamp–Massey algorithm might be a pipelined implementation of $2t$ cells, with the rth cell performing the rth iteration of the algorithm. This would be a high-speed decoder in which $2t$ Reed–Solomon sensewords are being decoded at the same time. As the last cell is performing the final iteration on the least-recent Reed–Solomon codeword still in the decoder, the first cell is performing the first iteration on the most-recent Reed–Solomon codeword in the decoder. This decoder has the same number of computational elements as $2t$ Berlekamp–Massey decoders working concurrently, but the data flow is different and perhaps simpler.

The Sugiyama algorithm, in contrast to the Berlekamp–Massey algorithm, begins with two polynomials of nonzero degree, one of degree $2t$, and one of degree $2t - 1$. At each iteration, the algorithm may decrease the degrees of the two polynomial iterates. The central computation of the Sugiyama algorithm at the ℓth iteration has the following form:

$$\begin{bmatrix} s(x) \\ t(x) \end{bmatrix} \leftarrow \begin{bmatrix} 0 & 1 \\ 1 & -Q^{(\ell)}(x) \end{bmatrix} \begin{bmatrix} s(x) \\ t(x) \end{bmatrix}.$$

The Sugiyama algorithm terminates with the locator polynomial $\Lambda(x)$ of degree ν. The polynomial $\Lambda(x)$ is the same locator polynomial as computed by the Berlekamp–Massey algorithm. The coefficients of the quotient polynomial $Q^{(\ell)}(x)$ are computed, one by one, by the division algorithm.

Because $Q^{(\ell)}(x)$ need not have degree 1, the structure of one computational step of the Sugiyama algorithm seems quite different from the structure of one computational step of the Berlekamp–Massey algorithm. Another difference in the two algorithms is that the Sugiyama algorithm has a variable number of iterations, while the Berlekamp–Massey algorithm has a fixed number of iterations. However, there are similarities at a deeper level. It is possible to recast the description of the Sugiyama algorithm to expose common elements in the structure of the two algorithms.

To restructure the Sugiyama algorithm, let d_ℓ denote the degree of $Q^{(\ell)}(x)$, and write

$$\begin{bmatrix} 0 & 1 \\ 1 & -Q^{(\ell)}(x) \end{bmatrix} = \begin{bmatrix} 0 & 1 \\ 1 & -Q_0^{(\ell)} \end{bmatrix} \begin{bmatrix} 1 & -Q_1^{(\ell)}x \\ 0 & 1 \end{bmatrix} \begin{bmatrix} 1 & -Q_2^{(\ell)}x^2 \\ 0 & 1 \end{bmatrix} \cdots \begin{bmatrix} 1 & -Q_{d_\ell}^{(\ell)}x^{d_\ell} \\ 0 & 1 \end{bmatrix}.$$

To multiply any vector by the matrix on the left side, multiply that vector, sequentially, by each matrix of the sequence on the right side. Indeed, this matrix factorization is easily seen to be a representation of the individual steps of the division algorithm. With this decomposition of the matrix on the right by the product of matrices on the left, the notion of an iteration can be changed so that each multiplication by one of these submatrices on the left is counted as one iteration. The iterations on this new, finer scale now have the form

$$\begin{bmatrix} s(x) \\ t(x) \end{bmatrix} \leftarrow \begin{bmatrix} \bar{\epsilon}_r & -\delta_r x^\ell \\ \epsilon_r & \bar{\epsilon}_r - \delta_r \epsilon_r \end{bmatrix} \begin{bmatrix} s(x) \\ t(x) \end{bmatrix},$$

where $\bar{\epsilon}_r = 1 - \epsilon_r$ and $\delta_r = Q_r^{(\ell)}$. The degrees of the polynomials decrease by one at each iteration, and now the Sugiyama algorithm has a fixed number of iterations. The two by two matrices now more closely resemble those of the Berlekamp–Massey algorithm.

A systolic implementation of the Sugiyama algorithm can be designed by defining one cell to perform the computation of one coefficient of $(s(x), u(x))$. Then δ_r must be provided as a global variable to all cells. This is strikingly different from the case of the Berlekamp–Massey algorithm because now, since δ_r arises naturally within one cell, it need not be computed as a global variable.

It is possible to make the similarity between the Sugiyama algorithm and the Berlekamp–Massey algorithm even stronger by redefining the polynomials to make each matrix contain only the first power of x. Let $u(x) = x^\ell \delta_r t(x)$. Then the iteration can be written as follows:

$$\begin{bmatrix} s(x) \\ u(x) \end{bmatrix} = \begin{bmatrix} \bar{\epsilon}_r & -\delta_r x \\ \epsilon_r & (\bar{\epsilon}_r - \delta_r \epsilon_r)x \end{bmatrix} \begin{bmatrix} s(x) \\ u(x) \end{bmatrix}.$$

Another change can be made by recalling that the Sugiyama algorithm terminates with a normalization step to put the result in the form of a monic polynomial. The coefficient $\delta_r = Q_r^{(\ell)}$ is the rth coefficient of the quotient polynomial iteration ℓ. If $t(x)$ is found to be monic, then δ_r is immediately available as a coefficient of $s(x)$.

3.11 The Welch–Berlekamp decoder

The Welch–Berlekamp algorithm provides yet another method for decoding Reed–Solomon codes. In contrast to many other decoding algorithms, and in correspondence with the code-domain Berlekamp–Massey algorithm of Section 3.8, the Welch–Berlekamp decoder provides a method for decoding directly from the code-domain syndromes rather than the transform-domain syndromes, as is the case for many other decoders.

The Welch–Berlekamp decoder for Reed–Solomon codes consists of the Welch–Berlekamp algorithm, discussed in Section 3.12, augmented by the additional steps that are described in this section. These additional steps prepare the senseword for the algorithm and interpret the result of the algorithm. The senseword is prepared by converting it to a polynomial $M(x)$ called the modified syndrome polynomial. This is a syndrome in altered form, which we will define below.

The purpose of this section is to recast the decoding problem in the form of the polynomial equation

$$\Lambda(x)M(x) = N(x) \quad (\text{mod } G(x)),$$

where $G(x)$ has the form[1]

$$G(x) = \prod_{\ell=0}^{2t-1}(x - X_\ell),$$

with all X_ℓ distinct, and where the unknown polynomials $\Lambda(x)$ and $N(x)$ satisfy $\deg \Lambda(x) \le t$ and $\deg N(x) < t$. These two polynomials are the error-locator polynomial, defined in earlier sections, and the modified error-evaluator polynomial, defined later. Although this polynomial equation will be developed in this section in terms of polynomials in the transform domain, it will be solved in the Section 3.12 by an algorithm that accepts the modified syndromes in the code domain.

Let $c(x)$ be a codeword polynomial from an (n, k) Reed–Solomon code with generator polynomial $g(x)$, having zeros at α^j for $d - 1$ consecutive values of j, and let

$$v(x) = c(x) + e(x),$$

where the error polynomial

$$e(x) = \sum_{\ell=1}^{\nu} e_{i_\ell} x^{i_\ell}$$

has weight ν. The code-domain syndrome polynomial, defined as

$$s(x) = R_{g(x)}[v(x)]$$
$$= R_{g(x)}[e(x)],$$

---

[1] Whereas the previously defined generator polynomial $g(x)$ has d^*-1 consecutive zeros in the transform domain, the new polynomial $G(x)$ has $d^* - 1$ consecutive zeros in the code domain.

is easy to compute from $v(x)$ by simple polynomial division.

The syndrome polynomial differs from $e(x)$ by the addition of a polynomial multiple of $g(x)$, which means that it can be written as

$$s(x) = \widehat{c}(x) + e(x)$$

for some other codeword polynomial $\widehat{c}(x)$. This equation then leads to a surrogate problem. Instead of finding $c(x)$ that is closest to $v(x)$, find $\widehat{c}(x)$ that is closest to $s(x)$. From $s(x)$ and $\widehat{c}(x)$, it is trivial to compute $e(x)$, then $c(x)$. For this reason, finding the surrogate codeword polynomial $\widehat{c}(x)$ is equivalent to finding $e(x)$.

The purpose of the forthcoming lemma is to provide an opening statement regarding the null space of an r by n Vandermonde matrix T, over the field F, as given by

$$T = \begin{bmatrix} 1 & 1 & \cdots & 1 \\ \beta_1^1 & \beta_2^1 & \cdots & \beta_n^1 \\ \beta_1^1 & \beta_2^2 & \cdots & \beta_n^2 \\ \vdots & \vdots & & \vdots \\ \beta_1^{r-1} & \beta_2^{r-1} & \cdots & \beta_n^{r-1} \end{bmatrix},$$

where $\beta_1, \ldots, \beta_n$ are distinct but arbitrary elements of the field F, and r is less than n. The null space of T consists of all vectors v such that $Tv = 0$.

The formula to be given in the lemma is suggestive of the Forney formula, but is actually quite different. The proof is based on the well known Lagrange interpolation formula:

$$f(x) = \sum_{\ell=0}^{n-1} f(\beta_\ell) \frac{\Pi_{\ell' \neq \ell}(x - \beta_{\ell'})}{\Pi_{\ell' \neq \ell}(\beta_\ell - \beta_{\ell'})}.$$

This can be written as

$$f(x) = \sum_{\ell=0}^{n-1} \frac{g(x)}{(x - \beta_\ell)} \frac{f(\beta_\ell)}{g'(\beta_\ell)},$$

where $g(x) = \Pi_{\ell=0}^{r-1}(x - \beta_\ell)$.

Lemma 3.11.1 *Over the field F, let v be any vector of blocklength n in the null space of the r by n Vandermonde matrix T with r less than n. There exists a polynomial $N(x)$ over F of degree at most $n - r$ such that*

$$v_i = \frac{N(\beta_i)}{g'(\beta_i)}$$

for $i = 0, \ldots, n - 1$, where $g(x) = \Pi_{\ell=0}^{r-1}(x - \beta_\ell)$.

Proof: The null space of T is the set of v such that $Tv = 0$. The null space is a subspace of dimension $n - r$ of the n-dimensional vector space over F. For each polynomial $N(x)$ of degree less than $n - r$, let

$$v_i = \frac{N(\beta_i)}{g'(\beta_i)}$$

for $i = 0, \ldots, n - 1$. Because $g(x)$ has no double zeros, $g'(\beta_i)$ is nonzero. Additional columns with additional, distinct β_i can be appended to make T into a full rank Vandermonde matrix, which will have an inverse. This means that each such $N(x)$ must produce a unique v. Thus, the space of polynomials $N(x)$ of degree less than $n - r$ and the null space of T have the same dimension.

To complete the proof, we only need to show that all such v are in the null space of T. The Lagrange interpolation formula,

$$f(x) = \sum_{\ell=0}^{n-1} f(\beta_\ell) \frac{\Pi_{\ell' \neq \ell}(x - \beta_{\ell'})}{\Pi_{\ell' \neq \ell}(\beta_\ell - \beta_{\ell'})},$$

can be applied to $x^j N(x)$ for $j = 0, 1, \ldots, r - 1$ to write

$$x^j N(x) = \sum_{\ell=0}^{n-1} \frac{\beta_\ell^j N(\beta_\ell)}{g'(\beta_\ell)} \prod_{\ell' \neq \ell} (x - \beta_{\ell'}) \quad j = 0, 1, \ldots, r - 1.$$

But $\deg N(x) < n - r$, so, for $j = 0, \ldots, r - 1$, the polynomial $x^j N(x)$ has degree less than $n - 1$. Because the degrees of the polynomials on both sides of the equation must be the same, the coefficient of the monomial of degree $n - 1$ on the right must be zero. That is,

$$\sum_{\ell=0}^{n-1} \beta_\ell^j \frac{N(\beta_\ell)}{g'(\beta_\ell)} = 0$$

for $j = 0, 1, \ldots, r - 1$. Thus the vector with components $N(\beta_\ell)/g'(\beta_\ell)$ for $\ell = 0, \ldots, n - 1$ is in the null space of T, and the proof is complete. ∎

One consequence of the lemma is an unusual description of a Reed–Solomon code, which we digress to describe in the following corollary. The converse of the corollary is also true.

Corollary 3.11.2 *Let C be an $(n, k, r + 1)$ Reed–Solomon code over F with spectral zeros at $j = 0, \ldots, r - 1$. Let*

$$G(x) = \prod_{\beta_i \in S} (x - \beta_i),$$

where S is any set of m distinct elements of F, with m at least as large as r. For each polynomial $N(x)$ over F of degree less than $m - r$, and for each i at which $G(\beta_i) = 0$, define

$$c_i = \frac{N(\beta_i)}{G'(\beta_i)}.$$

Otherwise, define $c_i = 0$. Then c is a codeword of the Reed–Solomon code C.

Proof: If $m = r$, the proof is immediate because then $N(x) = 0$ and the only such codeword is the all-zero codeword. More generally, recall that c is a Reed–Solomon codeword that has all its nonzero values confined to locations in the set of locations S if and only if $\sum_{i=S} \omega^{ij} c_i = 0$ for $j = 0, \ldots, r - 1$. The corollary then follows from the lemma. ∎

Theorem 3.11.3 establishes the Welch–Berlekamp key equation for $\Lambda(x)$, which can be solved by the algorithm given in Section 3.12. It expresses the error-locator polynomial $\Lambda(x)$ in terms of the modified syndrome polynomial, $M(x)$, which is defined in the proof, and an additional polynomial, $N(x)$, which we will call the modified error-evaluator polynomial.

Theorem 3.11.3 *The error-locator polynomial $\Lambda(x)$ satisfies the polynomial equation*

$$\Lambda(x)M(x) = N(x) \quad (\bmod\ G(x)),$$

where $M(x)$ is the modified syndrome polynomial and the polynomial $N(x)$ (the modified error-evaluator polynomial) has degree at most $t - 1$.

Proof: Start with the key equation

$$\Lambda(x)S(x) = \Gamma(x) \quad (\bmod\ x^{2t}).$$

Because $\deg \Gamma(x) < \nu$, this allows us to write

$$\sum_k \Lambda_k S_{j-k} = 0 \quad \text{for} \quad k = \nu, \nu + 1, \ldots, 2t - 1.$$

The left side can be manipulated as follows:

$$\sum_{k=0}^{n-1} \Lambda_k S_{j-k} = \sum_{k=0}^{n-1} \Lambda_k \sum_{i=0}^{n-1} s_i \omega^{i(j-k)} = \sum_{i=0}^{n-1} s_i \left(\sum_{k=0}^{n-1} \Lambda_k \omega^{-ik} \right) \omega^{ij} = \sum_{i=0}^{n-1} s_i \lambda_i \omega^{ij}.$$

Using Lemma 3.11.1 with $\beta_i = \omega^i$, we can write

$$\lambda_i s_i = \frac{N(\omega^i)}{g'(\omega^i)} \quad i = 0, \ldots, r - 1.$$

Define the *modified code-domain syndrome* as $m_i = s_i g'(\omega^i)$, and note that m_i equals zero whenever s_i equals zero. Define the modified syndrome polynomial $M(x)$ as the transform-domain polynomial corresponding to the code-domain vector $\boldsymbol{m}$. Now we can write

$$\lambda_i m_i - N(\omega^i) = 0 \quad i = 0, \dots, r - 1.$$

But for any vector $\boldsymbol{v}$ of blocklength n, $v_i = 0$ for $i = 0, \dots, r - 1$ if and only if

$$V(x) = 0 \quad (\mathrm{mod}\ G(x)).$$

Therefore

$$\Lambda(x) M(x) = N(x) \quad (\mathrm{mod}\ G(x)),$$

as asserted in the theorem.

It remains to show that the modified error-evaluator polynomial $N(x)$ has degree at most $t - 1$. Define $g^*(x)$ and $N^*(x)$ by $g(x) = g^*(x)\mathrm{GCD}(\Lambda(x), g(x))$ and $N(x) = N^*(x)\mathrm{GCD}(\Lambda(x), g(x))$. Then

$$
\begin{aligned}
\deg N(x) &= \deg N^*(x) + \deg g(x) - \deg g^*(x) \\
&\le \deg W(x) + \deg g^*(x) - r - 1 + \deg g(x) - \deg g^*(x) \\
&= \deg W(x) - 1 \\
&\le t - 1.
\end{aligned}
$$

This completes the proof of the theorem. ■

3.12   The Welch–Berlekamp algorithm

The *Welch–Berlekamp algorithm* has been developed to solve the decoding equation formulated in the preceding section. In this section, we shall describe the Welch–Berlekamp algorithm more generally as a fast method to solve a certain polynomial equation, regardless of the origin of the polynomial equation.

The Welch–Berlekamp algorithm is an algorithm for solving the polynomial equation

$$\Lambda(x) M(x) = N(x) \quad (\mathrm{mod}\ G(x))$$

for polynomials $N(x)$ and $\Lambda(x)$ of least degree, where $M(x)$ is a known polynomial and

$$G(x) = \prod_{\ell=1}^{2t} (x - X_\ell)$$

is a known polynomial with distinct linear factors. This is the form of the polynomial equation derived in Section 3.11. This is the case in which we are interested. In contrast, recall that the Berlekamp–Massey algorithm is an algorithm for solving the equation

$$\Lambda(x)S(x) = \Gamma(x) \quad (\mathrm{mod}\ x^{2t}).$$

If all the constants X_ℓ in $G(x)$ are replaced by zeros, x^{2t} results, and the second polynomial equation is obtained (but with the notation $S(x)$ and $\Gamma(x)$ instead of $M(x)$ and $N(x)$). For this reason, the two problems might seem to be similar, but they are actually quite different because, in the first problem, the $2t$ factors of $G(x)$ must be distinct, while, in the second problem, the $2t$ factors of x^{2t} are the same.

Based on our experience with the Berlekamp–Massey algorithm, we may anticipate an algorithm that consists of solving two equations simultaneously,

$$N^{(r)}(x) = \Lambda^{(r)}(x)M(x) \quad (\mathrm{mod}\ G^{(r)}(x)),$$

$$A^{(r)}(x) = B^{(r)}(x)M(x) \quad (\mathrm{mod}\ G^{(r)}(x)),$$

where, for $r = 1, \ldots, 2t$,

$$G^{(r)}(x) = \prod_{\ell=1}^{r} (x - X_\ell).$$

We will find that the Welch–Berlekamp algorithm has this form. The rth iteration begins with the primary polynomials $N^{(r-1)}(x)$ and $A^{(r-1)}(x)$ computed by the $(r-1)$th iteration and augments these polynomials and others to compute the new polynomials required of the rth iteration. Eventually, at iteration $n - k$, the solution provided by the Welch–Berlekamp algorithm is the solution to the original problem.

The algorithm will be developed by using the mathematical notion of a *module*. A module is similar to a vector space, except that a module is defined over a ring rather than over a field. Consequently, the properties of a module are weaker and more general than those of a vector space. Although, in general, a module is not a vector space, the elements of a module might informally be called *vectors* because a more specific term is not in common use. In this section, the "ring of scalars" for the modules is the polynomial ring $F[x]$.

We shall actually solve the more general polynomial equation

$$\Lambda(x)M(x) + N(x)H(x) = 0 \quad (\mathrm{mod}\ G(x))$$

for $\Lambda(x)$ and $N(x)$, where the polynomials $M(x)$ and $H(x)$ are known and

$$G(x) = \prod_{\ell=1}^{n-k} (x - \omega^\ell),$$

where, now, $X_\ell = \omega^\ell$.

If $H(x) = -1$, the equation to be solved reduces to the Welch–Berlekamp key equation, which is the particular equation in which we are interested. However, by allowing $H(x)$ to be a polynomial, a recursive structure will be found for a module of solutions. Then we can choose from the module a particular solution for which $H(x) = -1$.

We shall need to refer to a few facts from the theory of modules. In particular, although every vector space has a basis, a module need not have a basis. The property that the underlying ring must have to ensure the existence of a basis for the module is that the ring should be a principal ideal ring. For any field F, the ring $F[x]$ is always a principal ideal ring, so a module over $F[x]$ does have a basis. This is the case in which we are interested. In contrast, $F[x, y]$ is not a principal ideal ring, so a module over $F[x, y]$ need not have a basis.

The set of all solutions, $(\Lambda(x)\ N(x))$, of the equation

$$\Lambda(x)M(x) + N(x)H(x) = 0 \quad (\bmod\ G(x))$$

is easily seen to be a module over $F[x]$. We denote this module by $\mathcal{M}$. In fact, we will actually compute this module $\mathcal{M}$ of all solutions $(\Lambda(x)\ N(x))$ of this equation by computing a basis for the module. There must be a basis for the module because $F[x]$ is a principal ideal ring. Accordingly, we will describe the module in terms of a basis. In particular, to find $(\Lambda(x)\ N(x))$, we shall construct a basis for this module. Then we need only look within the module to extract the particular element $(\Lambda(x)\ N(x))$ that satisfies the necessary conditions on the solution.

Let $[\psi_{11}(x)\ \psi_{12}(x)]$ and $[\psi_{21}(x)\ \psi_{22}(x)]$ be any two vectors that form a basis for the module $\mathcal{M}$. Then each solution of the equation can be written as a combination of these basis vectors. This means that $[\Lambda(x)\ N(x)]$ can be expressed as follows:

$$[\Lambda(x)\ N(x)] = [a(x)\ b(x)] \begin{bmatrix} \psi_{11}(x) & \psi_{12}(x) \\ \psi_{21}(x) & \psi_{22}(x) \end{bmatrix},$$

where $a(x)$ and $b(x)$ are coefficients forming the linear combination of the basis vectors.

Given the polynomials $M(x)$ and $H(x)$, let

$$M^*(x) = M(x)/\mathrm{GCD}[M(x), H(x)],$$

$$H^*(x) = H(x)/\mathrm{GCD}[M(x), H(x)].$$

Then, because $M^*(x)$ and $H^*(x)$ are coprime polynomials, the extended euclidean algorithm assures that the equation

$$e^*(x)M^*(x) + h^*(x)H^*(x) = 1$$

is satisfied by some pair of polynomials $(e^*(x), h^*(x))$. It is easy to see that the two vectors $(G(x)e^*(x)\ G(x)h^*(x))$ and $(-M^*(x)\ H^*(x))$ form a basis for the module $\mathcal{M}$.

Proposition 3.12.1   *The two vectors*

$$b_1 = (G(x)e^*(x) \quad G(x)h^*(x))$$

and

$$b_2 = (-H^*(x) \quad M^*(x))$$

form a basis for the module $\{(\Lambda(x) \quad N(x))\}$ *defined by the equation*

$$\Lambda(x)M(x) + N(x)H(x) = 0,$$

where $M(x)$ *and* $H(x)$ *are given.*

Proof: Direct inspection shows that the two vectors solve the polynomial equation and are linearly independent, so they form a basis for a module of dimension 2.   ■

Proposition 3.12.2   *For any basis of the module* $\mathcal{M}$,

$$\det \begin{bmatrix} \psi_{11}(x) & \psi_{12}(x) \\ \psi_{21}(x) & \psi_{22}(x) \end{bmatrix} = \gamma G(x)$$

for some constant γ, *where the rows of the matrix are the basis vectors.*

Proof: The determinant of the matrix formed by these two basis vectors is clearly $G(x)$. Any other basis can be written in terms of this basis as follows:

$$\begin{bmatrix} \psi_{11}(x) & \psi_{12}(x) \\ \psi_{21}(x) & \psi_{22}(x) \end{bmatrix} = \begin{bmatrix} a_{11}(x) & a_{12}(x) \\ a_{21}(x) & a_{22}(x) \end{bmatrix} \begin{bmatrix} G(x)e^*(x) & G(x)h^*(x) \\ -H^*(x) & M^*(x) \end{bmatrix}.$$

The determinant of the first matrix on the right side is a polynomial. Because this transformation between bases can be inverted, the reciprocal of that determinant is also a polynomial. Because these two polynomials must be reciprocals, they must each be a scalar. Hence the determinant of the left side is a scalar multiple of $G(x)$.   ■

To construct the recursive algorithm, we shall solve a sequence of smaller problems. For $i = 1, \ldots, n-k$, let $\mathcal{M}^{(i)}$ be the module over $F[x]$ of all solutions $[\Lambda^{(i)}(x) \quad N^{(i)}(x)]$ for the equation

$$\Lambda^{(i)}(x)M(x) + N^{(i)}(x)H(x) = 0 \qquad (\mathrm{mod}\ G^{(i)}(x)),$$

where

$$G^{(i)}(x) = \prod_{\ell=1}^{i} (x - \omega^\ell).$$

This gives a nested chain of modules,

$$\mathcal{M}^{(1)} \supset \mathcal{M}^{(2)} \supset \mathcal{M}^{(3)} \supset \cdots \supset \mathcal{M}^{(n-k)} = \mathcal{M}.$$

For $i = 1$, this reduces to

$$\Lambda^{(1)}(x)M(x) + N^{(1)}(x)H(x) = 0 \quad (\bmod \ x - \omega).$$

We shall regard the four polynomials $\Lambda(x)$, $M(x)$, $N(x)$, and $H(x)$ that appear in the equation to be solved as representing transform-domain vectors, so that we may define the code-domain vectors as $\lambda_i = (1/n)\Lambda(\alpha^{-i})$, $v_i = (1/n)N(\alpha^{-i})$, $\mu_i = (1/n)M(\alpha^{-i})$, and $h_i = (1/n)H(\alpha^{-i})$. Then, in the code domain, the equation to be solved becomes

$$\lambda_i \mu_i + v_i h_i = 0 \quad i = 1, \ldots, n - k,$$

where the vectors $\boldsymbol{\mu}$ and $\boldsymbol{h}$ are known, and the vectors $\boldsymbol{\lambda}$ and $\boldsymbol{v}$ are to be computed such that this equation is satisfied. It is easy to see what is required in the code domain for each component, but we must give the solution in the transform domain. However, we will develop an algorithm that begins with the code-domain variables $\boldsymbol{\mu}$ and $\boldsymbol{h}$ and ends with the transform-domain variables $\Lambda(x)$ and $N(x)$.

The first iteration, for $i = 1$, is to form the module of all vectors $(\Lambda(x) \ N(x))$ such that, for a given $\boldsymbol{\mu}$ and $\boldsymbol{h}$, the equation

$$\lambda_1 \mu_1 + v_1 h_1 = 0$$

is satisfied. Suppose that μ_1 is nonzero. Then a basis for the module $\mathcal{M}^{(1)}$ consists of two vectors of polynomials given by

$$\boldsymbol{b}_1^{(1)}(x) = (-v_1 \quad \mu_1),$$
$$\boldsymbol{b}_2^{(1)}(s) = (1 - \omega^1 x \quad 0).$$

It is easy to see that, for $i = 1$, the two vectors $\boldsymbol{b}_1^{(1)}(x)$ and $\boldsymbol{b}_2^{(1)}(x)$ span the set of solution vectors $(\Lambda(x) \ N(x))$ to the expression $\Lambda(x)M(x) + N(x)H(x) = 0 \quad (\bmod \ 1 - x\omega^1)$. Thus for any polynomial coefficients $a_1(x)$ and $a_2(x)$, we have the following solution:

$$[\Lambda(x) \quad N(x)] = [a_1(x) \ a_2(x)] \begin{bmatrix} -h_1 & s_1 \\ 1 - x\omega^1 & 0 \end{bmatrix}.$$

To verify the solution, let

$$[\lambda_i \ v_i] = \left[\frac{1}{n}\Lambda(\omega^{-i}) \quad \frac{1}{n}N(\omega^{-i}) \right],$$

(1) Initialize:

$$\Psi^{(0)} = \begin{bmatrix} 1 & 1 \\ 0 & 1 \end{bmatrix}.$$

(2) Choose $j\ell$ such $\mu_{j\ell}^{(\ell-1)}$ is nonzero. Halt if there is no such $j\ell$.

(3) For $k = 1, \ldots, n$, set

$$\begin{bmatrix} \mu_k^{(\ell)} \\ \lambda_k^{(\ell)} \end{bmatrix} = \begin{bmatrix} -h_j^{(\ell-1)} & \mu_j^{(\ell-1)} \\ (\omega^{k-1} - \omega^{j-1}) & 0 \end{bmatrix} \begin{bmatrix} \mu_k^{(\ell-1)} \\ \lambda_k^{(\ell-1)} \end{bmatrix}.$$

(4)

$$\begin{bmatrix} \Psi_{11}^{(\ell)} & \Psi_{12}^{(\ell)} \\ \Psi_{21}^{(\ell)} & \Psi_{22}^{(\ell)} \end{bmatrix} = \begin{bmatrix} h_k^{(\ell-1)} & \mu_k^{(\ell-1)} \\ (x - \omega^{j-1}) & 0 \end{bmatrix} \begin{bmatrix} \Psi_{11}^{(\ell-1)} & \Psi_{12}^{(\ell-1)} \\ \Psi_{21}^{(\ell-1)} & \Psi_{22}^{(\ell-1)} \end{bmatrix}.$$

(5) Increment ℓ. Then go to step (2).

Figure 3.6. Welch–Berlekamp algorithm.

so

$$\lambda_i \mu_i + v_i h_i = \frac{1}{n} a_1(\omega^{-i})[-h_i \mu_i + \mu_i h_i] + \frac{1}{n} a_2(\omega^{-i})(1 - \omega^{-i}\omega^1)$$

$$= 0 \quad \text{for } i = 1.$$

so the first iteration indeed is correct.

Now we turn to the next iteration. Because of the nesting of modules, any solution to the equation for $i = 2, \ldots, n - k$ must be in the module $\mathcal{M}^{(i)}$. Define

$$\begin{bmatrix} \mu_i^{(1)} \\ \lambda_i^{(1)} \end{bmatrix} = \begin{bmatrix} -h_1 & \mu_1 \\ \omega^{-1} - \omega^{-i} & 0 \end{bmatrix} \begin{bmatrix} \mu_i \\ \lambda_i \end{bmatrix} \quad i = 2, 3, \ldots, n - k,$$

noting that the bottom row is not all zero. The two rows are linearly independent if s_i is nonzero.

The Welch–Berlekamp algorithm continues in this way as shown in Figure 3.6.

Problems

3.1 A linear map from one vector space to another is a map that satisfies

$$f(a\boldsymbol{v}_1 + b\boldsymbol{v}_2) = af(\boldsymbol{v}_1) + bf(\boldsymbol{v}_2),$$

where a and b are any elements of the underlying field. Show the mapping that takes each senseword v to the closest error word e cannot be a linear map. Consequently, a decoder must contain nonlinear functions.

3.2 Periodically repeat the first eight symbols of the Fibonacci sequence to give the following sequence:

$$1, 1, 2, 3, 5, 8, 13, 21, 1, 1, 2, 3, 5, 8, 13, 21, 1, 1, 2, \ldots$$

Use the Berlekamp–Massey algorithm to compute the minimum-length linear recursion that produces the above sequence.

3.3 Prove that if the sequence $S = (S_0, S_1, \ldots, S_{r-1})$ has linear complexity L', and the sequence $T = (T_0, T_1, \ldots, T_{r-1})$ has linear complexity L'', then the sequence $S + T$ has linear complexity not larger than $L' + L''$.

3.4 Show that the syndromes of a BCH code, given by

$$S_j = \sum_{j=0}^{n-1} \omega^{i(j+j_0)} v_i \quad j = 1, \ldots, 2t,$$

form a *sufficient statistic* for computing e. That is, show that no information about e which is contained in v is lost by the replacement of v by the set of syndromes.

3.5 (a) Design an encoder for the nonlinear $(15, 8, 5)$ code, discussed in Section 2.11, by modifying a nonsystematic encoder for a BCH $(15, 5, 7)$ code.

(b) Devise a decoder for the same code by augmenting the binary form of the Berlekamp–Massey algorithm with one extra iteration.

(c) Using part (a), derive a code-domain decoder for this code.

3.6 (a) The $(15, 11, 5)$ Reed–Solomon code with $j_0 = 1$ is defined over the field $GF(16)$ and constructed with the primitive polynomial $p(x) = x^4 + x + 1$. Let a set of syndromes be $S_1 = \alpha^4$, $S_2 = 0$, $S_3 = \alpha^8$, and $S_4 = \alpha^2$. Find the generator polynomial of the code.

(b) Find the error values using the Peterson–Gorenstein–Zierler decoder.

(c) Repeat by using the Berlekamp–Massey algorithm.

3.7 The nonlinear $(15, 8, 5)$ code, discussed in Section 2.11, can be decoded by using the Berlekamp–Massey algorithm for binary codes for three iterations ($r = 1$, 3, and 5) and then choosing A and B so that $\Delta_5 = 0$. By working through the iterations of this algorithm, set up an equation that A and B must solve.

3.8 A $(255, 223)$ Reed–Solomon code over $GF(256)$ is to be used to correct any pattern of ten or fewer errors and to detect any pattern of more than ten errors. Describe how to modify the Berlekamp–Massey algorithm to accomplish this. How many errors can be detected with certainty?

3.9   Let m be even and let r be such that

$$\mathrm{GCD}(2^r \pm 1, 2^{m/2} - 1) = 1.$$

A generalized extended Preparata code of blocklength $n = 2^m$ consists of the set of binary codewords described by the pairs $c = (a, b)$, satisfying the following conditions:

   (i) a and b each has an even number of ones;
   (ii) $A_1 = B_1$;
   (iii) $A_s + A_1^s = B_s$, where $s = 2^r + 1$.

Find the number of codewords and the minimum distance of such a code.

3.10 The irreducible polynomial $x^{20} + x^3 + 1$ is used to construct $GF(2^{20})$ with $\alpha = x$ primitive. A Reed–Solomon code is given over this field with $j_0 = 1$, distance 5, and blocklength $2^{20} - 1 = 1\,048\,575$. Suppose that the syndromes of a given senseword are $S_1 = v(\alpha) = x^4$, $S_2 = v(\alpha^2) = x^8$, $S_3 = v(\alpha^3) = x^{12} + x^9 + x^6$, and $S_4 = v(\alpha^4) = x^{16}$.

   (a) Are the errors in the subfield $GF(2)$? Why?
   (b) How many errors are there? Why?
   (c) Find the error-locator polynomial $\Lambda(x)$.
   (d) Find the location of the error or errors. Find the magnitudes of the error or errors.

3.11 Improve the double-error-correcting decoder for the binary BCH code with Hamming distance 6, described in this chapter, to detect triple errors as well. Repeat for the extended Preparata codes with Lee distance 6.

3.12 Are the code-domain syndrome and the transform-domain syndrome of a cyclic code related by the Fourier transform? How is one related to the other?

3.13 Describe how the code-domain Berlekamp–Massey algorithm can be augmented to compute a (frequency-domain) error-locator polynomial $\Lambda(x)$ directly from the code-domain syndromes by also running a shadow copy of the iteration in the transform domain.

Notes

The popularity of Reed–Solomon codes and other cyclic codes can be attributed, in large part, to the existence of good decoding algorithms. By specializing to the class of Reed–Solomon codes and BCH codes, very powerful and efficient algorithms can be developed. These are the specialized methods of locator decoding, which are quite effective for codes of this class. Peterson (1960) made the first important step in the development of these locator decoding algorithms when he introduced the error-locator polynomial as a pointer to the error pattern, and so replaced a seemingly intractable nonlinear problem by an attractive linear problem.

Locator decoding was discussed in the context of nonbinary codes by Gorenstein and Zierler (1961). In its original form, locator decoding involved inverting matrices of size t. Berlekamp (1968) introduced a faster form of this computation, for which Massey (1969) later gave a simpler formulation and an appealing development in the language of shift registers. Forney (1965), and later Horiguchi (1989) and Koetter (1997), developed attractive methods of computing the error magnitudes. Blahut (1979) reformulated the family of locator decoding algorithms within the Fourier transform methods of signal processing. He also recast the Berlekamp–Massey algorithm in the code domain so that syndromes need not be computed. Welch and Berlekamp (1983) patented another decoding algorithm in the code domain that also eliminates the need to compute syndromes. Further work in this direction was published by Berlekamp (1996). Dabiri and Blake (1995) reformulated the Welch–Berlekamp algorithm in the language of modules, using ideas from Fitzpatrick (1995), with the goal of devising a systolic implementation. Our discussion of the Welch–Berlekamp algorithm follows this work as described in Dabiri's Ph.D. thesis (Dabiri, 1996). The Welch–Berlekamp algorithm is sometimes viewed as a precursor and motivation for the Sudan (1997) family of decoding algorithms that is described in Chapter 4.

4 Within or Beyond the Packing Radius

The geometric structure of a code over a finite field consists of a finite set of points in a finite vector space with the separation between any two points described by the Hamming distance between them. A linear code has the important property that every codeword sees the same pattern of other codewords surrounding it. A tabulation of all codeword weights provides a great deal of information about the geometry of a linear code. The number of codewords of weight w in a linear code is equal to the number of codewords at distance w from an arbitrary codeword.

Given any element, which we regard geometrically as a point, of the vector space, not necessarily a codeword, the task of decoding is to find the codeword that is closest to the given point. The (componentwise) difference between the given point and the closest codeword is the (presumed) error pattern. A *bounded-distance decoder* corrects all error patterns of weight not larger than some fixed integer τ, called the *decoding radius*. A bounded-distance decoder usually uses a decoding radius equal to the packing radius t, though this need not always be true. In this chapter, we shall study both the case in which τ is smaller than t and the case in which τ is larger than t, though the latter case has some ambiguity. In many applications, a bounded-distance decoder is preferred to a complete decoder. In these applications, the limited decoding distance of a bounded-distance decoder can actually be a strength because, in these applications, a *decoding error* is much more serious than a *decoding failure*. Then the packing inefficiency of the code becomes an advantage. When the number of actual errors exceeds the packing radius, it is much more likely that the bounded-distance decoder will fail to decode rather than produce an incorrect codeword. In other applications, a decoding failure may be as undesirable as a decoding error. Then a complete decoder will be indicated. In fact, however, it is usually not necessary to use anything even close to a complete decoder; a decoder could perform usefully well beyond the packing radius and still fall far short of a complete decoder.

For a large code, the packing radius may be an inadequate descriptor of the true power of the code to correct errors because many error patterns of weight much larger than the packing radius t are still uniquely decodable – some decodable error patterns may have weight comparable to the covering radius. Of course, if τ is any integer larger than the packing radius t of the code C, then there are some error patterns of weight τ that cannot be uniquely decoded. The point here is that, in many large codes with large τ, these ambiguous codewords may be so scarce that they have little effect on the probability of decoding error in a bounded-distance decoder. The reason that bounded-distance decoders for this case have not been widely used is that efficient decoding algorithms have not been available.

A *false neighbor* of error pattern e is a nonzero codeword c that is at least as close to e as the all-zero codeword. An error pattern e is uniquely and correctly decodable if and only if it has no false neighbors.

Much of this chapter will be spent studying the weight and distance structure of Reed–Solomon codes and the implications for decoding. Other than for Reed–Solomon codes, very little can be said about the weight and distance structure of most linear codes unless the code is small enough to do a computer search. One strong statement that can be made involves a set of identities known as the MacWilliams equation. This equation relates the weight distribution of a code to the weight distribution of its dual code. The MacWilliams equation may be useful if the dual code is small enough for its weight distribution to be found by direct methods such as computer search. We will end the chapter by deriving the MacWilliams equation.

4.1   Weight distributions

For any code $\mathcal{C}$, the number of codewords at distance ℓ from a given codeword c, in general, depends on the codeword c. A linear code, however, is invariant under any vector-space translation of the code that places another codeword at the origin. This means that the number of other codewords at distance ℓ from a particular codeword is independent of the choice of codeword. Every codeword of a linear code sees exactly the same number of other codewords at distance ℓ from itself. This number is denoted A_ℓ.

If a linear block code has a minimum distance $d_{\min}$, then we know that at least one codeword of weight $d_{\min}$ exists, and no codewords of smaller nonzero weight exist. Sometimes, we are not content with this single piece of information; we wish to know how many codewords have weight $d_{\min}$ and what is the distribution of the weights of the other codewords. For example, in Table 2.3, we gave a list of codeword weights for the $(23, 12, 7)$ binary Golay code. For any small code, it is possible to find a similar table of all the weights by exhaustive search. But exhaustive search is intractable for most codes of interest. Instead, analytical techniques can be employed, if they can be found. Since even the minimum distance is unknown for many codes, it is clear that, in general, such analytical techniques will be difficult to find.

Let A_ℓ denote the number of codewords of weight ℓ in an (n, k) linear code. The $(n + 1)$-dimensional vector, with components A_ℓ for $\ell = 0, \ldots, n$, is called the *weight distribution* of the code. Obviously, if the minimum distance is $d_{\min} = d$, then

$A_0 = 1, A_1, \ldots, A_{d-1}$ are all zero and A_d is not zero. To say more than this, we will need to do some work.

The weight distribution tells us a great deal about the geometric arrangement of codewords in $GF(q)^n$. A sphere of radius $d_{\min}$ centered on the all-zero codeword contains exactly $A_{d_{\min}}$ other codewords; all of them are on the surface of the sphere. For example, a $(31,15,17)$ Reed–Solomon code over $GF(32)$ has 8.22×10^9 codewords of weight 17. A sphere around the origin of radius 17 has 8.22 billion codewords on its surface. There are even more codewords on the surface of a sphere of radius 18, and so forth. The weight distribution gives the number of codewords on each such sphere, and so reveals a great deal about the geometry of these points. There may be other questions that can be posed about the geometric arrangement of codewords that are not answered by the weight distribution.

Describing the weight distribution of a code analytically is a difficult problem, and this has not been achieved for most codes. For the important case of the Reed–Solomon codes (or any maximum-distance code), an analytical solution is known. This section will provide a formula for the weight distribution of a maximum-distance code. The formula is obtained using the fact stated in Theorem 2.1.2 that, in a maximum-distance code, the values in any $n - k$ places are forced by the values in the other k places.

For an arbitrary linear code, we will not be able to give such a formula. It is clear from the proof of Theorem 2.1.2 that, if the code is not a maximum-distance code, then it is not true that any set of k places may be used as designated places. This statement applies to all nontrivial binary codes because, except for the repetition codes and the simple parity-check codes, no binary code is a maximum-distance code.

For a maximum-distance code, we can easily compute the number of codewords of weight $d = d_{\min}$. Such a codeword must be zero in exactly $n - d$ components. Theorem 2.1.2 states that for a maximum-distance code, any set of $k = n - d + 1$ components of a codeword uniquely determines that codeword. Partition the set of integers from 0 to $n - 1$ into two sets, $\mathcal{T}_d$ and $\mathcal{T}_d^c$, with $\mathcal{T}_d$ having d integers. Consider all codewords that are zero in those places indexed by the integers in $\mathcal{T}_d^c$. Pick one additional place. This additional place can be assigned any of q values. Then $n - d + 1$ codeword components are fixed; the remaining $d - 1$ components of the codeword are then determined as stated in Theorem 2.1.2. Hence there are exactly q codewords for which any given set of $n - d$ places is zero. Of these, one is the all-zero codeword and $q - 1$ are of weight d. The $n - d$ locations at which a codeword of weight d is zero, as indexed by elements of $\mathcal{T}_d^c$, can be chosen in $\binom{n}{d}$ ways, so we have

$$A_d = \binom{n}{d}(q - 1),$$

because there are $q - 1$ nonzero codewords corresponding to each of these zero patterns.

To find A_ℓ for $\ell > d$, we use a similar, but considerably more complicated, argument. This is done in proving the following theorem.

Theorem 4.1.1 *The weight distribution of a maximum-distance (n, k, d) linear code over $GF(q)$ is given by $A_0 = 1$, $A_\ell = 0$ for $\ell = 1, \ldots, d - 1$, and, for $\ell \geq d$,*

$$A_\ell = \binom{n}{\ell}(q - 1)\sum_{j=0}^{\ell-d}(-1)^j\binom{\ell - 1}{j}q^{\ell-d-j}.$$

Proof: That this weight is zero for $\ell < d$ follows from the definition of d for $\ell < d$. The proof of the theorem for $\ell \geq d$ is divided into three steps as follows.

Step (1) Partition the set of integers from zero to $n - 1$ into two sets, $\mathcal{T}_\ell$ and $\mathcal{T}_\ell^c$, with $\mathcal{T}_\ell$ having ℓ integers, and consider only codewords that are equal to zero in those places indexed by the integers in $\mathcal{T}_\ell^c$ and are nonzero otherwise. Let M_ℓ be the number of such codewords of weight ℓ. We shall prove that M_ℓ is given by

$$M_\ell = (q - 1)\sum_{j=0}^{\ell-d}(-1)^j\binom{\ell - 1}{j}q^{\ell-d-j}.$$

Then, because M_ℓ does not depend on $\mathcal{T}_\ell$, we have, for the total code,

$$A_\ell = \binom{n}{\ell}M_\ell.$$

The expression for M_ℓ will be proved by developing an implicit relationship, for ℓ greater than d, between M_ℓ and $M_{\ell'}$ for ℓ' less than ℓ.

Choose a set of $n - d + 1$ designated components as follows. All of the $n - \ell$ components indexed by the integers in $\mathcal{T}_\ell^c$ are designated components, and any $\ell - d + 1$ of the components indexed by the integers in $\mathcal{T}_\ell$ are also designated components. Recall that the components indexed by $\mathcal{T}_\ell^c$ have been set to zero. By arbitrarily specifying the latter $\ell - d + 1$ components, not all zero, we get $q^{\ell-d+1} - 1$ nonzero codewords, all of weight at most ℓ.

From the set of ℓ places indexed by $\mathcal{T}_\ell$, we can choose any subset of ℓ' places. There will be $M_{\ell'}$ codewords of weight ℓ' whose nonzero components are confined to these ℓ' places. Hence,

$$\sum_{\ell'=d}^{\ell}\binom{\ell}{\ell'}M_{\ell'} = q^{\ell-d+1} - 1.$$

This recursion implicitly gives M_{d+1} in terms of M_d, M_{d+2} in terms of M_d and M_{d+1}, and so forth. Next, we will solve the recursion to give an explicit formula for M_ℓ.

Step (2)   In this step, we will rearrange the equation stated in the theorem into a form more convenient to prove. Treat q as an indeterminate for the purpose of manipulating the equations as polynomials in q. Define the notation

$$\left[\sum_{n=-N_1}^{N_2} a_n q^n \right] = \sum_{n=0}^{N_2} a_n q^n$$

as an operator, keeping only coefficients of nonnegative powers of q. Note that this is a linear operation. With this convention, the expression to be proved can be written as follows:

$$M_\ell = (q-1) \left[q^{-(d-1)} \sum_{j=0}^{\ell-1} (-1)^j \binom{\ell-1}{j} q^{\ell-1-j} \right].$$

The extra terms included in the sum correspond to the negative powers of q and do not contribute to M_ℓ. Now we can collapse the summation by using the binomial theorem to write

$$M_\ell = (q-1) \left[q^{-(d-1)} (q-1)^{\ell-1} \right].$$

Step (3)   To finish the proof, we will show that the expression for M_ℓ derived in step (2) solves the recursion derived in step (1). Thus

$$\sum_{\ell'=d}^{\ell} \binom{\ell}{\ell'} M_{\ell'} = \sum_{\ell'=0}^{\ell} \binom{\ell}{\ell'} M_{\ell'}$$

$$= (q-1) \sum_{\ell'=0}^{\ell} \binom{\ell}{\ell'} \left[q^{-(d-1)} (q-1)^{\ell'-1} \right]$$

$$= (q-1) \left[q^{-(d-1)} (q-1)^{-1} \sum_{\ell'=0}^{\ell} \binom{\ell}{\ell'} (q-1)^{\ell'} \right]$$

$$= (q-1) \left[q^{-d} \left(1 - \frac{1}{q} \right)^{-1} q^\ell \right]$$

$$= (q-1) \left[\sum_{i=0}^{\infty} q^{\ell-d-i} \right]$$

$$= (q-1) \sum_{i=0}^{\ell-d} q^{\ell-d-i}$$

$$= q^{\ell-d+1} - 1,$$

as was to be proved.   ■

Corollary 4.1.2   *The weight distribution of an (n, k) maximum-distance code over $GF(q)$ is given by $A_0 = 1$, $A_\ell = 0$ for $\ell = 1, \ldots, d - 1$, and, for $\ell \geq d$,*

$$A_\ell = \binom{n}{\ell} \sum_{j=0}^{\ell-d} (-1)^j \binom{\ell}{j} (q^{\ell-d+1-j} - 1).$$

Proof: Use the identity

$$\binom{\ell}{j} = \binom{\ell-1}{j} + \binom{\ell-1}{j-1}$$

to rewrite the equation to be proved as follows:

$$A_\ell = \binom{n}{\ell} \sum_{j=0}^{\ell-d} (-1)^j \left[\binom{\ell-1}{j} + \binom{\ell-1}{j-1} \right] (q^{\ell-d+1-j} - 1)$$

$$= \binom{n}{\ell} \left[\sum_{j=0}^{\ell-d} (-1)^j \binom{\ell-1}{j} (qq^{\ell-d-j} - 1) \right.$$

$$\left. - \sum_{j=1}^{\ell-d+1} (-1)^{j-1} \binom{\ell-1}{j-1} (q^{\ell-d+1-j} - 1) \right].$$

Now replace j by i in the first term and $j - 1$ by i in the second term to write

$$A_\ell = \binom{n}{\ell} (q - 1) \sum_{i=0}^{\ell-d} (-1)^i \binom{\ell-1}{i} q^{\ell-d-i}.$$

The last line is the statement of Theorem 4.1.1, which completes the proof of the theorem.   ∎

The above corollary is useful for calculating the weight distribution of a Reed–Solomon code. As an example, the weight distribution of the $(31, 15, 17)$ Reed–Solomon code over $GF(32)$ is shown in Table 4.1. Even for small Reed–Solomon codes such as this one, the number of codewords of weight ℓ can be very large. This explains why it is not practical, generally, to find the weight distribution of a code by simple enumeration of the codewords.

Table 4.1. *Approximate weight distribution for the (31,15,17) Reed–Solomon code*

ℓ	A_ℓ
0	1
1–16	0
17	8.22×10^9
18	9.59×10^{10}
19	2.62×10^{12}
20	4.67×10^{13}
21	7.64×10^{14}
22	1.07×10^{16}
23	1.30×10^{17}
24	1.34×10^{18}
25	1.17×10^{19}
26	8.37×10^{19}
27	4.81×10^{20}
28	2.13×10^{21}
29	6.83×10^{21}
30	1.41×10^{22}
31	1.41×10^{22}

4.2   Distance structure of Reed–Solomon codes

A Reed–Solomon code is a highly structured arrangement of q^k points in the vector space $GF(q)^n$. The most important geometrical descriptor of the code is the minimum distance (or the packing radius). Indeed, the large minimum distance of these codes is one property responsible for the popularity of the Reed–Solomon codes. As for any code, a sphere of radius $d_{\min} - 1$ about any codeword does not contain another codeword. In addition, spheres about codewords of radius not larger than the packing radius t do not intersect.

Because we know the weight distribution of a Reed–Solomon code, we know exactly how many Reed–Solomon codewords are in any sphere centered on a codeword. The number of codewords within a sphere of radius τ about any codeword is given in terms of the weight distribution as $\sum_{\ell=0}^{\tau} A_\ell$. If τ equals t (or is smaller than t), then the only codeword in the sphere is the codeword at the center of the sphere.

The volume of a sphere of radius τ about any codeword is the number of points of $GF(q)^n$ in that sphere. This is

$$V = \sum_{\ell=0}^{\tau} (q-1)^\ell \binom{n}{\ell},$$

because there are $\binom{n}{\ell}$ ways of choosing ℓ places and there are $(q-1)^\ell$ ways of being different from the codeword in all of these places. If τ is the packing radius t, then any two such spheres about codewords are disjoint.

To appreciate the practical aspects of this comment, consider the $(256, 224, 33)$ extended Reed–Solomon code over the field $GF(256)$, which has packing radius 16. There are $q^k V = 256^{224} \sum_{\ell=0}^{16} 255^\ell \binom{256}{\ell}$ points inside the union of all decoding spheres, and there are 256^{256} points in the vector space $GF(256)^{256}$. The ratio of these two numbers is 2.78×10^{-14}. This is the ratio of the number of sensewords decoded by a bounded-distance decoder to the number of sensewords decoded by a complete decoder. A randomly selected word will fall within one of the decoding spheres with this probability. Thus, with extremely high probability, the randomly selected word will fall between the decoding spheres and will be declared uncorrectable. Even though the disjoint decoding spheres cannot be made larger, almost none of the remaining space is inside a decoding sphere. Our intuition, derived largely from packing euclidean spheres in three-dimensional real vector space, is a poor guide to the 224-dimensional vector subspace of $GF(256)^{256}$. If the decoding spheres are enlarged to radius $t+1$, then they will intersect. A sphere will have one such intersection for each codeword at minimum distance. For the $(256, 224, 33)$ extended Reed–Solomon code, a decoding sphere of radius 17 will intersect with $255 \binom{256}{33}$ other such decoding spheres. But there are $255^{17} \binom{256}{17}$ words on the surface of a sphere of radius 17. The ratio of these numbers is on the order of 10^{-23}. Thus ambiguous sensewords are sparse on the surface of that decoding sphere.

Clearly, we may wish to tinker with our notion of a bounded-distance decoder by attempting to decode partially to a radius larger than the packing radius. We can visualize making the decoding spheres larger, but with dimples in the directions of the nearest neighboring codewords. Although decoding to a unique codeword cannot be guaranteed when the decoding radius exceeds the packing radius, for a large code most sensewords only a small distance beyond the packing radius have a unique nearest codeword and so can be uniquely decoded.

A far more difficult task is to find how many codewords are in a sphere of radius τ, centered on an arbitrary point $\boldsymbol{v}$ of $GF(q)^n$. Of course, if τ is not larger than t, then there cannot be more than one codeword in such a sphere, but there may be none. If τ is larger than t, then the answer to this question will depend on the particular $\boldsymbol{v}$, as is suggested by Figure 4.1. This figure shows the decoding situation from the point of view of the sensewords. Two sensewords are marked by an $\times$ in the figure, and spheres of radius τ are drawn about each senseword. One senseword has two codewords within distance τ; the other has only one codeword within distance τ. For a given value of τ, we may wish to count how many $\boldsymbol{v}$ have ℓ codewords in a sphere of radius τ about $\boldsymbol{v}$. A more useful variation of this question arises if the weight of $\boldsymbol{v}$ is specified. For a given value of τ, how many $\boldsymbol{v}$ of weight w have ℓ codewords in a sphere of radius τ about $\boldsymbol{v}$?

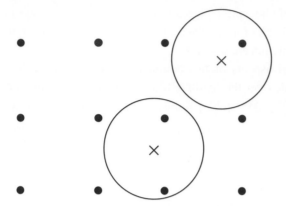

Figure 4.1. Oversized spheres about sensewords.

Equivalently, for a given value of τ, how many v, lying on a sphere of radius w about the all-zero codeword, have ℓ codewords in a sphere of radius τ about v?

For any v, the *proximate set of codewords* is the set of all codewords that are closest to v. There must be at least one codeword c in the set of proximate codewords. There can be multiple proximate codewords only if they all are at the same distance.

A *false neighbor* of an arbitrary vector v is any nonzero codeword in the set of proximate codewords. If the weight of v is at most t, then v has no false neighbors. For any τ larger than t, some v of weight τ will have a false neighbor, though most will not. We would like to know how many such v have a false neighbor. This question will be partially answered in the following sections.

4.3   Bounded-distance decoding

A *bounded-distance decoder* is one that decodes all patterns of τ or fewer errors for some specified integer τ. In Chapter 3, we discussed bounded-distance decoders that decode up to the packing radius t, which is defined as the largest integer smaller than $d_{min}/2$. Figure 4.2 shows the decoding situation from the point of view of the codewords. In this figure, spheres of radius t are drawn about each codeword. These spheres do not intersect, but they would intersect if the spheres were enlarged to radius $t + 1$. A senseword that lies within a decoding sphere is decoded as the codeword at the center of that sphere. A senseword that lies between spheres is flagged an uncorrectable. Because Hamming distance is symmetric, a sphere of radius t drawn about any senseword would contain at most one codeword. The illustration in Figure 4.2, drawn in euclidean two-dimensional space, does not adequately show the situation in n dimensions. In n dimensions, even though the radius of the spheres is equal to the packing

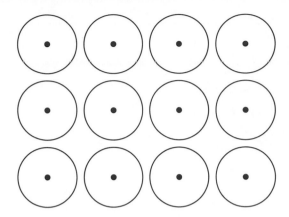

Figure 4.2. Decoding up to the packing radius.

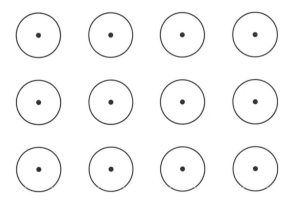

Figure 4.3. Decoding to less than the packing radius.

radius t, the region between the spheres is much larger than the region within the sphere.

To reduce the probability of incorrect decoding, the decoding spheres can be made smaller, as in Figure 4.3, but this will make the probability of correct decoding smaller as well. The decoding will be correct if the senseword lies in the decoding sphere about the correct codeword.

To increase the probability of correct decoding, the decoding spheres can be made larger, as shown in Figure 4.4. The decoding spheres will overlap if their common radius is larger than the packing radius. This results in a decoder known as a *list decoder* because there can be more than one decoded codeword. An alternative to the list decoder, in which the decoding regions are no longer spheres, is shown in Figure 4.5. In this decoding situation, the senseword is decoded as the closest codeword provided that codeword is within Hamming distance τ of the senseword.

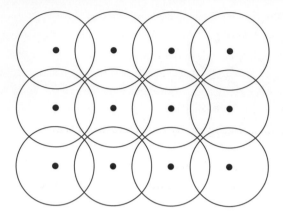

Figure 4.4. List decoding.

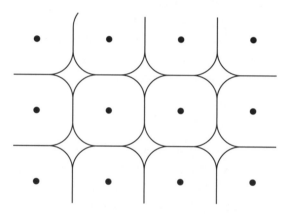

Figure 4.5. Decoding beyond the packing radius.

4.4   Detection beyond the packing radius

A bounded-distance decoder corrects all error patterns up to a specified weight τ, usually chosen to be equal to the packing radius t. If τ is equal to t, then the bounded-distance decoder will correct every error pattern for which the number of errors is not larger than t. If this decoder is presented with an error pattern in which the number of errors is larger than t, then it will not correct the error pattern. The decoder may then sometimes decode incorrectly and may sometimes fail to decode. The usual requirement is that the decoder must detect the error whenever a senseword lies outside the union of all decoding spheres. Such a senseword is said to have an *uncorrectable error pattern*, a property that depends on the chosen value of τ. It is important to investigate how a particular decoding algorithm behaves in the case of an uncorrectable error pattern.

For example, the Peterson algorithm first inverts a t by t matrix of syndromes, and that matrix does not involve syndrome S_{2t-1}. If the determinant of this matrix and all submatrices are equal to zero, the decoding will be completed without ever using syndrome S_{2t-1}. An error pattern for which all syndromes except S_{2t-1} are zero will be decoded as the all-zero error word, even though this cannot be the correct error pattern if S_{2t-1} is nonzero. Thus error patterns that are beyond the packing radius, yet not in a false decoding sphere, may be falsely decoded by the Peterson decoder.

One obvious way to detect an uncorrectable error pattern after the decoding is complete is to verify that the decoder output is, indeed, a true codeword and is within distance t of the decoder input. This final check is external to the central decoding algorithm. There are other, less obvious, checks that can be embedded within the decoding algorithm.

An uncorrectable error pattern in a BCH codeword can be detected when the error-locator polynomial does not have the number of zeros in the error-locator field equal to its degree, or when the degree of $\Lambda(x)$ is larger than τ, or when the error pattern has one or more components not in the symbol field $GF(q)$. All of these cases can be recognized by observing properties of the recursively computed error spectrum:

$$E_j = -\sum_{k=1}^{t} \Lambda_k E_{j-k}.$$

The error pattern will be in $GF(q)$ if and only if the error spectrum over $GF(q^m)$ satisfies the conjugacy constraint $E_j^q = E_{((qj))}$ for all j. If this condition is tested as each E_{qj} is computed, whenever E_j is already known, then an uncorrectable error pattern may be detected and the computation can be halted. For a Reed–Solomon code, $n = q - 1$, so this test is useless because then $((qj)) = j$ and the test only states the obvious condition that $E_j^q = E_j$ in $GF(q)$.

The following theorem states that $\deg \Lambda(x)$ is equal to the number of zeros of $\Lambda(x)$ in $GF(q)$ if and only if E_j is periodic with its period n dividing $q - 1$. By using this test, $\Lambda(x)$ need not be factored to find the number of its zeros.

If either of the two conditions fails, that is, if

$$E_j^q \neq E_{((qj))} \quad \text{for some } j$$

or

$$E_{((n+j))} \neq E_j \quad \text{for some } j,$$

then a pattern with more than t errors has been detected.

Theorem 4.4.1   *Let ω be an element of order n, a divisor of $q - 1$. In the field $GF(q)$, suppose that $\Lambda(x)$ with degree τ, at most equal to $n/2$, is the smallest degree polynomial* *for which*

$$E_j = -\sum_{k=1}^{\tau} \Lambda_k E_{j-k}$$

for $j = \tau, \ldots, \tau + n - 1$. The number of distinct powers of ω that are zeros of $\Lambda(x)$ in $GF(q)$ is equal to $\deg \Lambda(x)$ if and only if $E_{n+j} = E_j$ for $j = \tau, \ldots, \tau + n - 1$.

Proof: If $E_{n+j} = E_j$ holds for $j = \tau, \ldots, \tau + n - 1$, then the recursion requires that $E_{n+j} = E_j$ must then hold for all j. This can be written as $(x^n - 1)E(x) = 0$. The recursion also implies that $\Lambda(x)E(x) = 0$. Then, by the proof of Theorem 1.7.1, $\Lambda(x)E(x) = 0 \pmod{x^n - 1}$. Thus $\Lambda(x)$ divides $x^n - 1$, and so all of its zeros must also be distinct zeros of $x^n - 1$.

To prove the converse, observe that if the number of distinct powers of ω that are zeros of $\Lambda(x)$ is equal to the degree of $\Lambda(x)$, then $\Lambda(x)$ divides $x^n - 1$. But then $\Lambda(x)E(x) = 0 \pmod{x^n - 1}$, so the recursion must satisfy $E_{n+j} = E_j$ for $j = \tau, \ldots, \tau + n - 1$. ∎

4.5   Detection within the packing radius

To reduce the probability of a false decoding, a bounded-distance decoder may be designed to correct only up to τ errors, where τ is strictly smaller than the packing radius t. Then every pattern of more than t errors, but less than $d_{min} - \tau$ errors, can be detected, but not corrected. It is very easy to modify the Berlekamp–Massey algorithm to decode Reed–Solomon codes for this purpose. The modified Berlekamp–Massey algorithm processes 2τ syndromes in the usual way, and so makes 2τ iterations to generate $\Lambda(x)$. If at most τ errors occurred, then, after 2τ iterations, $\Lambda(x)$ will be a polynomial of degree at most τ, and the computation of $\Lambda(x)$ is complete. To verify this, the iterations of the algorithm are continued to process the remaining $d_{min} - 1 - 2\tau$ syndromes. If at most τ errors occurred, the algorithm will not attempt to update $\Lambda(x)$ during any of the remaining $2t - 2\tau$ iterations. If the degree of $\Lambda(x)$ is larger than τ, or if the algorithm does attempt to update $\Lambda(x)$ during any of these remaining $2t - 2\tau$ iterations, then there must be more than τ errors. In such a case, the senseword can be flagged as having more than τ errors.

The following argument shows that, provided ν is less than $d_{min} - \tau$, every pattern of ν errors will be detected by this procedure. We are given syndromes S_j for $j = 0, \ldots, d_{min} - 2$. Therefore we have the syndromes S_j, for $j = 0, \ldots, \tau + \nu - 1$. If we had 2ν syndromes, S_j for $j = 0, \ldots, 2\nu - 1$, then we could correct ν errors. Suppose a genie could give us these extra syndromes, S_j, for $j = \tau + \nu, \ldots, 2\nu - 1$. Then we could continue the Berlekamp–Massey algorithm to compute an error-locator polynomial whose degree equals the number of errors, v. The Berlekamp–Massey algorithm contains a rule for updating L, which says that if the recursion $(\Lambda(x), L)$ is to be updated and $2L < r$, then L is replaced by $r - L$. But at iteration 2τ, by assumption $L_{2\tau} \leq \tau$, and δ_r equals zero for $r = 2\tau + 1, \ldots, \tau + v$. Thus L is not updated by the Berlekamp–Massey algorithm before iteration $\tau + v + 1$. Therefore

$$L_{2v} \geq (\tau + v + 1) - L_{2\tau}$$
$$\geq (\tau + v + 1) - \tau$$
$$= v + 1.$$

But then deg $\Lambda(x) = L_{2v} \geq v + 1$, which is contrary to the assumption that there are at most v errors.

4.6 Decoding with both erasures and errors

The decoder for a Reed–Solomon code may be designed both to fill erasures and correct errors. This decoder is used with a channel that makes both erasures and errors. Hence a senseword now consists of channel input symbols, some of which may be in error, and blanks that denote erasures.

To decode a senseword with errors and erasures, it is necessary to find a codeword that differs from the senseword in the fewest number of places. This will be the correct codeword provided the number of errors v and the number of erasures ρ satisfy

$$2v + \rho + 1 \leq d_{\min}.$$

The task of finding the error-locator polynomial now becomes a little more complicated. We must find the error-locator polynomial, even though some symbols of the senseword are erased. To do this, we will devise a way to mask off the erased symbols so that the errors can be corrected as if the erasures were not there.

Suppose the ρ erasures are at locations $i_1, i_2, \ldots, i_\rho$. At the positions with these known indices, the senseword v_i has blanks, which initially we will fill with zeros. Define the erasure vector as that vector of length n having component f_{i_ℓ} for $\ell = 1, \ldots, \rho$ equal to the erased symbol; in all other locations, $f_i = 0$. Then

$$v_i = c_i + e_i + f_i \qquad i = 0, \ldots, n - 1,$$

where e_i is the error value and f_i is the erased value.

Let ψ be any vector that is zero at every erasure location and nonzero at every nonerasure location. We can suppress the values of the erasure components in v by means of a componentwise multiplication of $\boldsymbol{\psi}$ and $\boldsymbol{v}$. Thus, for $i = 0, \ldots, n-1$, let

$$\overline{v}_i = \psi_i v_i$$
$$= \psi_i(c_i + e_i + f_i)$$
$$= \psi_i c_i + \psi_i e_i.$$

Because ψ_i is zero at the erasure locations, the values of the erased symbols are not relevant to this equation. Define the modified codeword by $\overline{c}_i = \psi_i c_i$ and the modified error word by $\overline{e}_i = \psi_i e_i$. The modified senseword becomes

$$\overline{v}_i = \overline{c}_i + \overline{e}_i \quad i = 0, \ldots, n-1.$$

This equation puts the problem into the form of a problem already solved, provided there are enough syndromes.

To choose an appropriate $\boldsymbol{\psi}$, define the *erasure-locator polynomial*

$$\Psi(x) = \prod_{\ell=1}^{\rho}(1 - x\omega^{i_\ell}) = \sum_{j=0}^{n-1} \Psi_j x^j,$$

where the indices i_ℓ for $\ell = 1, \ldots, \rho$ point to the erasure locations. The inverse Fourier transform of the vector $\boldsymbol{\Psi}$ has component ψ_i equal to zero whenever i is an erasure location and ψ_i is not equal to zero otherwise. This is the ψ_i that was required earlier.

The equation

$$\overline{v}_i = \psi_i c_i + \overline{e}_i$$

becomes

$$\overline{V} = \boldsymbol{\Psi} * C + \overline{E}$$

in the transform domain. Because $\Psi(x)$ is a polynomial of degree ρ, the vector $\boldsymbol{\Psi}$ is nonzero only in a block of $\rho + 1$ consecutive components (from $j = 0$ to $j = \rho$). Because $\boldsymbol{c}$ is a codeword of a Reed–Solomon code, C is zero in a defining set of $d_{\min} - 1$ cyclically consecutive components. Therefore the nonzero components of C lie in a cyclically consecutive block of length at most $n - d_{\min} + 1$. The convolution $\boldsymbol{\Psi} * C$ has its nonzero components within a block consisting of at most $n - d_{\min} + 1 + \rho$ cyclically consecutive components. Thus the convolution $\boldsymbol{\Psi} * C$ is zero in a block of $d_{\min} - 1 - \rho$ consecutive components. This means that if ν is any integer satisfying

$$\nu \leq (d_{\min} - 1 - \rho)/2,$$

then it is possible to decode ν errors in $\overline{c}$.

Any errors-only Reed–Solomon decoder will recover $\bar{c}$ from $\bar{v}$ if v satisfies this inequality. In particular, one may use any procedure that computes the error-locator polynomial $\overline{\Lambda}(x)$ from $\bar{v}$. Once the error-locator polynomial $\overline{\Lambda}(x)$ is known, it can be combined with the erasure-locator polynomial. Define the error-and-erasure locator polynomial $\Lambda(x)$ as follows:

$$\Lambda(x) = \Psi(x)\overline{\Lambda}(x).$$

The error-and-erasure locator polynomial now plays the same role that the error-locator polynomial played before. The zeros of $\Lambda(x)$ point to the locations that have either errors or erasures. Because λ_i is zero if i is the location of either an error or an erasure, $\lambda_i(e_i + f_i) = 0$. The convolution theorem then leads to

$$\Lambda * (E + F) = 0,$$

where the left side is a cyclic convolution. Because $\Lambda_0 = 1$ and $\Lambda_j = 0$ for $j > v + \rho$, this can be written in the form of a cyclic recursion:

$$E_j + F_j = -\sum_{k=1}^{v+\rho} \Lambda_k (E_{((j-k))} + F_{((j-k))}).$$

In this way, the sum $E_j + F_j$ is computed for all j. Because

$$C_j = V_j - (E_j + F_j) \quad j = 0, \ldots, n-1,$$

the rest of the decoding is straightforward.

4.7   Decoding beyond the packing radius

Techniques that decode a Reed–Solomon code a small distance beyond the BCH bound can be obtained by forcing the Berlekamp–Massey algorithm to continue beyond $n - k$ iterations. After the algorithm has completed $n - k$ iterations, the $n - k$ syndromes have all been used, and no more syndromes are available. The decoding algorithm can then be forced to continue analytically, leaving the missing syndromes as unknowns, and the computation of the locator polynomial becomes a function of these unknowns. The unknowns are then selected to obtain a smallest-weight error pattern, provided it is unique, in the symbol field of the code. If the error pattern is not unique, then the unknowns can be selected in several ways and a list of all codewords at equal distance from the senseword can be obtained. Because the complexity of this procedure increases very quickly as one passes beyond the packing radius of the code, only a limited penetration beyond the packing radius is possible in this way.

First, consider a Reed–Solomon code with a defining set $\{\omega^0, \ldots, \omega^{2t-1}\}$ in the field $GF(q)$. We wish to form the list of all codewords within distance τ of the senseword $\boldsymbol{v}$. If τ is larger than t, then there will be some $\boldsymbol{v}$ for which there are at least two codewords on the decoded list. For other $\boldsymbol{v}$, for the same τ, there will be no codewords on the decoded list. If the error pattern has a weight only a little larger than t, then usually there will be exactly one codeword on the list.

Any polynomial $\Lambda(x)$ of degree ν, with ν distinct zeros in $GF(q)$ and $\Lambda_0 = 1$, is an error-locator polynomial for the error pattern if

$$\sum_{j=0}^{n-1} \Lambda_j S_{r-j} = 0 \quad r = \nu, \ldots, 2t-1.$$

If such a polynomial of the smallest degree has a degree at most t, then it is the polynomial produced by the Berlekamp–Massey algorithm. Even when there are more than t errors, the polynomial of smallest degree may be unique, and the senseword can be uniquely decoded whenever that unique polynomial can be found. If the polynomial of smallest degree is not unique, then there are several possible error patterns, all of the same weight, that agree with the senseword. To force the Berlekamp–Massey algorithm beyond the packing radius to radius τ, one can introduce $2(\tau - t)$ additional syndromes as unknowns. Then solve for $\Lambda(x)$ in terms of these unknown syndromes and choose the unknowns to find all polynomials $\Lambda(x)$ with $\deg \Lambda(x)$ distinct zeros in the locator field $GF(q)$. Each of these is a valid locator polynomial that produces a unique error spectrum. The complexity of this approach of forcing missing syndromes is proportional to $q^{2(\tau-t)}$, so it is impractical if τ is much larger than t, even if q is small.

Other cyclic codes may be decoded beyond the packing radius – or at least beyond the BCH bound – in the same way. For an arbitrary cyclic code, it is sometimes true that the packing radius is larger than the BCH radius. Many such codes do not have good performance, and we need not worry about decoding those. There are a few such cases, however, where the codes are good. The $(23, 12, 7)$ binary Golay code, for which the BCH bound is only 5, is one example. Another example is the $(127, 43, 31)$ binary BCH code, which has a designed distance of only 29. The method of forcing the Berlekamp–Massey algorithm beyond the designed distance can be used for these codes.

If the packing radius is larger than the BCH radius and the number of errors is not larger than the packing radius, then there is a unique locator polynomial, $\Lambda(x)$, that satisfies all the syndromes, even if the syndromes are noncontiguous. The binary Golay code is a case in point. Although the Berlekamp–Massey algorithm would be a poor choice to decode the binary Golay code – much better algorithms are available – it is an instructive example of the technique of syndrome filling. Suppose that the senseword has three errors. The syndromes are at the zeros of the generator polynomial. Because the code is a binary code, syndromes with even index are squares of syndromes occurring earlier in the same cycle of the index sequence. Only the syndromes with odd index need be used in the decoding (as was discussed in Section 3.5), and S_1, S_3, S_9, S_{13} are the only syndromes with odd index. To form the locator polynomial $\Lambda(x)$ for a pattern of three errors in $GF(2)$, the Berlekamp–Massey algorithm requires syndromes S_1, S_3, and S_5; all are elements of the locator field $GF(2^{11})$. Syndrome S_5, an element of $GF(2^{11})$, is missing, but can be found by trial and error. There is only one way to assign a value to S_5 such that the linear recursion given by

$$S_j = -\sum_{k=1}^{3} \Lambda_k S_{j-k},$$

formed by the Berlekamp–Massey algorithm, correctly produces the syndromes S_9 and S_{13}. However, trying all 2^{11} possibilities for S_5, even in a symbolic way, is rather clumsy and not satisfactory, so we will not develop this method further for the Golay code.

Later, in Chapter 12, we shall see that a generalization of syndrome filling to two-dimensional codes becomes quite attractive in the context of hyperbolic codes and hermitian codes.

4.8   List decoding of some low-rate codes

We shall now study the topic of bounded-distance decoding beyond the packing radius from a fresh point of view. Given the senseword v, the task is to find all codewords c such that the Hamming distance between v and c is not larger than some given τ that is larger than the packing radius t. Because we have chosen τ larger than t, the decoded codeword need not be unique. Depending on the particular senseword, it may be that no codewords are decoded, or that only one codeword is decoded, or that several codewords are decoded.

An (n, k) narrow-sense Reed–Solomon code can be described as the set of codewords of blocklength n whose spectra are described by polynomials $C(x)$ of degree at most $k - 1$. Let $\alpha_0, \alpha_1, \ldots, \alpha_{n-1}$ be n distinct elements of the field $GF(q)$. Then the code is given by

$$\mathcal{C} = \{(C(\alpha_0), C(\alpha_1), \ldots, C(\alpha_{n-1})) \mid C(x) \in GF(q)[x], \deg C(x) < k < n\}.$$

Under this formulation, the Reed–Solomon codewords are written as

$$c = (C(\alpha_0), C(\alpha_1), \ldots, C(\alpha_{n-1})).$$

If $\alpha_0, \alpha_1, \ldots, \alpha_{n-1}$ are all of the nonzero elements of $GF(q)$, then $\mathcal{C}$ is a primitive cyclic Reed–Solomon code. If $\alpha_0, \alpha_1, \ldots, \alpha_{n-1}$ are some, but not all, of the nonzero elements of $GF(q)$, then $\mathcal{C}$ is a punctured Reed–Solomon code. If $\alpha_0, \alpha_1, \ldots, \alpha_{n-1}$ are all of the elements of $GF(q)$, including the zero element, then $\mathcal{C}$ is a singly extended Reed–Solomon code.

The decoding algorithm of this section recovers the spectrum polynomial from the senseword $\boldsymbol{v}$. The recovery of the correct spectrum polynomial $C(x)$ from the senseword $\boldsymbol{v}$ is equivalent to the recovery of the correct codeword $\boldsymbol{c}$ from the senseword $\boldsymbol{v}$. From this point of view, the traditional decoding problem can be restated as follows. Whenever the number of errors is less than, or equal to, the packing radius t, then find the *unique* polynomial $C(x) \in GF(q)[x]$ such that the vector $\boldsymbol{c} = (C(\alpha_0), C(\alpha_1), \ldots, C(\alpha_{n-1}))$ and the senseword $\boldsymbol{v}$ differ in at most t positions. That is, find a polynomial $C(x)$ of degree less than k such that

$$\|\{i \mid C(\alpha_i) \neq v_i \ i = 0, \ldots, n-1\}\| \leq t,$$

where $\|\mathcal{S}\|$ denotes the cardinality of the set $\mathcal{S}$. Clearly, this is an alternative statement of the task of bounded-distance decoding.

In contrast, our decoding task in this section is the task of *list decoding*. The task now is to find *all* polynomials $C(x)$ such that

$$\|\{i \mid C(\alpha_i) \neq v_i \ i = 0, \ldots, n-1\}\| \leq \tau,$$

where τ is an integer larger than t. An equivalent statement of this condition is

$$\|\{i \mid C(\alpha_i) = v_i \ i = 0, \ldots, n-1\}\| > n - \tau.$$

This condition may be satisfied by a single $C(x)$, by several $C(x)$, or by none.

The *Sudan decoder* is a list decoder for low-rate Reed–Solomon codes that, given any senseword $\boldsymbol{v}$, will compute all codewords $\boldsymbol{c}$ such that the Hamming distance $d_H(\boldsymbol{c}, \boldsymbol{v})$ is not larger than some specified integer τ, called the *Sudan radius*, provided that τ has not been chosen too large. More directly, the Sudan decoder finds every spectrum polynomial $C(x)$ corresponding to a codeword, $\boldsymbol{c}$, within Hamming distance τ of the senseword $\boldsymbol{v}$. The Sudan decoder is a version of a bounded-distance decoder with a radius larger than t, and so sometimes it gives more than one codeword as its output. By saying that the decoder can correct up to τ errors, we mean that if the number of errors is less than or equal to τ, then the decoder can find *all* spectrum polynomials $C(x)$ over $GF(q)$ for which the corresponding codeword satisfies the distance condition.

For any positive integers a and b, the *weighted degree* (or the (a, b)-weighted degree) of the monomial $x^{j'} y^{j''}$ is $aj' + bj''$. The weighted degree of the polynomial $v(x, y)$ is the largest weighted degree of any monomial appearing in a term of $v(x, y)$ with a nonzero coefficient. The weighted degree of the polynomial $v(x, y)$ is denoted $\deg^{(a,b)} v(x, y)$. The weighted degree can be used to put a partial order, called the *weighted order*, on the polynomials by partially ordering the polynomials by the values of their weighted degrees. (The partial order becomes a total order if a supplementary rule is given for breaking ties.)

The Sudan decoder for an (n, k) Reed–Solomon code is based on finding and factoring a certain bivariate polynomial $Q(x, y)$, called the *Sudan polynomial*. The complexity of the algorithm is dominated by the complexity of two tasks: that of finding $Q(x, y)$ and that of factoring $Q(x, y)$. These tasks will not be regarded as part of the theory of the Sudan decoder itself, but require the availability of other algorithms for those tasks that the Sudan decoder can call.

Given the senseword v, by letting $x_i = \alpha_i$ and $y_i = v_i$ define n points (x_i, y_i), for $i = 0, \ldots, n - 1$, in the affine plane $GF(q)^2$. The x_i are n distinct elements of $GF(q)$ paired with the n components $y_i = v_i$ of the senseword v, so the n points (x_i, y_i) are distinct points of the plane. There exist nonzero bivariate polynomials

$$Q(x, y) = \sum_{j' j''} Q_{j' j''} x^{j'} y^{j''}$$

over $GF(q)$ with zeros at each of these n points. These are the polynomials that satisfy $Q(x_i, y_i) = 0$ for $i = 0, \ldots, n - 1$. Because $Q(x_i, y_i) = 0$ for $i = 0, \ldots, n - 1$, we have a set of n linear equations for the unknown coefficients $Q_{j' j''}$. The polynomial $Q(x, y)$ that has the required zeros must exist if the bidegree is constrained so that the number of coefficients is not too large. We shall require such a bivariate polynomial $Q(x, y)$ for which $\deg^{(1, k-1)} Q(x_i, y_i) < n - \tau$, where k is the dimension of the Reed–Solomon code. The weighted degree is chosen less than $n - \tau$ to guarantee the existence of a nonzero solution for the set of unknown $Q_{j' j''}$. Then we will show that every spectrum polynomial $C(x)$ satisfying appropriate distance conditions for an (n, k) Reed–Solomon code can be extracted from the Sudan polynomial $Q(x, y)$.

Theorem 4.8.1 (Sudan theorem)   *Let $Q(x, y)$ be a nonzero bivariate polynomial for which*

$$\deg^{(1, k-1)} Q(x, y) < n - \tau$$

that satisfies $Q(x_i, y_i) = 0$ for $i = 0, \ldots, n-1$. Then for any $C(x)$ of degree at most $k-1$, the polynomial $y - C(x)$ is a factor of $Q(x, y)$ if the vector $c = (C(x_0), \ldots, C(x_{n-1}))$ is within Hamming distance τ of the vector $y = (y_0, \ldots, y_{n-1})$.

Proof: Let $C(x)$ be the spectrum polynomial of any codeword c, whose Hamming distance from the senseword v is at most τ. This means that $C(x_i) \neq y_i$ for at most τ values of i, or, equivalently, $C(x_i) = y_i$ for at least $n - \tau$ values of i.

Because $Q(x_i, y_i) = 0$ for $i = 0, \ldots, n - 1$ and $C(x_i) = y_i$ for at least $n - \tau$ values of i, we have $Q(x_i, C(x_i)) = 0$ for at least $n - \tau$ values of i. But, for any $C(x)$ with degree less than k, $Q(x, C(x))$ is a univariate polynomial in only x, and

$$\deg Q(x, C(x)) \leq \deg^{(1,k-1)} Q(x, y)$$

$$< n - \tau.$$

A nonzero polynomial in one variable cannot have more zeros than its degree. Because $Q(x, C(x))$ does have more zeros than its largest possible degree, it must be the zero polynomial.

Now view $Q(x, y)$ as a polynomial in the ring $GF(q)[x][y]$. This polynomial, which we now denote $Q_x(y)$, is a polynomial in y with its coefficients in $GF(q)[x]$. Then we have that $Q_x(C(x))$ is identically zero. Because $GF(q)[x]$ is a ring with identity, the division algorithm for rings with identity implies that because $C(x)$ is a zero of $Q_x(y)$, then $y - C(x)$ is a factor of $Q(x, y)$. This is the statement that was to be proved. ∎

The Sudan theorem leads to the structure of the Sudan decoder. This decoder is a list decoder consisting of three stages. The input to the Sudan decoder is the senseword v. The senseword v is represented as a set of points in the plane, given by $\{(\alpha_i, v_i) \mid i = 0, \ldots, n - 1\}$, which we write as $\{(x_i, y_i) \mid i = 0, \ldots, n - 1\}$, where $x_i = \alpha_i$ and $y_i = v_i$.

Step (1) Find any nonzero bivariate polynomial $Q(x, y)$ over $GF(q)$ such that $Q(x_i, y_i) = 0$ for all $i = 1, \ldots, n$, and $\deg^{(1,k-1)} Q(x, y) < n - \tau$.

Step (2) Factor the bivariate polynomial $Q(x, y)$ into its irreducible bivariate factors over $GF(q)$.

Step (3) List all polynomials $C(x)$ whose degrees are less than k for which $y - C(x)$ is a factor of $Q(x, y)$ and $C(x_i) \neq y_i$ for at most τ values of i.

The Sudan theorem justifies this procedure if the polynomial $Q(x, y)$ with the required weighted degree exists and can be found. This polynomial will always exist if the decoding radius τ is not too large. One way to determine a value of τ for which a suitable Sudan polynomial exists is to choose any integer m smaller than $k - 1$ and any integer ℓ such that

$$(k - 1) \binom{\ell + 1}{2} + (m + 1)(\ell + 1) > n.$$

Then choose τ satisfying $m + \ell(k - 1) < n - \tau$. We shall see that this choice of τ assures that the needed Sudan polynomial $Q(x, y)$, exists and can be computed. Specifically, we will show that the specified conditions on m and ℓ, together with the condition on the weighted degree of $Q(x, y)$, combine to ensure the existence of the bivariate polynomial $Q(x, y)$ with the required properties. The Sudan theorem then states that because $Q(x, y)$ has these required properties, every polynomial $C(x)$ for which $\|\{i \mid C(x_i) \neq y_i\}\| \leq \tau$ corresponds to a factor of $Q(x, y)$ with the form $y - C(x)$. By finding the factors of this form, one finds the nearby codewords. These are the codewords c for which $d_H(v, c) \leq \tau$.

To prove the first claim, define

$$Q(x, y) = \sum_{j''=0}^{\ell} \sum_{j'=0}^{m+(\ell-j'')(k-1)} Q_{j'j''} x^{j'} y^{j''}.$$

Then

$$\deg^{(1,k-1)} Q(x, y) \leq j''(k-1) + m + (\ell - j'')(k-1)$$
$$= m + \ell(k-1)$$
$$< n - \tau.$$

The number of unknown coefficients $Q_{j'j''}$ is equal to the number of terms in the double sum defining $Q(x, y)$. For a fixed j'', the inner sum has $m + (\ell - j'')(k-1) + 1$ terms. Then

$$\sum_{j''=0}^{\ell} [(\ell - j'')(k-1) + m + 1] = \sum_{i=0}^{\ell} i(k-1) + \sum_{i=0}^{\ell} (m+1)$$
$$= (k-1)\binom{\ell+1}{2} + (m+1)(\ell+1).$$

Therefore $Q(x, y)$ has exactly $(k-1)\binom{\ell+1}{2} + (m+1)(\ell+1)$ unknown coefficients, which is larger than n because of how m and ℓ were chosen. On the other hand, the set of equations $Q(x_i, y_i) = 0$ for $i = 0, \dots, n-1$ yields a set of n linear equations involving the more than n coefficients $Q_{j'j''}$. This set of linear equations can be expressed as a matrix equation

$$MQ = 0.$$

By forming a vector Q composed of the unknown coefficients $Q_{j'j''}$, and a matrix M with elements $x_i^{j'} y_i^{j''}$, the number of unknowns is larger than the number of equations. This means that the number of rows of M is smaller than the number of columns of M.

Hence at least one nonzero solution exists for the set of $Q_{j'j''}$. This provides the required $Q(x, y)$.

4.9   Bounds on the decoding radius and list size

The Sudan theorem leads to the Sudan decoder. We have considered the Sudan decoder only in its most basic form; more advanced versions are known. The Sudan theorem also leads to statements about the performance of the Sudan decoder in terms of the relationship between the Sudan decoding radius, the list size, and the code rate. These statements include the observation that the Sudan radius reduces to the packing radius for high-rate Reed–Solomon codes. The Sudan theorem also leads to inferences about the distance structure of a low-rate Reed–Solomon code at distances larger than the packing radius.

The Sudan polynomial must have at least $n + 1$ monomials. At the same time, its $(1, k - 1)$-weighted degree should be made small. In the previous section, the polynomial

$$Q(x, y) = \sum_{j''=0}^{\ell} \sum_{j'=0}^{m+(\ell-j'')(k-1)} Q_{j'j''} x^{j'} y^{j''}$$

was used as the Sudan polynomial. This polynomial has a total of $(k - 1)\binom{\ell+1}{2} + (m + 1)(\ell + 1)$ monomials. In this section, we will look more carefully at the Sudan polynomial and define it slightly differently.

The Sudan theorem says that to correct up to τ errors, one should use a Sudan polynomial $Q(x, y)$ with $\deg^{(1,k-1)} Q(x, y) < n - \tau$. To make τ large, we must make $\deg^{(1,k-1)} Q(x, y)$ small without violating the constraints in the definition of the Sudan polynomial. To see how best to choose the monomials of $Q(x, y)$, in Figure 4.6 we show the bivariate monomials $x^{j'} y^{j''}$ arranged in the order of increasing $(1, k - 1)$-weighted degree, which is defined as $j' + (k - 1)j''$. The ℓth row consists of those

$1, x, x^2, \ldots, x^{k-2},$

$x^{k-1}, y, x^k, xy, x^{k+1}, x^2y, \ldots, x^{2(k-1)-1}, x^{k-2}y,$

$x^{2(k-1)}, x^{k-1}y, y^2, \ldots, x^{3(k-1)-1}, x^{2(k-1)-1}y, x^{k-2}y^2,$

$\vdots$

$x^{(r-1)(k-1)}, x^{(r-2)(k-1)}y, \ldots, y^{r-1}, \ldots, x^{r(k-1)-1}, x^{(r-1)(k-1)-1}y, \ldots, x^{k-2}y^{r-1},$

$\vdots$

Figure 4.6. Bivariate monomials in $(1, k - 1)$-weighted graded order.

monomials whose $(\ell, k - 1)$-weighted degree is smaller than $\ell(k - 1)$, but not smaller than $(\ell - 1)(k - 1)$. This means that the monomial $x^{i'} y^{i''}$ is placed before the monomial $x^{j'} y^{j''}$ if $i' + i''(k - 1) < j' + j''(k - 1)$ or if $i' + i''(k - 1) = j' + j''(k - 1)$ and $i'' < j''$. Groups of monomials with the same $(1, k - 1)$-weighted degree are clustered in Figure 4.6 by highlighting each cluster with an underline. For all monomials on the ℓth row, the number of monomials in the same cluster is ℓ. The total number of monomials on the ℓth row is exactly $\ell(k - 1)$. The total number of monomials in the first ℓ rows is $\sum_{i=0}^{\ell} i(k - 1) = \binom{\ell}{2}(k - 1)$.

To make the Sudan radius τ large, the $(1, k - 1)$-weighted degree should be made small. Thus to obtain the required $Q(x, y)$ with the fewest monomials, one should pick the $n + 1$ monomials appearing first in the ordered list of bivariate monomials as given in Figure 4.6. The $(1, k - 1)$-weighted degree of the $(n + 1)$th monomial is the largest $(1, k - 1)$-weighted degree of any linear combination of these monomials. This elementary method of determining the number of terms needed in $Q(x, y)$ results in simpler expressions for the Sudan decoding radius, and for the bound on the number of list-decoded codewords.

Before we give the general expression, we shall work out several examples of the exact expression for the Sudan decoding radius τ and the upper bound on the list size.

Example Choose the set of all monomials whose $(1, k - 1)$-weighted degree is less than $2(k - 1)$. There are $3(k - 1)$ such monomials. These are the monomials in the first two rows of Figure 4.6. If $3(k - 1) > n$, one can form a linear combination of these $3(k-1)$ monomials in the first two rows to form a bivariate polynomial, $Q(x, y)$, passing through n points (x_i, y_i) for $i = 1, \ldots, n$ and satisfying $\deg^{(1,k-1)} Q(x, y) < 2(k - 1)$. Because this $Q(x, y)$ has a y degree equal to 1, the Sudan polynomial can have at most one factor of the form $y - C(x)$.

Let $\deg^{(1,k-1)} Q(x, y) = M$. By assumption, the leading monomial of $Q(x, y)$ is on the second row of Figure 4.6. There are $k - 1$ clusters on the first row, each cluster with a single monomial. There are $k - 1$ clusters on the second row, each cluster with two monomials. Of these $k - 1$ clusters on the second row, $M + 1 - (k - 1)$ clusters have terms with degree M or less, and so appear in $Q(x, y)$. Therefore, there are $k - 1 + 2(M + 1 - (k - 1))$ monomials appearing in a polynomial $Q(x, y)$ of degree M. We require that the number of monomials be larger than n. Therefore

$$k - 1 + 2(M + 1 - (k - 1)) > n,$$

which leads to

$$M > \frac{n + k - 3}{2}.$$

But $M = n - \tau - 1$, so the Sudan radius can be obtained as

$$\tau = n - 1 - M$$
$$< n - 1 - \frac{n + k - 3}{2}$$
$$= \frac{n - k + 1}{2} = \frac{d_{min}}{2}.$$

Therefore τ is not larger than the packing radius of the code.

The statement $3(k - 1) > n$ is equivalent to the statement $(k - 1)/n > 1/3$. Thus we see that for an (n, k) Reed–Solomon code with rate $k/n > 1/3 + 1/n$, the Sudan decoder can find at most one codeword within a Hamming distance equal to the packing radius of the code, which is no better than the performance of the conventional locator decoding algorithms.

Example Consider the set of all monomials whose $(1, k - 1)$-weighted degree is less than $3(k - 1)$. There are $6(k - 1)$ such monomials as listed in Figure 4.6. So if $6(k - 1) > n$, then there exists a Sudan polynomial, $Q(x, y)$, for which $\deg^{(1,k-1)} Q(x, y) < 3(k - 1)$ and $Q(x_i, y_i) = 0$ for $i = 1, \ldots, n$. Because this $Q(x, y)$ has y degree equal to 2, it can have only two factors of the form $y - C(x)$.

Again, let $\deg^{(1,k-1)} Q(x, y) = M$. By assumption, the leading monomial of $Q(x, y)$ is on the third row of Figure 4.6, so M is not smaller than $2(k - 1)$ and not larger than $3(k-1)$. Thus, referring to Figure 4.6, there are $k-1$ clusters on the first row, each cluster with a single monomial. There are $k-1$ clusters on the second row, each cluster with two monomials. The number of clusters taken from the third row is $M - 2(k - 1) + 1$, each cluster with three monomials. Therefore, the total number of monomials with $(1, k-1)$-weighted degree not larger than M is $(k - 1) + 2(k - 1) + 3(M - 2(k - 1) + 1)$. We require that the number of monomials be larger than n. Therefore

$$M > \frac{n + 3k - 6}{3}.$$

But $M = n - \tau - 1$, so the Sudan radius can be obtained as

$$\tau = \left\lceil \frac{2n - 3k + 3}{3} \right\rceil - 1 = \frac{2d_{min} - (k - 1)}{3} - 1,$$

which is larger than the packing radius if $2(k - 1) < d_{min}$.

The inequality $6(k - 1) > n$ can be expressed as $(k - 1)/n > 1/6$. We conclude that if an (n, k) Reed–Solomon code has the rate $k/n > 1/6 + 1/n$, then the Sudan decoder can decode at most two codewords with up to τ errors provided $1/6 < (k-1)/n \leq 1/3$.

In particular, the $(256, 64, 193)$ extended Reed–Solomon code over $GF(256)$ has a packing radius equal to 96 and a Sudan radius equal to 107. The Sudan decoder can correct up to 107 errors. Whenever there are 96 or fewer errors, the decoder will produce one codeword. When there are more than 96 errors, but not more than 107 errors, the decoder may sometimes produce two codewords, but this is quite rare. Even if there are 107 errors, the decoder will almost always produce only one codeword. We can conclude that there are at most two codewords of a $(256, 64, 193)$ Reed–Solomon code within Hamming distance 107 of any vector in the vector space $GF(256)^{256}$. This means that any sphere of radius 107 about any point of the space will contain at most two codewords of an extended $(256, 65, 193)$ Reed–Solomon code.

In general, one may consider the set of all bivariate monomials whose $(1, k - 1)$-weighted degree is less than $\ell(k - 1)$. By generalizing the above examples, one can determine the relationship between the rate of the Reed–Solomon code and the largest number of codewords that the Sudan decoder can produce. Specifically, we have the following proposition.

Proposition 4.9.1 *For any (n, k) Reed–Solomon code and any integer ℓ larger than 1, if*

$$\frac{2n}{\ell(\ell + 1)} + 1 < k \le \frac{2n}{\ell(\ell - 1)} + 1,$$

then there are at most $\ell - 1$ codewords up to the Sudan decoding radius τ, which is

$$\tau = \left\lceil \frac{(\ell - 1)(2n - \ell k + \ell)}{2\ell} \right\rceil - 1.$$

Proof: To ensure that the bivariate polynomial $Q(x, y)$ has more than n free coefficients, and has $(1, k - 1)$-weighted degree M, select at least the first $n + 1$ monomials of Figure 4.6. The total number of monomials on the ℓth row is $\ell(k - 1)$, and the total number of monomials on the first ℓ rows is $\binom{\ell}{2}(k - 1)$. It is necessary to choose $M \ge (\ell - 1)(k - 1)$, because, otherwise, the number of unknown coefficients cannot be greater than n. It is sufficient to choose $M < \ell(k-1)$. Thus the number of monomials with $(1, k - 1)$-weighted degree not larger than M is given by

$$\frac{\ell(\ell - 1)}{2}(k - 1) + \ell(M - (\ell - 1)(k - 1) + 1) > n,$$

where the first term on the left is the number of monomials in the first $\ell - 1$ rows in Figure 4.6 and $M - (\ell - 1)(k - 1) + 1$ is the number of clusters in the ℓth row that have monomials of degree M or less. Thus

$$M > \frac{n}{\ell} + \frac{(\ell - 1)(k - 1)}{2} - 1.$$

Substituting $M = n - \tau - 1$, we obtain

$$\tau < \frac{(\ell - 1)(2n - \ell k + \ell)}{2\ell},$$

or, equivalently,

$$\tau = \left\lceil \frac{(\ell - 1)(2n - \ell k + \ell)}{2\ell} \right\rceil - 1.$$

By the properties imposed on $Q(x, y)$, all codewords within Hamming distance τ of any vector in the vector space $GF(q)^n$ correspond to factors of $Q(x, y)$ with the form $y - C(x)$. But the $(1, k - 1)$-weighted degree of $Q(x, y)$ is less than $\ell(k - 1)$, which implies that if $Q(x, y)$ is regarded as a polynomial in y over the ring $GF(q)[x]$, then its degree is at most $\ell - 1$. Such a polynomial can have at most $\ell - 1$ factors of the form $y - C(x)$. Therefore for a Reed–Solomon code whose rate satisfies the inequality of the proposition, the Sudan decoder can find at most $\ell - 1$ codewords. This completes the proof of the proposition. ∎

Except for a few cases, the bound on the list size in Proposition 4.9.1 is equal to the y degree of the optimal $Q(x, y)$. Note, however, that the first $\ell - 1$ monomials in the ℓth row of Figure 4.6 have a y degree at most $\ell - 2$, which is why the bound is not always tight.

Let $N_{\max}$ be the largest number of Reed–Solomon codewords in any sphere of Sudan radius τ. By an argument similar to the proof of the above proposition, we immediately have the following corollary.

Corollary 4.9.2 *For any (n, k) Reed–Solomon code such that*

$$\frac{2n}{\ell(\ell + 1)} + 1 \; < \; k \leq \frac{2n}{\ell(\ell - 1)} + 1,$$

then $N_{\max}$ equals $\ell - 1$ if

$$\frac{k}{n} \leq \frac{2}{\ell(\ell - 1)} + \frac{1}{n}\left(1 - \frac{2}{\ell}\right),$$

and otherwise equals $\ell - 2$.

Proof: If $n \geq [\ell(\ell - 1)/2](k - 1) + \ell - 1$, then the y degree of the optimal $Q(x, y)$ is $\ell - 1$ and otherwise it is $\ell - 2$. ∎

Although the corollary allows the possibility that $N_{\max} = \ell - 2$, the case that $N_{\max} = \ell - 1$ is much more common. For example, among all Reed–Solomon codes of blocklength 256 and dimension $k \geq 27$, the upper bound fails to be tight for only one code, namely when k is 86.

Given a blocklength n, we can easily find the range of k for various values of the integer ℓ. This can be seen in Table 4.2.

The Sudan theorem can be used to provide a statement about the geometrical configuration of codewords in a Reed–Solomon code, which is given in the following proposition.

Table 4.2. *Code rate versus r*

ℓ	Code rate	Range of k	$N_{\max}$
2	$\dfrac{k}{n} > \dfrac{1}{3} + \dfrac{1}{n}$	$\dfrac{n+3}{3} < k$	1
3	$\dfrac{1}{6} + \dfrac{1}{n} < \dfrac{k}{n} \leq \dfrac{1}{3} + \dfrac{1}{n}$	$\dfrac{n+1}{3} < k \leq \dfrac{n+3}{3}$	1
		$\dfrac{n+6}{6} < k \leq \dfrac{n+1}{3}$	2
4	$\dfrac{1}{10} + \dfrac{1}{n} < \dfrac{k}{n} \leq \dfrac{1}{6} + \dfrac{1}{n}$	$\dfrac{n+3}{6} < k \leq \dfrac{n+6}{6}$	2
		$\dfrac{n+10}{10} < k \leq \dfrac{n+3}{6}$	3

Proposition 4.9.3 *Suppose that the integers $n > k > 1$ satisfy*

$$\frac{\ell}{2} \leq \frac{n}{k-1} < \frac{\ell+1}{2}$$

for an integer $\ell \geq 2$. Then any sphere of a radius less than $(\ell - 1)(2n - \ell k + \ell)/2\ell$ about any point of the vector space $GF(q)^n$ contains at most $\ell - 1$ codewords of any (n, k) Reed–Solomon code over $GF(q)$.

Proof: If an (n, k) Reed–Solomon code has parameters satisfying the stated inequality, then it has a rate satisfying the inequality in the corollary, and the conclusion follows from the previous proposition.   ∎

4.10   The MacWilliams equation

The weight distribution of a maximum-distance code is given in Section 4.1. For codes that are not maximum-distance codes, we do not have anything like Theorem 4.1.1. For small n, the weight distribution can be found by a computer search, but for large n this becomes impractical quickly. In general, we do not know the weight distribution of a code of moderate blocklength.

The strongest tool we have is an expression of the relationship between the weight distribution of a linear code and the weight distribution of its dual code–an expression known as the *MacWilliams equation*. The MacWilliams equation holds for any linear code over a finite field. The MacWilliams equation also holds for linear codes over certain rings, in particular $\mathbf{Z}_4$, provided the notions of an inner product and a dual code are appropriately defined.

It is clear that a linear code, $\mathcal{C}$, implicitly determines its dual code, $\mathcal{C}^{\perp}$, and so the weight distribution of $\mathcal{C}^{\perp}$ is implicit in $\mathcal{C}$. The MacWilliams equation makes this connection of weight distributions explicit. It completely describes the relationship between the weight distribution of the code C and the weight distribution of the dual code $C^{\perp}$.

Before we can derive the MacWilliams equation, we need to introduce the ideas of the intersection and direct sum of two subspaces of a vector space and to prove some properties.

Let U and V be any two linear subspaces of F^n. Then $U \cap V$, called the *intersection* of U and V, denotes the set of vectors that are in both U and V; and $U + V$, called the *direct sum*, denotes the set of all linear combinations $a\boldsymbol{u} + b\boldsymbol{v}$, where $\boldsymbol{u}$ and $\boldsymbol{v}$ are in U and V, respectively, and a and b are scalars. Both $U \cap V$ and $U + V$ are subspaces of F^n.

Theorem 4.10.1

$$\dim[U \cap V] + \dim[U + V] = \dim[U] + \dim[V].$$

Proof: A basis for $U \cap V$ has $\dim[U \cap V]$ vectors. Because $U \cap V$ is contained in both U and V, this basis can be extended to a basis for U by adding $\dim[U] - \dim[U \cap V]$ more basis vectors, none of which are in V. Similarly, it can be extended to a basis for V by adding $\dim[V] - \dim[U \cap V]$ more basis vectors, none of which are in U. All of these basis vectors taken together form a basis for $U + V$. That is,

$$\dim[U + V] = \dim[U \cap V] + (\dim[U] - \dim[U \cap V]) + (\dim[V] - \dim[U \cap V]),$$

from which the theorem follows. ■

Theorem 4.10.2

$$U^{\perp} \cap V^{\perp} = (U + V)^{\perp}.$$

Proof: U is contained in $U + V$, and thus $(U + V)^{\perp}$ is contained in $U^{\perp}$. Similarly, $(U + V)^{\perp}$ is contained in $V^{\perp}$. Therefore $(U + V)^{\perp}$ is contained in $U^{\perp} \cap V^{\perp}$. On the other hand, write an element of $U + V$ as $a\boldsymbol{u} + b\boldsymbol{v}$, and let $\boldsymbol{w}$ be any element of $U^{\perp} \cap V^{\perp}$. Then $\boldsymbol{w} \cdot (a\boldsymbol{u} + b\boldsymbol{v}) = 0$, and thus $U^{\perp} \cap V^{\perp}$ is contained in $(U + V)^{\perp}$. Hence the two are equal. ■

Let A_{ℓ} for $\ell = 0, \ldots, n$ and B_{ℓ} for $\ell = 0, \ldots, n$ be the weight distributions of the linear code C and its dual code $C^{\perp}$, respectively. Define the weight polynomials

$$A(x) = \sum_{\ell=0}^{n} A_{\ell} x^{\ell} \qquad \text{and} \qquad B(x) = \sum_{\ell=0}^{n} B_{\ell} x^{\ell}.$$

The following theorem relates these two polynomials.

Theorem 4.10.3 (MacWilliams)   *The weight polynomial $A(x)$ of an (n, k) linear code over $GF(q)$ and the weight polynomial of its dual code are related as follows:*

$$q^k B(x) = [1 + (q-1)x]^n A\left(\frac{1-x}{1+(q-1)x}\right).$$

Proof: The proof will be in two parts. In part (1), we shall prove that

$$\sum_{i=0}^{n} B_i \binom{n-i}{m} = q^{n-k-m} \sum_{j=0}^{n} A_j \binom{n-j}{n-m}$$

for $m = 0, \ldots, n$. In part (2), we shall prove that this equates to the condition of the theorem.

Part (1)   For a given m, partition the integers from zero to $n-1$ into two subsets, $\mathcal{T}_m$ and $\mathcal{T}_m^c$, with set $\mathcal{T}_m$ having m elements. In the vector space $GF(q)^n$, let V be the m-dimensional subspace, consisting of all vectors that have zeros in components indexed by the elements of $\mathcal{T}_m^c$. Then $V^\perp$ is the $(n-m)$-dimensional subspace consisting of all vectors that have zeros in components indexed by the elements of $\mathcal{T}_m$.

Because

$$(\mathcal{C} \cap V)^\perp = \mathcal{C}^\perp + V^\perp,$$

we can write

$$\dim[\mathcal{C}^\perp + V^\perp] = n - \dim[\mathcal{C} \cap V].$$

On the other hand,

$$\dim[\mathcal{C}^\perp + V^\perp] = (n-k) + (n-m) - \dim[\mathcal{C}^\perp \cap V^\perp].$$

Equating these yields

$$\dim[\mathcal{C}^\perp \cap V^\perp] = \dim[\mathcal{C} \cap V] + n - k - m.$$

For each choice of $\mathcal{T}_m$, there are $q^{\dim[\mathcal{C} \cap V]}$ vectors in $\mathcal{C} \cap V$ and $q^{\dim[\mathcal{C}^\perp \cap V^\perp]}$ vectors in $\mathcal{C}^\perp \cap V^\perp$. Consider $\{\mathcal{T}_m\}$ to be the collection of all such $\mathcal{T}_m$. Enumerate the vectors in each of the $\mathcal{C} \cap V$ that can be produced from some subset $\mathcal{T}_m$ in the collection $\{\mathcal{T}_m\}$. There will be $\sum_{\{\mathcal{T}_m\}} q^{\dim[\mathcal{C} \cap V]}$ vectors in the enumeration, many of them repeated appearances. Similarly, an enumeration of all vectors in each $\mathcal{C}^\perp \cap V^\perp$ produced from $\mathcal{T}_m$ in $\{\mathcal{T}_m\}$ is given by

$$\sum_{\{\mathcal{T}_m\}} q^{\dim[\mathcal{C}^\perp \cap V^\perp]} = q^{n-k-m} \sum_{\{\mathcal{T}_m\}} q^{[\mathcal{C} \cap V]}.$$

To complete part (1) of the proof, we must evaluate the two sums in the equation. We do this by counting how many times a vector of weight j in C shows up in a set $C \cap V$. A vector of weight j is in $C \cap V$ whenever the j positions fall within the m positions in which vectors in V are allowed to be nonzero, or, equivalently, whenever the $n - m$ positions where vectors in V must be zero fall within the $n - j$ zero positions of the codeword. There are $\binom{n-j}{n-m}$ choices for the $n - m$ zero components, and thus the given codeword of weight j shows up in $\binom{n-j}{n-m}$ sets. There are A_j codewords of weight j. Therefore

$$\sum_{\{\mathcal{T}_m\}} q^{\dim[C \cap V]} = \sum_{j=0}^{n} A_j \binom{n-j}{n-m}.$$

Similarly, we can count the vectors in $C^{\perp} \cap V^{\perp}$. The earlier equation then becomes

$$\sum_{i=0}^{n} B_i \binom{n-i}{m} = q^{n-k-m} \sum_{j=0}^{n} A_j \binom{n-j}{n-m}.$$

Because m is arbitrary, the first part of the proof is complete.

Part (2) Starting with the conclusion of part (1), write the polynomial identity as follows:

$$\sum_{m=0}^{n} y^m \sum_{i=0}^{n} B_i \binom{n-i}{m} = \sum_{m=0}^{n} y^m q^{n-k-m} \sum_{j=0}^{n} A_j \binom{n-j}{n-m}.$$

Interchange the order of the summations:

$$\sum_{i=0}^{n} B_i \sum_{m=0}^{n-i} \binom{n-i}{m} y^m = q^{n-k} \sum_{j=0}^{n} A_j \sum_{m=0}^{n} \binom{n-j}{n-m} \left(\frac{y}{q}\right)^n \left(\frac{q}{y}\right)^{n-m},$$

recalling that $\binom{n-i}{m} = 0$ if $m > n - i$. Using the binomial theorem, this becomes

$$\sum_{i=0}^{n} B_i (1+y)^{n-i} = q^{n-k} \sum_{j=0}^{n} A_j \left(\frac{y}{q}\right)^n \left(1 + \frac{q}{y}\right)^{n-j}.$$

Finally, make the substitution $y = (1/x) - 1$ to get

$$q^k x^{-n} \sum_{i=0}^{n} B_i x^i = q^n \sum_{j=0}^{n} A_j \left(\frac{1-x}{xq}\right)^n \left(\frac{1 + (q-1)x}{1-x}\right)^{n-j}$$

or

$$q^k \sum_{i=0}^{n} B_i x^i = (1 + (q-1)x)^n \sum_{j=0}^{n} A_j \left(\frac{1-x}{1+(q-1)x} \right)^j,$$

which completes the proof of the theorem.   ∎

We close this section with a simple application of Theorem 4.10.3. By explicitly listing the codewords, we can see that the weight distribution of the Hamming $(7,4)$ code is given by

$$(A_0, A_1, \ldots, A_7) = (1, 0, 0, 7, 7, 0, 0, 1),$$

and thus

$$A(x) = x^7 + 7x^4 + 7x^3 + 1.$$

The dual code is the binary cyclic code known as the *simplex code*. Its generator polynomial

$$g(x) = x^4 + x^3 + x^2 + 1$$

has zeros at α^0 and α^1. The weight polynomial $B(x)$ of the simplex code is given by

$$2^4 B(x) = (1+x)^7 A \left(\frac{1-x}{1-x} \right)$$
$$= (1-x)^7 + 7(1+x)^3 (1-x)^4 + 7(1+x)^4 (1-x)^3 + (1+x)^7.$$

This reduces to

$$B(x) = 7x^4 + 1.$$

The weight distribution of the $(7,3)$ simplex code consists of one codeword of weight 0 and seven codewords of weight 4.

Problems

4.1   Show that the number of codewords of weight ℓ in an (n, k, d) Reed–Solomon code is equal to the number of vectors C of length n having linear complexity ℓ and $C_{j_0} = C_{j_0+1} = \cdots = C_{j_0+d-2} = 0$.

4.2   Use the MacWilliams equation and Table 2.4 to compute the weight distribution of the dual of the $(24, 12, 8)$ extended Golay code. Why is the answer obvious?

4.3   Verify that the sequence of terms A_ℓ in the weight distribution formula for a maximum-distance (n, k) code sums to q^k. Why must this be so?

4.4   What is the decoding radius for the Sudan decoder for a $(7, 3)$ Reed–Solomon code?

4.5   Consider a $(64, 59, 5)$ Reed–Solomon code over $GF(64)$.

   (a) How many codewords have weight 5?
   (b) How many codewords have weight 6?
   (c) How many error patterns have weight 3?
   (d) What fraction of error patterns of weight 3 is undetected?
   (e) What is the probability that a random senseword will be decoded?

4.6   (a) What is the area of a circle of radius 1 in R^2? What is the area of a circle of radius $1 - \epsilon$? What fraction of the area of a unit circle lies within ϵ of the circle's edge?
   (b) What is the volume of a sphere of radius 1 in R^3? What is the volume of a sphere of radius $1 - \epsilon$? What fraction of the volume of a unit sphere lies within ϵ of the surface?
   (c) Repeat this exercise in R^n. What fraction of the hypervolume of a hypersphere lies within ϵ of the surface in the limit as n goes to infinity?

4.7   (a) Two circles of radius 1 in R^2 have their centers separated by distance $2 - \epsilon$, where ϵ is positive. What is the area of the overlap as a fraction of the area of a circle?
   (b) Two spheres of radius 1 in R^3 have their centers separated by distance $2 - \epsilon$, where ϵ is positive. What is the area of overlap as a fraction of the volume of a sphere?
   (c) Repeat this exercise in R^n.

4.8   (a) Form a cap of a unit circle in the euclidean plane by placing a line perpendicular to a radius halfway between the origin and the circumference. What is the ratio of the area of this cap to the area of the circle?
   (b) Repeat this calculation for a unit sphere.
   (c) What can be said about the corresponding calculation for a hypersphere of dimension n in the limit as n goes to infinity?

4.9   (a) Partition a unit square in R^2 into four equal squares. Draw the maximum circle in each of the four squares. Draw another circle between and touching the four circles. What is the radius of this circle?
   (b) Partition a unit cube in R^3 into eight equal cubes. Draw the maximum sphere in each of the eight cubes. Draw another sphere between and touching the eight spheres. What is the radius of this sphere?
   (c) Repeat this exercise in R^n. Does the central sphere ever have a radius larger than $1/2$? What is the radius of the central sphere in the limit as n goes to infinity?

4.10   Let $C(x)$ be a univariate polynomial over F, and let $Q(x, y)$ be a bivariate polynomial over F such that $Q(x, C(x)) = 0$. Prove that $y - C(x)$ is a factor of $Q(x, y)$.

4.11   If two binary codes have the same weight distribution, do they necessarily have the same distance distribution? Give two $(5, 2, 2)$ binary codes with the same weight distributions and different distance distributions.

4.12   A *distance-invariant code* is one for which the distribution of distances from one codeword to other codewords does not depend on the specified codeword.

(a) Is every linear code a distance-invariant code?

(b) Is the $(15, 8, 5)$ nonlinear code given in Section 2.11 a distance-invariant code?

(c) Prove that the Gray image of a linear code over $\mathbf{Z}_4$ is always a distance-invariant code.

4.13   A $(256, 256 - 2t)$ Reed–Solomon code over the field $GF(257)$ is used to encode datawords of eight-bit symbols into codewords of eight-bit symbols. Whenever a check symbol takes on the value 256, it is replaced by the eight-bit symbol 0. How many codewords are encoded with a single error? How many codewords are encoded with ν errors? How does this reduce the performance of the code?

4.14   Describe how to use the Sudan decoder for decoding Reed–Solomon codes on a channel that makes both errors and erasures.

4.15   Describe how to use the Sakata algorithm (described in Section 8.4) to compute a basis for the ideal of all bivariate polynomials that has zeros at n given points of the bicyclic plane (see also Problem 8.11).

Notes

The formula for the weight distribution of a maximum distance code was first published in a laboratory report by Assmus, Mattson, and Turyn (1965), and was independently discovered the following year by Forney (1966) and by Kasami, Lin, and Peterson (1966). MacWilliams derived her equation in 1963 (MacWilliams, 1963). Blahut (1984) reformulated erasure decoding so that it is a simple preliminary initialization phase of the Berlekamp–Massey algorithm. The Sudan approach to list decoding first appeared in 1997, and was improved in Guruswami and Sudan (1999). A simple description of the structure and performance was given by W. Feng (Feng, 1999; Feng and Blahut, 1998). Because of the complicated polynomial computations, the original Sudan decoder did not initially appear in a mature form suitable for practical implementation. Later work by Koetter, and by Roth and Ruckenstein (2000), simplified these computations and made the decoder more attractive.

5 Arrays and the Two-Dimensional Fourier Transform

An *array* $\boldsymbol{v} = [v_{i'i''}]$ is a doubly indexed set of elements from any given alphabet. The alphabet may be a field F, and this is the case in which we are interested. We will be particularly interested in arrays over the finite field $GF(q)$. An array is a natural generalization of a sequence; we may refer to an array as a two-dimensional sequence or, with some risk of confusion, as a two-dimensional vector.

An array may be finite or infinite. We are interested in finite n' by n'' arrays, and in those infinite arrays $[v_{i'i''}]$ that are indexed by nonnegative integer values of the indices i' and i''. An infinite array is *periodic* if integers n' and n'' exist such that $v_{i'+n',i''+n''} = v_{i'i''}$. Any finite array can be made into a doubly periodic infinite array by periodically replicating it on both axes.

The notion of an array leads naturally to the notion of a bivariate polynomial; the elements of the array $\boldsymbol{v}$ are the coefficients of the bivariate polynomial $v(x, y)$. Accordingly, we take the opportunity in this chapter to introduce bivariate polynomials and some of their basic properties. The multiplication of bivariate polynomials is closely related to the two-dimensional convolution of arrays. Moreover, the evaluation of bivariate polynomials, especially bivariate polynomials over a finite field, is closely related to the two-dimensional Fourier transform.

This chapter is restricted to two-dimensional arrays, bivariate polynomials, and the two-dimensional Fourier transform. However, it is possible to define an array in more than two dimensions. Indeed, nearly everything in this and subsequent chapters generalizes to more than two dimensions, much of it in a very straightforward manner, although there may be a few pitfalls along the way. However, for concreteness, we prefer to stay in two dimensions so that the ideas are more accessible.

5.1 The two-dimensional Fourier transform

If the field F (or an extension of the field F) contains elements β and γ of order n' and n'', respectively, then the n' by n'' array $\boldsymbol{v}$ has a *bispectrum* $\boldsymbol{V}$, which is another n' by n'' array whose components are given by the following two-dimensional

Fourier transform:

$$V_{j'j''} = \sum_{i'=0}^{n'-1} \sum_{i''=0}^{n''-1} \beta^{i'j'} \gamma^{i''j''} v_{i'i''} \qquad \begin{matrix} j' = 0, \ldots, n'-1 \\ j'' = 0, \ldots, n''-1. \end{matrix}$$

The two-dimensional Fourier transform relationship will be represented by a doubly shafted arrow,

$$v \Leftrightarrow V,$$

instead of the singly shafted arrow used for the one-dimensional Fourier transform.

The two-dimensional Fourier transform can be written as follows:

$$V_{j'j''} = \sum_{i'=0}^{n'-1} \beta^{i'j'} \left[\sum_{i''=0}^{n''-1} \gamma^{i''j''} v_{i'i''} \right],$$

or as

$$V_{j'j''} = \sum_{i''=0}^{n''-1} \gamma^{i''j''} \left[\sum_{i'=0}^{n'-1} \beta^{i'j'} v_{i'i''} \right].$$

These expressions are arranged to emphasize that several copies of the one-dimensional Fourier transform are embedded within the two-dimensional Fourier transform. Thus the first rearrangement suggests an n''-point one-dimensional Fourier transform on each row, followed by an n'-point one-dimensional Fourier transform on each column. The second rearrangement suggests an n'-point one-dimensional Fourier transform on each column followed by an n''-point one-dimensional Fourier transform on each row. Because each of the one-dimensional Fourier transforms can be inverted by the inverse one-dimensional Fourier transform, it is apparent that the inverse two-dimensional Fourier transform is given by

$$v_{i'i''} = \frac{1}{n'} \frac{1}{n''} \sum_{j'=0}^{n'-1} \sum_{j''=0}^{n''-1} \beta^{-i'j'} \gamma^{-i''j''} V_{j'j''} \qquad \begin{matrix} i' = 0, \ldots, n'-1 \\ i'' = 0, \ldots, n''-1. \end{matrix}$$

The field elements n' and n'' in the denominator are understood to be the sum of n' ones and of n'' ones, respectively, in the field F.

When $n' = n'' = n$, it is simplest – though not necessary – to choose the same element for β and γ. Then with ω an element of order n, we write the two-dimensional Fourier transform as

$$V_{j'j''} = \sum_{i'=0}^{n-1} \sum_{i''=0}^{n-1} \omega^{i'j'} \omega^{i''j''} v_{i'i''} \qquad \begin{matrix} j' = 0, \ldots, n-1 \\ j'' = 0, \ldots, n-1, \end{matrix}$$

and the inverse two-dimensional Fourier transform as

$$v_{i'i''} = \frac{1}{n^2} \sum_{j'=0}^{n-1} \sum_{j''=0}^{n-1} \omega^{-i'j'} \omega^{-i''j''} V_{j'j''} \qquad \begin{array}{l} i' = 0, \ldots, n-1 \\ i'' = 0, \ldots, n-1. \end{array}$$

The inverse two-dimensional Fourier transform is very similar to the two-dimensional Fourier transform. This similarity can be emphasized by writing it in the following form:

$$v_{((n'-i'))((n''-i''))} = \frac{1}{n^2} \sum_{j'=0}^{n'-1} \sum_{j''=0}^{n''-1} \omega^{i'j'} \omega^{i''j''} V_{j'j''}.$$

Accordingly, we define the *reciprocal array* v as the n' by n'' array with components

$$\tilde{v}_{i'i''} = v_{((n'-i'))((n''-i''))}.$$

The reciprocal array is formed simply by reversing all rows and reversing all columns.

Alternatively, one can write the inverse two-dimensional Fourier transform as follows:

$$v_{i'i''} = \frac{1}{n^2} \sum_{j'=0}^{n'-1} \sum_{j''=0}^{n''-1} \omega^{i'j'} \omega^{i''j''} V_{((n'-j'))((n''-j''))}.$$

Accordingly, we define the *reciprocal bispectral array* as the n' by n'' array with components

$$\tilde{V}_{j'j''} = V_{((n'-j'))((n''-j''))}.$$

The relationship $\tilde{v} \Leftrightarrow \tilde{V}$ then follows immediately.

5.2   Properties of the two-dimensional Fourier transform

Many useful properties of the two-dimensional Fourier transform carry over from the one-dimensional Fourier transform. Such properties include linearity, modulation, translation, and the convolution property. The translation property, for example, says that the array $[v_{((i'-\ell'))((i''-\ell''))}]$ transforms into the array $[V_{j'j''} \beta^{\ell'j'} \gamma^{\ell''j''}]$. The meaning of the double parentheses notation should be apparent; in the first index, the double parentheses denote modulo n', and in the second they denote modulo n''. A list of the properties follows.

(1) Inversion:

$$v_{i'i''} = \frac{1}{n'n''} \sum_{j'=0}^{n'-1} \sum_{j''=0}^{n''-1} \beta^{-i'j'} \gamma^{-i''j''} V_{j'j''} \qquad \begin{matrix} i' = 0, \dots, n'-1 \\ i'' = 0, \dots, n''-1, \end{matrix}$$

where

$$n' = 1 + 1 + 1 + \cdots + 1(n' \text{ terms})$$
$$n'' = 1 + 1 + 1 + \cdots + 1(n'' \text{ terms}).$$

(2) Linearity:

$$\lambda v + \mu v' \Leftrightarrow \lambda V + \mu V'.$$

(3) Modulation:

$$[v_{i'i''} \beta^{i'\ell'} \gamma^{i''\ell''}] \Leftrightarrow [V_{((j'+\ell'))((j''+\ell''))}].$$

(4) Translation:

$$[v_{((i'-\ell'))((i''-\ell''))}] \Leftrightarrow [V_{j'j''} \beta^{\ell'j'} \gamma^{\ell''j''}].$$

(5) Convolution property:

$$e = f ** g \quad \Leftrightarrow \quad E_{j'j''} = F_{j'j''} G_{j'j''} \qquad \begin{matrix} j' = 0, \dots, n'-1 \\ j'' = 0, \dots, n''-1, \end{matrix}$$

where

$$[f ** g]_{i'i''} = \sum_{\ell'=0}^{n'-1} \sum_{\ell''=0}^{n''-1} f_{((i'-\ell'))((i''-\ell''))} g_{\ell'\ell''}.$$

(6) Polynomial zeros: the bivariate polynomial $v(x,y) = \sum_{i'=0}^{n'-1} \sum_{i''=0}^{n''-1} v_{i'i''} x^{i'} y^{i''}$ has a zero at $(\beta^{j'}, \gamma^{j''})$ if and only if $V_{j'j''} = 0$. Likewise the bivariate polynomial $V(x,y) = \sum_{j'=0}^{n'-1} \sum_{j''=0}^{n''-1} V_{j'j''} x^{j'} y^{j''}$ has a zero at $(\beta^{-i'}, \gamma^{-i''})$ if and only if $v_{i'i''} = 0$.

(7) Linear complexity: the weight of the square array v is equal to the cyclic complexity of its two-dimensional Fourier transform V.

(8) Reciprocation:

$$\left[v_{((n'-i')),i''} \right] \Leftrightarrow \left[V_{((n'-j')),j''} \right],$$
$$\left[v_{i',((n''-i''))} \right] \Leftrightarrow \left[V_{j',((n''-j''))} \right].$$

(9) Twist property: suppose $n' = n'' = n$. Then

$$[v_{i',((i''+bi'))}] \Leftrightarrow [V_{((j'-bj'')),j''}],$$

where the indices are cyclic (modulo n).

Most of these properties are immediate counterparts of properties of the one-dimensional Fourier transform. The linear complexity property, however, is not straightforward; it will take some effort in Chapter 7 to explain the linear complexity and the cyclic complexity of a two-dimensional array.

The twist property has no counterpart in one dimension. The twist property says that if the i'th row of a square n by n array is cyclically shifted by bi' places, then the j''th column of the Fourier transform is cyclically shifted by $-bj''$ places. The twist property is proved by defining the new array v' with the components given by

$$v'_{i'i''} = v_{i',((i''+bi'))},$$

which has the following Fourier transform:

$$V'_{j'j''} = \sum_{i'=0}^{n-1}\sum_{i''=0}^{n-1} \omega^{i'j'}\omega^{i''j''} v_{i',((i''+bi'))}$$

$$= \sum_{i'=0}^{n-1}\sum_{i''=0}^{n-1} \omega^{i'(j'-bj'')}\omega^{(i''+bi')j''} v_{i',((i''+bi'))}.$$

But

$$\sum_{i''=0}^{n-1} \omega^{(i''+bi')j''} v_{i',((i''+bi'))} = \sum_{i''=0}^{n-1} \omega^{i''j''} v_{i'i''}$$

because the offset by bi' simply amounts to a rearrangement of the terms, and this does not change the sum. Therefore

$$V'_{j'j''} = \sum_{i'=0}^{n-1}\sum_{i''=0}^{n-1} \omega^{i'(j'-bj'')}\omega^{i''j''} v_{i'i''}$$

$$= V_{((j'-bj'')),j''},$$

as was to be proved.

5.3   Bivariate and homogeneous trivariate polynomials

A *bivariate monomial* is a *term* of the form $x^{i'} y^{i''}$. The *degree* of the monomial $x^{i'} y^{i''}$ is $i' + i''$. For fixed positive integers a and b, the *weighted degree* of the monomial $x^{i'} y^{i''}$ is $ai' + bi''$. A *bivariate polynomial*, $v(x, y)$, is a linear combination of distinct bivariate monomials. The *coefficient* of the *term* $v_{i' i''} x^{i'} y^{i''}$ is the field element $v_{i' i''}$. The *bi-index* of the term $v_{i' i''} x^{i'} y^{i''}$ is the pair of integers (i', i''). The *degree* (or *total degree*) of the polynomial $v(x, y)$, denoted $\deg v(x, y)$, is the largest degree of any monomial appearing in a term of $v(x, y)$ with a nonzero coefficient. The *weighted degree* of the polynomial $v(x, y)$, denoted $\deg^{(a,b)} v(x, y)$, is the largest weighted degree of any monomial appearing in a term of $v(x, y)$ with a nonzero coefficient.

The bivariate polynomial $v(x, y)$ may be regarded as a polynomial in x whose coefficients are polynomials in y. The degree of this univariate polynomial is called the x degree of $v(x, y)$. The x degree and the y degree, taken together, form the pair (s_x, s_y), called the *componentwise degree* of $v(x, y)$. The degree of the polynomial cannot be larger than $s_x + s_y$, but it need not equal $s_x + s_y$ because the polynomial $v(x, y)$ need not include the monomial $x^{s_x} y^{s_y}$.

In Chapter 7, we will study ways to put an order on the monomials. Only then can we define the notions of leading term and monic polynomial, as well as another notion that will be introduced called the *bidegree* of a bivariate polynomial. The degree, the componentwise degree, and the bidegree of $v(x, y)$ will be denoted $\deg v(x, y)$, compdeg $v(x, y)$, and bideg $v(x, y)$, respectively.

The bivariate polynomial $v(x, y)$ is *reducible* over the field F if $v(x, y) = a(x, y)b(x, y)$ for some polynomials $a(x, y)$ and $b(x, y)$, both over the field F, and neither of which has degree 0. A nonconstant polynomial that is not reducible is called an *irreducible polynomial*. If a polynomial is not reducible in the field F, then it may be reducible when viewed in a sufficiently large algebraic extension of the field F. The nonconstant polynomial $v(x, y)$ over the field F is called *absolutely irreducible* if it is not reducible in any algebraic extension of the field F. The polynomials $a(x, y)$ and $b(x, y)$, if they exist, are *factors* of $v(x, y)$, and are called *irreducible factors* if they themselves are irreducible. Any nonconstant polynomial can be written uniquely as a product of its irreducible factors, possibly repeated. We will state this formally as the unique factorization theorem for bivariate polynomials after giving the definition of a monic bivariate polynomial in Section 7.2.

The point $(\beta, \gamma) \in F^2$ is called a *zero* or *affine zero* (or *bivariate zero*) of the polynomial $v(x, y)$ if $v(\beta, \gamma) = 0$, where

$$v(\beta, \gamma) = \sum_{i'} \sum_{i''} v_{i' i''} \beta^{i'} \gamma^{i''}.$$

The bivariate polynomial $v(x, y)$ over the field F is also a bivariate polynomial over an extension field of F, so it may have zeros over the extension field that are not in the given field. When it is necessary to emphasize that the zeros are those in the field of the polynomial, they may be called *rational points*, or *rational zeros*, of the polynomial.

The point (β, γ) is called a *singular point* of the polynomial $v(x, y)$ if it is a zero of $v(x, y)$, and, moreover, the partial derivatives evaluated at β, γ satisfy

$$\frac{\partial v(\beta, \gamma)}{\partial x} = 0$$

$$\frac{\partial v(\beta, \gamma)}{\partial y} = 0.$$

(A formal partial derivative of a polynomial is defined in the same way as a formal derivative.) A nonsingular point of the bivariate polynomial $v(x, y)$ is called a *regular point* of $v(x, y)$. Passing through the regular affine point (β, γ) is the line

$$(x - \beta)\frac{\partial v(\beta, \gamma)}{\partial x} + (y - \gamma)\frac{\partial v(\beta, \gamma)}{\partial y} = 0,$$

called the *tangent line* to $v(x, y)$ at the point (β, γ).

We shall want to define a nonsingular polynomial as one with no singular points. Before doing this, however, we must enlarge the setting to the projective plane. In a certain sense, which will become clear, the bivariate polynomial $v(x, y)$ may want to have some additional zeros at "infinity." However, a field has no point at infinity, so these zeros are "invisible." To make the extra zeros visible, we will change the bivariate polynomial into a homogeneous trivariate polynomial and enlarge the affine plane to the projective plane. A homogeneous polynomial is a polynomial in three variables:

$$v(x, y, z) = \sum_{i'} \sum_{i''} \sum_{i'''} v_{i'i''i'''} x^{i'} y^{i''} z^{i'''},$$

for which $v(\lambda x, \lambda y, \lambda z) = \lambda^i v(x, y, z)$ for some i. This means that every term of a trivariate homogeneous polynomial has the same degree; if the degree is i, then $v_{i'i''i'''} = 0$ unless $i' + i'' + i''' = i$. Therefore $v_{i'i''i'''} = v_{i'i''(i-i'-i'')}$. The original polynomial can be recovered by setting z equal to 1.

The projective plane is the set of points (x, y, z) with the requirement that the rightmost nonzero component is a 1. Thus we can evaluate the trivariate homogeneous polynomial at points of the form $(x, y, 1)$, $(x, 1, 0)$, and $(1, 0, 0)$. The set of points of the form $(x, y, 1)$ forms a copy of the affine plane that is contained in the projective plane. The other points – those with $z = 0$ – are the points at infinity. A *projective zero* of the homogeneous polynomial $v(x, y, z)$ is a point of the projective plane (β, γ, δ) such that $v(\beta, \gamma, \delta) = 0$. Projective zeros of the form $(\beta, \gamma, 1)$ are also affine zeros.

The formal reason for introducing projective space is that the zeros of bivariate polynomials can be described in a more complete way. In the projective plane, all zeros become visible. This is important to recognize because a linear transformation of variables can send some zeros off to infinity, or pull zeros back from infinity. The number of polynomial zeros in the projective plane does not change under a linear transformation of variables. The practical reason for introducing the projective plane is that these zeros at infinity can be useful in some applications. In particular, these extra zeros will be used in later chapters to extend the blocklength of a code.

Any bivariate polynomial $v(x, y)$ can be changed into a trivariate homogeneous polynomial by inserting an appropriate power of z into each term to give each monomial the same degree. Thus the *Klein polynomial*

$$v(x, y) = x^3 y + y^3 + x$$

becomes

$$v(x, y, z) = x^3 y + y^3 z + z^3 x.$$

The original Klein polynomial is recovered by setting $z = 1$.

The *hermitian polynomial*

$$v(x, y) = x^{q+1} + y^{q+1} - 1$$

becomes

$$v(x, y, z) = x^{q+1} + y^{q+1} - z^{q+1}.$$

The original hermitian polynomial is recovered by setting $z = 1$.

We can now define a nonsingular polynomial. The bivariate polynomial $v(x, y)$ over the field F is called a *nonsingular bivariate polynomial* (or a *regular polynomial* or a *smooth polynomial*) if it has no singular points anywhere in the projective plane over any extension field of F. The polynomial is nonsingular if

$$\frac{\partial v(x, y, z)}{\partial x} = 0,$$

$$\frac{\partial v(x, y, z)}{\partial y} = 0,$$

and

$$\frac{\partial v(x, y, z)}{\partial z} = 0$$

are not satisfied simultaneously at any point (x, y, z) in any extension field of F at which $v(x, y, z) = 0$.

Associated with every polynomial is a positive integer called the *genus* of the polynomial. The genus is an important invariant of a polynomial under a linear transformation of variables. For an arbitrary polynomial, the genus can be delicate to define. Since we shall deal mostly with irreducible nonsingular polynomials, we will define the genus only for this case. (The alternative is to give a general definition. Then the following formula, known as the *Plücker formula*, becomes a theorem.)

Definition 5.3.1   *The genus of a nonsingular bivariate polynomial of degree d is the integer*

$$g = \frac{1}{2}(d-1)(d-2) = \binom{d-1}{2}.$$

5.4   Polynomial evaluation and the Fourier transform

Many properties of cyclic codes follow directly from properties of the Fourier transform. Likewise, many properties of bicyclic and epicyclic codes, which we shall study in later chapters, follow from properties of the two-dimensional Fourier transform.

The Fourier transform has been described as the evaluation of the polynomial $v(x)$ on the n powers of an nth root of unity. Let ω be an element of order n in the field F. Then

$$V_j = v(\omega^j)$$

$$= \sum_{i=0}^{n-1} v_i \omega^{ij},$$

which can be regarded either as the polynomial $v(x)$ evaluated at ω^j, or as the jth component of the Fourier transform V. If F is the finite field $GF(q)$ and ω is an element of order $q - 1$, then every nonzero element of the field is a power of ω, so the Fourier transform can be regarded as the evaluation of $v(x)$ at every nonzero element of the field.

The Fourier transform fails to evaluate $v(x)$ at the zero of the field. This exception could be viewed as a slight weakness of the Fourier transform in a finite field. We have seen, however, that it is straightforward to append one additional component to the Fourier transform by evaluating $v(x)$ at the point at zero. We have also seen that a second component can be appended by evaluating $v(x)$ at the point at infinity of the projective line.

The two-dimensional Fourier transform can be described in like fashion, as the evaluation of a bivariate polynomial on the n^2 pairs of powers of an nth root of unity.

Let the array $\boldsymbol{v}$ be represented as the bivariate polynomial $v(x, y)$, given by[1]

$$v(x, y) = \sum_{i'=0}^{n'-1} \sum_{i''=0}^{n''-1} v_{i'i''} x^{i'} y^{i''}.$$

The components of the Fourier transform can be written as follows:

$$V_{j'j''} = v(\beta^{j'}, \gamma^{j''})$$

$$= \sum_{i'=0}^{n'-1} \sum_{i''=0}^{n''-1} v_{i'i''} \beta^{i'j'} \gamma^{i''j''},$$

which can be regarded either as the polynomial $v(x, y)$ evaluated at the point $(\beta^{j'}, \gamma^{j''})$, or as the (j', j'')th component of the bispectrum V.

In some situations, the evaluation of bivariate polynomials on an n by n array is a more convenient description. In other instances, the Fourier transform is more convenient to work with. If F is the finite field $GF(q)$, and ω is an element of order $n = q - 1$, then every nonzero element of the field is a power of ω, so the two-dimensional Fourier transform can be regarded as the evaluation of $v(x, y)$ at every pair of nonzero elements of $GF(q)$. There are $(q-1)^2$ such points. The Fourier transform fails to evaluate $v(x, y)$ at those $2q - 1$ points at which either x or y is the zero element of the field.

Thus we are confronted with a choice: either use the two-dimensional Fourier transform – enjoying all its properties – on an array of $(q - 1)^2$ points, or use polynomial evaluation to form a larger array of q^2 points. In the language of coding theory, either use the two-dimensional Fourier transform to form bicyclic codes of blocklength $(q - 1)^2$, with the bicyclic set of automorphisms, or use polynomial evaluation to extend the code to blocklength q^2, thus spoiling the bicyclic structure. We will take an ambivalent position regarding this issue, speaking sometimes in terms of the Fourier transform, which produces an array of $(q - 1)^2$ components, and speaking sometimes in terms of polynomial evaluation, which produces an array of q^2 components.

The inverse two-dimensional Fourier transform also can be viewed as the evaluation of the bivariate polynomial. Let the array V be represented as the polynomial $V(x, y)$, given by

$$V(x, y) = \sum_{j'=0}^{n'-1} \sum_{j''=0}^{n''-1} V_{j'j''} x^{j'} y^{j''}.$$

---

[1] We may regard this polynomial to be an element of the ring $F[x, y]/\langle x^{n'} - 1, y^{n''} - 1\rangle$.

Then the components of the inverse two-dimensional Fourier transform can be written

$$v_{i'i''} = \frac{1}{n'n''} V(\beta^{-i'}, \gamma^{-i''}).$$

Consequently, whenever it suits our convenience, we can express the inverse two-dimensional Fourier transform in this way as the evaluation of a spectral polynomial $V(x, y)$. In particular, if $n' = n'' = n$,

$$v_{i'i''} = \frac{1}{n^2} V(\omega^{-i'}, \omega^{-i''}),$$

where $\omega = \beta = \gamma$.

5.5   Intermediate arrays

We have already noted that the two-dimensional Fourier transform can be written either as

$$V_{j'j''} = \sum_{i'=0}^{n'-1} \beta^{i'j'} \left[\sum_{i''=0}^{n''-1} \gamma^{i''j''} v_{i'i''} \right],$$

or as

$$V_{j'j''} = \sum_{i''=0}^{n''-1} \gamma^{i''j''} \left[\sum_{i'=0}^{n'-1} \beta^{i'j'} v_{i'i''} \right].$$

The first rearrangement has an inner sum that describes an n''-point one-dimensional Fourier transform on each row of the array v. The second rearrangement has an inner sum that describes an n'-point one-dimensional Fourier transform on each column of the array v. Accordingly, define the *intermediate array* $w = [w_{j'i''}]$ by

$$w_{j'i''} = \sum_{i'=0}^{n'-1} \beta^{i'j'} v_{i'i''}$$

and the intermediate array $W = [W_{i'j''}]$ by

$$W_{i'j''} = \sum_{i''=0}^{n''-1} \gamma^{i''j''} v_{i'i''}.$$

This suggests the following diagram:

$$v \leftrightarrow W$$

$$\updownarrow \qquad \updownarrow$$

$$w \leftrightarrow V,$$

where a horizontal arrow denotes a one-dimensional Fourier transform relationship along every row of the array and a vertical arrow denotes a one-dimensional Fourier transform relationship along every column of the array. The (two-dimensional) bispectrum V can be obtained from v first by computing W, then computing V, or first by computing w, then computing V. The rows of the array W are the (one-dimensional) spectra of the rows of the array v. The columns of the array w are the spectra of the columns of the array v.

The intermediate arrays w and W are themselves (nearly) related by the two-dimensional Fourier transform

$$w_{j'i''} = \frac{1}{n} \sum_{i'=0}^{n'-1} \beta^{i'j'} \gamma^{-i''j''} W_{i'j''}.$$

Except for the factor of $1/n$, this is exactly the form of the Fourier transform with β and γ^{-1} as the elements of order n' and n''. Even the factor of $1/n$ can be suppressed by redefining w.

One consequence of this observation is that various properties relating an array to its two-dimensional Fourier transform, as discussed in Section 5.2, also apply to the relationship between the intermediate arrays w and W. This corresponds to a mixture of properties on the original arrays. For example, the array v can be cyclically shifted on one axis and modulated on the other axis. Then, by the properties of the Fourier transform, V is modulated on one axis and cyclically shifted on the other axis.

5.6   Fast algorithms based on decimation

A *fast algorithm* for a computation is a procedure that significantly reduces the number of additions and multiplications needed for the computation compared with the natural way to do the computation. A *fast Fourier transform* is a computational procedure for the n-point Fourier transform that uses about $n \log n$ multiplications and about $n \log n$ additions in the field of the Fourier transform. We shall describe fast Fourier transform algorithms that exist whenever n is composite. These fast algorithms are closely related to the two-dimensional Fourier transform.

A two-dimensional Fourier transform can arise as a rearrangement of a one-dimensional Fourier transform. Such rearrangements are called *decimation algorithms* or *fast Fourier transform algorithms*. The term "decimation algorithm" refers to the method of breaking a large Fourier transform into a combination of small Fourier transforms. The term fast Fourier transform refers to the computational efficiency of these algorithms. The following paragraphs describe the *Good–Thomas decimation algorithm* and the *Cooley–Tukey decimation algorithm*, which arise as fast algorithms for computing a one-dimensional Fourier transform.

The *Good–Thomas decimation algorithm* uses the chinese remainder theorem, which is an elementary statement of number theory, to convert a one-dimensional Fourier transform of composite blocklength $n = n'n''$ into an n' by n'' two-dimensional Fourier transform, provided n' and n'' are coprime. Because n' and n'' are coprime, integers N' and N'' exist such that $N'n' + N''n'' = 1$. Let the index i be replaced by $i' = i(\bmod n')$ and $i'' = i(\bmod n'')$. Let the index j be replaced by $j' = N''j(\bmod n')$ and $j'' = N'j(\bmod n'')$. The chinese remainder theorem produces the following inverse relationships:

$$i = N''n''i' + N'n'i'' \pmod n$$

and

$$j = n''j' + n'j'' \pmod n.$$

Therefore

$$\omega^{ij} = \beta^{i'j'}\gamma^{i''j''},$$

where $\beta = \omega^{N''n''^2}$ has order n' and $\gamma = \omega^{N'n'^2}$ has order n''. Therefore, by defining the two-dimensional arrays, also called v and V, in terms of the vectors v and V as

$$v_{i'i''} = v_{((N''n''i'+N'n'i''))}$$

and

$$V_{j'j''} = V_{((n''j'+n'j''))},$$

we obtain the following the expression:

$$V_{j'j''} = \sum_{i'=0}^{n'-1}\sum_{i''=0}^{n''-1} \beta^{i'j'}\gamma^{i''j''}v_{i'i''}.$$

The original one-dimensional Fourier transform now has been expressed in a form exactly the same as a two-dimensional Fourier transform.

The Cooley–Tukey decimation algorithm is an alternative algorithm that converts a one-dimensional Fourier transform of composite blocklength $n = n'n''$ into a variation of a two-dimensional Fourier transform. In this case, the factors n' and n''

need not be coprime. Let

$$i = i' + n'i''i' = 0, \ldots, n' - 1; \quad i'' = 0, \ldots, n'' - 1;$$
$$j = n''j' + j''j' = 0, \ldots, n' - 1; \quad j'' = 0, \ldots, n'' - 1.$$

Then, because $\omega^{n'n''} = 1$,

$$\omega^{ij} = \beta^{i'j'}\omega^{i'j''}\gamma^{i''j''},$$

where $\beta = \omega^{n''}$ and $\gamma = \omega^{n'}$. Therefore by defining the two-dimensional arrays

$$v_{i'i''} = v_{((i'+n'i''))} \quad \text{and} \quad V_{j'j''} = V_{((n''j'+j''))},$$

also called v and V, we have the following expression:

$$V_{j'j''} = \sum_{i'=0}^{n'-1} \beta^{i'j'} \left[\omega^{i'j''} \sum_{i''=0}^{n''-1} \gamma^{i''j''} v_{i'i''} \right].$$

This nearly has the form of a two-dimensional Fourier transform, but is spoiled by the appearance of $\omega^{i'j''}$. The original one-dimensional Fourier transform now has been expressed in a form almost the same as a two-dimensional Fourier transform. Thus the Cooley–Tukey decimation algorithm is less attractive than the Good–Thomas decimation algorithm, but its great advantage is that it can be used even when n' and n'' are not coprime.

5.7   Bounds on the weights of arrays

The pattern of zeros in the bispectrum of a nonzero (two-dimensional) array gives information about the Hamming weight of the array just as the pattern of zeros in the spectrum of a nonzero (one-dimensional) vector gives information about the Hamming weight of the vector. We shall use the pattern of zeros in the bispectrum of an array to bound the weight of the array.

A special case is an array in which the number of rows and the number of columns are coprime. Then the chinese remainder theorem can be used to turn the two-dimensional Fourier transform into a one-dimensional Fourier transform. In this way, various bounds on the weight of sequences become bounds on the weight of such arrays. In this section, we are concerned, instead, with bounds on the weight of a square two-dimensional array based on the pattern of zeros in its bispectrum. We will convert bounds on the weight of sequences into bounds on the weight of the arrays. We will also develop bounds on the weight of arrays directly.

One way to derive distance bounds based on bispectral zeros for a general array is simply to take the product of the one-dimensional bounds on the weight of vectors, given in Section 1.8. Thus the BCH bound on the weight of a vector can be used to give a bound relating the weight of an array to the pattern of zeros in the bispectrum of that array.

BCH product bound   Any nonzero n by n array v whose bispectrum V has a consecutive columns equal to zero, and b consecutive rows equal to zero, must have a weight at least equal to $(a + 1)(b + 1)$.

Proof: Let W be the intermediate array obtained by computing the inverse Fourier transform of every column of V. Each column of W is either everywhere zero or, by the BCH bound, has a weight at least $b + 1$. Thus if v is nonzero, W has at least $b + 1$ nonzero rows. Every such nonzero row has at least a consecutive zeros. Because v is the row-wise inverse Fourier transform of W, v has at least $a + 1$ nonzero components in every nonzero row, and there are at least $b + 1$ nonzero rows. ∎

BCH dual product bound   Any nonzero n by n array v whose bispectrum V is zero in an a by b subarray must have weight at least $\min(a + 1, b + 1)$.

Proof: Suppose, without loss of generality, that a is the number of columns in the subarray of zeros. Because v is nonzero, the bispectrum V is nonzero. If the a consecutive columns passing through the subarray are not all zero, then there is at least one nonzero column in V with at least b consecutive zeros. The BCH bound then asserts that there are at least $b + 1$ nonzero rows. Otherwise, there are at least a consecutive nonzero columns, and so at least one nonzero row with at least a consecutive zeros. The BCH bound then asserts that there are at least $a + 1$ nonzero columns, and so at least $a + 1$ nonzero elements. ∎

For an example of the BCH product bound, consider the nonzero array v whose bispectrum V, written as an array, has a zero pattern of the form

$$
V = \begin{bmatrix}
 & 0 & 0 & 0 & & & & \\
0 & 0 & 0 & 0 & 0 & 0 & 0 & 0 \\
0 & 0 & 0 & 0 & 0 & 0 & 0 & 0 \\
 & 0 & 0 & 0 & & & & \\
 & 0 & 0 & 0 & & & & \\
 & 0 & 0 & 0 & & & & \\
 & 0 & 0 & 0 & & & & \\
 & 0 & 0 & 0 & & & &
\end{bmatrix}.
$$

Then v must have weight at least 12 because the bispectrum has three consecutive columns of zeros and two consecutive rows of zeros. To repeat explicitly the argument for this example, observe that the BCH bound implies that the inverse Fourier transform of each nonzero row has weight at least 4, so the intermediate array has at least four nonzero columns and two consecutive zeros in each such column. Then, again by the BCH bound, v has at least three nonzero elements in each nonzero column, and there are at least four nonzero columns.

For an example of the BCH dual product bound, consider the nonzero array v whose bispectrum has a zero pattern of the form

$$
V = \begin{bmatrix}
0 & 0 & 0 & & \\
0 & 0 & 0 & & \\
& & & & \\
& & & & \\
& & & & \\
& & & &
\end{bmatrix}.
$$

There are three columns with two consecutive zeros. If these columns are all zero, there is a nonzero row with three consecutive zeros, so the weight of v is at least 4. If at least one of the three columns is not all zero, then there are at least three nonzero rows in v, so v has weight at least 3.

The BCH product bound is not the only such product relationship between the pattern of bispectral zeros and the weight of an array. One could also state a Hartmann–Tzeng product bound and a Roos product bound in the same way.

Next, we will give a statement regarding a single run of consecutive bispectral zeros in either the row direction or the column direction. To be specific, we will describe these zeros as lying consecutively in the row direction. The statement remains the same if the consecutive bispectral zeros lie in the column direction. A similar statement can be given even with the bispectral zeros lying along a generalized "knight's move."

BCH bispectrum property   Any two-dimensional array v of weight $d - 1$ or less, with $d - 1$ consecutive zeros in some row of its bispectrum V, has zeros for every element of that row of the bispectrum.

Proof: Without loss of generality, suppose that V has $d - 1$ consecutive zeros in its first row. Because v has weight at most $d - 1$, the intermediate array w, obtained by taking the Fourier transform of each column of v, has at most $d - 1$ nonzero columns, and so at most $d - 1$ nonzero elements in the first row. Because the Fourier transform of the first row of w has $d - 1$ consecutive zeros, the first row of w, and of V, must all be zero. ∎

For example, if the array v has bispectrum V containing four zeros in a row, as follows:

$$
V = \begin{bmatrix} 0 & 0 & 0 & 0 & & & \\ & & & & & & \\ & & & & & & \\ & & & & & & \end{bmatrix},
$$

then either v has at least five nonzero columns, or the entire first row of V is zero. In the latter case, if v is nonzero, it has at least one nonzero column of weight at least 2. No more than this can be concluded.

The BCH bispectrum property can be combined with the twist property of the two-dimensional Fourier transform to show that any two-dimensional square array v of weight $d - 1$ or less, with $d - 1$ consecutive zeros in a diagonal of its bispectrum V, has zeros in every element of that diagonal. This can be generalized further to place the consecutive zeros on various definitions of a generalized diagonal. In effect, placing consecutive zeros on any "straight line" leads to an appropriate generalization of the BCH bound.

The BCH bispectrum condition finds its strength when it is applied simultaneously in the row direction of an array and the column direction, as described next. Indeed, then it includes the product bound as a special case.

Truncated BCH product bound   Any nonzero n by n array whose bispectrum V has a consecutive columns each equal to zero in $a^2 + 2a$ rows and a consecutive rows each equal to zero in $a^2 + 2a$ consecutive columns must have weight at least $(a + 1)^2$.

For example, suppose that the array v has bispectrum V containing two consecutive rows with eight consecutive zeros and two consecutive columns with eight consecutive zeros. Such a bispectrum is given by

$$
V = \begin{bmatrix} 0 & 0 & 0 & 0 & 0 & 0 & 0 & 0 \\ 0 & 0 & 0 & 0 & 0 & 0 & 0 & 0 \\ 0 & 0 & & & & & & \\ 0 & 0 & & & & & & \\ 0 & 0 & & & & & & \\ 0 & 0 & & & & & & \\ 0 & 0 & & & & & & \\ 0 & 0 & & & & & & \end{bmatrix}.
$$

Suppose that the array v has weight at most 8. Then the intermediate array, formed by taking the Fourier transform of each column of v, has at most eight nonzero columns, so each row of the intermediate array has at most eight nonzero elements. Any such row

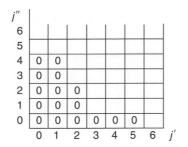

Figure 5.1. Pattern of spectral zeros forming a cascade set.

that has eight consecutive zeros in its Fourier transform must, by the BCH bound, be all zero. Thus the top two rows, because they have eight consecutive zeros, must actually be all zero. Similarly, the two columns of given zeros are also all zero. Therefore this reduces to the product bound. Hence if v is not all zero, it has weight at least 9.

It is reasonable now to ask whether the pattern of required spectral zeros can be further reduced without changing the bound on minimum distance. In fact, we shall see that the bispectral array

$$V = \begin{bmatrix} 0 & 0 & 0 & 0 & 0 & 0 & 0 & 0 \\ 0 & 0 & 0 & 0 & & & & \\ 0 & 0 & & & & & & \\ 0 & 0 & & & & & & \\ 0 & & & & & & & \\ 0 & & & & & & & \\ 0 & & & & & & & \\ 0 & & & & & & & \end{bmatrix},$$

If nonzero, corresponds to a vector v of weight at least 9. This is a consequence of the multilevel bound, which will be described next.

The multilevel bound will be given with the bispectral zeros arranged in a pattern known as a *cascade set*. Such a set is best described if the bispectrum is displayed against a pair of coordinate axes, rather than as a matrix, so that the pattern will appear with the rows indexed from bottom to top. Figure 5.1 shows a set of zeros clustered in such a pattern.

Definition 5.7.1 *The cascade set Δ is a proper subset of N^2 with the property that if $(k', k'') \in \Delta$, then $(j', j'') \in \Delta$ whenever $j' \leq k'$ and $j'' \leq k''$.*

Figure 5.2 is an illustration of a typical cascade set.

A cascade set always has this form of a descending stairway with a finite number of steps of varying integer-valued rises and runs. A cascade set is completely defined

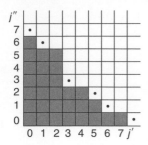

Figure 5.2. Typical cascade set.

by specifying its *exterior corners*. These are the elements of N^2 marked by dots in Figure 5.2.

Definition 5.7.2 *The cascade hull of a set of points in N^2 is the smallest cascade set that contains the given set of points.*

The zeros of a bispectrum, in totality or in part, may form a cascade set. An example of a bispectrum, V, can be constructed with its zeros forming the cascade set shown in Figure 5.2. Any set of zeros of V that forms a cascade set leads to the following bound on the weight of a nonzero array v.

Multilevel bound　Any nonzero n by n array v, whose set of bispectral zeros contains a cascade set Δ that includes all (j', j'') such that $(j' + 1)(j'' + 1) < d$, has weight at least d.

Proof: The bispectrum V of the array v can be converted to the array v in two steps, described as follows:

$$v$$

$$\uparrow$$

$$w \leftarrow V.$$

The horizontal arrow denotes a Fourier transform along each row of V producing the intermediate array w; the vertical arrow denotes a Fourier transform along each column of w producing the array v. The BCH componentwise bound will be used to bound the number of nonzero columns in w, and then the BCH componentwise bound will be used a second time to bound the number of nonzero entries in each nonzero column of v.

Let $d^{(j')}$ denote the BCH bound on the weight of the j'th row of w, as determined by the number of consecutive zeros in that row of V. If row zero is not all zero, then it has weight at least $d^{(0)}$. Then there are at least $d^{(0)}$ nonzero columns in v, each of weight at least 1. If row zero, instead, is everywhere zero, but row one is not everywhere zero, then v has at least $d^{(1)}$ nonzero columns. Each nonzero column of v has weight at least 2 because each column of w is zero in row zero. Similarly, if all rows of w before row j'' are zero, and row j'' is not zero, then there are at least $d^{(j'')}$ nonzero columns in v. Thus each nonzero column has weight at least $j'' + 1$, because each such column of w has j' consecutive zeros in the initial j'' rows.

Because the array is not zero everywhere, one of the assumptions about the number of all-zero leading rows must be true, so one of the bounds holds. The weight of the array v is not smaller than the smallest such bound. That is,

$$\text{wt } v \geq \min_{j'=0,\dots,n-1} [(j'+1)d^{(j')}].$$

But

$$d^{(j')} = \min_{j' | (j',j'') \notin \Delta} (j''+1).$$

Combining these yields

$$\text{wt } v \geq \min_{(j',j'') \notin \Delta} [(j'+1)(j''+1)] \geq d,$$

as was to be proved.   ■

To see that the bound is tight, let $g_{j'}(x)$ be a polynomial with spectral zeros at all j smaller than j' and let $g_{j''}(x)$ be a polynomial with spectral zeros at all j smaller than j''. Then the product $g_{j'}(x)g_{j''}(y)$ has weight $(j'+1)(j''+1)$.

A special case of the multilevel bound is known as the *hyperbolic bound*.

Hyperbolic bound   Any nonzero n by n array, whose set of bispectral zeros is a hyperbolic set given by

$$\mathcal{A} = \{(j',j'') \mid (j'+1)(j''+1) < d\},$$

must have weight at least d.

To complete this section, there is one more bound that we will include. To prove this bound, we must anticipate facts, such as *Bézout's theorem*, that are not discussed until Chapter 7. We shall call this bound the *weak Goppa bound* because the same techniques will be used in Chapter 9 to give a stronger bound, known as the *Goppa bound*.

Weak Goppa bound   Any nonzero n by n two-dimensional array v over F, whose n by n bispectrum V over F is zero for $j' + j'' > J$, has weight at least $n^2 - nJ$.

Proof: An n by n bispectrum exists only if F contains an element of order n. Therefore the bound is vacuous either if $F = GF(2)$ (because then $n = 1$), or if $J \geq n$, so we can ignore these cases.

A statement equivalent to the theorem is that $V(x, y)$, a nonzero bivariate polynomial of degree at most J over the field F, has at most nJ zeros of the form $(\omega^{-i'}, \omega^{-i''})$, where ω is an element of order n in F. Regard the array V as a polynomial $V(x, y)$ of degree at most J. To prove the bound, it suffices to find an irreducible polynomial, $G(x, y)$, of degree n that has a zero at every point of the form $(\omega^{-i'}, \omega^{-i''})$. Bézout's theorem asserts that $V(x, y)$ has at most nJ zeros in common with such a $G(x, y)$, so $V(x, y)$ can have at most nJ zeros of the form $(\omega^{-i'}, \omega^{-i''})$. Therefore v, because it has n^2 components, has weight at least $n^2 - nJ$.

Let

$$G(x, y) = x^n - 1 + \beta y^n - \beta,$$

where $\beta \neq 1$ if the characteristic of the field is 2; otherwise, $\beta = 1$. The polynomial $G(x, y)$ has degree n and has the required zeros at every $(\omega^{-i'}, \omega^{-i''})$. The three partial derivatives of the homogeneous trivariate polynomial $G(x, y, z)$ are

$$\frac{\partial G(x, y, z)}{\partial x} = nx^{n-1}; \quad \frac{\partial G(x, y, z)}{\partial y} = \beta n y^{n-1}; \quad \frac{\partial G(x, y, z)}{\partial z} = -(1 + \beta)nz^{n-1}.$$

If the characteristic of the field is p, then n divides $p^m - 1$; therefore n is not zero in F. Because the three partial derivatives are all equal to zero only if $x = y = z = 0$, which is not a point of the curve, the polynomial is nonsingular. Therefore according to a theorem to be given (as Theorem 9.1.1) in Section 9.1, the polynomial is irreducible, so $V(x, y)$ and $G(x, y)$ can have no common polynomial factor. The proof is complete because Bézout's theorem now says that $V(x, y)$ has at most nJ zeros on the bicyclic plane. ∎

If the field is the finite field $GF(q)$ with the characteristic p, a slightly stronger statement can be obtained by proving that the polynomial

$$G'(x, y) = x^q - x + y^q - y$$

is irreducible. This polynomial $G'(x, y)$ has all the bicyclic zeros of $G(x, y)$ and additional zeros whenever either x or y (or both) equals zero. Because $q = 0$ in $GF(q)$, the partial derivatives of $G'(x, y)$ reduce to

$$\frac{\partial G'(x, y)}{\partial x} = \frac{\partial G'(x, y)}{\partial y} = -1 \neq 0,$$

so the polynomial is nonsingular and so is irreducible. The nonzero polynomial $V(x, y)$ of degree at most J can have at most q zeros in common with $G'(x, y)$. Therefore $V(x, y)$ must have at least $q^2 - qJ$ nonzeros in the affine plane.

Problems

5.1 Do absolutely irreducible univariate polynomials with degree larger than 1 exist? Do absolutely irreducible bivariate polynomials with degree larger than 1 exist?

5.2 Let n' and n'' be coprime. Given an (n', k') cyclic code with generator polynomial $g'(x)$ and an (n'', k'') cyclic code with generator polynomial $g''(x)$, answer the following.

(a) Prove that the product code formed from these two cyclic codes is equivalent to an $(n'n'', k'k'')$ cyclic code.

(b) What is the minimum distance?

(c) Find the generator polynomial in terms of $g'(x)$ and $g''(x)$.

5.3 Rearrange the components of the one-dimensional vector v of blocklength 35 into a two-dimensional array, and rearrange the components of its spectrum V into another two-dimensional array so that the two arrays are related by a two-dimensional Fourier transform.

5.4 Let a' and n be coprime and let a'' and n be coprime. Show that any two-dimensional array v of weight $d - 1$ or less, for which $V_{j_0' + a'k, j_0'' + a''k} = 0$ for $k = 1, \ldots, d - 1$, also satisfies $V_{j_0' + a'k, j_0'' + a''k} = 0$ for $k = 0, \ldots, n - 1$.

5.5 Set up the equations for computing an n' by n'' two-dimensional Fourier transform, where n' and n'' are coprime, as a one-dimensional Fourier transform.

5.6 State and prove the two-dimensional conjugacy constraint

$$V_{j'j''}^q = V_{((qj')),((qj''))}$$

for the two-dimensional Fourier transform in the field $GF(q)$.

5.7 Is the polynomial

$$p(x, y) = x^{17} + y^{16} + y$$

singular or nonsingular?

5.8 A two-dimensional cyclic convolution, denoted $e = f * *g$, is given by

$$e_{i'i''} = \sum_{\ell'=0}^{n-1} \sum_{\ell''=0}^{n-1} f_{((i'-\ell'))((i''-\ell''))} g_{\ell'\ell''}.$$

State and prove the two-dimensional convolution theorem.

5.9 An *elliptic polynomial* over the field F is a polynomial of the form

$$y^2 + a_1 xy + a_3 y = x^3 + a_2 x^2 + a_4 x + a_6.$$

(An elliptic curve is the set of rational zeros of an elliptic polynomial.) What is the genus of a nonsingular elliptic polynomial?

5.10 How many zeros does the Klein polynomial have in the affine plane over $GF(8)$?

5.11 (a) Prove that the Klein polynomial over any field of characteristic 2 is nonsingular.

   (b) Prove that the hermitian polynomials over any field of characteristic 2 are nonsingular.

5.12 For any field F, let $F^\circ[x, y] = F[x, y]/\langle x^n - 1, y^n - 1 \rangle$, with typical element $p(x, y) = \sum_{i'=0}^{n-1} \sum_{i''=0}^{n-1} p_{i'i''} x^{i'} y^{i''}$. The *transpose* of the element $p(x, y)$ of $F^\circ[x, y]$ is the element $p^{\mathsf{T}}(x, y) = p(y, x)$. The *reciprocal* of the element $p(x, y)$ is the element $\tilde{p}(x, y) = x^{n-1} y^{n-1} p(x^{-1}, y^{-1})$.

   For any field F, let $\boldsymbol{p} = [p_{i'i''} \ i' = 0, \ldots, n-1; i'' = 0, \ldots, n-1]$ be an n by n array of elements of F. The *transpose* of the n by n array $\boldsymbol{p}$ is the array $\boldsymbol{p}^{\mathsf{T}} = [p_{i''i'}]$. The *reciprocal* of the n by n array $\boldsymbol{p}$ is the array $\tilde{\boldsymbol{p}} = [p_{n-1-i', n-1-i''}]$.

   A polynomial of $F^\circ[x, y]$, with *defining set* $\mathcal{A} \subset \{0, \ldots, n-1\}^2$, is the polynomial $p(x, y)$ with coefficient $p_{i'i''}$ equal to zero for all $(i', i'') \in \mathcal{A}$. Let $\mathcal{A}_J = \{(i', i'') | i' + i'' > J\}$. Prove that if $p(x, y)$ has defining set $\mathcal{A}_J$ and $\overline{p}(x, y)$ has defining set $\mathcal{A}_J^c$, then $p^{\mathsf{T}}(x, y)$ has defining set $\mathcal{A}_{2n-3-J}$.

Notes

The two-dimensional Fourier transform is widely used in the literature of two-dimensional signal processing and image processing. Most of the properties of the two-dimensional Fourier transform parallel properties of the one-dimensional Fourier transform and are well known. The role of the two-dimensional Fourier transform in coding theory and in the bounds on the weight of arrays, such as the BCH componentwise bound, was discussed by Blahut (1983).

A statement very similar to the weak Goppa bound appears in the computer science literature under the name *Schwartz's lemma*. This lemma, published in 1980, is used in the field of computer science for probabilistic proofs of computational complexity (Schwartz, 1980). The hyperbolic bound appears in the work of Saints and Heegard (1995).

The Good–Thomas decimation algorithm (Good, 1960; Thomas, 1963) is well known in signal processing as a way to use the chinese remainder theorem to change a one-dimensional Fourier transform of composite blocklength into a two-dimensional Fourier transform. The Good–Thomas decimation algorithm is closely related to the decomposition of a cyclic code of composite blocklength with coprime factors into a bicyclic code.

6   The Fourier Transform and Bicyclic Codes

Given the field F, the vector space F^n exists for every positive integer n, and a linear code of blocklength n is defined as any vector subspace of F^n. Subspaces of dimension k exist in F^n for every integer $k \leq n$. In fact, very many subspaces of dimension k exist. Each subspace has a minimum Hamming weight, defined as the smallest Hamming weight of any nonzero vector in that subspace. We are interested in those subspaces of dimension k over $GF(q)$ for which the minimum Hamming weight is large.

In the study of F^n and its subspaces, there is no essential restriction on n. This remark is true in the finite field $GF(q)$ just as in any other field. However, in the finite field, it is often useful to index components of the vector space $GF(q)^n$ by the elements of the field $GF(q)$, when $n = q$, or by the nonzero elements of the field $GF(q)$, when $n = q - 1$. The technique of using the elements of $GF(q)$ to index the components of the vector over $GF(q)$ is closely related both to the notion of a cyclic code and to polynomial evaluation. The essential idea of using nonzero field elements as indices can be extended to blocklength $n = (q - 1)^2$ by indexing the components of the vector v by pairs of nonzero elements of $GF(q)$. Then the vector v is displayed more naturally as a two-dimensional array.

A two-dimensional array can be rearranged into a one-dimensional vector by placing its rows side by side, or by placing its columns top to bottom. Thus an $n = n'n''$ code can be constructed as a subset of the set of n' by n'' arrays. Such a code is sometimes called a *two-dimensional code* because codewords are displayed as two-dimensional arrays. Of course, if it is a linear code, the code as a vector space will have a dimension, denoted k, but in a different sense – in the sense of a code. A two-dimensional code is also called a *bivariate code* when codewords are regarded as bivariate polynomials.

6.1   Bicyclic codes

One class of two-dimensional codes is the class of bivariate codes called *two-dimensional cyclic codes* or *bicyclic codes*. A bicyclic code may be defined by the property that the two-dimensional code is invariant under both a cyclic shift in the row direction and a cyclic shift in the column direction. Bicyclic codes can also be defined in terms of the two-dimensional Fourier transform.

Recall that we may describe a cyclic code in terms of its spectrum using the terminology of the Fourier transform,

$$C_j = \sum_{i=0}^{n-1} \omega^{ij} c_i \quad j = 0, \ldots, n-1,$$

where ω is an element of order n of the finite field $GF(q)$. A cyclic code is defined as the set of vectors of length n with a fixed set of components of the Fourier transform C equal to zero.

In a similar way, a bicyclic code is defined in terms of the two-dimensional Fourier transform,

$$C_{j'j''} = \sum_{i'=0}^{n'-1} \sum_{i''=0}^{n''-1} \beta^{i'j'} \gamma^{i''j''} c_{i'i''},$$

where β and γ are elements of $GF(q)$ of order n' and n'', respectively. An n' by n'' bicyclic code consists of the set of n' by n'' arrays c with a fixed set of components of the two-dimensional Fourier transform C equal to zero. This fixed set of bi-indices is called the *defining set* $\mathcal{A}$ of the bicyclic code. In the language of polynomials, a bicyclic code is the set of bivariate polynomials $c(x, y)$ of componentwise degree at most $(n' - 1, n'' - 1)$, such that

$$C_{j'j''} = 0 \quad (j', j'') \in \mathcal{A},$$

where $C_{j'j''} = c(\beta^{j'}, \gamma^{j''})$. Clearly, a bicyclic code is a linear code.

To define an $(n'n'', k)$ code on the bicyclic plane over $GF(q)$, let n' and n'' be divisors of $q - 1$. Select a set of $n'n'' - k$ components of the two-dimensional n' by n'' Fourier transform to be a (two-dimensional) defining set, denoted $\mathcal{A}$, and constrain these $n'n'' - k$ components of the bispectrum to be zero. Figure 6.1 provides an example of a defining set in an array with $n' = n'' = 7$. The remaining k components of the bispectrum can be filled with any k symbols from $GF(q)$, and the two-dimensional inverse Fourier transform then gives the codeword corresponding to those given symbols. Indeed, this assignment could be the encoding rule, with the k unconstrained components of C filled with the k data symbols.

The same bicyclic code can be stated in either of two ways. One way is as follows:

$$C = \{c = [c_{i'i''}] \mid c(\omega^{j'}, \omega^{j''}) = 0 \text{ for } (j', j'') \in \mathcal{A}\}$$

0	0					
				0		
				0		
			0			
		0				
	0	0				

Figure 6.1. Defining set in two dimensions.

(which is in the form $cH^T = 0$); the other is

$$
\mathcal{C} = \left\{ c = \left[\frac{1}{n'n''} C(\omega^{-i'}, \omega^{-i''}) \right] \mid C_{j'j''} = 0 \text{ for } (j',j'') \in \mathcal{A} \right\}
$$

(which is in the form $c = aG$).

Recall that a cyclic code is a set of vectors with the property that it is closed under cyclic shifts. Likewise, a bicyclic code is a set of n' by n'' arrays that is closed under both cyclic shifts in the row direction and cyclic shifts in the column direction. Once again, the bicyclic codes take their name from this property, although we do not regard the property as especially important. Rather, the important property of the bicyclic codes is the relationship between the code-domain representation and the transform-domain representation, as defined by the Fourier transform.

The two-dimensional Fourier transform can be computed in either order:

$$
C_{j'j''} = \sum_{i'=0}^{n'-1} \beta^{i'j'} \sum_{i''=0}^{n''-1} \gamma^{i''j''} c_{i'i''}
$$

$$
= \sum_{i''=0}^{n''-1} \gamma^{i''j''} \sum_{i'=0}^{n'-1} \beta^{i'j'} c_{i'i''}.
$$

Define two intermediate arrays b and B as follows:

$$
b_{j'i''} = \sum_{i'=0}^{n'-1} \beta^{i'j'} c_{i'i''};
$$

$$
B_{i'j''} = \sum_{i''=0}^{n''-1} \gamma^{i''j''} c_{i'i''}.
$$

Then C can be computed from either b or B as follows:

$$C_{j'j''} = \sum_{i''=0}^{n''-1} \gamma^{i''j''} b_{j'i''};$$

$$C_{j'j''} = \sum_{i'=0}^{n'-1} \beta^{i'j'} B_{i'j''}.$$

This set of equations can be represented by the following diagram:

$$
\begin{array}{ccc}
c & \longleftrightarrow & B \\
\updownarrow & & \updownarrow \\
b & \longleftrightarrow & C,
\end{array}
$$

where a horizontal arrow denotes a one-dimensional Fourier transform relationship along every row of the array and a vertical arrow denotes a one-dimensional Fourier transform relationship along every column of the array. The rows of the array B are the spectra of the rows of c (viewed as row codewords). The columns of b are the spectra of the columns of c (viewed as column codewords). Because b and B, in effect, are also related by a two-dimensional Fourier transform, one might also regard b as a codeword (of a different, noncyclic code) and B as its bispectrum.

It is well known that if n' and n'' are coprime, then an n' by n'' bicyclic code is equivalent to a cyclic code. To form the bicyclic codewords from the cyclic codewords, simply read down the extended diagonal. Because $\text{GCD}[n', n''] = 1$, the extended diagonal $\{(i(\bmod n'), i(\bmod n'')) \mid i = 0, \ldots, n - 1\}$ passes once through every element of the array. Likewise, a cyclic code of blocklength $n'n''$ is equivalent to a bicyclic code. One way to map from the cyclic code into the bicyclic code is simply to write the components of the cyclic code, in order, down the extended diagonal of the n' by n'' array. This relationship between the cyclic form and the bicyclic form of such a code, when n' and n'' are coprime, can be formally described by the chinese remainder theorem. The relationship between the one-dimensional spectrum and the two-dimensional spectrum can be described by the Good–Thomas algorithm. Specifically, the codeword index i is replaced by $i' = i(\bmod n')$ and $i'' = i(\bmod n'')$. The index i can be recovered from (i', i'') by using the expression

$$i = N''n''i' + N'n'i'' \ (\bmod n),$$

where the integers N' and N'', sometimes called *Bézout coefficients*, are those satisfying $N'n' + N''n'' = 1$. Further, the bispectrum indices are given by

$$j' = N''j \ (\bmod n')$$

and

$$j'' = N'j \pmod{n''}.$$

In this way, any cyclic code of blocklength $n = n'n''$ with n' and n'' coprime can be represented as a bicyclic code.

6.2   Codes on the affine plane and the projective plane

A *primitive bicyclic code* over $GF(q)$ has blocklength $n = (q-1)^2$. By appending additional rows and columns, a linear code of blocklength $n = q^2$ can be described in a natural way as an extended bicyclic code. We shall describe such codes more directly, and more elegantly, in terms of the evaluation of bivariate polynomials.

The *affine plane* over the finite field $GF(q)$, denoted $GF(q)^2$, consists of the set of all pairs of elements of $GF(q)$. The *bicyclic plane* over the finite field $GF(q)$, denoted $GF(q)^{*2}$, is the set of all pairs of nonzero elements of $GF(q)$. The bicyclic plane has the structure of a torus. The *projective plane* over the finite field $GF(q)$, denoted $GF(q)^{2+}$ or $\boldsymbol{P}^2(GF(q))$, is the set of triples (β, γ, δ) of elements of $GF(q)$ such that the rightmost nonzero element of the triple is a one. The point $(0,0,0)$ is not part of the projective plane. Thus by going from the affine plane into the projective plane, the points (β, γ) are replaced by the points $(\beta, \gamma, 1)$, and new points $(\beta, 1, 0)$ and $(1, 0, 0)$ are created. Each point with $z = 0$ is called a *"point at infinity."* The set of points at infinity is called the *"line at infinity."* The set of points of the projective plane that are not points at infinity forms a copy of the affine plane within the projective plane. The points of the affine plane are called *affine points*.

The projective plane has more points than the affine plane or the bicyclic plane, but it also has a more cumbersome structure. The bicyclic plane has fewer points than the affine plane or the projective plane, but it has the simplest structure, which is the structure of a torus. Often, it is helpful to think in terms of the projective plane, even though the applications may be in the affine plane or the bicyclic plane. Other times, it is simpler to think in terms of the bicyclic plane.

Let $C(x, y)$ be a bivariate polynomial of componentwise degree at most $(n-1, n-1)$. We can regard the coefficients of $C(x, y)$ as a bispectrum $\boldsymbol{C}$, with components $C_{j'j''}$ for $j' = 0, \ldots, n-1$ and $j'' = 0, \ldots, n-1$. The array $\boldsymbol{c}$ is obtained by the two-dimensional inverse Fourier transform

$$c_{i'i''} = \frac{1}{n^2} \sum_{j'=0}^{n-1} \sum_{j''=0}^{n-1} C_{j'j''} \omega^{-i'j'} \omega^{-i''j''},$$

which is the same as the array obtained by evaluating the polynomial $C(x, y)$ at all pairs of reciprocal powers of ω,

$$c_{i'i''} = \frac{1}{n^2} C(\omega^{-i'}, \omega^{-i''}).$$

Evaluating bivariate polynomials in this way is slightly stronger, in one sense, than is the two-dimensional Fourier transform, because one can also evaluate $C(x, y)$ at the points with $x = 0$ or $y = 0$. The array c then has q^2 components. To define a code on the affine plane, choose a fixed set of bi-indices as the defining set $\mathcal{A}$. Let

$$\mathcal{S} = \{C(x, y) \mid C_{j'j''} = 0 \text{ for } (j', j'') \in \mathcal{A}\}.$$

The code on the affine plane over $GF(q)$ is defined as

$$\mathcal{C} = \left\{ c = \frac{1}{n^2}[C(\beta, \gamma)] \quad \beta, \gamma \in GF(q) \mid C(x, y) \in \mathcal{S} \right\}.$$

Thus polynomial $C(x, y)$ is evaluated at every point of the affine plane. The bicyclic code, then, is the restriction of the affine code to the bicyclic plane.

To extend the code by $q + 1$ additional components, define the code on the projective plane. Replace $C(x, y)$ by the homogeneous trivariate polynomial $C(x, y, z)$ of the form

$$C(x, y, z) = \sum_{j'=0}^{q-1} \sum_{j''=0}^{q-1} C_{j'j''} x^{j'} y^{j''} z^{J-j'-j''},$$

where J is the largest degree of any $C(x, y)$ in S. Redefine $\mathcal{S}$ as a set of homogeneous trivariate polynomials,

$$\mathcal{S} = \{C(x, y, z) \mid C_{j'j''} = 0 \text{ for } (j', j'') \in \mathcal{A}\}.$$

The code in the projective plane is defined as

$$\mathcal{C} = \left\{ c = \frac{1}{n^2}[C(\beta, \gamma, \delta)] \mid C(x, y, z) \in \mathcal{S} \right\},$$

where (β, γ, δ) ranges over the points of the projective plane. Because the projective plane has $q^2 + q + 1$ points, the extended code has blocklength $n = q^2 + q + 1$. The blocklength of the code on the projective plane is larger than the blocklength of the code on the affine plane, which is q^2.

6.3   Minimum distance of bicyclic codes

The weight of an individual n by n array is related to the pattern of zeros of its two-dimensional Fourier transform, as was studied in Section 5.7. We can choose the pattern of zeros to ensure that the weight of the array is large. This relationship can be used to define a code as the set of all arrays with a given set of bispectral zeros. Statements relating $d_{\min}$ to the defining set $\mathcal{A}$ can be made directly from the bounds on the weight of an array that were given in Section 5.7.

Two examples of bicyclic codes are *product codes* and *dual-product codes*. The defining set of a product code consists of all elements of selected rows and all elements of selected columns of the array. The defining set of a dual-product code is the complement of the defining set of a product code. Figure 6.2 shows examples of defining sets for a product code and a dual-product code. On the left, the defining set gives a product code. It is the product of two cyclic codes; the defining set consists of rows and columns of the array. On the right, the defining set gives the dual of a product code, which, by the BCH dual-product bound given in Section 5.7, has minimum distance 4.

A two-dimensional code that is designed to fit the BCH product bound is called a *BCH product code*, or, if the symbol field and the locator field are the same, a *Reed–Solomon product code*. The bispectrum of a $(225, 169, 9)$ Reed–Solomon product code over $GF(16)$ is illustrated in Figure 6.3.

The product code illustrated in Figure 6.3 is the product of two $(15, 13, 3)$ Reed–Solomon codes over $GF(16)$. Each component code has minimum distance 3, so the product code has minimum distance 9. To see the strength of the truncated BCH product bound of Section 5.7, consider reducing the defining set of this example. The bispectrum has two consecutive rows equal to zero, and two consecutive columns equal to zero. But the truncated BCH product bound says that to ensure the weight of a vector is at least 9, it is enough to have only eight consecutive zeros in each of these rows and columns. This means that there is a $(225, 197, 9)$ code over $GF(16)$ with the defining set shown in Figure 6.4.

Figure 6.2. Examples of defining sets.

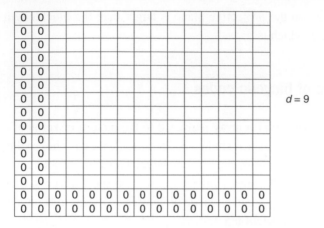

Figure 6.3. Defining set for a $(225, 169, 9)$ Reed–Solomon product code.

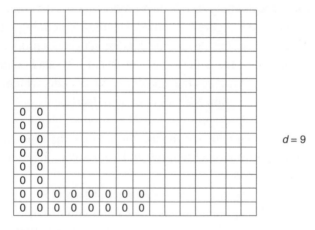

Figure 6.4. Defining set for a $(225, 197, 9)$ code.

The dual of a product code, called a dual-product code, can also be studied with the aid of the dual-product bound. In the two-dimensional array of bispectral components, choose an a by b rectangular subarray as the two-dimensional defining set of the code. Any codeword of weight $\min(a, b)$ or less must have a bispectrum that is zero everywhere in any horizontal or vertical stripe passing through the rectangle of check frequencies. This implies, in turn, that the bispectrum is zero everywhere; therefore the codeword is the all-zero codeword. Consequently,

$$d_{\min} \geq 1 + \min(a, b) = \min(a + 1, b + 1).$$

Hence the example gives a $(49, 45, 3)$ code over $GF(8)$. The binary subfield-subcode is a $(49, 39)$ $d \geq 3$ code. A dual-product code does not have a large minimum distance.

The bispectral zeros of the dual-product code can always be chosen so that the defining set is given by

$$\mathcal{A} = \{(j',j'') \mid j' = 0,\ldots,a-1; j'' = 0,\ldots,b-1\},$$

which is a cascade set. Now the minimum distance can be seen to be a consequence of the multilevel bound,

$$d_{\min} \geq \min_{(j',j'')\notin\mathcal{A}} (j'+1)(j''+1),$$

which reduces to the expression $d_{\min} \geq \min(a+1,b+1)$ given earlier. Any cascade set can be regarded as the union of rectangles, so a cascade defining set can be regarded as a union of rectangular defining sets. In this way, the multilevel bound is then seen to be a generalization of the dual-product bound.

For example, if $\mathcal{C}_1$ and $\mathcal{C}_2$ are each a dual-product code with bispectral zeros in sets $\mathcal{A}_1$ and $\mathcal{A}_2$, defined as above, and $\mathcal{C} = \mathcal{C}_1 \cap \mathcal{C}_2$, then this code has bispectral zeros for all (j',j'') in the set $\mathcal{A} = \mathcal{A}_1 \cup \mathcal{A}_2$, which again is a cascade set. The minimum distance satisfies

$$d_{\min} \geq \min_{(j',j'')\notin\mathcal{A}} (j'+1)(j''+1)$$

by the multilevel bound.

In the following two sections, we shall discuss two examples of bicyclic codes, namely the hyperbolic codes and other bicyclic codes based on the BCH bound. The minimum distances of these codes are not noteworthy, though they may have other desirable attributes. Two-dimensional codes with good minimum distances can be obtained by *puncturing* or *shortening*. In Chapter 10, we shall discuss a powerful method of puncturing (or shortening) that uses a bivariate polynomial to define a set of points in the plane, which will define the components of a punctured or shortened bicyclic code.

6.4   Bicyclic codes based on the multilevel bound

The two examples of the defining sets of bicyclic codes that we have illustrated in Section 6.3 are both cascade sets. A *cascade code* is a two-dimensional code whose defining set is a cascade set. A general example of such a defining set is shown

Figure 6.5. Defining set for a cascade code.

in Figure 6.5. The bispectrum of the cascade code corresponding to the cascade set of Figure 6.5 has the following form:

$$
C = \begin{bmatrix}
0 & 0 & 0 & 0 & 0 & C_{0,5} & C_{0,6} \\
0 & 0 & 0 & 0 & 0 & C_{1,5} & C_{1,6} \\
0 & 0 & 0 & C_{2,3} & C_{2,4} & C_{2,5} & C_{2,6} \\
0 & C_{3,1} & C_{3,2} & C_{3,3} & C_{3,4} & C_{3,5} & C_{3,6} \\
0 & C_{4,1} & C_{4,2} & C_{4,3} & C_{4,4} & C_{4,5} & C_{4,6} \\
0 & C_{5,1} & C_{5,2} & C_{5,3} & C_{5,4} & C_{5,5} & C_{5,6} \\
C_{6,0} & C_{6,1} & C_{6,2} & C_{6,3} & C_{6,4} & C_{6,5} & C_{6,6}
\end{bmatrix} .
$$

Standard matrix notation requires that the row with $j' = 0$ be written at the top, and that $C_{j'j''}$ be the entry in row j' and column j''. In contrast, the illustration of the cascade set, with indices arranged in the usual pattern of a cartesian coordinate system, shows a reflection of the visual pattern of zeros.

The inverse Fourier transform of any row of this matrix is a Reed–Solomon codeword. The set of inverse Fourier transforms of any row of all such matrices is a Reed–Solomon code. The BCH bound gives the minimum weight of each of these Reed–Solomon codes. The intermediate array, then, consists of rows that are codewords of different Reed–Solomon codes. The multilevel bound can then be obtained by applying the BCH bound to each column of the intermediate array.

A bicyclic code designed to exploit the multilevel bound

$$
d_{\min} \geq \min_{(j',j'') \notin A} (j' + 1)(j'' + 1)
$$

is called a *hyperbolic code*. The name derives from the fact that $(x + 1)(y + 1) = d$ is the equation of a hyperbola.

Definition 6.4.1 *A hyperbolic code with designed distance d is a bicyclic code with defining set given by*

$$
A = \{(j',j'') \mid (j' + 1)(j'' + 1) < d\}.
$$

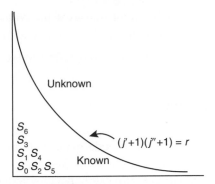

Figure 6.6. Syndromes for a hyperbolic code.

0					
0					
0					
0	0				
0	0	0			
0	0	0	0	0	0

Figure 6.7. Defining set for a hyperbolic code.

The defining set of a hyperbolic code is bounded by a hyperbola, as illustrated in Figure 6.6.

Proposition 6.4.2 *The minimum distance d of a hyperbolic code is at least as large as its designed distance.*

Proof: An obvious combination of the multilevel bound with the statement of the theorem yields

$$d_{\min} \geq \min_{j' j'' \notin A} (j' + 1)(j'' + 1)$$

$$= \min_{(j'+1)(j''+1) \geq d} (j' + 1)(j'' + 1)$$

$$= d,$$

which proves the proposition. ∎

For example, the defining set for a $(49, 35, 7)$ hyperbolic code over $GF(8)$, with $d = 7$, is shown in the bispectrum of Figure 6.7. This hyperbolic code, when judged solely by dimension and minimum distance, is inferior to the $(63, 51, 7)$ BCH code over $GF(8)$. The comparison is less clear for decoders that decode beyond the minimum distance. The hyperbolic code also has the minor feature that for a code of blocklength

$(q-1)^2$ over $GF(q)$, the computations of the decoding algorithm are in the symbol field $GF(q)$; it is not necessary to introduce an extension field for the decoding algorithm.

6.5   Bicyclic codes based on the BCH bound

A bivariate code may be preferred to a long univariate code even if its minimum distance is less. This is because the decoding complexity may be much less, and the code may be able to correct a great many error patterns well beyond the packing radius, which more than compensates for the smaller packing radius. A bicyclic code may be designed by repeatedly using the BCH bispectrum property, described in Section 5.7. The defining set may be rather irregular, consisting of the union of enough consecutive runs in various directions that, taken together, ensure the minimum distance of the code through repeated application of the BCH bispectrum property.

An example of such a code is the $(49, 39, 5)$ code over $GF(8)$, whose bispectrum is given in Figure 6.8. The defining set of this code is as follows:

$$\mathcal{A} = \{(0, 1), (0, 2), (0, 3), (0, 4), (1, 2), (2, 2), (2, 3), (3, 2), (3, 4), (6, 3)\}.$$

To see that the minimum distance of this code is at least 5, suppose that a codeword c has weight less than 5. Then we can use the BCH bispectrum property of Section 5.7 to conclude that the top row of the bispectrum is zero everywhere, as in part (a) of Figure 6.9. Then we conclude that column two is zero everywhere, as in part (b) of Figure 6.9. Next, we conclude that the general diagonal $(j', j'') = (0, 1) + (1, 6)j$ is zero everywhere, as in part (c) of Figure 6.9. We continue in this way to conclude that the general diagonal $(j', j'') = (2, 2) + (1, 4)j$ is zero everywhere, as in part (d) of Figure 6.9. Continuing with the steps, as shown in parts (e), (f), and beyond, we eventually find that the entire array is zero. Hence if the weight of c is less than 5, then c is the all-zero array. Thus the minimum distance of this code is at least 5.

A senseword $v = c + e$ for a bicyclic code with defining set $\mathcal{A}$ is decoded by evaluating $v(x, y)$ at $(\omega^{j'}, \omega^{j''})$ for $(j', j'') \in \mathcal{A}$. The *two-dimensional syndrome* of the

0	0	0	0		
0					
0	0				
0			0		
		0			

Figure 6.8. Defining set for a $(49, 39, 5)$ code.

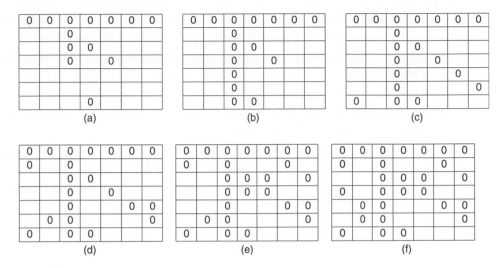

Figure 6.9. Inferring the spectrum from a few of its components.

error pattern $e_{i'i''}$ is defined as

$$S_{j'j''} = \sum_{i'=0}^{n-1} \sum_{i''=0}^{n-1} e_{i'i''} \omega^{i'j'} \omega^{i''j''} \quad (j',j'') \in \mathcal{A},$$

which can be computed from the two-dimensional senseword v. The task, then, is to recover the error pattern from the two-dimensional syndromes.

A pattern of v errors can be described in terms of row locators, column locators, and error magnitudes. The ℓth error, lying in row i'_ℓ and column i''_ℓ, has a row locator, defined as $X_\ell = \omega^{i'_\ell}$, and a column locator, defined as $Y_\ell = \omega^{i''_\ell}$. The ℓth error magnitude is defined as $Z_\ell = e_{i'_\ell i''_\ell}$. Then the (j',j'') syndrome can be written more compactly as

$$S_{j'j''} = \sum_{\ell=1}^{v} Z_\ell X_\ell^{j'} Y_\ell^{j''} \quad (j',j'') \in \mathcal{A}.$$

This set of nonlinear equations can always be solved for X_ℓ, Y_ℓ, and Z_ℓ for $\ell = 1, \ldots, v$ when v is not larger than the packing radius t. Tractable algorithms are known only for special $\mathcal{A}$, as in the previous example.

Thus suppose that we choose the defining set $\mathcal{A}$ to include $\{(1,1)\}, (1,2), \ldots, (1,2t)\}$. The syndromes are

$$S_{11} = Z_1 X_1 Y_1 + Z_2 X_2 Y_2 + \cdots + Z_v X_v Y_v = E_{11},$$
$$S_{12} = Z_1 X_1 Y_1^2 + Z_2 X_2 Y_2^2 + \cdots + Z_v X_v Y_v^2 = E_{12},$$
$$\vdots$$
$$S_{1,2t} = Z_1 X_1 Y_1^{2t} + Z_2 X_2 Y_2^{2t} + \cdots + Z_v X_v Y_v^{2t} = E_{1,2t}.$$

With the substitution $W_\ell = Z_\ell X_\ell$, this set of equations is familiar from the decoding of a Reed–Solomon code. There is a difference, however; here the Y_ℓ need not be distinct because several errors might occur in the same column. It is a simple matter, however, to observe that the terms with the same Y_ℓ combine to obtain a similar set of equations with possibly smaller ν, and this smaller set also satisfies $\nu \le t$. Then the Berlekamp–Massey algorithm, followed by recursive extension, will yield $E_{11}, E_{12}, \ldots, E_{1n}$, the entire first row of the error bispectrum.

In general, if $\nu \le t$ errors occur, then, for any integers j_0', j_0'' and a', a'', the syndromes $S_{j_0'+a'k, j_0''+a''k}$ for $k = 1, \ldots, 2t$ uniquely determine the entire line of syndromes $S_{j_0'+ak, j_0''+a'k}$ for $k = 1, \ldots, n$. To compute this line of syndromes, let $\overline{Y}_{\ell'}$ for $\ell' = 1, \ldots, \nu'$ denote the distinct terms $X_\ell^{a'} Y_\ell^{a''}$ over ℓ. Then

$$S_{j_0'+a'k, j_0''+a''k} = \sum_{\ell=1}^{\nu} Z_\ell X_\ell^{j_0'+a'k} Y_\ell^{j_0''+a''k}$$

$$= \sum_{\ell'=1}^{\nu'} \overline{X}_{\ell'} \overline{Y}_{\ell'}^{k} \qquad k = 1, \ldots, 2t,$$

where $\nu' \le \nu \le t$, and $\overline{X}_{\ell'}$ denotes the sum of the factors multiplying $\overline{Y}_{\ell'}^{k}$ in each equation. The Berlekamp–Massey algorithm, followed by recursive extension, will produce all the syndromes in this line. In particular, any $2t$ consecutive syndromes in a straight line (horizontal, vertical, or at any angle) can be extended to all syndromes in that line. Repeated applications of this process complete the decoding.

6.6   The (21, 12, 5) bicyclic BCH code

To obtain a binary double-error-correcting BCH code of blocklength 21, choose $\omega = \alpha^3$ in the extension field $GF(64)$. Then ω has order 21, so we have a Fourier transform of blocklength 21. The conjugacy classes modulo 21 of ω are as follows:

$\{\omega^0\}$,

$\{\omega^1, \omega^2, \omega^4, \omega^8, \omega^{16}, \omega^{11}\}$,

$\{\omega^3, \omega^6, \omega^{12}\}$,

$\{\omega^5, \omega^{10}, \omega^{20}, \omega^{19}, \omega^{17}, \omega^{13}\}$,

$\{\omega^7, \omega^{14}\}$,

$\{\omega^9, \omega^{18}, \omega^{15}\}$,

and these partition the spectral components into the following chords:

$\{C_0\}$,

$\{C_1, C_2, C_4, C_8, C_{16}, C_{11}\}$,

$\{C_3, C_6, C_{12}\}$,

$\{C_5, C_{10}, C_{20}, C_{19}, C_{17}, C_{13}\}$,

$\{C_7, C_{14}\}$,

$\{C_9, C_{18}, C_{15}\}$.

To satisfy the BCH bound for a distance-5 code, choose C_1 and C_3 equal to zero, which makes all conjugates of C_1 and C_3 also equal to zero. The other spectral components are arbitrary, except for the conjugacy constraint that $C_j^2 = C_{2j}$. This constraint implies that $C_0 \in GF(2)$, $C_5 \in GF(64)$, $C_7 \in GF(4)$, and $C_9 \in GF(8)$. Otherwise, these components can be specified arbitrarily. All other spectral components are determined by the conjugacy constraints.

The $(21, 12, 5)$ BCH code can also be described as a bicyclic code. As such, it is cyclic in both the row direction and the column direction. Then it has the form of a set of three by seven, two-dimensional binary arrays. The codeword c, with indices i' and i'', is a three by seven, two-dimensional array of the form

$$c = \begin{bmatrix} c_{0,0} & c_{0,1} & c_{0,2} & c_{0,3} & c_{0,4} & c_{0,5} & c_{0,6} \\ c_{1,0} & c_{1,1} & c_{1,2} & c_{1,3} & c_{1,4} & c_{1,5} & c_{1,6} \\ c_{2,0} & c_{2,1} & c_{2,2} & c_{2,3} & c_{2,4} & c_{2,5} & c_{2,6} \end{bmatrix}.$$

The bispectrum C of the bicyclic codeword c has the form

$$C = \begin{bmatrix} C_{0,0} & C_{0,1} & C_{0,2} & C_{0,3} & C_{0,4} & C_{0,5} & C_{0,6} \\ C_{1,0} & C_{1,1} & C_{1,2} & C_{1,3} & C_{1,4} & C_{1,5} & C_{1,6} \\ C_{2,0} & C_{2,1} & C_{2,2} & C_{2,3} & C_{2,4} & C_{2,5} & C_{2,6} \end{bmatrix}.$$

The bispectrum is in the field $GF(64)$, because that field is the smallest field containing both an element of order 3 and an element of order 7.

To obtain a bicyclic binary BCH double-error-correcting code of blocklength 21, let α be a primitive element in the extension field $GF(64)$, and let $\beta = \alpha^{21}$ and $\gamma = \alpha^9$. Then (β, γ) has *biorder* three by seven, so we have a three by seven two-dimensional Fourier transform. Because the code is binary, the components of the bispectrum satisfy the two-dimensional conjugacy constraint:

$$C_{j'j''}^2 = C_{((2j')),((2j''))}.$$

This constraint breaks the bispectrum into two-dimensional conjugacy classes. The first index is interpreted modulo 7; the second, modulo 3. The two-dimensional conjugacy classes modulo 7 by 3 can be used to partition the components of the array C into two-dimensional chords, just as the one-dimensional conjugacy classes modulo 21 can be used to partition the components of the vector C into one-dimensional chords. The two-dimensional and one-dimensional chords are as follows:

$\{C_{0,0}\}$,

$\{C_{1,1}, C_{2,2}, C_{1,4}, C_{2,1}, C_{1,2}, C_{2,4}\}$,

$\{C_{0,3}, C_{0,6}, C_{0,5}\}$,

$\{C_{1,3}, C_{2,6}, C_{1,5}, C_{2,3}, C_{1,6}, C_{2,5}\}$,

$\{C_{0,1}, C_{0,2}, C_{0,4}\}$,

$\{C_{1,0}, C_{2,0}\}$,

$\{C_0\}$,

$\{C_1, C_2, C_4, C_8, C_{16}, C_{11}\}$,

$\{C_3, C_6, C_{12}\}$,

$\{C_5, C_{10}, C_{20}, C_{19}, C_{17}, C_{13}\}$,

$\{C_9, C_{18}, C_{15}\}$,

$\{C_7, C_{14}\}$.

The entries in the two columns are equivalent, related by the chinese remainder theorem, and portray two ways of representing the same 21-point vector.

The conjugacy constraint implies that $C_{0,0} \in GF(2)$, $C_{0,1} \in GF(8)$, $C_{1,1} \in GF(64)$, $C_{1,3} \in GF(64)$, $C_{1,0} \in GF(4)$, and $C_{0,3} \in GF(8)$. All other bispectral components are implied by the conjugacy constraint. The resulting code is a $(21, 12, 5)$ bicyclic binary BCH code. To satisfy the BCH bound for a distance-5 code, we shall choose $C_{1,1} = C_{0,3} = 0$, which makes all conjugates of $C_{1,1}$ and $C_{0,3}$ zero also. The other bispectral components are arbitrary, except that they must satisfy the conjugacy constraint. The bispectrum, then, can be rewritten as follows:

$$C = \begin{bmatrix} C_{0,0} & C_{0,1} & C_{0,1}^2 & 0 & C_{0,1}^4 & 0 & 0 \\ C_{1,0} & 0 & 0 & C_{1,3} & 0 & C_{1,3}^4 & C_{1,3}^{16} \\ C_{1,0}^2 & 0 & 0 & C_{1,3}^8 & 0 & C_{1,3}^{32} & C_{1,3}^2 \end{bmatrix}$$

$$= \begin{bmatrix} C_0 \\ C_1 \\ C_2 \end{bmatrix}.$$

The two-dimensional codeword c is obtained by taking a two-dimensional inverse Fourier transform of C. This consists of taking either the seven-point inverse Fourier transform of all rows of c, then taking the three-point inverse Fourier transform of all columns, or first taking the three-point inverse Fourier transform of all columns of c, then taking the seven-point inverse Fourier transform of all rows.

A superficial inspection of the bispectrum C immediately tells us much about the structure of a codeword. The BCH bound says that because there are four consecutive zeros down the extended diagonal ($C_{1,1} = C_{2,2} = C_{0,3} = C_{1,4} = 0$), the weight of a nonzero codeword is at least 5. Furthermore, unless $C_{0,1}$ is zero, there are at least three nonzero rows in a codeword, because then there are two consecutive zeros in a

Table 6.1. *Weight distribution of the* $(21, 12, 5)$ *BCH code*

ℓ	A_ℓ
0 or 21	1
1 or 20	0
2 or 19	0
3 or 18	0
4 or 17	0
5 or 16	21
6 or 15	168
7 or 14	360
8 or 13	210
9 or 12	280
10 or 11	1008

nonzero column of the bispectrum. Finally, because the top row of C is the spectrum of a Hamming codeword, the column sums of c must form a Hamming codeword. In particular, if c has odd weight, either three or seven columns must have an odd number of ones. Therefore a codeword of weight 5 has exactly three columns with an odd number of ones.

We will show later that if c is a bicyclic BCH codeword of weight 5, then every row of this array has odd weight, and if c is a codeword of weight 6, then there are two rows of the array that have odd weight. This means that appending a check sum to each row triply extends the code to a $(24, 12, 8)$ code.

The weight distribution of the $(21, 12, 5)$ binary BCH code is given in Table 6.1.

6.7 The Turyn representation of the (21, 12, 5) BCH code

The binary $(21, 12, 5)$ bicyclic BCH code can be represented as a linear combination of three binary $(7, 4, 3)$ Hamming codewords. This representation, known as the *Turyn representation*, is given by the concatenation of three sections,

$$c = | \, c_0 \, | \, c_1 \, | \, c_2 \, |,$$

where

$$\begin{bmatrix} c_0 \\ c_1 \\ c_2 \end{bmatrix} = \begin{bmatrix} 1 & 0 & 1 \\ 1 & 1 & 0 \\ 1 & 1 & 1 \end{bmatrix} \begin{bmatrix} b_0 \\ b_1 \\ b_2 \end{bmatrix}.$$

The vectors b_1 and b_2 are any two – possibly the same – codewords from the binary $(7, 4, 3)$ Hamming code, with spectra given by

$$B_1 = [B_{1,0} \ 0 \ 0 \ B_{1,3} \ 0 \ B_{1,3}^4 \ B_{1,3}^2],$$
$$B_2 = [B_{2,0} \ 0 \ 0 \ B_{2,3} \ 0 \ B_{2,3}^4 \ B_{2,3}^2],$$

and b_0 is any codeword from the reciprocal binary $(7, 4, 3)$ Hamming code, with spectrum

$$B_0 = [B_{0,0} \ B_{0,1} \ B_{0,1}^2 \ 0 \ B_{0,1}^4 \ 0 \ 0].$$

The components $B_{0,0}, B_{1,0}$, and $B_{2,0}$ are elements of $GF(2)$, while the components $B_{0,1}, B_{1,3}$, and $B_{2,3}$ are elements of $GF(8)$.

The three Hamming codewords can be recovered from c as follows:

$$\begin{bmatrix} b_0 \\ b_1 \\ b_2 \end{bmatrix} = \begin{bmatrix} 1 & 0 & 1 \\ 1 & 1 & 0 \\ 1 & 1 & 1 \end{bmatrix}^{-1} \begin{bmatrix} c_0 \\ c_1 \\ c_2 \end{bmatrix} = \begin{bmatrix} 1 & 1 & 1 \\ 1 & 0 & 1 \\ 0 & 1 & 1 \end{bmatrix} \begin{bmatrix} c_0 \\ c_1 \\ c_2 \end{bmatrix}.$$

A hint of the Turyn representation appears in the structure of the bispectrum C of the bicyclic BCH code given in Section 6.7 as

$$C = \begin{bmatrix} C_{0,0} & C_{0,1} & C_{0,1}^2 & 0 & C_{0,1}^4 & 0 & 0 \\ C_{1,0} & 0 & 0 & C_{1,3} & 0 & C_{1,3}^4 & C_{1,3}^{16} \\ C_{1,0}^2 & 0 & 0 & C_{1,3}^8 & 0 & C_{1,3}^{32} & C_{1,3}^2 \end{bmatrix} = \begin{bmatrix} C_0 \\ C_1 \\ C_2 \end{bmatrix}.$$

The top row, denoted C_0, is the spectrum of a reciprocal $(7, 4, 3)$ Hamming codeword. The middle and bottom rows, denoted C_1 and C_2, resemble the spectrum of a $(7, 4, 3)$ Hamming codeword, except the elements are from the field $GF(64)$ instead of $GF(8)$.

To put the rows of the array C in the form of Hamming spectra, we will write $GF(4)$ and $GF(64)$ as extensions of $GF(2)$ and $GF(8)$, respectively. Let β be a zero of the polynomial $x^2 + x + 1$, which is irreducible over $GF(2)$. Then

$$GF(4) = \{a + \beta b \mid a, b \in GF(2)\}$$

and

$$GF(64) = \{a + \beta b \mid a, b \in GF(8)\},$$

where a can be called the "real part" and b can be called the "imaginary part" of the element of $GF(4)$ or of $GF(64)$. Then each element of C_1 and C_2 can be broken into a real part and an imaginary part, sum that

$$
C = \begin{bmatrix} C_0 \\ C_1 \\ C_2 \end{bmatrix}
$$

$$
= \begin{bmatrix} C_{0R} \\ C_{1R} \\ C_{2R} \end{bmatrix} + \beta \begin{bmatrix} 0 \\ C_{1I} \\ C_{2I} \end{bmatrix}.
$$

Let C_1^8 denote a row whose elements, componentwise, are the eighth powers of the elements of C_1. This row is then equal to row C_2; so we have

$$
C_2 = C_1^8 = (C_{1R} + \beta C_{1I})^8 = C_{1R}^8 + \beta^8 C_{1I}^8
$$
$$
= C_{1R} + \beta^2 C_{1I},
$$

where we have used the facts that $\beta^3 = 1$ and $a^8 = a$ for any element a of $GF(8)$. Therefore

$$
C = \begin{bmatrix} C_0 \\ C_1 \\ C_2 \end{bmatrix} = \begin{bmatrix} C_0 \\ C_{1R} + \beta C_{1I} \\ C_{1R} + \beta^2 C_{1I} \end{bmatrix} = \begin{bmatrix} 1 & 0 & 0 \\ 0 & 1 & \beta \\ 0 & 1 & \beta^2 \end{bmatrix} \begin{bmatrix} C_0 \\ C_{1R} \\ C_{1I} \end{bmatrix}.
$$

Next, referring to the diagram

$$
\begin{array}{ccc}
c & \longleftrightarrow & B \\
\updownarrow & & \updownarrow \\
b & \longleftrightarrow & C,
\end{array}
$$

we will show that B is an array of three rows, each of which is a Hamming codeword spectrum.

First, compute the three-point inverse Fourier transform of each column of C. Thus

$$
\begin{bmatrix} B_0 \\ B_1 \\ B_2 \end{bmatrix} = \begin{bmatrix} 1 & 1 & 1 \\ 1 & \beta^{-1} & \beta^{-2} \\ 1 & \beta^{-2} & \beta^{-1} \end{bmatrix} \begin{bmatrix} C_0 \\ C_1 \\ C_2 \end{bmatrix}.
$$

From the previous derivation, this becomes

$$
\begin{bmatrix} B_0 \\ B_1 \\ B_2 \end{bmatrix} = \begin{bmatrix} 1 & 1 & 1 \\ 1 & \beta^2 & \beta \\ 1 & \beta & \beta^2 \end{bmatrix} \begin{bmatrix} C_0 \\ C_1 \\ C_1^8 \end{bmatrix}
$$

$$
= \begin{bmatrix} 1 & 1 & 1 \\ 1 & \beta^2 & \beta \\ 1 & \beta & \beta^2 \end{bmatrix} \begin{bmatrix} 1 & 0 & 0 \\ 0 & 1 & \beta \\ 0 & 1 & \beta^2 \end{bmatrix} \begin{bmatrix} C_0 \\ C_{1R} \\ C_{1I} \end{bmatrix}
$$

$$
= \begin{bmatrix} 1 & 0 & 1 \\ 1 & 1 & 0 \\ 1 & 1 & 1 \end{bmatrix} \begin{bmatrix} C_0 \\ C_{1R} \\ C_{1I} \end{bmatrix}.
$$

Now the components of vectors C_{1R} and C_{1I} are in $GF(8)$ and are the spectra of binary Hamming codewords, and C_0 is the spectrum of a reciprocal binary Hamming codeword. Finally, take the inverse Fourier transform of each row vector on both sides of this equation. This is the horizontal arrow in the above diagram of Fourier transforms. This yields

$$
\begin{bmatrix} c_0 \\ c_1 \\ c_2 \end{bmatrix} = \begin{bmatrix} 1 & 0 & 1 \\ 1 & 1 & 0 \\ 1 & 1 & 1 \end{bmatrix} \begin{bmatrix} b_0 \\ b_1 \\ b_2 \end{bmatrix},
$$

as was asserted earlier.

6.8   The (24, 12, 8) bivariate Golay code

There are two cyclic Golay codes: the $(23, 12, 7)$ binary cyclic Golay code and the $(11, 6, 5)$ ternary cyclic Golay code, each of which can be extended by one symbol. We will consider only the $(23, 12, 7)$ binary cyclic Golay code and the $(24, 12, 8)$ extended binary Golay code. There are many ways of constructing the extended Golay code. The $(24, 12, 8)$ extended binary Golay code is traditionally obtained by extending the $(23, 12, 7)$ binary Golay code by a single bit. Here we will give an original method that constructs a $(24, 12, 8)$ code by appending one check bit to each row of the $(21, 12, 5)$ binary bicyclic BCH code. It is obliquely related to the Turyn representation of the Golay code. The triply extended binary BCH codeword has the following form:

$$
c^+ = \begin{bmatrix} c_0^+ \\ c_1^+ \\ c_2^+ \end{bmatrix} = \begin{bmatrix} c_{0,0} & c_{0,1} & c_{0,2} & c_{0,3} & c_{0,4} & c_{0,5} & c_{0,6} & c_0^+ \\ c_{1,0} & c_{1,1} & c_{1,2} & c_{1,3} & c_{1,4} & c_{1,5} & c_{1,6} & c_1^+ \\ c_{2,0} & c_{2,1} & c_{2,2} & c_{2,3} & c_{2,4} & c_{2,5} & c_{2,6} & c_2^+ \end{bmatrix}
$$
$$
= \begin{bmatrix} 1 & 0 & 1 \\ 1 & 1 & 0 \\ 1 & 1 & 1 \end{bmatrix} \begin{bmatrix} b_0^+ \\ b_1^+ \\ b_2^+ \end{bmatrix},
$$

where the plus superscript denotes an overall binary check symbol on each row.

As follows from the previous section, in this representation b_1^+ and b_2^+ are extended Hamming codewords of blocklength 8, and b_0^+ is an extended reciprocal Hamming codeword of blocklength 8. To show that the triply extended code has minimum weight 8, we will show that every codeword of the $(21, 12, 5)$ binary bicyclic BCH code of weight 5 must have three rows of odd weight, that every codeword of weight 6 must have two rows of odd weight, and (obviously) every codeword of weight 7 must have at least one row of odd weight. Then we can conclude that the $(24, 12, 8)$ binary triply extended BCH code is the $(24, 12, 8)$ extended binary Golay code because, as we have said but not proved, only one linear $(24, 12, 8)$ binary code exists. Thus we will conclude that the $(24, 12, 8)$ binary triply extended BCH code, with any single component deleted, becomes a cyclic code under a suitable permutation.

The bispectrum of the $(21, 12, 5)$ bicyclic BCH code is given by

$$
C = \begin{bmatrix} C_0 \\ C_1 \\ C_2 \end{bmatrix} = \begin{bmatrix} C_{0,0} & C_{0,1} & C_{0,1}^2 & 0 & C_{0,1}^4 & 0 & 0 \\ C_{1,0} & 0 & 0 & C_{1,3} & 0 & C_{1,3}^4 & C_{1,3}^{16} \\ C_{1,0}^2 & 0 & 0 & C_{1,3}^8 & 0 & C_{1,3}^{32} & C_{1,3}^2 \end{bmatrix}.
$$

Note that each row of the bispectrum C individually satisfies a Gleason–Prange condition. Each row either has zeros at all indices that are equal to a nonzero square modulo p, or it has zeros at all indices that are not equal to a nonzero square modulo p, where $p = 7$.

As an example of C, we write one codeword bispectrum of the $(21, 12, 5)$ bicyclic BCH code by setting $C_7 = 0$ and $C_0 = C_5 = C_9 = 1$. Then

$$
C = \begin{bmatrix} 1 & 1 & 1 & 0 & 1 & 0 & 0 \\ 0 & 0 & 0 & 1 & 0 & 1 & 1 \\ 0 & 0 & 0 & 1 & 0 & 1 & 1 \end{bmatrix}.
$$

Take the three-point inverse Fourier transform of each column, then take the seven-point inverse Fourier transform of each row to obtain the codeword c:

$$C = \begin{bmatrix} 1 & 1 & 1 & 0 & 1 & 0 & 0 \\ 0 & 0 & 0 & 1 & 0 & 1 & 1 \\ 0 & 0 & 0 & 1 & 0 & 1 & 1 \end{bmatrix}$$

$$\downarrow$$

$$\begin{bmatrix} 1 & 1 & 1 & 0 & 1 & 0 & 0 \\ 1 & 1 & 1 & 1 & 1 & 1 & 1 \\ 1 & 1 & 1 & 1 & 1 & 1 & 1 \end{bmatrix} \rightarrow \begin{bmatrix} 0 & 0 & 0 & 1 & 0 & 1 & 1 \\ 1 & 0 & 0 & 0 & 0 & 0 & 0 \\ 1 & 0 & 0 & 0 & 0 & 0 & 0 \end{bmatrix} = c.$$

Codeword c has weight 5, and each row has odd weight. Moreover, all 21 bicyclic translates of

$$c = \begin{bmatrix} 0 & 0 & 0 & 1 & 0 & 1 & 1 \\ 1 & 0 & 0 & 0 & 0 & 0 & 0 \\ 1 & 0 & 0 & 0 & 0 & 0 & 0 \end{bmatrix}$$

are codewords. Because there are only 21 minimum-weight codewords, this accounts for all minimum-weight codewords.

Each of the three rows of C, denoted C_0, C_1, and C_2, is the spectrum of a codeword over $GF(4)$, which codewords we denote by b_0, b_1, and b_2. Thus we have the following Fourier transform relationship:

$$b_0 \longleftrightarrow C_0,$$

$$b_1 \longleftrightarrow C_1,$$

$$b_2 \longleftrightarrow C_2.$$

Because C_0, C_1, and C_2 individually satisfy a Gleason–Prange condition, b_0, b_1, and b_2 can each be rearranged by using the Gleason–Prange permutation, thereby producing three new valid $GF(4)$ codewords. This means that the columns of the array

$$b = \begin{bmatrix} b_0 \\ b_1 \\ b_2 \end{bmatrix},$$

triply extended along rows, can be rearranged by the Gleason–Prange permutation to produce another triply extended array that corresponds to another codeword spectrum which also satisfies a Gleason–Prange condition. But the columns of b are simply the three-point Fourier transforms of the columns of the codeword c. If the columns of c are permuted by the Gleason–Prange permutation, then so are the columns of b. Because this permutation of b produces an array corresponding to another valid codeword, this permutation of c also produces another valid codeword.

Theorem 6.8.1 *The binary $(21, 12, 5)$ bicyclic BCH code, triply extended, is equivalent to the $(24, 12, 8)$ extended Golay code.*

Proof: The set of such triply extended codewords forms a linear code. Because the blocklength is increased by 3, and, as we will show, the minimum weight has been increased by 3, this is actually a $(24, 12, 8)$ code. It is equivalent to the extended binary Golay code. We will prove only that the code has distance 8, accepting the fact that the binary Golay code is unique. We must prove that bicyclic BCH codewords of weight 5, 6, or 7 will always have at least three, two, or one ones, respectively, in the three extension bits.

Because each row of C satisfies the Gleason–Prange condition, the Gleason–Prange permutation of the set of columns of c produces another codeword of the triply extended bicyclic BCH code. By using both a cyclic shift in the row direction of the bicyclic code and the Gleason–Prange permutation on the columns of the triply extended code, an automorphism of the triply extended code can be produced under which the extension column is interchanged with any chosen column. The new extension column can then be deleted to obtain another codeword of the bicyclic BCH code. We will show that whenever a codeword of the extended code has a weight less than 8, a column of that codeword can be deleted to obtain a codeword of the $(21, 12, 5)$ bicyclic BCH code with weight less than 5. Because such a codeword does not exist in that BCH code, the extended code can have no codeword of weight less than 8.

To this end, recall that every nonzero codeword c of the bicyclic BCH code has weight at least 5 (by the BCH bound) and the triply extended codeword must have even weight, so it has weight at least 6. Moreover, the column sum of a bicyclic BCH codeword c is a binary Hamming codeword, denoted b_0, and the column sum of the triply extended codeword $c^{\mid}$ is an extended binary Hamming codeword, denoted $b_0^{\mid}$, and so has weight 0, 4, or 8. If c^+ has weight 6, then the column sum b_0^+ must have either weight 0 or weight 4. Hence it has at least one column with at least two ones. Using the Gleason–Prange permutation, one such column of weight 2 can be moved to the extension position and deleted to give a bicyclic codeword of weight 4. Because there is no bicyclic codeword of weight 4, the extended code can have no codeword of weight 6. Hence every nonzero codeword has weight not smaller than 8. ∎

We have concluded that this is the binary $(24, 12, 8)$ extended Golay code because only one linear binary $(24, 12, 8)$ code exists. Thus we come to the rather unexpected conclusion that the triply extended BCH $(21, 12, 5)$ code with any component deleted, under a suitable permutation, becomes the cyclic Golay code. The number of codewords of weight 8 in the extended code is the sum of the numbers of codewords of weights 5,

6, 7, and 8 in the BCH $(21, 12, 5)$ code. Thus there are 759 codewords of weight 8 in the triply extended code.

6.9   The (24, 14, 6) Wagner code

The $(24, 14, 6)$ *Wagner code* is a linear, binary code that was discovered more than 40 years ago by computer search. The Wagner code claims no close relatives and does not appear as an example of any special class of codes. It has no $(23, 14, 5)$ cyclic subcode. The literature of the Wagner code is not extensive, and no simple construction of it is known to date. We shall construct the Wagner code in this section.

Together, the Golay code and the Wagner code provide the following pair of linear binary codes:

$(24, 14, 6)$ and $(24, 12, 8)$.

Thus we might ask: "Whose cousin is the Wagner code?" hoping that the Golay code might be the answer. However, the extended Golay code cannot be a subcode of the Wagner code because the $(23, 12, 8)$ Golay code is a perfect code, and the minimum weight of any of its cosets is at most 3. The contrast between the two codes is even more evident from the comparison of the weight distribution of the Wagner code with the weight distribution of the Golay code, shown in Table 6.2 for the full Wagner code and for a punctured Wagner code.

Of course, we can always say that the Wagner code is a subcode of the union of certain cosets of the Golay code. This statement, however, says almost nothing because the whole vector space is the union of cosets of the Golay code.

We will construct the Wagner code as a concatenated code, which is a code inside a code. Consider the $(8, 4, 4)$ code C_1 over $GF(4)$ with the generator matrix

$$G_1 = \begin{bmatrix} 1 & 0 & 0 & 0 & 1 & 0 & \alpha & \alpha \\ 0 & 1 & 0 & 0 & \alpha & 1 & 0 & \alpha \\ 0 & 0 & 1 & 0 & \alpha & \alpha & 1 & 0 \\ 0 & 0 & 0 & 1 & 0 & \alpha & \alpha & 1 \end{bmatrix},$$

where α is a primitive element of $GF(4) = \{0, 1, \alpha, 1 + \alpha\}$. An inspection verifies that the minimum distance of code C_1 is 4. The code C_1 could be regarded as a $(16, 8, 4)$ code over $GF(2)$ by replacing each element of G_1 by a two by two matrix:

replace 0 by $\begin{bmatrix} 0 & 0 \\ 0 & 0 \end{bmatrix}$, 1 by $\begin{bmatrix} 1 & 0 \\ 0 & 1 \end{bmatrix}$, and α by $\begin{bmatrix} 0 & 1 \\ 1 & 1 \end{bmatrix}$.

But this is not the code we want. We want, instead, a $(24, 8, 8)$ code.

Table 6.2. *Comparison of weight distributions*

ℓ	A_ℓ			
	Wagner code		Golay code	
	$(23, 14, 5)$	$(24, 14, 6)$	$(23, 12, 7)$	$(24, 12, 8)$
0 or 23/24	1	1	1	1
1 or 22/23	0	0	0	0
2 or 21/22	0	0	0	0
3 or 20/21	0	0	0	0
4 or 19/20	0	0	0	0
5 or 18/19	84	0	0	0
6 or 17/18	252	336	0	0
7 or 16/17	445	0	253	0
8 or 15/16	890	1335	506	759
9 or 14/15	1620	0	0	0
10 or 13/14	2268	3888	0	0
11 or 12/13	2632	0	1288	0
12	–	5264	–	2576

Toward this end, let $\mathcal{C}_2$ be the $(3, 2, 2)$ code over $GF(2)$ with generator matrix given by

$$G_2 = \begin{bmatrix} 1 & 0 & 1 \\ 0 & 1 & 1 \end{bmatrix}.$$

The concatenation of codes $\mathcal{C}_1$ and $\mathcal{C}_2$ is the binary $(24, 8, 8)$ code $\mathcal{C}_{12}$, in which each symbol of a codeword of $\mathcal{C}_1$ is regarded as a pair of bits, and those two bits are encoded by $\mathcal{C}_2$. To find the 8 by 24 generator matrix G_{12} for code $\mathcal{C}_{12}$, replace each zero by

$$\begin{bmatrix} 0 & 0 & 0 \\ 0 & 0 & 0 \end{bmatrix},$$

replace each one by

$$\begin{bmatrix} 1 & 0 & 1 \\ 0 & 1 & 1 \end{bmatrix},$$

and replace each α by

$$\begin{bmatrix} 0 & 1 & 1 \\ 1 & 1 & 0 \end{bmatrix}.$$

(The third matrix is obtained by multiplying the columns of G_2, regarded as elements of $GF(4)$, by α.) The resulting generator matrix for $\mathcal{C}_{12}$ is given by

$$
G_{12} = \begin{bmatrix}
1 & 0 & 1 & 0 & 0 & 0 & 0 & 0 & 0 & 0 & 0 & 0 & 1 & 0 & 1 & 0 & 0 & 0 & 0 & 1 & 1 & 0 & 1 & 1 \\
0 & 1 & 1 & 0 & 0 & 0 & 0 & 0 & 0 & 0 & 0 & 0 & 1 & 1 & 0 & 0 & 0 & 1 & 1 & 0 & 1 & 1 & 0 \\
0 & 0 & 0 & 1 & 0 & 1 & 0 & 0 & 0 & 0 & 0 & 0 & 1 & 1 & 1 & 0 & 1 & 0 & 0 & 0 & 0 & 1 & 1 \\
0 & 0 & 0 & 0 & 1 & 1 & 0 & 0 & 0 & 0 & 0 & 1 & 1 & 0 & 0 & 1 & 1 & 0 & 0 & 0 & 1 & 1 & 0 \\
0 & 0 & 0 & 0 & 0 & 0 & 1 & 0 & 1 & 0 & 0 & 0 & 0 & 1 & 1 & 0 & 1 & 1 & 1 & 0 & 1 & 0 & 0 & 0 \\
0 & 0 & 0 & 0 & 0 & 0 & 0 & 1 & 1 & 0 & 0 & 0 & 1 & 1 & 0 & 1 & 1 & 0 & 0 & 1 & 1 & 0 & 0 & 0 \\
0 & 0 & 0 & 0 & 0 & 0 & 0 & 0 & 0 & 1 & 0 & 1 & 0 & 0 & 0 & 0 & 1 & 1 & 0 & 1 & 1 & 1 & 0 & 1 \\
0 & 0 & 0 & 0 & 0 & 0 & 0 & 0 & 0 & 1 & 1 & 0 & 0 & 0 & 1 & 1 & 0 & 1 & 1 & 0 & 0 & 1 & 1
\end{bmatrix}.
$$

The dual code of $\mathcal{C}_{12}$, denoted $\mathcal{C}_{12}^{\perp}$, is the code with a check matrix equal to G_{12}. It is easy to see that $\mathcal{C}_{12}^{\perp}$ has minimum distance 3 because the first three columns of G_2 are linearly dependent. Therefore $\mathcal{C}_{12}^{\perp}$ is a $(24, 16, 3)$ code. In fact, there are 24 codewords of weight 4. To obtain a code $\mathcal{C}'$ with minimum distance 6, we will expurgate all codewords of odd weight or of weight 4. Words of odd weight are eliminated by appending a single parity-check equation to H. Words of weight 4 are eliminated by noting that such words have a single 1 in either the first 12 positions or the last 12 positions.

Finally, the Wagner code is defined as a subcode of the code $\mathcal{C}_{12}^{\perp}$. It is defined by the check matrix H, consisting of G_{12} augmented by two additional rows, one row of weight 12 with ones in the first 12 columns, and one row of weight 12 with ones in the last 12 columns. Thus

$$
H = \begin{bmatrix}
1 & 0 & 1 & 0 & 0 & 0 & 0 & 0 & 0 & 0 & 0 & 0 & 1 & 0 & 1 & 0 & 0 & 0 & 0 & 1 & 1 & 0 & 1 & 1 \\
0 & 1 & 1 & 0 & 0 & 0 & 0 & 0 & 0 & 0 & 0 & 0 & 1 & 1 & 0 & 0 & 0 & 1 & 1 & 0 & 1 & 1 & 0 \\
0 & 0 & 0 & 1 & 0 & 1 & 0 & 0 & 0 & 0 & 0 & 0 & 1 & 1 & 1 & 0 & 1 & 0 & 0 & 0 & 0 & 1 & 1 \\
0 & 0 & 0 & 0 & 1 & 1 & 0 & 0 & 0 & 0 & 0 & 1 & 1 & 0 & 0 & 1 & 1 & 0 & 0 & 0 & 1 & 1 & 0 \\
0 & 0 & 0 & 0 & 0 & 0 & 1 & 0 & 1 & 0 & 0 & 0 & 0 & 1 & 1 & 0 & 1 & 1 & 1 & 0 & 1 & 0 & 0 & 0 \\
0 & 0 & 0 & 0 & 0 & 0 & 0 & 1 & 1 & 0 & 0 & 0 & 1 & 1 & 0 & 1 & 1 & 0 & 0 & 1 & 1 & 0 & 0 & 0 \\
0 & 0 & 0 & 0 & 0 & 0 & 0 & 0 & 0 & 1 & 0 & 1 & 0 & 0 & 0 & 0 & 1 & 1 & 0 & 1 & 1 & 1 & 0 & 1 \\
0 & 0 & 0 & 0 & 0 & 0 & 0 & 0 & 0 & 1 & 1 & 0 & 0 & 0 & 1 & 1 & 0 & 1 & 1 & 0 & 0 & 1 & 1 \\
1 & 1 & 1 & 1 & 1 & 1 & 1 & 1 & 1 & 1 & 1 & 1 & 0 & 0 & 0 & 0 & 0 & 0 & 0 & 0 & 0 & 0 & 0 & 0 \\
0 & 0 & 0 & 0 & 0 & 0 & 0 & 0 & 0 & 0 & 0 & 0 & 1 & 1 & 1 & 1 & 1 & 1 & 1 & 1 & 1 & 1 & 1 & 1
\end{bmatrix}.
$$

This H gives a $(24, 14)$ code. The last two rows of H, taken together, eliminate all codewords of $\mathcal{C}_{12}$ with odd weight. This eliminates all codewords of weight 3 or 5. The last row eliminates all codewords with odd weight in the last 12 bits. The next-to-last row eliminates all codewords with odd weight in the first 12 bits. This eliminates all codewords of weight 4. Because all codewords of weight 3, 4, or 5 are eliminated, the minimum distance of the Wagner code is 6.

6.10   Self-dual codes

The $(23, 12, 7)$ binary Golay code is a very special code, because it is the only nontrivial perfect binary code other than the binary Hamming codes. The $(24, 12, 8)$ extended binary Golay code is a self-dual code that comes from the Golay code, and the $(8, 4, 4)$ extended Hamming code is a self-dual code that comes from the Hamming code. (Recall that a self-dual code is a code that satisfies $\mathcal{C} = \mathcal{C}^{\perp}$.) It is natural to look for other good self-dual, binary codes. For brevity, we will mention here only those good self-dual codes whose blocklengths are multiples of 8, especially those whose blocklengths are multiples of 24. Thus for multiples of 24, one might hope to find binary linear codes with the following parameters: $(24, 12, 8)$, $(48, 24, 12)$, and $(72, 36, 16)$. Of these the $(24, 12, 8)$ and the $(48, 24, 12)$ binary self-dual codes do exist. The second is an extended binary quadratic residue code. However, the quadratic residue code of blocklength 72 is a $(72, 36, 12)$ code, so it is not the conjectured code. In fact, it is not known whether a $(72, 36, 16)$ binary code exists, whether linear or nonlinear. This is a long-standing and straightforward unsolved question of coding theory.

The parameters of some selected binary self-dual codes are shown in Table 6.3. The codes that were selected to put in this table are those that have a blocklength equal to a multiple of 8 and are also the best codes known of their blocklength. A few of the quadratic residue codes, but not all, satisfy both of these conditions and are listed in the table. However, all quadratic residue codes are self-dual codes, and so would be listed in a more extensive table.

Table 6.3 was deliberately formed to be rather suggestive and to call for conjectures regarding missing entries. However, although the code parameters suggest a pattern, the underlying codes are very different, at least in their conventional representation. What deeper pattern may be hidden within this list of codes, if any, remains undiscovered.

Table 6.3. *Parameters of some binary self-dual codes*

n	k	d	
8	4	4	Hamming code
16	8	4	
24	12	8	Golay code
32	16	8	quadratic residue code
48	24	12	quadratic residue code
56	28	12	double circulant code
64	32	12	double circulant code
80	40	16	quadratic residue code
96	48	16	Feit code

Problems

6.1 Prove that the $(24, 12, 8)$ extended Golay code has 759 codewords of weight 8.

6.2 Prove that no binary $(24, 14, 6)$ code is the union of cosets of the binary $(24, 12, 8)$ extended Golay code.

6.3 Let $\mathcal{C}$ be an $(n, n - 1)$ simple binary parity-check code, and let $\mathcal{C}^3$ be the $(n^3, (n - 1)^3)$ binary code obtained as the three-dimensional product code, using $\mathcal{C}$ as each component code.
   (a) How many errors can $\mathcal{C}^3$ correct?
   (b) Give two error patterns of the same weight that have the same syndrome, and so are uncorrectable.

6.4 Using the Turyn representation, describe the construction of a $(72, 36, d)$ code from the $(24, 12, 8)$ Golay code and its reciprocal. What is d?

6.5 Let $\mathcal{C}'$ be the $(24, 12, 8)$ binary Golay code constructed by the Turyn representation, and let $\mathcal{C}''$ be the $(24, 4, 4)$ code obtained by padding the $(8, 4)$ binary Hamming code with 16 zeros. What are the possible weights of $c' + c''$, where $c' \in \mathcal{C}'$ and $c'' \in \mathcal{C}''$.

6.6 Construct a binary $(18, 7, 7)$ linear code as a triply extended bicyclic BCH code.

6.7 The $(21, 15, 3)$ bicyclic binary BCH code is obtained by setting spectral component $C_{1,1}$ and all its conjugates equal to zero. The two-dimensional spectrum of the codeword c is given by

$$
C = \begin{bmatrix}
C_{0,0} & C_{0,1} & C_{0,2} & C_{0,3} & C_{0,4} & C_{0,5} & C_{0,6} \\
C_{1,0} & 0 & 0 & C_{1,3} & 0 & C_{1,5} & C_{1,6} \\
C_{2,0} & 0 & 0 & C_{2,3} & 0 & C_{2,5} & C_{2,6}
\end{bmatrix}.
$$

This code has minimum distance equal to at least 3. Because there are two consecutive zeros in several columns, the BCH bound says that, unless such a column is everywhere zero, then there are three nonzero rows in the codeword c. Can this code be triply extended by a check on each row to obtain a $(24, 15, 6)$ code?

6.8 By viewing it as a bicyclic code, prove that the cyclic code of blocklength 21 and defining set $\{0, 1, 3, 7\}$ has minimum weight 8. How does this compare to the BCH bound?

6.9 The Turyn construction of the binary Golay code can be used to construct a $(72, 36, 12)$ code by replacing the three extended Hamming codes by three Golay codes.
   (a) Show that this code has dimension 36 and minimum distance 12.
   (b) What is the minimum distance of the $(69, 36)$ BCH code with spectral zeros at C_1 and C_{23}. Can it be triply extended to form a $(72, 36, 12)$ code?

6.10   What is the minimum distance of the $(93, 48)$ narrowsense binary BCH code? Can this code be triply extended to form a binary $(96, 48, 16)$ code? What is the relationship to the Turyn construction?

6.11   A 15 by 15 binary bicyclic code over $GF(16)$ has the defining set

$$\mathcal{A} = \{j' = 1, 2 \text{ and } j'' = 1, 2, \ldots, 8\} \cup \{j' = 1, 2, \ldots, 8 \text{ and } j'' = 1, 2\}$$

containing 28 elements. What is the complete defining set of this code? What is the minimum distance? What is the dimension?

6.12   Give a definition of a fast Reed–Solomon code in a form that anticipates the use of the Good–Thomas fast Fourier transform. How might this simplify the encoder and decoder?

6.13   Can the $(49, 39, 5)$ code over $GF(8)$, specified by Figure 6.9, be extended by four symbols to produce a $(53, 43, 5)$ code? How? What can be said beyond this?

6.14   (a) Can the Turyn representation be used to construct the Wagner code from three linear codes of blocklength 8?

    (b) Can the Turyn representation be used to construct a $(48, 36, 6)$ code from three linear codes of blocklength 16?

6.15   The dual of the $(2^m - 1, 2^m - 1 - m)$ binary Hamming code is a $(2^m - 1, m)$ code, called a *simplex code* (or *first-order Reed–Muller code*). Is the dual of the $(2^m, 2^m - 1 - m)$ extended binary Hamming code equivalent to the extended simplex code? If not, what is the relationship?

6.16   A nonlinear binary code with blocklength 63 is given by the set of vectors c of blocklength 63 whose spectra C satisfy $C_1 = C_3 = C_5 = 0$, $C_7 = A$, and $C_9 = B$, where $A \in \{1, \alpha^7, \alpha^{14}, \alpha^{21}, \alpha^{28}, \alpha^{35}, \alpha^{42}, \alpha^{49}, \alpha^{56}\}$ and $B \in \{1, \alpha^9, \alpha^{18}, \alpha^{27}, \alpha^{36}, \alpha^{45}, \alpha^{54}\}$.

    (a) Is the code cyclic?

    (b) How many codewords does this code have?

    (c) What is the minimum distance of the code?

    (d) How does this code compare to the $(63, 35, 9)$ BCH code?

6.17   Show that the binary Gleason–Prange theorem can be extended to arrays with rows of length $p+1$. That is, if the rows of arrays v and u are related by a Gleason–Prange permutation, and if each row of the two-dimensional Fourier transform V satisfies a Gleason–Prange condition, then the corresponding row of the two-dimensional Fourier transform satisfies the same Gleason–Prange condition.

Notes

Some two-dimensional codes of simple structure have been found to be useful in applications. These are the interleaved codes and the product codes. These codes are used, not because their minimum distances are attractive, but because their implementations are affordable and because of good burst-correcting properties. Decoders for interleaved codes routinely correct many error patterns beyond their minimum distances. A general investigation of the structure of two-dimensional bicyclic codes can be found in Ikai, Kosako, and Kojima (1974) and Imai (1977). In general, the study of two-dimensional codes has not produced codes whose minimum distances are noteworthy. The exceptions are the two-dimensional bicyclic codes whose component blocklengths are coprime, but these codes are equivalent to one-dimensional cyclic codes. Therefore two-dimensional codes are not highly valued by those who judge codes only by n, k, and d.

The construction, herein, of the Golay code as a triply extended, two-dimensional BCH code seems to be original. It is related to an observation of Berlekamp (1971). Other than some recent work by Simonis (2000), the Wagner code has been largely ignored since its discovery (Wagner, 1965). Duursma, in unpublished work, provided the construction of the Wagner code as a concatenated code that is given in this chapter.

The terms "codes on the bicyclic plane," "codes on the affine plane," and "codes on the projective plane" were selected to continue and parallel the classification begun in Chapter 2. This classification will be completed in Chapter 9.

7 Arrays and the Algebra of Bivariate Polynomials

An array, $v = [v_{i'i''}]$, defined as a doubly indexed set of elements from a given alphabet, was introduced in Chapter 5. There we studied the relationship between the two-dimensional array v and its two-dimensional Fourier transform V. In this chapter, further properties of arrays will be developed by drawing material from the subject of commutative algebra, but enriching this material for our purposes and presenting some of it from an unconventional point of view.

The two-dimensional array v can be represented by the bivariate polynomial $v(x, y)$, so we can study arrays by studying bivariate polynomials, which is the theme of this chapter. The polynomial notation provides us with a convenient way to describe an array. Many important computations involving arrays can be described in terms of the addition, subtraction, multiplication, and division of bivariate polynomials. Although n-dimensional arrays also can be studied as n-variate polynomials, in this book we shall treat only two-dimensional arrays and bivariate polynomials.

As the chapter develops, it will turn heavily toward the study of ideals, zeros of ideals, and the relationship between the number of zeros of an ideal and the degrees of the polynomials in any set of polynomials that generates the ideal. A well known statement of this kind is *Bézout's theorem*, which bounds the number of zeros of an ideal generated by two polynomials.

7.1　Polynomial representations of arrays

An n' by n'' array, $v = [v_{i'i''}]$, over the field F can be represented as a bivariate polynomial $v(x, y)$, given by

$$v(x, y) = \sum_{i'=0}^{n'-1} \sum_{i''=0}^{n''-1} v_{i'i''} x^{i'} y^{i''}.$$

For a square array, which is our usual case, we will set $n = n' = n''$.

The set of bivariate polynomials over the field F is closed under addition and multiplication. It is a ring. The ring of bivariate polynomials over the field F is conventionally denoted $F[x, y]$. The ring of bivariate polynomials modulo $x^n - 1$ and modulo $y^n - 1$ is a quotient ring, which is conventionally denoted $F[x, y]/\langle x^n - 1, y^n - 1\rangle$. We also use the simpler notation $F^\circ[x, y]$ for this quotient ring. In the quotient ring $F^\circ[x, y]$, a multiplication product is reduced by setting $x^n = 1$ and $y^n = 1$.

The *ideal I* in $F[x, y]$ is a nonempty subset of $F[x, y]$ that is closed under addition of its elements and is closed under multiplication by any bivariate polynomial. Thus for a subset I to be an ideal in $F[x, y]$, $f(x, y) + g(x, y)$ must be in I if both $f(x, y)$ and $g(x, y)$ are in I, and $f(x, y)p(x, y)$ must be in I if $p(x, y)$ is any bivariate polynomial in $F[x, y]$ and $f(x, y)$ is any polynomial in I. An ideal of $F[x, y]$ is called a *proper ideal* if it is not equal to $\{0\}$ or $F[x, y]$. An ideal of $F[x, y]$ is called a *principal ideal* if there is one element of the ideal such that every element of the ideal is a multiple of that element.

If $g_1(x, y), g_2(x, y), \ldots, g_n(x, y)$ are any bivariate polynomials over F, the set of their polynomial combinations is written as follows:

$$I = \left\{ \sum_{\ell=1}^{n} a_\ell(x, y) g_\ell(x, y) \right\},$$

where the $a_\ell(x, y)$ are arbitrary polynomials in $F[x, y]$. It is easy to see that the set I forms an ideal. The polynomials $g_1(x, y), g_2(x, y), \ldots, g_n(x, y)$ are called *generators* of the ideal I, and, taken together, the *generator polynomials* form a *generator set*, denoted $\mathcal{G} = \{g_1(x, y), \ldots, g_n(x, y)\}$. This ideal is conventionally denoted as

$$I = \langle g_1(x, y), \ldots, g_n(x, y)\rangle,$$

or as $I(\mathcal{G})$. We shall see that every ideal of $F[x, y]$ can be generated in this way. In general, an ideal does not have a unique generator set; an ideal may have many different generator sets. A principal ideal can always be generated by a single polynomial, but it may also be generated by other generator sets containing more than one polynomial.

Recall that for any field F, not necessarily algebraically closed, the affine plane over F consists of the set $F^2 = \{(x, y) \mid x \in F, y \in F\}$. A *zero* (or *affine zero*) of the polynomial $v(x, y)$ is the pair (β, γ) of elements of F such that $v(\beta, \gamma) = 0$. Thus an affine zero of $v(x, y)$ is a point of the affine plane. The set of affine zeros of the polynomial $v(x, y)$ is a set of points in the affine plane. A *zero* (or *affine zero*) of an ideal I in $F[x, y]$ is a point of the affine plane that is a zero of every element of I. The set of affine zeros of the ideal I, denoted $\mathcal{Z}(I)$, is a set of points in the affine plane. It is equal to the set of common zeros of any set of generator polynomials for I. The set of common affine zeros of a set of irreducible multivariate polynomials is called a *variety* or an *affine variety*. An affine variety in the plane formed by a single irreducible bivariate polynomial is called a *plane affine curve*.

In the ring $F[x, y]$, the *reciprocal polynomial* $\tilde{v}(x, y)$ of the bivariate polynomial $v(x, y)$ is defined as

$$\tilde{v}(x, y) = x^{s_x} y^{s_y} v(x^{-1}, y^{-1}),$$

where (s_x, s_y) is the componentwise degree of $v(x, y)$. In the ring $F^\circ[x, y] = F[x, y]/\langle x^n - 1, y^n - 1 \rangle$, the *reciprocal polynomial* of $v(x, y)$ is defined as follows:

$$\tilde{v}(x, y) = x^{n-1} y^{n-1} v(x^{-1}, y^{-1}).$$

This reciprocal polynomial corresponds to the reciprocal array $\tilde{v}$ with elements

$$\tilde{v}_{i'i''} = v_{(n-1-i'),(n-1-i'')}.$$

The context will convey which form of the reciprocal polynomial is to be understood in any discussion.

7.2   Ordering the elements of an array

The elements of an array have two indices, both nonnegative integers. This double index (j', j'') is called the *bi-index* of the indexed element. We shall have many occasions to rearrange the elements of the two-dimensional array (either a finite array or an infinite array) into a one-dimensional sequence so that we can point to them, one by one, in a fixed order. This is called a *total order* on the elements of the array. A total order on any set is an ordering relationship that can be applied to any pair of elements in that set. The total order is expressed as $(j', j'') \prec (k', k'')$, meaning that (j', j'') comes before (k', k'') in the total order.

When the total order has been specified, it is sometimes convenient to represent the double index (j', j'') simply by the single index j, meaning that (j', j'') is the jth entry in the total order. Then we have two possible meanings for addition. By $j + k$, we do not mean the $(j + k)$th entry in the total order; by $j + k$, we always mean $(j' + k', j'' + k'')$. Likewise, by $j - k$, we mean $(j' - k', j'' - k'')$. Occasionally, we mildly violate this rule by writing $r - 1$ to index the term before the rth term in a sequence. In this case, subtraction is not at the component level. The context, and the fact that 1 is not in the form of a bi-index, will make it clear when this is the intended meaning.

A total order on an array of indices implies a total order on the monomials $x^{j'} y^{j''}$ of a bivariate polynomial. In particular, a total order determines the *leading monomial* $x^{s'} y^{s''}$ and the *leading term* $v_{s's''} x^{s'} y^{s''}$ of the bivariate polynomial $v(x, y)$. This is the unique monomial for which (s', s'') is greater than (j', j'') for any other monomial $x^{j'} y^{j''}$ with nonzero coefficient. The bi-index (s', s'') of the leading monomial of $v(x, y)$ is called the *bidegree* of $v(x, y)$, and is denoted bideg $v(x, y)$. (Recall that the *degree* of the polynomial is defined as $s' + s''$, and the *componentwise degree* (s_x, s_y) is defined separately for the x and y variables.) The bidegree of a polynomial cannot be determined until a total order is specified.

There are many ways of defining a total order. We shall limit the possible choices to those total orders that respect multiplication by monomials. This means that if the array $[v_{i'i''}]$ is represented by the polynomial $v(x, y)$, then the coefficients of the polynomial retain the same relative order when the polynomial $v(x, y)$ is multiplied by the monomial $x^a y^b$. In particular, if

$$(j', j'') \prec (k', k''),$$

then for any positive integers a and b we require that

$$(j' + a, j'' + b) \prec (k' + a, k'' + b).$$

Total orders that satisfy this condition are called *monomial orders* or *term orders*. The two most popular monomial orders are the lexicographic order and the graded order.

The *lexicographic order* is defined as $(j', j'') \prec (k', k'')$ if $j'' < k''$, or if $j'' = k''$ and $j' < k'$. The lexicographic order is usually unsatisfactory for infinite arrays, but is perfectly suitable for polynomials.

For example, the indices of a three by three array, arranged in increasing lexicographic order, are as follows:

$$(0, 0), (1, 0), (2, 0), (0, 1), (1, 1), (2, 1), (0, 2), (1, 2), (2, 2).$$

The nine elements of the array v, listed with indices in increasing lexicographic order, are as follows:

$$v_{00}, v_{10}, v_{20}, v_{01}, v_{11}, v_{21}, v_{02}, v_{12}, v_{22}.$$

The corresponding bivariate polynomial $v(x, y)$, with terms arranged so their indices are in decreasing lexicographic order, is given by

$$v(x, y) = v_{22}x^2y^2 + v_{12}xy^2 + v_{02}y^2 + v_{21}x^2y + v_{11}xy + v_{01}y + v_{20}x^2 + v_{10}x + v_{00}.$$

In particular, with the lexicographic order specified, the nonzero polynomial $v(x, y)$ has a monomial of largest degree, namely x^2y^2 if v_{22} is nonzero, and a *leading coefficient*, namely v_{22}. The bidegree of $v(x, y)$ in the lexicographic order is $(2, 2)$. If v_{22} were equal to 1, $v(x, y)$ would be an example of a monic bivariate polynomial in lexicographic order.

Note that in our definition of the lexicographic order the monomial x precedes the monomial y (reading from the right). As an alternative, the definition could be inverted so that y precedes x. A serious disadvantage of the lexicographic order is that, for an infinite array, a monomial can be preceded by an infinite number of other monomials. We shall avoid using the lexicographic order, preferring instead an order in which every monomial is preceded only by a finite number of other monomials.

The *graded order* (or graded lexicographic order) is defined as $(j', j'') \prec (k', k'')$ if $j' + j'' < k' + k''$, or if $j' + j'' = k' + k''$ and $j'' < k''$. The indices of a three by three array, arranged in increasing graded order, are as follows:

$$(0,0), (1,0), (0,1), (2,0), (1,1), (0,2), (2,1), (1,2), (2,2).$$

The nine elements of the array, listed with indices in increasing graded order, are as follows:

$$v_{00}, v_{10}, v_{01}, v_{20}, v_{11}, v_{02}, v_{21}, v_{12}, v_{22}.$$

The polynomial $v(x, y)$, with terms arranged with indices in decreasing graded order, is given by

$$v(x, y) = v_{22}x^2y^2 + v_{12}xy^2 + v_{21}x^2y + v_{02}y^2 + v_{11}xy + v_{20}x^2 + v_{01}y + v_{10}x + v_{00}.$$

The bidegree of $v(x, y)$ in the graded order is $(2, 2)$.

The polynomial $v(x, y)$ has the same leading term, namely $v_{22}x^2y^2$, for both the lexicographic order and the graded order, provided v_{22} is nonzero. If v_{22} and v_{12} are both zero, however, then in the lexicographic order the leading term of the polynomial would be $v_{02}y^2$, while in the graded order the leading term would be $v_{21}x^2y$. Thus before determining the leading term it is necessary to specify the ordering rule.

Another total order on indices (or monomials) that is useful is the weighted order. Let a and b be fixed positive integers. The *weighted order* (or *weighted graded order*) is defined as $(j', j'') \prec (k', k'')$ if $aj' + bj'' < ak' + bk''$, or if $aj' + bj'' = ak' + bk''$ and $j'' < k''$.

For example, with $a = 3$ and $b = 2$, the indices of a three by three array, arranged in increasing weighted order, are as follows:

$$(0, 0), (0, 1), (1, 0), (0, 2), (1, 1), (2, 0), (0, 3), (1, 2), (2, 1), (3, 0), \ldots$$

The bidegree of $v(x, y)$ in the weighted order is $(3, 0)$. Note that $(2, 0)$ appears before $(0, 3)$ in this weighted order. The polynomial $v(x, y)$, with leading coefficient v_{30} and written with terms in decreasing weighted order, is given by

$$v(x, y) = v_{30}x^3 + v_{21}x^2y + v_{12}xy^2 + v_{20}x^2 + v_{03}y^3 + v_{11}xy$$
$$+ v_{02}y^2 + v_{10}x + v_{01}y + v_{00}.$$

The bidegree of $v(x, y)$ in the weighted order is $(3, 0)$. Note that the positions of x^2 and y^3 in this weighted order are not as they would be in the graded order.

The bidegree of a polynomial depends on the choice of total order. When we wish to be precise, we shall speak of the *lexicographic bidegree*, the *graded bidegree*, or the *weighted bidegree*. The weighted bidegree is not the same as the *weighted degree*, defined in Sections 4.8 and 5.3. For a specified a and b, the weighted degree, denoted $\deg^{(a,b)} v(x, y)$, is the largest value of $aj' + bj''$ for any monomial with a nonzero coefficient.

For example, the polynomial

$$v(x, y) = x^5 y^2 + x^3 y^3 + xy^4$$

has $\deg v(x, y) = 7$, $\text{compdeg } v(x, y) = (5, 4)$, and $\text{bideg } v(x, y) = (5, 2)$ in the graded order, while in the lexicographic order, $\text{bideg } v(x, y) = (1, 4)$. In the weighted order with $a = 2$ and $b = 3$, the weighted bidegree and the weighted degree are expressed as $\text{bideg } v(x, y) = (5, 2)$ and $\deg^{(a,b)} v(x, y) = 16$.

An example of an order that is *not* a total order is the *division order*, which is denoted $j \lessdot k$ or $(j', j'') \lessdot (k', k'')$, meaning that $j' < k'$ and $j'' < k''$. The nonstrict form of this inequality is denoted $j \leqq k$, or $(j', j'') \leqq (k', k'')$, meaning that $j' \leq k'$ and $j'' \leq k''$. (We do not define a notation for strict inequality on only one component.) The division order is not a total order, because there are some pairs (j', j'') and (k', k'') that cannot be compared in the division order. This is called a *partial order*. Note, for example, that $(3, 7)$ and $(4, 2)$ cannot be compared by using the division order. A simple illustration will show that $j \lessdot k$ is not the opposite of $j \gtrdot k$. The first inequality means $j' < k'$ *and* $j'' < k''$. Its opposite is $j' \geq k'$ or $j'' \geq k''$. The second inequality means that $j' \geq k'$ and $j'' \geq k''$.

The division order on indices is closely related to a division order on monomials. The monomial $x^{j'} y^{j''}$ comes before $x^{k'} y^{k''}$ in the division order if the monomial $x^{j'} y^{j''}$ divides, as polynomials, the monomial $x^{k'} y^{k''}$. In terms of the exponents, this becomes, $(j', j'') \leqq (k', k'')$ if $j' \leq k'$ and $j'' \leq k''$.

The contrast between the division order and the graded order is shown in Figure 7.1. The highlighted region in Figure 7.1(a) shows the set of (j', j'') such that $(j', j'') \leqq (5, 3)$. The highlighted region in Figure 7.1(b) shows the set of (j', j'') such that $(j', j'') \preceq (5, 3)$, where $(j', j'') \preceq (k', k'')$ means that Figure 7.1(a) is equal to, or smaller than, Figure 7.1(b) in the graded order. These (j', j'') satisfy the sequence

$$(0, 0) \preceq (1, 0) \preceq (0, 1) \preceq (2, 0) \preceq (1, 1) \preceq \cdots \preceq (6, 2) \preceq (5, 3).$$

The shaded region in Figure 7.1(a) is contained in the shaded region in part (b) because $(j', j'') \preceq (k', k'')$ if $(j', j'') \leqq (k', k'')$.

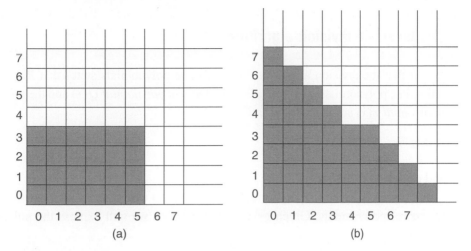

Figure 7.1. Division order and graded order.

The division order can be used to restate the definition of a cascade set as follows: The *cascade set* $\mathcal{P}$ is a proper subset of $\mathbf{N}^2$ with the property that if $(k', k'') \in \mathcal{P}$ and $(j', j'') \leq (k', k'')$, then $(j', j'') \in \mathcal{P}$.

For any monomial order, the degrees of $a(x, y)$ and $b(x, y)$ add under polynomial multiplication $a(x, y)b(x, y)$, as do the bidegrees and the componentwise degrees. Thus if $a(x, y)$ and $b(x, y)$ have bidegrees (s', s'') and (r', r''), respectively, then $a(x, y)b(x, y)$ has bidegree $(s' + r', s'' + r'')$ and degree $s' + s'' + r' + r''$. If $a(x, y)$ and $b(x, y)$ have componentwise degrees (s_x, s_y) and (r_x, r_y), respectively, then $a(x, y)b(x, y)$ has componentwise degree $(s_x + r_x, s_y + r_y)$.

A monic polynomial has been defined as a bivariate polynomial $v(x, y)$ whose leading coefficient $v_{s's''}$ is 1. Note that a polynomial may lose its status as a monic polynomial if the choice of total order is changed because, then, the leading term may change. A monic irreducible polynomial is called a *prime polynomial*. Any nonconstant polynomial $v(x, y)$ can be written uniquely (up to the order of the factors) as the product of a field element and its prime polynomial factors $p_\ell(x, y)$, some perhaps raised to a power m_ℓ. Thus

$$v(x, y) = \beta \prod_{\ell=1}^{N} p_\ell(x, y)^{m_\ell}.$$

This statement – that every bivariate polynomial has a unique factorization – is called the *unique factorization theorem* (for bivariate polynomials). In general, any ring that satisfies the unique factorization theorem is called a *unique factorization ring*, so the ring of bivariate polynomials over a field is an example of a unique factorization ring.

7.3   The bivariate division algorithm

Recall that the division algorithm for univariate polynomials states that for any two univariate polynomials, $f(x)$ and $g(x)$, the *quotient polynomial $Q(x)$* and the *remainder polynomial $r(x)$* are unique and satisfy

$$f(x) = Q(x)g(x) + r(x)$$

and $\deg r(x) < \deg g(x)$.

In the ring of bivariate polynomials, we shall want to divide simultaneously by two (or more) polynomials. Given $g_1(x, y)$ and $g_2(x, y)$, we will express the polynomial $f(x, y)$ as

$$f(x, y) = Q_1(x, y)g_1(x, y) + Q_2(x, y)g_2(x, y) + r(x, y),$$

where the *remainder polynomial $r(x, y)$* satisfies bideg $r(x, y) \preceq$ bideg $f(x, y)$, and no term of $r(x, y)$ is divisible by the leading term of either $g_1(x, y)$ or $g_2(x, y)$. Of course, because we speak of leading terms and the bidegree of polynomials, we are dealing with a fixed total order, which (unless otherwise specified) we shall always take to be the graded order.

The procedure we shall study mimics the steps of the division algorithm for univariate polynomials, reducing step by step the degree of a scratch polynomial, also called $f(x, y)$. At each step, if possible, the leading term of $f(x, y)$ is canceled by the leading term of either $g_1(x, y)$ or $g_2(x, y)$, multiplied by an appropriate monomial and coefficient. Otherwise, the leading term of $f(x, y)$ is assigned to the remainder polynomial $r(x, y)$. To make the procedure unambiguous, $g_1(x, y)$ is chosen as the divisor polynomial, whenever possible, in preference to $g_2(x, y)$.

The procedure will be made clear by an example. In $GF(2)$, let

$$f(x, y) = x^4 + x^3 y + xy^2 + xy + x + 1,$$
$$g_1(x, y) = x^3 + xy + 1,$$
$$g_2(x, y) = xy + y^2 + 1.$$

Initialize the quotient polynomials and the remainder polynomial as $Q_1(x, y) = Q_2(x, y) = r(x, y) = 0$.

Step (1) Multiplying $g_1(x, y)$ by x gives a leading term of x^4, which will cancel the leading term of $f(x, y)$. Then

$$f^{(1)}(x, y) = f(x, y) - xg_1(x, y)$$
$$= x^3 y + x^2 y + xy^2 + xy + 1,$$

$$Q_1^{(1)}(x, y) = x,$$

$$Q_2^{(1)}(x, y) = 0,$$

$$r^{(1)}(x, y) = 0.$$

Step (2) Multiplying $g_1(x, y)$ by y gives a leading term of $x^3 y$, which will cancel the leading term of $f^{(1)}(x, y)$. Then

$$f^{(2)}(x, y) = f^{(1)}(x, y) - y g_1(x, y)$$

$$= x^2 y + xy + y + 1,$$

$$Q_1^{(2)}(x, y) = x + y,$$

$$Q_2^{(2)}(x, y) = 0,$$

$$r^{(2)}(x, y) = 0.$$

Step (3) No monomial multiple of the leading term of $g_1(x, y)$ will cancel the leading term of $f^{(2)}(x, y)$. Multiplying $g_2(x, y)$ by x gives a leading term of $x^2 y$, which will cancel the leading term of $f^{(2)}(x, y)$. Then

$$f^{(3)}(x, y) = f^{(2)}(x, y) - x g_2(x, y)$$

$$= xy^2 + xy + x + y + 1,$$

$$Q_1^{(3)}(x, y) = x + y,$$

$$Q_2^{(3)}(x, y) = x,$$

$$r^{(3)}(x, y) = 0.$$

Step (4) Again, $g_1(x, y)$ cannot be used, but $g_2(x, y)$ multiplied by y can be used to cancel the leading term of $f^{(3)}(x, y)$. Then

$$f^{(4)}(x, y) = f^{(3)}(x, y) - y g_2(x, y)$$

$$= y^3 + xy + x + 1,$$

$$Q_1^{(4)}(x, y) = x + y,$$

$$Q_2^{(4)}(x, y) = x + y,$$

$$r^{(4)}(x, y) = 0.$$

Step (5) The leading term of $f^{(4)}(x, y)$ cannot be canceled by any multiple of $g_1(x, y)$ or $g_2(x, y)$, so it is assigned to the remainder polynomial. Then

$$f^{(5)}(x, y) = xy + x + 1,$$

$$Q_1^{(5)}(x, y) = x + y,$$

$$Q_2^{(5)}(x, y) = x + y,$$

$$r^{(5)}(x, y) = y^3.$$

Step (6) The leading term of $f^{(5)}(x, y)$ can be canceled by $g_2(x, y)$. Then

$$f^{(6)}(x, y) = f^{(4)}(x, y) - g_2(x, y)$$

$$= y^2 + x,$$

$$Q_1^{(6)}(x, y) = x + y,$$

$$Q_2^{(6)}(x, y) = x + y + 1,$$

$$r^{(5)}(x, y) = y^3.$$

In the final two steps, y^2 then x will be assigned to the remainder polynomial, because they cannot be canceled by any multiple of $g_1(x, y)$ or $g_2(x, y)$.

The result of the division algorithm for this example is as follows:

$$f(x, y) = Q_1(x, y)g_1(x, y) + Q_2(x, y)g_2(x, y) + r(x, y)$$

$$= (x + y)g_1(x, y) + (x + y + 1)g_2(x, y) + y^3 + y^2 + x.$$

Note that bideg $[(x + y)g_1(x, y)] \preceq$ bideg $f(x, y)$, and bideg $[(x + y + 1)g_2(x, y)] \preceq$ bideg $f(x, y)$.

The same procedure can be used with more than (or less than) two divisor polynomials, $g_\ell(x, y)$ for $\ell = 1, \ldots, L$, to compute a set of L quotient polynomials and one remainder polynomial, as stated in the following theorem.

Theorem 7.3.1 (division algorithm for bivariate polynomials) *Let $\mathcal{G} = \{g_\ell(x, y) \mid \ell = 1, \ldots, L\}$ be a set of bivariate polynomials from $F[x, y]$. Then every $f(x, y)$ can be written as follows:*

$$f(x, y) = Q_1(x, y)g_1(x, y) + \cdots + Q_L(x, y)g_L(x, y) + r(x, y),$$

where

$$\text{bideg } r(x, y) \preceq \text{bideg } f(x, y),$$

$$\text{bideg } [Q_\ell(x, y)g_\ell(x, y)] \preceq \text{bideg } f(x, y),$$

and no monomial of $r(x, y)$ is divisible by the leading monomial of any $g_\ell(x, y)$.

Proof: The proof can be obtained by formalizing the example given prior to the theorem. ∎

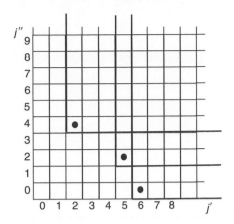

Figure 7.2. Removing quarter planes from the first quadrant.

Given an ordered set of polynomials, $\mathcal{G}$, the remainder polynomial will often be written as follows:

$r(x, y) = R_{\mathcal{G}}[f(x, y)].$

This is read as "$r(x, y)$ is the remainder, under division by $\mathcal{G}$, of $f(x, y)$." The condition that no monomial of $r(x, y)$ is divisible by the leading monomial of any $g_\ell(x, y)$ of $\mathcal{G}$ is illustrated in Figure 7.2 for a case with $L = 3$. Each filled circle in the figure represents the leading monomial of one of the $g_\ell(x, y)$. The set of monomials excluded by each $g_\ell(x, y)$ is highlighted as a quarter plane. No monomial of $r(x, y)$ can lie in any of these quarter planes.

In general, the decomposition of $f(x, y)$ given in Theorem 7.3.1 is not unique. The result of the division algorithm will depend, in general, on the order in which the $g_\ell(x, y)$ are listed; the remainder polynomial may be different if the $g_\ell(x, y)$ are permuted, and the quotient polynomials may also be different. In Section 7.4, we will show that if the polynomials of $\mathcal{G}$ form a certain preferred kind of set, known as a minimal basis for the ideal formed by $\mathcal{G}$, then the remainder polynomial $r(x, y)$ does not depend on the order in which the polynomials of $\mathcal{G}$ are listed.

Figure 7.3 shows the conditions, given in Theorem 7.3.1, on the monomials of $r(x, y)$ for a typical application of the division algorithm using the graded order. The open circle represents the leading monomial of $f(x, y)$. The upper staircase boundary of Figure 7.3 represents the condition that

$\text{bideg } r(x, y) \preceq \text{bideg } f(x, y),$

as required by the theorem. The monomials of $r(x, y)$ must be under this staircase. The solid circles represent the leading monomials of the $g_\ell(x, y)$. Theorem 7.3.1 requires that all monomials of the remainder polynomial $r(x, y)$ must lie in the set of those

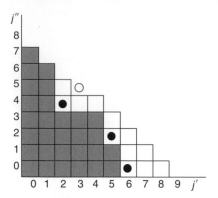

Figure 7.3. Conditions on the remainder polynomial.

monomials that are not divisible by the leading monomial of any $g_\ell(x, y)$. This requires that the possible monomials of the $r(x, y)$ correspond to a cascade set determined by the leading monomials of the $g_\ell(x, y)$. The combination of the two conditions means that the indices of all monomials of $r(x, y)$ lie in the shaded region of Figure 7.3.

We will close this section by describing a useful computation for forming, from any two polynomials $g_1(x, y)$ and $g_2(x, y)$, another polynomial that is contained in the ideal $\langle g_1(x, y), g_2(x, y) \rangle$, but in some sense is "smaller." To set up this computation, we shall introduce additional terminology. Let $\mathcal{G}$ be any set of polynomials in the ring of bivariate polynomials $F[x, y]$, with a fixed total order on the bi-indices. We need to coin a term for the set of bi-indices that are not the leading bi-index of any polynomial multiple of an element of the given set $\mathcal{G}$.

Definition 7.3.2 *The footprint, denoted $\Delta(\mathcal{G})$, of the set of bivariate polynomials given by*

$$\mathcal{G} = \{g_1(x, y), g_2(x, y), \ldots, g_n(x, y)\}$$

is the set of index pairs (j', j''), both nonnegative, such that $x^{j'} y^{j''}$ is not divisible by the leading monomial of any polynomial in $\mathcal{G}$.

Thus the complement of $\Delta(\mathcal{G})$ can be written as follows:

$$\Delta(\mathcal{G})^c = \{(j', j'') \mid (j', j'') = \text{bideg}\,[a(x, y)g_\ell(x, y)]; g_\ell(x, y) \in \mathcal{G}\}.$$

If any polynomial multiple of an element of $\mathcal{G}$ has leading bi-index (j', j''), then (j', j'') is not in $\Delta(\mathcal{G})$. Later, we will consider the footprint of the ideal generated by $\mathcal{G}$, which will be denoted $\Delta(I(\mathcal{G}))$. Although $\Delta(I(\mathcal{G}))$ is always contained in $\Delta(\mathcal{G})$, the two need not be equal to $\Delta(\mathcal{G})$. In Section 7.4, we will determine when the two footprints are equal.

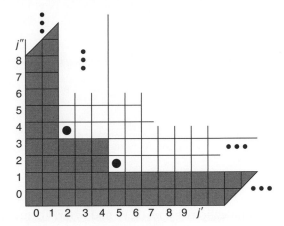

Figure 7.4. Footprint of $\{g_1(x,y), g_2(x,y)\}$.

The footprint of a set of polynomials is always a cascade set. An example of a footprint of set $\mathcal{G} = \{g_1(x,y), g_2(x,y)\}$ is shown in Figure 7.4. The squares containing the two solid circles correspond to the leading monomials of $g_1(x,y)$ and $g_2(x,y)$. The quadrant that consists of every square above and to the right of the first of these solid circles is not in the footprint of the set $\mathcal{G}$. The quadrant that consists of every square above and to the right of the second of these solid circles is not in the footprint of the set $\mathcal{G}$. The shaded squares are in the footprint, including all squares on the vertical strip at the left going to infinity and on the horizontal strip at the bottom going to infinity.

We shall now describe the promised useful operation, which we call a *conjunction*, that produces from two polynomials, $g_1(x,y)$ and $g_2(x,y)$ of set $\mathcal{G}$, another polynomial in the ideal $\langle g_1(x,y), g_2(x,y) \rangle$, whose leading monomial (unless it is the zero polynomial) is within the footprint of $\mathcal{G}$. This polynomial is called the *conjunction polynomial* of $g_1(x,y)$ and $g_2(x,y)$, and is denoted $b_{\mathcal{G}}^{g_1 g_2}(x,y)$. The conjunction polynomial will be found to be a linear polynomial combination of $g_1(x,y)$ and $g_2(x,y)$, and will be expressed as follows:

$$b_{\mathcal{G}}^{g_1 g_2}(x,y) = a_1(x,y)g_1(x,y) + a_2(x,y)g_2(x,y).$$

This means that the conjunction polynomial is in the ideal generated by $g_1(x,y)$ and $g_2(x,y)$.

Suppose, without loss of meaningful generality, that $g_1(x,y)$ and $g_2(x,y)$ are monic polynomials. Denote the leading monomial of $g_1(x,y)$ by $x^{s'} y^{s''}$; the bidegree of $g_1(x,y)$ is (s', s''). Denote the leading monomial of $g_2(x,y)$ by $x^{r'} y^{r''}$; the bidegree of $g_2(x,y)$ is (r', r''). The conjunction polynomial of $g_1(x,y)$ and $g_2(x,y)$ is defined as

$$b_{\mathcal{G}}^{g_1 g_2}(x,y) = R_{\mathcal{G}}[f(x,y)],$$

where

$$f(x, y) = m_1(x, y)g_1(x, y) - m_2(x, y)g_2(x, y),$$

and

$$m_1(x, y) = x^{r'} y^{r''} / \text{GCD}[x^{s'} y^{s''}, x^{r'} y^{r''}],$$
$$m_2(x, y) = x^{s'} y^{s''} / \text{GCD}[x^{s'} y^{s''}, x^{r'} y^{r''}].$$

The monomials $m_1(x, y)$ and $m_2(x, y)$ are chosen to "align" the polynomials $g_1(x, y)$ and $g_2(x, y)$ so that, after the multiplications by $m_1(x, y)$ and $m_2(x, y)$, the leading monomials are the same. Hence the monomials $m_1(x, y)$ and $m_2(x, y)$ are called *alignment monomials*. The alignment monomials can take several forms, such as 1 and $x^{s'-r'} y^{s''-r''}$, or $x^{r'-s'} y^{r''-s''}$ and 1, or $y^{r''-s''}$ and $x^{s'-r'}$, according to the signs of $r' - s'$ and $r'' - s''$.

Figure 7.5 illustrates the procedure for a case in which $s' \geq r'$ and $r'' \geq s''$. The leading monomials of $g_1(x, y)$ and $g_2(x, y)$ are denoted by two filled circles, located at coordinates (s', s'') and (r', r''). The asterisk in the illustration denotes the monomial of least degree that is divisible by the leading monomials of both $g_1(x, y)$ and $g_2(x, y)$. The shaded region is the intersection of the footprint of $\{g_1(x, y), g_2(x, y)\}$ and the set of remainders allowed by the division algorithm. The leading monomial of $f(x, y)$ is not greater than the monomial indicated by the open circle in the figure. The conjunction polynomial is the remainder polynomial and, unless it is identically zero, it has bidegree in the shaded region.

The expression for the division algorithm,

$$f(x, y) = Q_1(x, y)g_1(x, y) + Q_2(x, y)g_2(x, y) + R_G[f(x, y)],$$

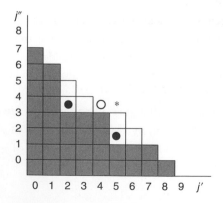

Figure 7.5. Possible bidegrees of the conjunction polynomial.

together with the definition of $f(x, y)$, show that we can write the conjunction polynomial in the form of a polynomial combination:

$$b_{\mathcal{G}}^{g_1 g_2}(x, y) = a_1(x, y) g_1(x, y) + a_2(x, y) g_2(x, y).$$

Thus the conjunction polynomial is in the ideal generated by $\mathcal{G}$. We have seen that, unless it is the zero polynomial, its leading monomial lies in $\Delta(\mathcal{G})$.

Now, unless it is zero, rename the conjunction polynomial $g_3(x, y)$ and append it to the set $\mathcal{G}$ to form a new set, again called $\mathcal{G}$. Unless $g_3(x, y)$ is equal to zero, the new set of polynomials $\mathcal{G} = \{g_1(x, y), g_2(x, y), g_3(x, y)\}$ has a footprint that is strictly smaller than the footprint of $\{g_1(x, y), g_2(x, y)\}$. To reduce the footprint further, the conjunction operation can again be applied, now to the pair $(g_1(x, y), g_3(x, y))$ and to the pair $(g_2(x, y), g_3(x, y))$, producing the new conjunction polynomials $g_4(x, y)$ and $g_5(x, y)$. These, if nonzero, can be further appended to the set $\mathcal{G}$, thereby reducing the footprint even further. Later, we shall examine this process rather closely. By finding conditions on the fixed points of this conjunction operation, we shall discover several important facts about sets of bivariate polynomials.

7.4   The footprint and minimal bases of an ideal

It is a well known fact that the number of zeros of the univariate polynomial $p(x)$ is not larger than the degree of $p(x)$. An indirect way of saying this is that the number of zeros of $p(x)$ is not larger than the number of nonnegative integers smaller than the leading bi-index of $p(x)$. This seems like a needlessly indirect way of stating this fact. It is, however, the form of the theorem that will generalize to two (or more) dimensions in a way that fits our needs.

We shall be interested in the number of affine zeros of an ideal in the ring of bivariate polynomials. This will have a close relationship to the footprint of the ideal. The footprint of an ideal is already defined because an ideal is a set of polynomials, and the footprint is defined for any set of polynomials. For emphasis and for the integrity of this section, however, we will repeat the definition specifically for an ideal.

Definition 7.4.1   *The footprint of the ideal $I \subset F[x, y]$, denoted $\Delta(I)$, is the set of index pairs (j', j''), both nonnegative, such that (j', j'') is not the leading bi-index of any polynomial in I.*

Thus, with regard to the ideal I, partition the set of all bivariate monomials into those monomials that are leading monomials of at least one polynomial of I and those monomials that are not the leading monomial of any polynomial of I. The footprint of I is the set of bi-indices that correspond to the monomials of the second set.

The *area* of a footprint is the number of points (j', j'') in the footprint. Possibly the area is infinite. It is obvious that the footprint is empty and that its area is zero if the ideal I is $F[x, y]$ itself. The converse of this statement – that the footprint is not empty except when the ideal I is $F[x, y]$ itself – follows from the fact, that if the footprint is empty, then the trivial monomial $x^0 y^0$ is in the ideal, as are all polynomial multiples of $x^0 y^0$.

If (j', j'') is not in the footprint of I and $(k', k'') \geqq (j', j'')$, then (k', k'') is not in the footprint of I either. This means that the footprint $\Delta(I)$ of any proper ideal I is a cascade set. A typical footprint is illustrated in Figure 7.6. Some special points outside the footprint, denoted by solid circles in the illustration, are called *exterior corners* of the footprint.

We will only study ideals in the ring of bivariate polynomials, although the generalization to the ring of multivariate polynomials is straightforward. For example, an illustration of a footprint of an ideal in the ring of trivariate polynomials is shown in Figure 7.7, which depicts how the notion of an exterior corner generalizes to three dimensions. Each exterior corner of the figure is indicated by a dot.

The footprint of a bivariate ideal is completely specified by its exterior corners. For each exterior corner, there is a polynomial in the ideal I with its leading monomial in

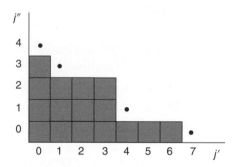

Figure 7.6. Typical bivariate footprint.

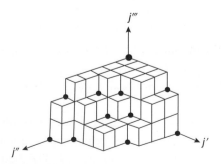

Figure 7.7. Typical trivariate footprint.

this exterior corner. Indeed, there must be a monic polynomial in I with this property. The monic polynomial, however, need not be unique. This is because the difference of two monic polynomials, both with their leading monomials in the same exterior corner of the footprint, can have a leading monomial that is not in the footprint.

This leads to the following definition.

Definition 7.4.2   *A minimal basis for the ideal I is a set of polynomials, $\{g_\ell(x,y)\} \subset I$, that consists of exactly one monic polynomial with a bidegree corresponding to each exterior corner of the footprint of I.*

Theorem 7.4.3   *The footprint of a minimal basis for the ideal $I \subset F[x,y]$ is the same as the footprint of I.*

Proof: For any set of polynomials $\mathcal{G} = \{g_1(x,y), \ldots, g_n(x,y)\}$, it is evident that

$$\Delta(\langle g_1(x,y), \ldots, g_n(x,y)\rangle) \subseteq \bigcap_{\ell=1}^{n} \Delta(\langle g_\ell(x,y)\rangle),$$

or, more concisely,

$$\Delta(I(\mathcal{G})) \subseteq \Delta(\mathcal{G}),$$

because every polynomial in $\mathcal{G}$ is also in $I(\mathcal{G})$. This inclusion holds with equality if $\mathcal{G}$ contains a minimal basis for the ideal it generates. This is evident because a minimal basis includes, for each exterior corner, one polynomial whose leading monomial lies in that exterior corner.   ∎

Thus a minimal basis is a set of monic polynomials of the ideal that has the same footprint as the entire ideal, and all polynomials of that set are essential to specifying the footprint.

The minimal basis of an ideal is not unique. We shall see that any minimal basis of I is a generator set for I. A minimal basis remains a generator set if it is enlarged by appending any other polynomials of I to it. Any set of polynomials of I that contains a minimal basis of an ideal I is called a *Gröbner basis*[1] or a *standard basis* of I. A Gröbner basis may be infinite, but we shall show in what follows that a minimal basis must be finite. This is the same as saying that the number of exterior corners in any cascade set must be finite. Loosely speaking, a descending staircase with an infinite number of steps at integer-valued coordinates cannot be contained in the first quadrant of the (j', j'') plane.

Theorem 7.4.4   *A minimal basis in $F[x,y]$ is finite.*

---

[1] A Gröbner basis is not required to be a minimal set. To respect the more common usage of the word basis, it might be better to use instead the terminology "Gröbner spanning set."

Proof: Partition the first quadrant of the (j', j'') plane into two parts, using the diagonal line $j' = j''$. This line must cross the boundary of the cascade set at some point, say at the point (k, k). Because steps in the footprint only occur at integer coordinates, there can be at most k exterior corners above this crossing point, and there can be at most k exterior corners below this crossing point. Hence there can be only a finite number of exterior corners, and, because there is one minimal basis polynomial for each exterior corner, there can be only a finite number of polynomials in a minimal basis. $\blacksquare$

Because the polynomials of a minimal basis are in one-to-one correspondence with the exterior corners of a cascade set, we will sometimes order the polynomials of a minimal basis by using the integer order on the exponent of y (or of x) in the leading monomial, calling this order the *staircase order*. This is *not* the same as ordering the minimal polynomials by using a specified total order on the leading monomials.

The set of polynomial combinations of the minimal polynomials, denoted

$$\langle g_\ell(x, y)\rangle = \left\{\sum_\ell a_\ell(x, y) g_\ell(x, y)\right\},$$

where the polynomial coefficients $a_\ell(x, y)$ are arbitrary, forms an ideal contained in the original ideal. The following theorem says that it is equal to the original ideal, so the minimal basis completely determines the ideal. Thus it is appropriate to call it a basis.

Theorem 7.4.5 *An ideal is generated by any of its minimal bases.*

Proof: Let $f(x, y)$ be any element of I, and let $\{g_\ell(x, y) \mid \ell = 1, \ldots, L\}$ be a minimal basis. Then by the division algorithm for bivariate polynomials we can write the following:

$$f(x, y) = a_1(x, y) g_1(x, y) + a_2(x, y) g_2(x, y) + \cdots + a_L(x, y) g_L(x, y) + r(x, y).$$

Because $f(x, y)$ and all $g_\ell(x, y)$ are in I, we conclude that the remainder polynomial $r(x, y)$ is in I also. Therefore the leading monomial of $r(x, y)$ is not an element of $\Delta(I)$. On the other hand, by the properties of the division algorithm, we know that the nonzero remainder polynomial $r(x, y)$ has a leading monomial that is not divisible by the leading monomial of any of the $g_\ell(x, y)$. Therefore, as implied by Theorem 7.4.3, if $r(x, y)$ is nonzero, its leading index must lie in $\Delta(I)$. The contradiction proves that $r(x, y)$ must be the zero polynomial. $\blacksquare$

In general, the remainder polynomial produced by the division algorithm may depend on the order of the polynomials in the set $\mathcal{G}$. The following theorem states that this cannot happen if $\mathcal{G}$ is a minimal basis.

Theorem 7.4.6 *Under division by the minimal basis $\mathcal{G} = \{g_\ell(x,y) \mid \ell = 1, \ldots, L\}$, any polynomial $f(x,y)$ has a unique remainder, independent of the order of the elements of $\mathcal{G}$.*

Proof: The division algorithm states that no term of the remainder polynomial $r(x,y)$ is divisible by the leading term of any $g_\ell(x,y)$. Thus the monomials of the remainder polynomial are in the footprint of the ideal, which, because $\mathcal{G}$ is a minimal basis, is equal to $\Delta(I(\mathcal{G}))$. Suppose that

$$f(x,y) = Q_1(x,y)g_1(x,y) + \cdots + Q_L(x,y)g_L(x,y) + r(x,y)$$

and

$$f(x,y) = Q_1'(x,y)g_1(x,y) + \cdots + Q_L'(x,y)g_L(x,y) + r'(x,y)$$

are two expressions generated by the division algorithm. Then

$$
\begin{aligned}
r(x,y) - r'(x,y) = {} & \left[Q_1(x,y) - Q_1'(x,y)\right]g_1(x,y) + \cdots \\
& + \left[Q_L(x,y) - Q_L'(x,y)\right]g_L(x,y).
\end{aligned}
$$

Therefore $r(x,y) - r'(x,y)$ is in the ideal $I(\mathcal{G})$. But no monomial of $r(x,y) - r'(x,y)$ is divisible by the leading monomial of any polynomial in the ideal. Hence $r(x,y) - r'(x,y) = 0$, and so $r(x,y) = r'(x,y)$. ∎

Although the remainder polynomial is unique under division by a minimal basis, the quotient polynomials need not be unique. The quotient polynomials may vary for different rearrangements of the polynomials of the minimal basis. In Section 7.5, we shall see how to choose the minimal basis so that the quotient polynomials are unique as well.

Theorem 7.4.7 (Hilbert basis theorem) Every ideal in the ring of bivariate polynomials is finitely generated.

Proof: Every ideal has a minimal basis, and a minimal basis always consists of a finite number of polynomials because a footprint has only a finite number of exterior corners. ∎

Although we have stated the bivariate case only, the Hilbert basis theorem also holds in the ring of n-variate polynomials, as do most of the notions we have discussed. Indeed, one can show that, if every ideal of the ring R is finitely generated, then every ideal of the ring $R[x]$ is finitely generated as well. The Hilbert basis theorem tells us that the set of zeros of an ideal in $F[x,y]$ is actually the set of common zeros of a finite number of polynomials. If these polynomials are irreducible, then these common zeros form a variety in the affine plane.

The following theorem says that any nested chain of ideals $I_1 \subseteq I_2 \subseteq I_3 \subseteq I_4 \subseteq \cdots \subseteq F[x, y]$ must eventually be constant. (Any ring with this property is called a *noetherian ring*.)

Corollary 7.4.8 (ascending chain condition)   *The ring $F[x, y]$ does not contain an infinite chain of properly nested ideals.*

Proof: Let $I_1 \subseteq I_2 \subseteq I_3 \subseteq \cdots$ be an infinite chain of nested ideals. We must show that eventually this chain is constant. Let

$$I = \bigcup_{\ell=1}^{\infty} I_\ell.$$

This is an ideal in $F[x, y]$, so it is generated by a minimal basis. For each generator polynomial in this finite set, there must be a value of ℓ at which that generator polynomial first appears in I_ℓ. There is only a finite number of such ℓ, so there is a largest. The largest such ℓ corresponds to an ideal I_ℓ that includes all generator polynomials of I. Therefore for this value of ℓ, $I_\ell = I$, and consequently $I_{\ell'} = I$ for all $\ell' \geq \ell$.   ∎

Note that a converse statement is also true: any ring that does not contain an infinite chain of properly nested ideals contains only rings that are finitely generated.

Definition 7.4.9   *If a minimal basis for an ideal in $F[x, y]$ consists only of monomials, then the basis is called a monomial basis, and the ideal it generates is called a monomial ideal.*

Corresponding to every ideal, $I \subset F[x, y]$, is a unique monomial ideal, denoted I^*, which is obtained by replacing each polynomial of a minimal basis of I by its leading monomial, then using these monomials as a basis to generate I^*. The footprint of I suffices to specify the monomial ideal I^*.

7.5   Reduced bases and quotient rings

The minimal basis of an ideal need not be unique; there can be many choices for a minimal basis of an ideal. Because we want to make the choice of basis unique, we constrain the basis further. The unique basis that we will now define is perhaps the most useful of the minimal bases of an ideal.

Definition 7.5.1   *A minimal basis is called a reduced basis if every nonleading monomial of each basis polynomial has a bi-index lying in the footprint of the ideal.*

Equivalently, no leading monomial of a basis polynomial in a reduced basis divides any monomial appearing in any other basis polynomial.

It is trivial to compute a reduced basis from a minimal basis because every invertible polynomial combination of the polynomials of a minimal basis contains a minimal basis. Simply list the polynomials of the minimal basis such that the leading monomials are in the total order. Then, as far as possible, subtract a monomial multiple of the last polynomial from each of the others to cancel monomials from other polynomials. Repeat this, in turn, for each polynomial higher in the list. (This process is similar to putting a matrix into reduced echelon form by gaussian elimination, hence the term "reduced basis.")

Theorem 7.5.2   *For each fixed monomial order, every nonzero ideal of $F[x, y]$ has exactly one reduced basis.*

Proof: Order the polynomials of any minimal basis of ideal I by their leading monomials; no two have the same leading monomial. Polynomial combinations of polynomials can be used to cancel a monomial that is a leading monomial of one polynomial from any other polynomial appearing on the list. Thus every ideal has at least one reduced basis.

To prove that the reduced basis is unique, let $\mathcal{G}$ and $\mathcal{G}'$ be two reduced bases for I. For each exterior corner of the footprint, each reduced basis will contain exactly one monic polynomial with its leading monomial corresponding to that exterior corner. Let $g_\ell(x, y)$ and $g'_\ell(x, y)$ be the monic polynomials of $\mathcal{G}$ and $\mathcal{G}'$, respectively, corresponding to the ℓth exterior corner of the footprint. By the division algorithm, their difference can be written as follows:

$$g_\ell(x, y) - g'_\ell(x, y) = Q_1(x, y)g_1(x, y) + \cdots + Q_L(x, y)g_L(x, y) + r(x, y).$$

Because all $g_\ell(x, y)$ are in the ideal, $r(x, y)$ is in the ideal also. But the leading monomial of $r(x, y)$ is not divisible by the leading monomial of any of the reduced polynomials. Thus

$$r(x, y) = R_\mathcal{G}[g_\ell(x, y) - g'_\ell(x, y)]$$
$$= 0.$$

By the definition of the reduced basis, all terms of $g_\ell(x, y) - g'_\ell(x, y)$ are elements of the footprint of I, and, hence, no term of $g_\ell(x, y) - g'_\ell(x, y)$ is divisible by the leading monomial of any $g_k(x, y)$. This implies that

$$g_\ell(x, y) - g'_\ell(x, y) = 0$$

for all ℓ, so the reduced basis is unique.   ∎

The theorem tells us that, in the graded order, there is a unique correspondence between reduced bases and ideals of $F[x, y]$. One way to specify an ideal is to specify

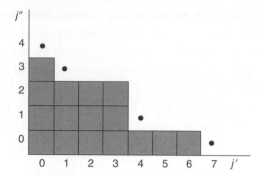

Figure 7.8. Leading monomials of a reduced basis.

its reduced basis. First, illustrate a footprint, as shown in Figure 7.8. Each black dot in an exterior corner corresponds to the leading monomial of one polynomial of the reduced basis. The coefficient of that monomial is a one. The other monomials of a basis polynomial must lie within the footprint, each paired with a coefficient from the field F to form one term of the polynomial. For an ideal with the footprint in the illustration, there are four polynomials in the reduced basis; these have the following form:

$$g_1(x, y) = x^7 \; + \; g_{60}^{(1)} x^6 \; + \; g_{50}^{(1)} x^5 \; + \; g_{40}^{(1)} x^4 \; + \cdots ;$$
$$g_2(x, y) = x^4 y + g_{32}^{(2)} x^3 y^2 \; + \; g_{40}^{(2)} x^4 \; + \; g_{31}^{(2)} x^3 y + \cdots ;$$
$$g_3(x, y) = x y^3 \; + \; g_{30}^{(3)} x^3 \; + \; g_{21}^{(3)} x^2 y + \; g_{12}^{(3)} x y^2 + \cdots ;$$
$$g_4(x, y) = y^4 \; + \; g_{30}^{(4)} x^3 \; + \; g_{21}^{(4)} x^2 y + \; g_{12}^{(4)} x y^2 + \cdots .$$

Each polynomial has one coefficient, possibly zero, for each gray square smaller than its leading monomial in the graded order. These coefficients are not completely arbitrary, however; some choices for the coefficients will not give a basis for an ideal with this footprint. Later, we will give a condition on these coefficients that ensures that the polynomials do form a reduced basis.

Looking at Figure 7.8, the following theorem should seem obvious.

Theorem 7.5.3 *If appending a new polynomial enlarges an ideal, then the footprint of the ideal becomes smaller.*

Proof: It is clear that after appending a new polynomial, the footprint must be contained in the original footprint, so it is enough to show that the footprint does not remain the same. Let I be the original ideal and I' the expanded ideal. Because the ideals are not equal, the reduced basis for I is not the same as the reduced basis for I'. Consider any polynomial of the reduced basis for I' not in the reduced basis for I. There can be only one monic polynomial in the ideal with the leading monomial of that reduced basis polynomial. Hence that exterior corner of Δ' is not an exterior corner of Δ. Hence $\Delta \neq \Delta'$, and the proof is complete. ∎

Now we turn our attention to certain polynomials that are not in the ideal I. Specifically, we will consider those polynomials whose monomials all lie inside the footprint. Polynomials with all monomials in Δ are not in I, because no polynomial of I has a leading monomial lying in the footprint of I. Nevertheless, the set of these polynomials can itself be given its own ring structure, defined in terms of the structure of I. This ring structure underlies the notion of a quotient ring.

The formal definition of a quotient ring uses the concept of an equivalence class. Every $I \subset F[x, y]$ can be used to partition $F[x, y]$ into certain subsets called *equivalence classes*. Given any $p(x, y) \in F[x, y]$, the equivalence class of $p(x, y)$, denoted $\{p(x, y)\}$, is the set given by

$$\{p(x, y) + i(x, y) \mid i(x, y) \in I\}.$$

The set of such sets, called a *quotient ring*, is denoted $F[x, y]/I$:

$$F[x, y]/I = \{\{p(x, y) + i(x, y) \mid i(x, y) \in I\}, p(x, y) \in F[x, y]\}.$$

Thus all elements of $F[x, y]$ that differ by an element of I are *equivalent*.

To give $F[x, y]/I$ a ring structure, we must define the ring operations. This means that we must define the addition and multiplication of equivalence classes, and then verify that these definitions do satisfy the axioms of a ring. Addition or multiplication of equivalence classes is quite straightforward in principle. Define the sum of equivalence classes $\{a(x, y)\}$ and $\{b(x, y)\}$ as the equivalence class $\{a(x, y) + b(x, y)\}$. Define the product of equivalence classes $\{a(x, y)\}$ and $\{b(x, y)\}$ as the equivalence class $\{a(x, y)b(x, y)\}$. It is straightforward to prove that the sum or product does not depend on the choice of representatives from the sets $\{a(x, y)\}$ and $\{b(x, y)\}$, so the definitions are proper.

A more concrete, and more computational, approach to the notion of a quotient ring is to define the elements of the quotient ring as individual polynomials, as follows. Let I be an ideal with the footprint Δ. The footprint of I is a cascade set Δ, which we use to define the support of a two-dimensional array of elements from the field F. To define the quotient ring, first define the set given by

$$F[x, y]/I = \{p(x, y) \mid p_{i'i''} = 0 \text{ if } (i', i'') \notin \Delta\}.$$

To form the polynomial $p(x, y)$ of $F[x, y]/I$, simply fill the cells of Δ with any coefficients $p_{i'i''}$. This array of coefficients defines a polynomial of $F[x, y]/I$. The coefficients are arbitrary, and each choice of coefficients gives one element of the quotient ring; there are no others. To reconcile this definition with the formal definition of $F[x, y]/I$, the individual polynomial $p(x, y)$ is regarded as the canonical representative of an equivalence class of that formal definition.

Addition in the quotient ring is defined as addition of polynomials. Under addition, $F[x, y]/I$ is a vector space, and the monomials that correspond to the cells of Δ are a basis for this vector space.

Multiplication in the quotient ring is defined as multiplication of polynomials modulo $\mathcal{G}$, where $\mathcal{G}$ is the reduced basis for the ideal I. Thus

$$a(x, y) \cdot b(x, y) = R_{\mathcal{G}}[a(x, y)b(x, y)].$$

The ring multiplication can be described in a very explicit way. To multiply $b(x, y)$ by x or y, the elements of the reduced basis $\mathcal{G}$ are used to "fold back" those terms of $xb(x, y)$ or $yb(x, y)$ that are outside the support of $F[x, y]/I$, which support is given by the footprint Δ. In the general case, when $b(x, y)$ is multiplied by $a(x, y)$, the polynomial product $a(x, y)b(x, y)$ produces terms whose indices lie outside the support of Δ. Those coefficients are then folded back, one by one, into Δ by subtracting, one by one, polynomial multiples of elements of $\mathcal{G}$ to cancel terms outside of Δ. Because $\mathcal{G}$ is a reduced basis for I, as a consequence of Theorem 7.4.6, it does not matter in which order the terms are folded back; the final result will be the same.

Every polynomial $g_\ell(x, y)$ of the reduced basis of the ideal I can be conveniently written as a sum of the form

$$g_\ell(x, y) = x^{m_\ell} y^{n_\ell} + g_\ell^-(x, y),$$

where the monomial $x^{m_\ell} y^{n_\ell}$ corresponds to an exterior corner of the footprint Δ, and $g_\ell^-(x, y)$ is an element of the quotient ring $F[x, y]/I$. Reduction modulo $\mathcal{G}$ simply consists of eliminating monomials outside the footprint, one by one, by setting $x^{m_\ell} y^{n_\ell} = -g_\ell^-(x, y)$ as needed and in any order. For this purpose, it may be convenient to order the polynomials of $\mathcal{G}$ by the staircase order such that m_ℓ is decreasing with ℓ. Then the steps of the reduction could be ordered by decreasing m_ℓ.

Theorem 7.5.4 *The quotient ring is a ring.*

Proof: Clearly, the set is closed under addition and closed under multiplication. To verify that it satisfies the properties of a ring, recall that, for a reduced basis, the equation

$$f(x, y) = Q_1(x, y)g_1(x, y) + \cdots + Q_L(x, y)g_L(x, y) + r(x, y)$$

for any $f(x, y)$ is satisfied by a unique $r(x, y)$. The associative law for multiplication is verified by setting $f(x, y) = a(x, y)b(x, y)c(x, y)$ and noting that the $r(x, y)$ solving this equation will be the same no matter how it is computed because $\mathcal{G}$ is a reduced basis. Thus

$$R_{\mathcal{G}}[a(x, y)R_{\mathcal{G}}[b(x, y)c(x, y)]] = R_{\mathcal{G}}[[R_{\mathcal{G}}[a(x, y)b(x, y)]]c(x, y)].$$

The other ring properties are verified by similar reasoning. ∎

7.6   The Buchberger theorem

As we have seen, a reduced basis $\mathcal{G}$ for an ideal can be used in two important constructions: it can be used to construct the ideal I as the set of polynomial combinations of
elements of $\mathcal{G}$, or it can be used to carry out multiplication in the quotient ring $F[x, y]/I$.
Because the reduced basis is so important, we want a test that can be used to recognize
a reduced basis. It is enough here to give a test to recognize a minimal basis, because
it is easy to recognize when a minimal basis is a reduced basis.

Theorem 7.6.1 (Buchberger)   *A set of polynomials, $\mathcal{G} = \{g_1(x, y), \ldots, g_L(x, y)\}$,
contains a minimal basis for the ideal I of $F[x, y]$ generated by $\mathcal{G}$ if and only if all
conjunction polynomials $b_{\mathcal{G}}^{g_\ell g_{\ell'}}(x, y)$, $\ell = 1, \ldots, L$ and $\ell' = 1, \ldots, L$, are equal to zero.*

Proof: The condition of the theorem can be seen to be necessary by repeating a
recurring argument. Suppose that $\mathcal{G}$ contains a minimal basis and that a conjunction
polynomial is not equal to zero. The conjunction polynomial is in the ideal I, so its
leading monomial is not in the footprint $\Delta(I)$. But the conjunction polynomial has a
leading monomial not divisible by the leading monomial of any $g_\ell(x, y)$ of $\mathcal{G}$. Therefore
since the conjunction polynomial is nonzero, its leading monomial is in the footprint
$\Delta(I)$. The contradiction proves that if $\mathcal{G}$ contains a minimal basis, then the conjunction
polynomial is zero.

To show that the condition is also sufficient will take some effort. We will show
that if all conjunction polynomials are zero, then $\mathcal{G}$ must contain a minimal basis by
showing that $\Delta(\mathcal{G}) = \Delta(I(\mathcal{G}))$.

Choose any polynomial $v(x, y)$ in the ideal I. Then because I is generated by $\mathcal{G}$, we
have the representation

$$v(x, y) = \sum_{\ell=1}^{L} h_\ell(x, y)g_\ell(x, y)$$

for some expansion polynomials $h_\ell(x, y)$, $\ell = 1, \ldots, L$. This representation for $v(x, y)$
need not be unique; in general, there will be many ways to choose the $h_\ell(x, y)$.

The bidegree of $v(x, y)$ satisfies

$$\text{bideg } v(x, y) \preceq \max_\ell \text{ bideg } h_\ell(x, y)g_\ell(x, y),$$

where, if equality does not hold, there must be a cancellation among the leading terms
of the $h_\ell(x, y)g_\ell(x, y)$. We will show that we can always choose the $h_\ell(x, y)$ so that this
inequality is satisfied with equality.

Let $\delta = \max_\ell \text{ bideg } h_\ell(x, y)g_\ell(x, y)$, and consider only those $h_\ell(x, y)$ for which
bideg $h_\ell(x, y)g_\ell(x, y) = \delta$. If bideg $v(x, y) \prec \delta$, cancellation must occur in the

leading monomial, so there must be at least two such terms, say $h_j(x, y)g_j(x, y)$ and $h_k(x, y)g_k(x, y)$. We shall focus our attention on the pair of polynomials $h_j(x, y)$ and $h_k(x, y)$, replacing them by new polynomials, of which at least one has a smaller bidegree. The leading monomials of $h_j(x, y)$ and $h_k(x, y)$ align $g_j(x, y)$ and $g_k(x, y)$ to have a common leading monomial so that the cancellation can occur. These leading monomials need not be the minimal monomials that align $g_j(x, y)$ and $g_k(x, y)$ when computing the conjunction polynomial of $g_j(x, y)$ and $g_k(x, y)$.

Because all conjunction polynomials are assumed to be zero, we can write

$$x^{a'_j} y^{a''_j} g_j(x, y) - x^{a'_k} y^{a''_k} g_k(x, y) = \sum_{\ell=1}^{L} Q_{jk\ell}(x, y)g_\ell(x, y)$$

with quotient polynomials $Q_{jk\ell}(x, y)$, where $x^{a'_j} y^{a''_j}$ and $x^{a'_k} y^{a''_k}$ are the minimal monomials that align $g_j(x, y)$ and $g_k(x, y)$. The left side has bidegree smaller than δ, as do all terms on the right side. From this equation, we can write $g_j(x, y)$ in terms of $g_k(x, y)$:

$$x^{b'} y^{b''} \left[x^{a'_j} y^{a''_j} g_j(x, y) \right] = x^{b'} y^{b''} \left[x^{a'_k} y^{a''_k} g_k(x, y) + \sum_{\ell=1}^{L} Q_{jk\ell}(x, y)g_\ell(x, y) \right],$$

where the terms in the sum on ℓ on the right do not contribute to the leading monomial, and where (b', b'') can be chosen so that the leading monomial on each side has bidegree δ. In other words, (b', b'') is chosen such that $x^{b'+a'_j} y^{b''+a''_j}$ and $x^{b'+a'_k} y^{b''+a''_k}$ are equal to the leading monomials of $h_j(x, y)$ and $h_k(x, y)$.

Now add and subtract a scalar multiple of the two sides of this equality from the earlier expression for $v(x, y)$ to write

$$v(x, y) = h_j(x, y)g_j(x, y) + \sum_{\ell \neq j} h_\ell(x, y)g_\ell(x, y)$$

$$= h_j(x, y)g_j(x, y) - Ax^{b'} y^{b''} x^{a'_j} y^{a''_j} g_j(x, y)$$

$$+ \sum_{\ell \neq j} h_\ell(x, y)g_\ell(x, y) + Ax^{b'} y^{b''} \left[x^{a'_k} y^{a''_k} g_k(x, y) + \sum_{\ell=1}^{L} Q_{jk\ell}(x, y)g_\ell(x, y) \right]$$

for some scalar A. The next step of the proof is to gather terms to rewrite this as follows:

$$v(x, y) = h'_j(x, y)g_j(x, y) + \sum_{\ell \neq j} h'_\ell(x, y)g_\ell(x, y),$$

where the new jth expansion polynomial is given by

$$h'_j(x, y) = h_j(x, y) - Ax^{b'+a'_j} y^{b''+a''_j} + Ax^{b'} y^{b''} Q_{jkj}(x, y),$$

and A has been chosen to cancel the leading monomial in the first two terms on the right so that the leading monomial of $h'_j(x, y)$ has a bidegree smaller than the bidegree of $h_j(x, y)$. The new kth expansion polynomial is given by

$$h'_k(x, y) = h_k(x, y) + Ax^{b'+a'_k}y^{b''+a''_k} + Ax^{b'}y^{b''}Q_{jkk}(x, y).$$

The other new expansion polynomials are

$$h'_\ell(x, y) = h_\ell(x, y) + Ax^{b'}y^{b''}Q_{jk\ell}(x, y) \quad \ell \neq j \text{ or } k.$$

For every ℓ other than j, $h_\ell(x, y)g_\ell(x, y)$ can have a leading monomial of bidegree δ only if it had it before, while for $\ell = j$ the bidegree of $h_j(x, y)g_j(x, y)$ is smaller. In this way, the number of expansion polynomials $h'_\ell(x, y)$ for which bideg $h'_\ell(x, y)g_\ell(x, y) = \delta$ is reduced by at least one. After this step, if there are two or more such terms remaining with bidegree equal to δ, the process can be repeated. If there are no such terms remaining, the value of δ is reduced, and the process is repeated for that new value of δ. The process will halt only if exactly one term remains for the current value of δ.

Thus, for the process to stop, we see that, eventually, we will obtain the polynomials $h_\ell(x, y)$ such that

$$\text{bideg } v(x, y) = \max_\ell \text{ bideg } h_\ell(x, y)g_\ell(x, y),$$

because there can be no cancellation in the leading monomial. From this equality, we see that the leading monomial of $v(x, y)$ is divisible by the leading monomial of one of the $g_\ell(x, y)$. Thus $\Delta(\mathcal{G}) = \Delta(I(\mathcal{G}))$, so, for each exterior corner of $\Delta(I(\mathcal{G}))$, $\mathcal{G}$ contains a polynomial with its leading monomial corresponding to that exterior corner. We conclude that the set $\mathcal{G}$ contains a minimal basis, and the proof is complete. ∎

The Buchberger theorem might be a bit daunting because it requires the computation of $L(L-1)/2$ conjunction polynomials. The following corollary presents the condition of the Buchberger theorem in a way that requires the computation of only $L-1$ conjunction polynomials. We shall see that with the polynomials arranged in staircase order (decreasing m_ℓ), the $\binom{L}{2}$ equalities required by the theorem reduce to only those $L-1$ equalities given by

$$R_{\mathcal{G}}[y^{n_{\ell+1}-n_\ell}g_\ell(x, y)] = R_{\mathcal{G}}[x^{m_\ell-m_{\ell+1}}g_{\ell+1}(x, y)]$$

for $\ell = 1, \ldots, L-1$. This simplification is closely tied to the footprint of the set of generator polynomials. Each leading monomial of a generator polynomial accounts for one exterior corner of the footprint, and the conjunction polynomial needs to be computed only for pairs of polynomials that correspond to neighboring exterior corners.

The notation $g(x, y) = x^m y^n + g^-(x, y)$ for a generator polynomial is used in this corollary so that the leading monomial can be displayed explicitly.

Corollary 7.6.2 *The set of monic polynomials $\mathcal{G} = \{g_\ell(x, y) = x^{m_\ell} y^{n_\ell} + g_\ell^-(x, y) \mid \ell = 1, \ldots, L\}$ is a reduced basis for an ideal with footprint Δ if and only if there is one leading monomial $x^{m_\ell} y^{n_\ell}$ for each exterior corner of Δ, all nonleading monomials are in Δ, and with the elements of $\mathcal{G}$ arranged in a staircase order,*

$$R_\mathcal{G}[y^{n_{\ell+1}-n_\ell} g_\ell^-(x, y)] = R_\mathcal{G}[x^{m_\ell - m_{\ell+1}} g_{\ell+1}^-(x, y)],$$

for $\ell = 1, \ldots, L-1$.

Proof: The polynomials are in a staircase order such that $m_0 > m_1 > m_2 > \cdots > m_L$ and $n_0 < n_1 < n_2 < \cdots < n_L$. Theorem 7.6.1 requires that

$$R_\mathcal{G}[y^{n_{\ell'}-n_\ell} g_\ell(x, y) - x^{m_\ell - m_{\ell'}} g_{\ell'}(x, y)] = 0$$

for all ℓ and $\ell' > \ell$. Because the leading monomials cancel, and the operation of forming the remainder modulo $\mathcal{G}$ can be distributed across addition, the Buchberger condition can be rewritten:

$$R_\mathcal{G}[y^{n_{\ell'}-n_\ell} g_\ell^-(x, y)] = R_\mathcal{G}[x^{m_\ell - m_{\ell'}} g_{\ell'}^-(x, y)]$$

for all ℓ and $\ell' > \ell$. To prove the theorem, we will assume only that this equation holds for $\ell = 1, \ldots, L-1$ and $\ell' = \ell+1$, and we will show that this implies that the equation holds for all ℓ and $\ell' > \ell$. In particular, we will show that because neighbors satisfy this condition, then second neighbors must satisfy this condition as well, and so on. This will be shown as a simple consequence of the fact that the $g_\ell^-(x, y)$ are all elements of $F[x, y]/I$. By reinterpreting the multiplications as operations in the quotient ring $F[x, y]/I$, the operator $R_\mathcal{G}$ becomes a superfluous notation and can be dropped. The proof, then, consists of the following simple calculations in $F[x, y]/I$.

We are given that

$$y^{n_{\ell+1}-n_\ell} g_\ell^-(x, y) = x^{m_\ell - m_{\ell+1}} g_{\ell+1}^-(x, y)$$

and

$$y^{n_\ell - n_{\ell-1}} g_{\ell-1}^-(x, y) = x^{m_{\ell-1} - m_\ell} g_\ell^-(x, y).$$

From these, we can write the following:

$$
\begin{aligned}
y^{n_{\ell+1}-n_\ell}[y^{n_\ell - n_{\ell-1}} g_{\ell-1}^-(x, y)] &= y^{n_{\ell+1}-n_\ell}[x^{m_{\ell-1}-m_\ell} g_\ell^-(x, y)] \\
&= x^{m_{\ell-1}-m_\ell}[y^{n_{\ell+1}-n_\ell} g_\ell^-(x, y)] \\
&= x^{m_{\ell-1}-m_\ell}[x^{m_\ell - m_{\ell+1}} g_{\ell+1}^-(x, y)].
\end{aligned}
$$

Therefore

$$y^{n_{\ell+1}-n_{\ell}-1}g_{\ell-1}^{-}(x,y) = x^{m_{\ell-1}-m_{\ell+1}}g_{\ell+1}^{-}(x,y).$$

Thus, if the condition of the corollary holds for neighbors, it also holds for second neighbors. In the same way, the condition can be verified for more distant neighbors, and the proof of the theorem is complete.   ∎

As an example of the theorem, consider the ideal $I = \langle x^3 + x^2 y + xy + x + 1, y^2 + y \rangle$ in the ring $GF(2)[x, y]$. By Buchberger's theorem, the set $\mathcal{G} = \{x^3 + x^2 y + xy + x + 1, y^2 + y\}$ contains a minimal basis because the conjunction polynomial given by

$$R_{\mathcal{G}}[y^2(x^3 + x^2 y + xy + x + 1) - x^3(y^2 + y)]$$

is equal to zero. Moreover, the set $\mathcal{G}$ has no subset that will generate the ideal, so the set itself is a minimal basis and, in fact, is the reduced basis for I. Thus the footprint of this ideal can be easily illustrated, as shown in Figure 7.9.

The theorem also tells us how to enlarge the ideal $\langle x^3 + x^2 y + xy + x + 1, y^2 + y \rangle$ to a new ideal, whose footprint is illustrated in Figure 7.10. To the original reduced basis, we simply adjoin a new polynomial, whose leading monomial is in the new exterior corner, and for it to be in the reduced basis of the new ideal, all nonleading terms of the new polynomial with nonzero coefficients must correspond to points of the new footprint. Thus the new polynomial appended to the basis $\mathcal{G}$ has the form

$$p(x, y) = x^2 y + ax^2 + bxy + cx + dy + e,$$

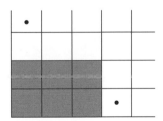

Figure 7.9. Footprint of $\{x^3 + x^2 y + xy + x + 1, \ y^2 + y\}$.

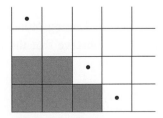

Figure 7.10. Footprint of an enlarged ideal.

where the unspecified coefficients a, b, c, d, and e are not arbitrary; they must be chosen to satisfy the Buchberger theorem. A straightforward computation of conjunction polynomials shows that there are exactly six such $p(x, y)$, all with coefficients in $GF(8)$, but we defer discussion of this point until Section 7.11 because then we will be prepared to give other insights as well.

We end this section with a rather long discussion of the important fact that the bivariate polynomials comprising any set $\mathcal{G}$ have no common polynomial factor if and only if the footprint of the ideal generated by $\mathcal{G}$ has a finite area. This fact is given as Theorem 7.6.4, whose proof rests on the unique factorization theorem. Theorem 7.6.4 can be read immediately, but instead we will preview the discussion of that theorem by first giving a simplified version with only two polynomials in the reduced basis.

Proposition 7.6.3 *Suppose that the set $\mathcal{G} = \{g_1(x, y), g_2(x, y)\}$ forms a minimal basis for the ideal $I \subset F[x, y]$. Then the footprint of $\mathcal{G}$ has a finite area if and only if the two polynomials of $\mathcal{G}$ do not have a common nontrivial polynomial factor.*

Proof: The footprint of the ideal has two exterior corners at (m_1, n_1) and (m_2, n_2), with $m_1 > m_2$ and $n_2 > n_1$. The footprint has infinite area unless both n_1 and m_2 are zero. Corresponding to the two exterior corners of $\mathcal{G}$ are the leading monomials of $g_1(x, y)$ and $g_2(x, y)$, given by $x^{m_1} y^{n_1}$ and $x^{m_2} y^{n_2}$. Clearly, if $g_1(x, y)$ and $g_2(x, y)$ have a common nontrivial polynomial factor, say $a(x, y)$, then n_1 and m_2 cannot both be zero, so the area is infinite.

Suppose now that $g_1(x, y)$ and $g_2(x, y)$ have no common nontrivial polynomial factor. We must show that n_1 and m_2 are both zero. By the Buchberger theorem,

$$R_{\mathcal{G}}[y^{n_2 - n_1} g_1(x, y) - x^{m_1 - m_2} g_2(x, y)] = 0,$$

which is equivalent to the equation

$$y^{n_2 - n_1} g_1(x, y) - x^{m_1 - m_2} g_2(x, y) = Q_1(x, y) g_1(x, y) + Q_2(x, y) g_2(x, y)$$

for some quotient polynomials $Q_1(x, y)$ and $Q_2(x, y)$. Thus

$$\left[y^{n_2 - n_1} - Q_1(x, y) \right] g_1(x, y) = \left[x^{m_1 - m_2} + Q_2(x, y) \right] g_2(x, y).$$

We can conclude, moreover, that $y^{n_2 - n_1}$ and $x^{m_1 - m_2}$ are the leading monomials of the two bracketed terms by the following argument. First, observe that the leading monomials on the left side of the previous equation cancel, so

$$\text{bideg} \left[y^{n_2 - n_1} g_1(x, y) - x^{m_1 - m_2} g_2(x, y) \right] \prec \text{bideg} \left[y^{n_2 - n_1} g_1(x, y) \right];$$
$$\text{bideg} \left[y^{n_2 - n_1} g_1(x, y) - x^{m_1 - m_2} g_2(x, y) \right] \prec \text{bideg} \left[x^{m_1 - m_2} g_2(x, y) \right].$$

Also observe that the division algorithm states that

$$\text{bideg} \left[Q_1(x,y)g_1(x,y)\right] \preceq \text{bideg} \left[y^{n_2-n_1}g_1(x,y) - x^{m_1-m_2}g_1(x,y)\right]$$
$$\prec \text{bideg} \left[y^{n_2-n_1}g_1(x,y)\right];$$
$$\text{bideg} \left[Q_2(x,y)g_2(x,y)\right] \preceq \text{bideg} \left[y^{n_2-n_1}g_1(x,y) - x^{m_1-m_2}g_2(x,y)\right]$$
$$\prec \text{bideg} \left[x^{m_1-m_2}g_2(x,y)\right].$$

Under any monomial order, the bidegrees of polynomials add under multiplication of polynomials, so the division algorithm requires that

$$\text{bideg} \left[Q_1(x,y)\right] \prec \text{bideg} \left[y^{n_2-n_1}\right];$$
$$\text{bideg} \left[Q_2(x,y)\right] \prec \text{bideg} \left[x^{m_1-m_2}\right],$$

as was claimed.

Now we can conclude that, because $g_1(x,y)$ and $g_2(x,y)$ have no common polynomial factor, the unique factorization theorem requires that the expression

$$\left[y^{n_2-n_1} - Q_1(x,y)\right]g_1(x,y) = \left[x^{m_1-m_2} + Q_2(x,y)\right]g_2(x,y)$$

can be partially factored as

$$[a(x,y)g_2(x,y)]g_1(x,y) = [a(x,y)g_1(x,y)]g_2(x,y)$$

for some common polynomial $a(x,y)$, where

$$a(x,y)g_2(x,y) = y^{n_2-n_1} - Q_1(x,y),$$
$$a(x,y)g_1(x,y) = x^{m_1-m_2} + Q_2(x,y).$$

The product of the leading monomials on the left of the first of these two equations must equal $y^{n_2-n_1}$, from which we conclude that the leading monomial of $a(x,y)$ does not depend on x. The product of the leading monomial on the left of the second of these two equations allows us to conclude that the leading monomial of $a(x,y)$ does not depend on y. Hence $a(x,y)$ has degree 0, and

$$x^{m_1}y^{n_1} + g_1^-(x,y) = x^{m_1-m_2} + Q_2(x,y),$$
$$x^{m_2}y^{n_2} + g_2^-(x,y) = y^{n_2-n_1} - Q_1(x,y).$$

Consequently, we conclude that if $g_1(x,y)$ and $g_2(x,y)$ have no common nontrivial polynomial factor $a(x,y)$, then m_2 and n_1 are both zero, and

$$g_1(x,y) = x^{m_1} + Q_2(x,y),$$
$$g_2(x,y) = y^{n_2} - Q_1(x,y).$$

as was to be proved. ∎

The general version of this proposition has more than two polynomials in a minimal basis of an ideal $I \subset F[x, y]$. Certainly, if any two polynomials of I have no common nontrivial polynomial factor, then those two polynomials generate an ideal whose footprint has finite area and contains the footprint of I. Hence $\Delta(I)$ must have finite area as well. However, the general situation is more subtle. It may be that, pairwise, generator polynomials do have a common polynomial factor but, jointly, do not have a common polynomial factor. An example of such a case is

$$\mathcal{G} = \{a(x, y)b(x, y), b(x, y)c(x, y), c(x, y)a(x, y)\}.$$

Then, any pair of generator polynomials generates an ideal whose footprint has infinite area, but the full set of generator polynomials generates an ideal whose footprint has finite area.

Before we prove the proposition for the case with an arbitrary number of polynomials in a minimal basis, we further preview the method of proof by considering an ideal with three polynomials in a minimal basis. Then the Buchberger theorem yields two equations:

$$y^{n_2-n_1} g_1(x, y) - x^{m_1-m_2} g_2(x, y) = \sum_{\ell=1}^{3} Q_\ell^{(1)}(x, y) g_\ell(x, y)$$

and

$$y^{n_3-n_2} g_2(x, y) - x^{m_2-m_3} g_3(x, y) = \sum_{\ell=1}^{3} Q_\ell^{(2)}(x, y) g_\ell(x, y).$$

Although the leading monomials of the terms on the left could be canceled, we choose not to cancel them. Instead, we will allow these monomials to migrate to different positions in the equations so that we can recognize the leading monomials within these equations.

Abbreviate $g_\ell(x, y)$ by g_ℓ and $Q_{\ell'}^{(\ell)}(x, y)$ by $Q_{\ell'}^{(\ell)}$, and eliminate $g_2(x, y)$ from these two equations, to obtain the following single equation:

$$\left[(x^{m_1-m_2} + Q_2^{(1)})Q_1^{(2)} - (y^{n_3-n_2} - Q_2^{(2)})(y^{n_2-n_1} - Q_1^{(1)}) \right] g_1$$
$$= \left[(x^{m_1-m_2} + Q_2^{(1)})(x^{m_2-m_3} + Q_3^{(2)}) - (y^{n_3-n_2} - Q_2^{(2)})Q_3^{(1)} \right] g_3.$$

The leading term on the left is the product of $y^{n_3-n_2}y^{n_2-n_1} = y^{n_3-n_1}$. The leading term on the right is the product $x^{m_1-m_2}x^{m_2-m_3} = x^{m_1-m_3}$. By gathering other terms into new polynomials $A_1(x, y)$ and $A_2(x, y)$, we may write this equation compactly as follows:

$$\left(y^{n_3-n_1} + A_1(x, y) \right) g_1(x, y) = (x^{m_1-m_3} + A_3(x, y)) g_3(x, y),$$

where only the leading monomial $y^{n_3-n_1}$ is written explicitly in the first term on the left, and only the leading monomial $x^{m_1-m_3}$ is written explicitly in the first term on the right. Consequently, again by the unique factorization theorem, we conclude that either the leading monomials of $g_1(x,y)$ and $g_3(x,y)$ must involve only x and y, respectively, or $g_1(x,y)$ and $g_3(x,y)$ have the nontrivial polynomial factor $a(x,y)$ in common.

To see that a common factor $a(x,y)$ of $g_1(x,y)$ and $g_3(x,y)$ must then also divide $g_2(x,y)$, note that the two equations of the Buchberger theorem imply that any common factor $a(x,y)$ of both $g_1(x,y)$ and $g_3(x,y)$ must be a factor of both $[x^{m_1-m_2}+Q_2^{(1)}(x,y)]g_2(x,y)$ and $[y^{n_3-n_2}-Q_2^{(2)}(x,y)]g_2(x,y)$, where $x^{m_1-m_2}$ and $y^{n_3-n_2}$ are the leading monomials of the bracketed terms. Because $a(x,y)$ cannot divide both bracketed terms, it must divide $g_2(x,y)$. Hence $a(x,y)$ divides all three polynomials of the reduced basis unless the leading monomial of $g_1(x,y)$ involves only x and the leading monomial of $g_3(x,y)$ involves only y.

The general theorem involves essentially the same proof, but the reduced basis now has L polynomials. To carry through the general proof, we will need to set up and solve a linear system of polynomial equations of the form $A(x,y)a(x,y) = b(x,y)$, where $A(x,y)$ is a matrix of polynomials, and $a(x,y)$ and $b(x,y)$ are vectors of polynomials. The matrix $A(x,y)$ is a matrix of polynomials, so it need not have an inverse within the ring of polynomials because $\det A(x,y)$ can have an inverse only if it is a polynomial of degree 0. However, for our needs, it will be enough to use *Cramer's rule* in the divisionless form:

$$(\det A(x,y))a_i(x,y) = \det A^{(i)}(x,y),$$

where $A^{(i)}(x,y)$ is the matrix obtained by replacing the ith column of $A(x,y)$ by the column vector $b(x,y)$ on the right side of the previous equation. The definition of a determinant applies as well to a matrix of polynomials. As usual, the determinant of the matrix $A(x,y)$, with elements $a_{ij}(x,y)$, is defined as

$$\det A(x,y) = \sum \xi_{i_1\dots i_n} a_{1i_1}(x,y)a_{2i_2}(x,y)\cdots a_{ni_n}(x,y),$$

where $i_1,i_2,\dots,i_n$ is a permutation of the integers $1,2,\dots,n$; the sum is over all possible permutations of these integers; and $\xi_{i_1\dots i_n}$ is ± 1, according to whether the permutation is even or odd. In particular, the product of all diagonal terms of $A(x,y)$ appears in $\det A(x,y)$, as does the product of all terms in the first extended off-diagonal, and so on.

Theorem 7.6.4   *Suppose that the set of bivariate polynomials $\mathcal{G}$ forms a minimal basis for the ideal in $F[x,y]$ that it generates. Then either the footprint of $\mathcal{G}$ has finite area, or the elements of $\mathcal{G}$ have a common nontrivial polynomial factor.*

Proof: Index the generator polynomials in staircase order. The footprint of the ideal has an exterior corner at (m_ℓ, n_ℓ), corresponding to the generator polynomial $g_\ell(x,y)$

for $\ell = 1, \ldots, L$. Clearly, if the generator polynomials have a nontrivial common polynomial factor, then n_1 and m_L cannot both be zero, so the footprint has infinite area.

The proof of the converse begins with the corollary to the Buchberger theorem, which states that

$$y^{n_{\ell+1}-n_\ell} g_\ell(x,y) - x^{m_\ell - m_{\ell+1}} g_{\ell+1}(x,y) = \sum_{\ell'=1}^{L} Q_{\ell'}^{(\ell)}(x,y) g_{\ell'}(x,y) \quad \ell = 1, \ldots, L-1.$$

Let $\delta_\ell = m_\ell - m_{\ell+1}$ and $\epsilon_\ell = n_{\ell+1} - n_\ell$, and abbreviate $g_\ell(x,y)$ as g_ℓ and $Q_{\ell'}^{(\ell)}(x,y)$ as $Q_{\ell'}^{(\ell)}$. The system of equations can be written in matrix form as follows:

$$
\begin{bmatrix}
y^{\epsilon_1} & -x^{\delta_1} & 0 & \cdots & 0 \\
0 & y^{\epsilon_2} & -x^{\delta_2} & & 0 \\
0 & 0 & y^{\epsilon_3} & \cdots & 0 \\
\vdots & & & & \vdots \\
0 & 0 & 0 & \cdots & -x^{\delta_{L-1}}
\end{bmatrix}
\begin{bmatrix}
g_1 \\ g_2 \\ g_3 \\ \vdots \\ g_L
\end{bmatrix}
$$

$$
=
\begin{bmatrix}
Q_1^{(1)} & Q_2^{(1)} & \cdots & Q_L^{(1)} \\
Q_1^{(2)} & Q_2^{(2)} & \cdots & Q_L^{(2)} \\
\vdots & & & \vdots \\
Q_1^{(L-1)} & Q_2^{(L-1)} & \cdots & Q_L^{(L-1)}
\end{bmatrix}
\begin{bmatrix}
g_1 \\ g_2 \\ g_3 \\ \vdots \\ g_L
\end{bmatrix},
$$

where the matrices have L columns and $L-1$ rows. This equation can be rearranged as

$$
M
\begin{bmatrix}
g_1 \\ g_1 \\ g_3 \\ \vdots \\ g_{L-1}
\end{bmatrix}
=
\begin{bmatrix}
Q_L^{(1)} \\
Q_L^{(2)} \\
\vdots \\
x^{\delta_{L-1}} + Q_L^{(L-1)}
\end{bmatrix}
g_L,
$$

where

$$
M =
\begin{bmatrix}
Q_1^{(1)} - y^{\epsilon_1} & x^{\delta_1} + Q_2^{(1)} & Q_3^{(1)} & \cdots & Q_{L-1}^{(1)} \\
Q_1^{(2)} & Q_2^{(2)} - y^{\epsilon_2} & x^{\delta_2} + Q_3^{(2)} & \cdots & \\
\vdots & & \vdots & & \vdots \\
Q_1^{(L-1)} & & & & Q_{L-1}^{(L-1)} - y^{\epsilon_{L-1}}
\end{bmatrix}
$$

is an $L-1$ by $L-1$ matrix of polynomials. The common term $g_L(x,y)$ has been factored out of every term of the vector on the right, so we can treat it separately. By Cramer's rule, we can write

$$(\det M(x, y))g_1(x, y) = (\det M^{(1)}(x, y))g_L(x, y).$$

Because the product of all diagonal terms of $M(x, y)$ forms one of the terms of $\det M(x, y)$, we can conclude further that $\prod_{\ell=1}^{L-1} y^{\epsilon_\ell}$ is the leading monomial of $\det M(x, y)$ because each factor in the product is the leading monomial of one term of the matrix diagonal. This leading monomial is equal to $y^{n_L - n_1}$. Moreover, $\prod_{\ell=1}^{L-1} x^{\delta_\ell}$ is the leading monomial of $\det M^{(1)}(x, y)$. This monomial is equal to $x^{m_1 - m_L}$. Thus, for some appropriate polynomials $A_1(x, y)$ and $A_L(x, y)$,

$$[y^{n_L - n_1} + A_1(x, y)]g_1(x, y) = [x^{m_1 - m_L} + A_L(x, y)]g_L(x, y).$$

From this, as in the proof of Proposition 7.6.3, we can conclude that if $g_1(x, y)$ and $g_L(x, y)$ have no common nontrivial polynomial factor, then the leading monomial of $g_1(x, y)$ is $x^{m_1 - m_L}$ and the leading monomial of $g_L(x, y)$ is $y^{n_L - n_1}$. This means that n_1 and m_L are zero, so

$$g_1(x, y) = x^{m_1} + g_1^-(x, y),$$

$$g_L(x, y) = y^{n_L} + g_L^-(x, y).$$

Therefore the footprint has finite area.

Our final task is to show that the common factor $a(x, y)$ must also divide any other generator polynomial $g_k(x, y)$. Consider the ideal $I_k = \langle g_1(x, y), g_k(x, y), g_L(x, y) \rangle$, generated by only these three polynomials. The ideal I contains the ideal I_k, so $\Delta(I)$ is contained in $\Delta(I_k)$. Let $\mathcal{G}_k = \{g_\ell^{(k)}(x, y)\}$ be a minimal basis for I_k. Each $g_\ell^{(k)}(x, y)$ is a linear combination of $g_1(x, y)$, $g_k(x, y)$, and $g_L(x, y)$. Moreover, since the conjunction polynomial associated with this set of generator polynomials must equal zero, we have

$$y^{n_k - n_1} g_1(x, y) - x^{m_1 - m_k} g_k(x, y) = \sum_\ell Q_\ell^{(1)}(x, y) g_\ell^{(k)}(x, y)$$

and

$$y^{n_L - n_k} g_k(x, y) - x^{m_k - m_L} g_L(x, y) = \sum_\ell Q_\ell^{(L)}(x, y) g_\ell^{(k)}(x, y).$$

But the $g_\ell^{(k)}(x, y)$ are polynomial combinations of $g_1(x, y)$, $g_k(x, y)$, and $g_L(x, y)$. Therefore any common polynomial factor of both $g_1(x, y)$ and $g_L(x, y)$ must also be a factor of both $[x^{m_1 - m_k} + A_1(x, y)]g_k(x, y)$ and $[y^{n_L - n_k} + A_2(x, y)]g_k(x, y)$ for some polynomials $A_1(x, y)$ and $A_2(x, y)$. We conclude that $a(x, y)$ is a factor of $g_k(x, y)$. The same argument holds for every k, so $a(x, y)$ divides all polynomials of $\mathcal{G}$. ∎

Let $f(x, y)$ and $g(x, y)$ be nonzero polynomials over the field F of degree m and degree n, respectively, and with no common polynomial factor. In Section 7.8, we shall discuss Bézout's theorem, which says that the number of common zeros of $f(x, y)$ and $g(x, y)$ in F^2 is at most mn. In the language of ideals, Bézout's theorem says that an ideal, $I = \langle f(x, y), g(x, y) \rangle$, generated by two coprime polynomials, has at most mn zeros in F^2. This may be viewed as a generalization to bivariate polynomials of the statement that a (univariate) polynomial of degree n over the field F has at most n zeros over F (or over any extension of F). In this section, we shall give a different generalization of this statement, formulated in the language of polynomial ideals, that counts exactly the number of affine zeros of certain ideals, not necessarily ideals defined by only two generator polynomials.

The notion of the footprint will be used in this section as a vehicle to pass the well known properties of linear vector spaces over to the topic of commutative algebra, where these properties become statements about the zeros of ideals. Though it will take several pages to complete the work of this section, the result of this work can be expressed succinctly: every proper ideal of $F[x, y]$ has at least one affine zero in an appropriate extension field, and the largest ideal with a given finite set of affine zeros in F^2 has a footprint with an area equal to the number of affine zeros in the set.

For any finite set of points $\mathcal{P}$ in the affine plane over F, let $I(\mathcal{P})$ be the set consisting of all bivariate polynomials in $F[x, y]$ having a zero at every point of $\mathcal{P}$. Then $I(\mathcal{P})$ is an ideal. It is the locator ideal for the points of $\mathcal{P}$.

Definition 7.7.1   *A locator ideal is an ideal in $F[x, y]$ with a finite number of affine zeros in F^2 contained in no larger ideal in $F[x, y]$ with this same set of affine zeros.*

Clearly, I is a locator ideal if and only if $I = I(\mathcal{Z}(I))$, where $\mathcal{Z}(I)$ is the finite set of rational affine zeros of I. It is apparent that the locator ideal for a given finite set of affine points of F^2 is the set of all bivariate polynomials whose zeros include these points. Obviously, the locator ideal for a given set of affine points is unique. For the finite field $GF(q)$, both $x^q - x$ and $y^q - y$ are always elements of every locator ideal.

The locator ideal is unique because if there were two different locator ideals with the same set of zeros, then the ideal generated by the union of those two ideals would be a larger ideal with the same set of zeros. Moreover, a locator ideal over a finite field can have no unnecessary zeros. This is because an unnecessary zero at (α^a, α^b), for example, could be eliminated by appending to the ideal the polynomial $p(x, y) = (\sum_i x^i \alpha^{-ia})(\sum_i y^i \alpha^{-ib})$, which has a zero everywhere except at (α^a, α^b). This remark is closely related to the statement that will be called the *discrete weak nullstellensatz* in Section 7.9.

The locator ideal in $F[x, y]$ is the largest ideal that has a given set of zeros in F^2. However, the word "maximal" cannot be used here because it has another meaning. A *maximal ideal* is a proper ideal that is not contained in a larger proper ideal. This might be, as in a finite field, a locator ideal with a single zero, but not a locator ideal with two zeros.

In later sections, we shall study the role of the locator ideal in "locating" the nonzeros of a given bivariate polynomial, $V(x, y)$. In this section, we shall develop the important fact that the number of affine zeros in F^2 of a locator ideal is equal to the area of its footprint. (Thus the *locator footprint* for a finite set of points has the same cardinality as the set of points.) This statement is a generalization of the statement that the minimal monovariate polynomial, having zeros at n specified points, has degree n.

Before proving this statement, we will provide an example that explains these notions. Consider the ideal $I = \langle x^3 + xy^2 + x + 1, y^2 + xy + y \rangle$ in the ring of bivariate polynomials over $GF(8)$, with monomials ordered by the graded order. We shall find the footprint and all affine zeros of this ideal, and we shall find that it is a locator ideal; there is no larger ideal with the same set of affine zeros. Even before we compute the footprint, it should be apparent that (in the graded order) the footprint of I is contained in the set of indices $\{(0, 0), (0, 1), (1, 0), (1, 1), (2, 0), (2, 1)\}$ because I is generated by polynomials with leading monomials x^3 and y^2. This quick computation says that the footprint has area at most 6. We shall see in a moment that the actual area of the footprint is 3.

To find the affine zeros of the ideal, $I = \langle x^3 + xy^2 + x + 1, y^2 + xy + y \rangle$, set

$$x^3 + xy^2 + x + 1 = 0,$$
$$y^2 + xy + y = y(y + x + 1) = 0.$$

The second equation says that either $y = 0$ or $y = x + 1$. But if $y = x + 1$, the first equation yields

$$x^3 + x(x + 1)^2 + x + 1 = 1 \neq 0.$$

Therefore zeros can occur only if $y = 0$, and the first equation becomes

$$x^3 + x + 1 = 0.$$

This has three zeros in the extension field $GF(8)$. Thus we have found three zeros in the affine plane $GF(8)^2$, which can be expressed as $(\alpha, 0)$, $(\alpha^2, 0)$, and $(\alpha^4, 0)$. In Section 7.8, we will find that three more zeros of this ideal, which are needed in that section to satisfy the equality form of Bézout's theorem, lie at infinity. The locator ideal, in contrast, considers only affine zeros, which is why we have that

$$\Delta(\langle x^3 + xy^2 + x + 1, y^2 + xy + y \rangle) = 3$$

for the ideal, but, for the set,

$$\Delta(\{x^3 + xy^2 + x + 1, y^2 + xy + y\}) = 6.$$

What is the footprint of the ideal $I = \langle x^3 + xy^2 + x + 1, y^2 + xy + y \rangle$? To answer, first find a minimal basis. Because

$$y(x^3 + xy^2 + x + 1) + (x^2 + xy + x)(y^2 + xy + y) = y,$$

we have that $y \in I$. Because $y^2 + xy + y$ is a multiple of y, the ideal can be written as follows:

$$I = \langle x^3 + xy^2 + x + 1, y \rangle.$$

The Buchberger theorem allows us to conclude that $\mathcal{G} = \{x^3 + xy^2 + x + 1, y\}$ is a minimal basis for I because $R_{\mathcal{G}}[y(x^3 + xy^2 + x + 1) - x^3 y] = 0$. (What is the reduced basis?)

The footprint of I is illustrated in Figure 7.11. The footprint has area 3, which is equal to the number of affine zeros of I in $GF(8)^2$.

If we choose to regard I as an ideal in, say, $GF(32)[x, y]$ instead of $GF(8)[x, y]$, we will find that there are no affine zeros over the ground field $GF(32)$, a field that does not contain $GF(8)$. The smallest extension field of $GF(32)$ that contains all of the affine zeros of I is $GF(2^{15})$ since this extension field of $GF(32)$ contains $GF(8)$, and $GF(8)$ actually is the field where the zeros lie. The reason for the discrepancy between the area of the footprint and the number of affine zeros in $GF(32)$ is that I is not a locator ideal in the ring of polynomials $GF(32)[x, y]$, so the statement does not apply. We also note that, even in the algebraic closure of $GF(32)$, or of $GF(8)$, this ideal has no additional zeros. Thus in the algebraic closure, $I(\mathcal{Z}(I)) = I$, where $\mathcal{Z}(I)$ denotes the set of zeros of I.

The conclusion of this example is actually quite general and underlies much of the sequel. It is restated below as a theorem, whose proof begins with the following proposition.

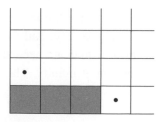

Figure 7.11. Footprint of $\langle x^3 + xy^2 + x + 1, y \rangle$.

Proposition 7.7.2   *For any $I \subset F[x, y]$, the dimension of the vector space $F[x, y]/I$ is equal to the area of the footprint of I.*

Proof: As a vector space, $F[x, y]/I$ is spanned by the set of monomials that are not the leading monomial of any polynomial in I. Because these monomials are linearly independent, they form a basis. There is one such monomial for each point of the footprint of I, so the number of monomials in the vector space basis is equal to the area of the footprint of I.   ■

Theorem 7.7.3   *The number of affine zeros over F of a locator ideal in $F[x, y]$ is equal to the area of the footprint of that ideal.*

Proof: Let I be a locator ideal in the ring $F[x, y]$ for a finite set of t points, $(\beta_\ell, \gamma_\ell)$, for $\ell = 1, \ldots, t$ in the affine plane F^2. Let $\{ p_i(x, y) \mid i = 1, \ldots, n\}$ be a (possibly infinite) basis for the vector space $F[x, y]/I$. Let $b(x, y)$ be any element of $F[x, y]/I$. It can be written as follows:

$$b(x, y) = \sum_{i=1}^{n} a_i p_i(x, y),$$

for some set of coefficients $\{a_i \mid i = 1, \ldots, n\}$, which can be arranged as the vector $\boldsymbol{a}$. Let $\boldsymbol{b}$ be the vector of blocklength t with components $b_\ell = b(\beta_\ell, \gamma_\ell)$ for $\ell = 1, \ldots, t$. Then the equation

$$b(\beta_\ell, \gamma_\ell) = \sum_{i=1}^{n} a_i p_i(\beta_\ell, \gamma_\ell)$$

can be represented as the following matrix equation:

$$\boldsymbol{b} = \boldsymbol{Pa},$$

where the t by n matrix $\boldsymbol{P}$ has the elements $P_{\ell i} = p_i(\beta_\ell, \gamma_\ell)$.

We now provide the proof of the theorem in three steps.

Step (1) This step shows the following. Let I be an ideal in the ring $F[x, y]$ having the finite set $\mathcal{B}$ of t zeros in the affine plane F^2. Then the dimension of $F[x, y]/I$ as a vector space over F is not smaller than t. That is, $n \geq t$, where $n = \dim F[x, y]/I$.

The proof of step (1) follows. Let $\mathcal{B} = \{(\beta_\ell, \gamma_\ell) \mid \ell = 1, \ldots, t\}$. It is an elementary fact that for any vector $\boldsymbol{b} = [b_1, b_2, \ldots, b_t]$ of length t, there is a polynomial $s(x, y) \in F[x, y]$ with $s(\beta_\ell, \gamma_\ell) = b_\ell$ for $\ell = 1, \ldots, t$. It is given by the *Lagrange interpolation formula*:

$$s(x, y) = \sum_{\ell=1}^{n} b_\ell \frac{\Pi_{\ell' \neq \ell}(x - \beta_{\ell'})(y - \gamma_{\ell'})}{\Pi_{\ell' \neq \ell}(\beta_\ell - \beta_{\ell'})(\gamma_\ell - \gamma_{\ell'})}.$$

Every polynomial in I has a zero at $(\beta_\ell, \gamma_\ell)$ for $\ell = 1, \ldots, t$. Therefore $s(x, y)$ maps into the polynomial $p(x, y)$ of the quotient ring $F[x, y]/I$ for which $p(\beta_\ell, \gamma_\ell) = s(\beta_\ell, \gamma_\ell)$. Thus we conclude that for every $\boldsymbol{b}$, there is a polynomial $p(x, y)$ in $F[x, y]/I$ such that $p(\beta_\ell, \gamma_\ell) = b_\ell$. The polynomial $p(x, y)$ has at most n nonzero coefficients, which can be arranged into a vector $\boldsymbol{a}$ of length n, regarding $x^{i'} y^{i''}$ for $(i', i'') \in \Delta$ as a basis for $F[x, y]/I$. Then the statement $p(\beta_\ell, \gamma_\ell) = \beta_\ell$ can be written $\boldsymbol{Pa} = \boldsymbol{b}$, where $\boldsymbol{P}$ is an n by t matrix with elements of the form $\beta_\ell^{i'} \gamma_\ell^{i''}$ for each $(i', i'') \in \Delta$. The set of all such $\boldsymbol{b}$ forms a t-dimensional vector space that is covered by $\boldsymbol{Pa}$, where $\boldsymbol{a}$ is an element of an n-dimensional vector space. Therefore $n \geq t$.

Step (2) This step shows the following. If $\mathcal{B}$ is a finite set of t points in the affine plane F^2 and $I \subset F[x, y]$ is the locator ideal of $\mathcal{B}$, then the dimension of $F[x, y]/I$, as a vector space over F, is not larger than the number of elements of $\mathcal{B}$. That is, $n \leq t$, where $n = \dim F[x, y]/I$.

The proof of step (2) follows. Let $b(x, y)$ be any polynomial in $F[x, y]/I$. If $b(\beta_\ell, \gamma_\ell) = 0$ for $\ell = 1, \ldots, t$, then $b(x, y) \in I$ because I is a locator ideal for $\mathcal{B}$. But the only element in both I and $F[x, y]/I$ is the zero polynomial. Therefore the null space of the map $\boldsymbol{Pa} = \boldsymbol{b}$ has dimension 0. This implies that $n \leq t$.

Step (3) The first two steps and the proposition combine to show that the area of the footprint of the locator ideal is equal to the number of zeros of the ideal.   ■

The line of reasoning used in this proof will be used later in Section 7.8 to give the affine form of Bézout's theorem. We will only need to show that if $f(x, y)$ and $g(x, y)$ are two elements of $F[x, y]$ of degree m and n, respectively, that have no factor in common, then the dimension of $F[x, y]/\langle f(x, y), g(x, y)\rangle$ is not larger than mn. Then Bézout's theorem will follow from step (1).

Corollary 7.7.4 *If the number of affine zeros of the ideal I is equal to the area of $\Delta(I)$, then I is the locator ideal for these zeros.*

Proof: We will show that if I is not the locator ideal for these zeros, then the footprint of I cannot have an area equal to the number of zeros of I. Let $\mathcal{Z}$ be the set of zeros of I. Certainly, I is contained in the locator ideal of $\mathcal{Z}$. If I is not the locator ideal of $\mathcal{Z}$, then additional polynomials must be appended to I to form the locator ideal. Then, by Theorem 7.5.3, the footprint must become smaller, so it cannot have an area equal to the number of zeros of I.   ■

Theorem 7.7.3 will be useful in many ways, and it underlies much of the sequel. Here we use it to draw several elementary conclusions regarding the number of zeros of an ideal.

First, we consider the set of common zeros of two distinct irreducible polynomials, or, equivalently, the set of zeros of the ideal generated by these two polynomials. It may seem rather obvious that two distinct irreducible polynomials can have only a finite number of common zeros. In the real field, this remark comes from the notion that the curves defined by the two polynomials cannot wiggle enough to cross each other infinitely often. This conclusion, however, requires a proof. We will find that this is a consequence of Theorem 7.7.3 by showing that two distinct irreducible polynomials must generate an ideal whose footprint has finite area.

For example, consider $I = \langle x^2y + xy + 1, xy^2 + 1 \rangle$ over the field $GF(2)$. Because

$$y^3(x^2y + xy + 1) + (xy^2 + y^2 + 1)(xy^2 + 1) = y^3 + y^2 + 1$$

and

$$y(x^2y + xy + 1) + (x + 1)(xy^2 + 1) = x + y + 1,$$

we can conclude that neither $(0, 3)$ nor $(1, 0)$ is in the footprint of I. Hence the footprint has a finite area (which is not larger than 6). From this, using Theorem 7.7.3, it is a short step to infer that the two polynomials have a finite number of common zeros. Indeed, $I = \langle x^2y + xy + 1, xy^2 + 1 \rangle$ can be regarded as an ideal over any extension of $GF(2)$, so the total number of affine zeros in any extension field is at most six.

Theorem 7.7.5 *Two bivariate polynomials with no common nontrivial polynomial factor have at most a finite number of common affine zeros in F^2.*

Proof: Let $\mathcal{G}$ be a minimal basis for the ideal I formed by the two polynomials $f(x, y)$ and $g(x, y)$. Then the polynomials of $\mathcal{G}$ can have no common nontrivial polynomial factor because, if they did, then $f(x, y)$ and $g(x, y)$ would also. Therefore the footprint of I' has a finite area, and so $\mathcal{G}$ has a finite number of common affine zeros in F, as do $f(x, y)$ and $g(x, y)$.   ∎

Corollary 7.7.6 *A bivariate[2] ideal cannot have an infinite number of zeros unless all polynomials in its reduced basis have a common polynomial factor.*

Proof: Suppose there is no common polynomial factor. Then each pair of generator polynomials has at most a finite number of common zeros. Any zero of the ideal is a zero of every pair of its generator polynomials, so, by the theorem, there are at most a finite number of zeros of the ideal.   ∎

Conversely, if all polynomials of a reduced basis have a common polynomial factor, then the ideal has an infinite number of zeros in the algebraic closure of the field because a single nontrivial bivariate polynomial has an infinite number of zeros. If all generators have a common polynomial factor, the ideal has all the zeros of this polynomial factor.

An ideal generated by the single polynomial $\langle g(x, y) \rangle$ in an algebraically closed field F must have many zeros in the affine plane (unless $g(x, y)$ has degree 0), because, for any $\beta \in F$, either $g(x, \beta)$ or $g(\beta, y)$ must be a univariate polynomial with at least one

---

[2] This is an example of a statement in the ring of bivariate polynomials that is not true in the ring of trivariate polynomials.

zero. Also, the ideal $\langle x - \beta, y - \gamma \rangle$, generated by two polynomials, clearly has a zero at (β, γ). The general form of this statement, that every proper ideal in $F[x, y]$ has at least one zero, seems quite plausible and easy to accept, but is actually deeper and trickier to prove than it seems. It will be proved in Section 7.9.

7.8   The Bézout theorem

A set of generator polynomials for an ideal of $F[x, y]$ determines the ideal, and so, indirectly, determines the number of zeros of the ideal. We want to know what can be said about the number of zeros of an ideal if the generator polynomials are not fully specified. For example, if we are given only the leading monomials of the generator polynomials, then what can we say about the number of zeros of the ideal? If, moreover, we are given that the polynomials have no common polynomial factor, then, as we shall see, we can bound the number of zeros of the ideal by using the bidegrees of the generator polynomials. If there are only two generator polynomials, this bound is simply stated and is known as the *Bézout theorem*.

The corresponding question for an ideal of $F[x]$ is elementary. The number of zeros of any univariate polynomial $p(x)$ over the field F is not larger than the degree of $p(x)$, and the number of common zeros of several polynomials is not larger than the smallest of their degrees. If the field F is an algebraically closed field, then the number of zeros of the polynomial $p(x)$ is exactly equal to the degree of $p(x)$, provided multiple zeros are counted as such. For example, a polynomial of degree m over the rational field need not have any zeros, but if it is regarded as a polynomial over the complex field, which is algebraically closed, then it will have exactly m zeros, provided multiple zeros are counted as such. We can always embed a field into an algebraically closed extension field. For example, the union of all extension fields is an algebraically closed extension field. In this sense, a univariate polynomial of degree m always has exactly m zeros.

A nonsingular univariate polynomial, which is defined as a univariate polynomial with the property that the polynomial and its derivative are not simultaneously zero at any point, has no multiple zeros. Then we have the following statement: in an algebraically closed field, a nonsingular polynomial of degree m has exactly m distinct zeros.

What is the corresponding statement for bivariate polynomials? We cannot give any general relationship between the degree of a bivariate polynomial and the number of its zeros. For example, over the real field, the polynomial

$$p(x, y) = x^2 + y^2 - 1$$

has an uncountably infinite number of zeros.

We can make a precise statement, however, about the number of simultaneous zeros of two polynomials. This is a generalization to bivariate polynomials of the familiar statement that the number of zeros of a univariate polynomial is not larger than its degree. The generalization, known as *Bézout's theorem*, says that two bivariate polynomials of degree m and n, respectively, that do not have a polynomial factor in common, have at most mn common zeros. The number of common zeros is equal to mn if the field is algebraically closed, the plane is extended to the projective plane, and the common multiple zeros are counted as such.

Thus Bézout's theorem tells us that as polynomials over $GF(2)$, the polynomial

$$f(x,y) = x^3 y + y^3 + x$$

and the polynomial

$$g(x,y) = x^2 + y + 1$$

have exactly eight common zeros in the projective plane, possibly repeated (in a sufficiently large extension field).

Theorem 7.8.1 (Bézout's theorem)   *Let $f(x,y)$ and $g(x,y)$ be bivariate polynomials over the field F of degree m and degree n, respectively, and with no common polynomial factor. In the projective plane over a sufficiently large extension field of F, the number of points at which $f(x,y)$ and $g(x,y)$ are both zero, counted with multiplicity, is equal to mn.*

Only the affine form of Bézout's theorem will be proved herein, this by an unconventional method. The affine form states that the number of common zeros in the affine plane of the two polynomials is *at most mn*. It will be proved as a corollary to a more general theorem at the end of the section.

For an example of Bézout's theorem, we will consider the two polynomials $x^3 + y^2 x + x^2 + 1$ and $y^2 + xy + y$ over $GF(2)$. The second polynomial can be factored as $y(y + x + 1)$, which equals zero if and only if $y = 0$ or $y = x + 1$. If $y = 0$, the first polynomial reduces to $x^3 + x^2 + 1$, which has three zeros in $GF(8)$. If $y = x+1$, the first polynomial reduces to $x^2 + x + 1$, which has two zeros in $GF(4)$. We will write these five zeros in $GF(64)$ because this is the smallest field that contains both $GF(4)$ and $GF(8)$. If α is an appropriate element of $GF(64)$, then $GF(4)^*$ is the orbit of α^{21}, and $GF(8)^*$ is the orbit of α^9. In terms of α, the zeros of $x^3 + x^2 + 1$ are at α^{27}, α^{45}, and α^{54}. In terms of α, the zeros of $x^2 + x + 1$ are at α^{21} and α^{42}. Thus the given pair of polynomials has five common affine zeros at $(\alpha^{27}, 0)$, $(\alpha^{45}, 0)$, $(\alpha^{54}, 0)$, $(\alpha^{21}, \alpha^{42})$, and $(\alpha^{42}, \alpha^{21})$. The sixth common zero, required by Bézout's theorem, cannot be an affine zero. To find it, we must write the polynomials in homogeneous form, as $x^3 + y^2 x + x^2 z + z^3$ and $y^2 + xy + yz$. (Recall that to view the zeros in the projective plane, one must replace $f(x,y)$ and $g(x,y)$ by the homogeneous trivariate polynomials $f(x,y,z)$ and

$g(x, y, z)$, and then the rightmost nonzero coordinate of (x, y, z) must be a one.) Then to find the points at infinity, set $z = 0$ and $y = 1$. This gives the polynomials $x^3 + x$ and $x + 1$, which have a common zero when $x = 1$. Thus the sixth common zero is at $(1, 1, 0)$. To conclude, in projective coordinates, the six zeros are at $(\alpha^{27}, 0, 1)$, $(\alpha^{45}, 0, 1)$, $(\alpha^{54}, 0, 1)$, $(\alpha^{21}, \alpha^{42}, 1)$, $(\alpha^{42}, \alpha^{21}, 1)$, and $(1, 1, 0)$.

In Section 7.7, we studied the ideal in $GF(2)[x, y]$ formed by the two polynomials $x^3 + y^2 x + x + 1$ and $y^2 + xy + y$, and we found only three affine zeros. Bézout's theorem tells us that the two polynomials have exactly six zeros in common. Where are the other three zeros? To make these zeros visible, write the polynomials in homogeneous trivariate form: $x^3 + y^2 x + xz^2 + z^3$ and $y^2 + xy + yz$. There is a zero at the projective point $(x, y, z) = (1, 1, 0)$. Moreover, to provide three zeros, this must be a multiple zero with multiplicity 3. Hence we have found the six zeros required by Bézout's theorem. These are $(\alpha, 0, 1)$, $(\alpha^2, 0, 1)$, $(\alpha^4, 0, 1)$, $(1, 1, 0)$, $(1, 1, 0)$, and $(1, 1, 0)$, where $\alpha \in GF(8)$. The zero of multiplicity 3 has been listed three times. Except for the fact that we have not given – nor will we give – a formal definition of the term multiplicity, this completes our examples of Bézout's theorem. (For the record, the multiplicity of zeros has the same intuitive interpretation as it does for polynomials of one variable, but is more delicate to define precisely.)

To avoid the issue of multiplicity of common zeros, several conditions are needed. We must first restrict $f(x, y)$ and $g(x, y)$ to be nonsingular polynomials, because then all affine zeros of each polynomial have multiplicity 1. Less obvious is the condition that $f(x, y)$ and $g(x, y)$ cannot have a point of tangency. At such a point, the pair of polynomials $(f(x, y), g(x, y))$ can have a multiple zero, even though each is nonsingular. This is a rather technical consideration we only hint at by considering the two points in R^2 at which a straight line intersects a circle. As the line is moved away from the center of the circle, the two points of intersection move closer together and eventually coalesce when the line becomes tangent to the circle. Thus the two zeros merge to become a double zero at the point of tangency.

A polynomial of degree 1 is given by $\ell(x, y) = ax + by + c$, where a and b are not both zero. This is the equation of a line. A polynomial of degree 1 and a polynomial of degree m have common zeros at not more than m points unless the polynomials share a common polynomial factor. Therefore if $f(x, y)$ is a bivariate polynomial of degree m with coefficients in the field F, and if $f(x, y)$ and the polynomial of degree 1, $\ell(x, y)$, have more than m common zeros, then $\ell(x, y)$ must divide $f(x, y)$, with division as bivariate polynomials.

Two lines intersect in one point. This remark leads to an easy interpretation of Bézout's theorem for two homogeneous polynomials $f(x, y)$ and $g(x, y)$, of degrees m and n, respectively. Simply divide the two bivariate homogeneous polynomials by y^m and y^n, respectively, and set $t = x/y$ to produce two univariate polynomials $f(t, 1)$ and $g(t, 1)$, respectively. Then in a sufficiently large extension field of F, $f(t, 1) = a \prod_{i=1}^{m} (t - \beta_i)$ and $g(t, 1) = b \prod_{i=1}^{n} (t - \gamma_j)$. This means that $f(x, y) =$

$a\prod_{i=1}^{m}(x - \beta_i y)$ and is zero on the line $x = \beta_i y$. Similarly, $g(x, y) = b\prod_{j=1}^{n}(x - \gamma_j y)$. For each i and j, $1 \leq i \leq m$, $1 \leq j \leq n$, the line $x - \beta_i y = 0$ intersects the line $x - \gamma_j y = 0$ at the origin. In this way, we see that two homogeneous bivariate polynomials have a common zero of multiplicity mn at the origin.

We shall provide proof of a restricted form of Bézout's theorem: that the number of affine zeros is *at most* mn; there may be other zeros at infinity, which we do not count. Our approach is unconventional. It is to show that the number of points in the footprint of the ideal, generated by two coprime polynomials, cannot be larger than the product of the degrees of the two polynomials. We already know that the number of affine zeros of a locator ideal is equal to the number of points in the footprint. Bézout's theorem follows by combining these two statements. However, rather than proceed with this plan directly, we will provide a more general statement regarding the number of zeros of an ideal when given only the bidegrees of a set of generator polynomials. Bézout's theorem then follows from this general statement.

Before starting the proof of the theorem, we will look carefully at some examples. Over $GF(2)$, let

$$f(x, y) = x^5 + x^4 y + x^2 y^3 + \cdots,$$
$$g(x, y) = x^2 y + xy^2 + y^3 + \cdots,$$

where the unstated terms are arbitrary monomials of smaller degree in the graded order. The polynomials have the leading monomials x^5 and $x^2 y$ in the graded order with $x \succ y$. First, note that, up to the leading term,

$$yf(x, y) + (x^3 + xy^2)g(x, y) = xy^5 + \cdots,$$

which has the leading monomial xy^5, and

$$(xy + y^2)f(x, y) + (x^4 + x^3 y + x^2 y^2 + xy^3 + y^4)g(x, y) - y^7 + \cdots,$$

which has the leading monomial y^7. Thus the ideal $\langle f(x, y), g(x, y) \rangle$ has polynomials with leading monomials $x^5, x^2 y, xy^5$, and y^7. It is possible that these are all of the leading monomials of the reduced basis, and so the footprint of the ideal $\langle f(x, y), g(x, y) \rangle$ may be as shown in Figure 7.12. Indeed, this will be the footprint if the unspecified terms of $f(x, y)$ and $g(x, y)$ happen to be such that Buchberger's theorem is satisfied. The area of the footprint of this ideal is 15, which is equal to the product of the degrees of $f(x, y)$ and $g(x, y)$. Because the area is 15, Theorem 7.7.3 implies that there are exactly 15 common affine zeros of $f(x, y)$ and $g(x, y)$ if and only if $\langle f(x, y), g(x, y) \rangle$ is a locator ideal. This is not an unimportant coincidence. It is an example of a general rule we need to understand. We will see that the area of the footprint is never larger than the product of the degrees of $f(x, y)$ and $g(x, y)$. Because the number of zeros of $\langle f(x, y), g(x, y) \rangle$

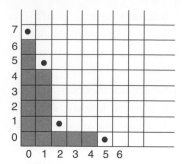

Figure 7.12. An insight into Bézout's theorem.

is not larger than the area of the footprint (with equality if this is a locator ideal), we will see that the number of common zeros is not larger than the product of the degrees.

A geometric insight into Bézout's theorem can be obtained from an inspection of Figure 7.12. Consider the white square at the intersection of the column of squares that runs through the lowest dot and the row of squares that runs through the second lowest dot. The diagonal (of slope -1) from this square, up and to the left, contains the leading monomial of the third basis polynomial. Similarly, the two middle dots determine another white square by their row/column intersection. The diagonal (of slope -1) through that white square contains the leading monomial of the fourth basis polynomial. In general, each pair of polynomials of the reduced basis of the ideal will imply another polynomial of the reduced basis with a leading monomial on, or below, the diagonal defined by the previously formed step.

To prove the affine form of Bézout's theorem, we will prove that the area of the footprint of the ideal $\langle f(x, y), g(x, y) \rangle$ is not larger than the product of the degrees of $f(x, y)$ and $g(x, y)$. We will generalize Bézout's theorem to a statement that the leading monomials of any set of generators of any ideal in $F[x, y]$ determine an upper bound on the number of affine zeros of the ideal. The affine form of Bézout's theorem is then obtained as a corollary to this more general statement by restricting it to only two polynomials. Before giving the proof, we will examine a few more simple cases.

First, suppose that the leading monomial of $g(x, y)$ is x^m and the leading monomial of $h(x, y)$ is y^n. If $g(x, y)$ and $h(x, y)$ form a minimal basis of $\langle g(x, y), h(x, y) \rangle$, and this is a locator ideal, then the area of the footprint is clearly mn, as shown in Figure 7.13. In this case, there are at most mn common affine zeros. If $g(x, y)$ and $h(x, y)$ do not form a minimal basis, or if $\langle g(x, y), h(x, y) \rangle$ is not a locator ideal, then the footprint of the ideal is contained within this rectangle, and so the area of the footprint is strictly smaller than mn. Therefore the ideal $\langle g(x, y), h(x, y) \rangle$ has fewer than mn zeros.

In the general case, although the polynomials $g(x, y)$ and $h(x, y)$ have degrees m and n, respectively, their leading monomials need not be x^m and y^n, respectively. In general, the leading monomials have the form $x^s y^{m-s}$ and $x^{n-r} y^r$, respectively. The footprint

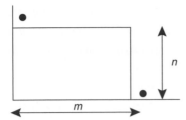

Figure 7.13. Area of a rectangular footprint.

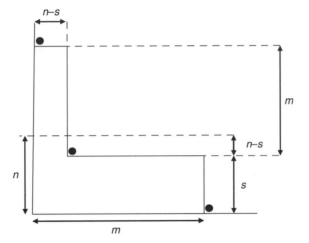

Figure 7.14. Illustrating the invariance of the area of a footprint.

of the set $\{g(x, y), h(x, y)\}$ is infinite, but if the two polynomials have no common polynomial factor, then the footprint of the ideal $\langle g(x, y), h(x, y) \rangle$ must be finite. This means that a minimal basis must include polynomials with leading monomials of the form $x^{m'}$ and $y^{n'}$.

The simplest such example is an ideal of the form $\langle g(x, y), h(x, y) \rangle$ that has three polynomials in its reduced basis, as indicated by the three dots in Figure 7.14. In this example, the leading monomial of $g(x, y)$ is x^m and the leading monomial of $h(x, y)$ is $y^s x^{n-s}$ with $s \neq n$; thus, both are monic polynomials. The nonzero polynomial

$$f(x, y) = y^s g(x, y) - x^{m-n+s} h(x, y),$$

which has a degree at most $m + s$, is in the ideal. Under division by $g(x, y)$ and $h(x, y)$, the remainder polynomial $r(x, y)$ has a degree at most $m + s$. Because we have specified that there are only three polynomials in the minimal basis, its leading monomial must involve only y. If the remainder polynomial has the largest possible degree, then its leading monomial is y^{m+s} and the footprint is as shown in Figure 7.14. It has the area $sm + m(n - s) = mn$, again equal to the product of the degrees of $g(x, y)$ and $h(x, y)$.

If the remainder polynomial has less than the largest possible degree, then the area of the footprint will be smaller than the product of the degrees of $g(x, y)$ and $h(x, y)$.

In general, this process of appending a conjunction polynomial to a set of polynomials may need to be repeated an arbitrary number of times, terminating with a minimal basis that must contain one polynomial, whose leading monomial is solely a power of y, and another, whose leading monomial is solely a power of x. The set of bidegrees of the polynomials computed in this way defines a cascade set. This cascade set must contain the footprint of the ideal; therefore its area is a bound on the area of the footprint. We will bound the area by finding a finite sequence of cascade sets whose areas are nonincreasing, the first of which has the area mn and the last of which contains the footprint.

Consider the cascade set Δ with the exterior corners at

$$\{(s'_1, s''_1), (s'_2, s''_2), (s'_3, s''_3), \ldots, (s'_M, s''_M)\},$$

with $s'_1 < s'_2 < s'_3 < \cdots < s'_M$ and $s''_1 > s''_2 > s''_3 > \cdots > s''_M$. If the exterior corners satisfy $s''_1 = 0$ and $s'_M = 0$, then the cascade set has a finite area. Otherwise, the cascade set has an infinite area. To facilitate the proof of the Bézout theorem, we define the *essential cascade set* $\widehat{\Delta}$ as the cascade set with the exterior corners at

$$\{(s'_1 + s''_1, 0), (s'_2, s''_2), (s'_3, s''_3), \ldots, (s'_{M-1}, s''_{M-1}), (0, s'_M + s''_M)\}.$$

The essential cascade set is obtained by moving the two extreme exterior corners to the boundary (along lines of slope -1). It has a finite area. A cascade set with finite area is equal to its essential cascade set. Define the *essential area* of a cascade set to be equal to the area of its essential cascade set. This is the shaded area of Figure 7.15. The essential area of a finite cascade set is equal to its area. The following theorem is a precursor of a generalized form of Bézout's theorem.

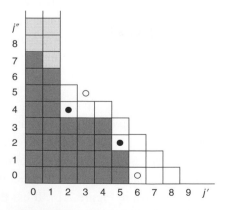

Figure 7.15. Division order and graded order.

Theorem 7.8.2 *The essential area of the footprint of a set of bivariate polynomials is nonincreasing under the appending of a conjunction polynomial of those polynomials.*

Proof: The proof involves only simple geometric reasoning. Because appending a polynomial to a set of polynomials cannot make the footprint larger, and so cannot make the essential cascade set larger, no proof is needed if the footprint has a finite area.

Suppose that the footprint has an infinite area, with exterior corners at (s_1', s_1''), $(s_2', s_2''), \ldots, (s_M', s_M'')$ $s_1'' > s_2'' > s_3''$, where $s_1' < s_2' < s_3' < \cdots < s_M'$ and s_1' and s_M'' are not both zero. Again, there is nothing to prove unless the new exterior corner is above the previous top corner or to the right of the previous bottom corner. To align polynomials with the leading monomials $x^{s_1'}y^{s_1''}$ and $x^{s_2'}y^{s_2''}$, with $s_1' < s_2'$ and $s_1'' > s_2''$, it is required that these polynomials be translated to have the leading monomial $x^{s_2'}y^{s_1''}$. Then, under division by $\mathcal{G}$, the degree of the remainder polynomial is at most $s_2' + s_1'' > s_1' + s_2''$. Hence the new exterior corner lies in the square formed by the top left corner and the bottom right corner, so the essential area does not increase. ∎

The proof of the following corollary is similar to the *Buchberger algorithm*, which is described in Chapter 8. The Buchberger theorem suggests that a minimal basis for an ideal can be obtained by repeatedly appending nonzero conjunction polynomials to the set of generator polynomials $\mathcal{G}$. Enlarge $\mathcal{G}$ by appending conjunction polynomials until Corollary 7.6.2 is satisfied. The essential area is finite and is nonincreasing when conjunction polynomials are appended, as is shown in Figure 7.16. This construction

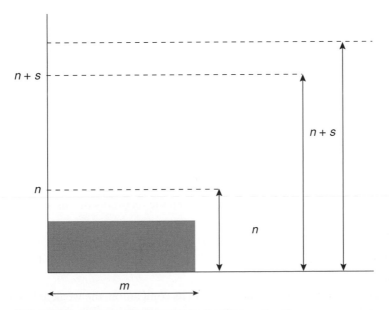

Figure 7.16. Changing the footprint by Buchberger iterations.

leads to the following corollary. If the ideal is generated by two coprime polynomials, then the corollary reduces to the affine form of Bézout's theorem.

Corollary 7.8.3 *The number of zeros of an ideal with no common nontrivial polynomial factor is not larger than the essential area of the footprint of any set of generator polynomials.*

Proof: Compute a minimal basis by appending conjunction polynomials as needed. The essential area of the footprint is nonincreasing under the operation of appending a conjunction polynomial, so the area of the footprint of the minimal basis is not larger than the essential area of the given set of generator polynomials. The process eventually stops, at which point the footprint is equal to its essential cascade set. Because the number of zeros of the ideal is not larger than the area of the footprint of the minimal basis, the corollary follows from the theorem.   ∎

Corollary 7.8.4 (Bézout's theorem)   *Over the field F, let $g(x, y)$ and $h(x, y)$ be bivariate polynomials of degree m and n, respectively, and with no common nontrivial polynomial factor. The footprint of $\langle g(x, y), h(x, y) \rangle$ has an area not larger than mn.*

Proof: This is a special case of the previous corollary. The polynomials have leading monomials of the form $x^{m-r}y^r$ and $x^{n-s}y^s$, respectively. Hence the footprint of $\{g(x, y), h(x, y)\}$ has exterior corners at $(m - r, r)$ and $(n - s, s)$, so the essential area of the footprint is mn.   ∎

7.9   Nullstellensätze

We shall want to know when a given ideal is the largest ideal with its set of zeros. The answer to this question is given by a theorem due to Hilbert, known as the *nullstellensatz*, one of the gems of nineteenth century mathematics. For any ideal I in $F[x, y]$, let $\mathcal{Z}(I)$ denote the set of affine zeros of I. We may assume that the field F is large enough to contain all the affine zeros of I. We may even dismiss this concern from the discussion by demanding that only algebraically closed fields be studied. However, we are mostly interested in finite fields, and these are never algebraically closed. Furthermore, in a finite field with $q = n + 1$ elements, it is often desirable to work in the ring $F^{\circ}[x, y] = F[x, y]/\langle x^n - 1, y^n - 1 \rangle$, rather than $F[x, y]$. Then the theorems of this section become almost trivial.

For any set of points $\mathcal{P}$ in the affine plane over F, or some extension of F, let $I(\mathcal{P})$ denote the ideal in $F[x, y]$ consisting of all bivariate polynomials that have zeros at every point of $\mathcal{P}$.

We shall discuss both the nullstellensatz and its companion, the weak nullstellensatz, but first we will give discrete versions of the weak nullstellensatz and the nullstellensatz reformulated for ideals of the quotient ring $F^\circ[x, y] = F[x, y]/\langle x^n - 1, y^n - 1 \rangle$. (Ideals in a quotient ring are defined, just as are ideals in $F[x, y]$.) The discrete versions will actually be more useful for some of our needs. The discrete versions of these theorems can be seen as rather obvious consequences of the convolution theorem, an elementary property of the two-dimensional Fourier transform, and are immediate observations to the field of signal processing.

For any ideal I in $F^\circ[x, y] = F[x, y]/\langle x^n - 1, y^n - 1 \rangle$, as before, let $\mathcal{Z}^\circ(I)$ denote the set of bicyclic zeros of I in the smallest extension of F that contains an element ω of order n, provided such an extension field exists. More simply, we may assume that F is chosen large enough so that it contains an element ω of order n. In any case, the bicyclic zeros of I are the zeros of the form $(\omega^{j'}, \omega^{j''})$.

Theorem 7.9.1 (discrete weak nullstellensatz) *If F contains an element ω of order n, then every proper ideal of the quotient ring $F[x, y]/\langle x^n - 1, y^n - 1 \rangle$ has at least one zero in the bicyclic plane.*

Proof: The bicyclic plane is the set of all points of the form $(\omega^{j'}, \omega^{j''})$. Let I be an ideal of $F^\circ[x, y] = F[x, y]/\langle x^n - 1, y^n - 1 \rangle$ with no bicyclic zeros. Every polynomial $s(x, y)$ of I can be represented by an n by n array of coefficients, denoted s. Thus the ring $F[x, y]/\langle x^n - 1, y^n - 1 \rangle$ can be regarded as the set $\{s\}$ of all n by n arrays under componentwise addition and bicyclic convolution. Accordingly, we may regard $F^\circ[x, y]$ as the set $\{s\}$ of such arrays and I as an ideal of $F^\circ[x, y]$. Every such s has an n by n two-dimensional Fourier transform S with components given by $S_{j'j''} = \sum_{j'} \sum_{j''} \omega^{i'j'} \omega^{i''j''} s_{i'i''}$. Thus the Fourier transform maps the ring $F^\circ[x, y]$ into the set of Fourier transform arrays $\{S\}$ and maps the ideal I into an ideal $I^* \subset \{S\}$, which is the transform-domain representation of the ideal I. The defining properties of an ideal, under the properties of the Fourier transform, require that the ideal $I^* \subset \{S\}$ is closed under componentwise addition, and that the componentwise product of any element of I^* with any array A is an element of I^*. Thus the set I^* itself is an ideal of the ring of all n by n arrays under componentwise addition and multiplication.

If I has no bicyclic zeros of the form $(\omega^{j'}, \omega^{j''})$, then, for every bi-index (r', r''), one can choose an array $S^{(r',r'')}$ in I^* that is nonzero at the component with bi-index (r', r''). For any such $S^{(r',r'')}$, one can choose the array $A^{(r',r'')}$, nonzero only where $(j', j'') = (r', r'')$, so that

$$A_{j'j''}^{(r',r'')} S_{j'j''}^{(r',r'')} = \begin{cases} 1 & (j', j'') = (r', r'') \\ 0 & (j', j'') \neq (r', r''). \end{cases}$$

Thus for every (r', r''), there is an element of I, denoted $\delta^{(r',r'')}(x, y)$, such that

$$\delta^{(r',r'')}(\omega^{j'}, \omega^{j''}) = \begin{cases} 1 & \text{if}(j', j'') = (r', r'') \\ 0 & \text{if}(j', j'') \neq (r', r''). \end{cases}$$

The set of all such $\delta^{(r',r'')}(\omega^{j'}, \omega^{j''})$ forms a basis for the vector space of all n by n arrays. Therefore I^* contains all n by n arrays. Thus I is equal to $F^\circ[x, y]$, so I is not a proper ideal. ∎

It is now clear that if F has an element ω of order n, then the ideals of $F^\circ[x, y] = F[x, y]/\langle x^n - 1, y^n - 1 \rangle$ can be expressed in terms of the Fourier transform in a simple way. Let I be any ideal of $F^\circ[x, y]$ and let $\mathcal{Z}^\circ(I)$ denote the zeros of the ideal I in the discrete bicyclic plane $\{(\omega^{j'}, \omega^{j''})\}$ over F. The following theorem says that if F has an element ω of order n, then any ideal I of $F^\circ[x, y]$ is completely characterized by its set of zeros $\mathcal{Z}^\circ$ in the bicyclic plane over F.

Theorem 7.9.2 (discrete nullstellensatz) *If F has an element of order n, and J is an ideal of the ring $F^\circ[x, y] = F[x, y]/\langle x^n - 1, y^n - 1 \rangle$, then $I(\mathcal{Z}^\circ(J)) = J$.*

Proof: Let ω be an element of F of order n. For any element $s(x, y)$ of $F^\circ[x, y]$, let $S_{j'j''} = s(\omega^{j'}, \omega^{j''})$ be the (j', j'')th component of its Fourier transform. Following the line of the proof of Theorem 7.9.1, for each $(\omega^{r'}, \omega^{r''}) \notin \mathcal{Z}^\circ(I)$ there is an element $s^{(r',r'')}(x, y)$ of J for which its Fourier transform $S_{j'j''}^{r'r''}$ is nonzero at the component with bi-index (j', j'') equal to (r', r''). Moreover, there is an element $a(x, y)$ of $F^\circ[x, y]$ such that $A_{j'j''} = a(\omega^{j'}, \omega^{j''})$ is nonzero only at (r', r''). In particular, $a(x, y)$ can be chosen such that $a(x, y)s(x, y)$ has the following Fourier transform:

$$\delta^{(r',r'')}(\omega^{j'}, \omega^{j''}) = \begin{cases} 1 & \text{if } (j', j'') = (r', r'') \\ 0 & \text{if } (j', j'') \neq (r', r''). \end{cases}$$

This means that for every $(\omega^{r'}, \omega^{r''}) \notin \mathcal{Z}^\circ(I)$, there is a function $\delta^{(r'r'')}(x, y)$ in J such that

$$\delta^{(r',r'')}(\omega^{j'}, \omega^{j''}) = \begin{cases} 1 \text{ if } (j', j'') = (r', r'') \\ 0 \text{ if } (j', j'') \neq (r', r''). \end{cases}$$

Clearly, the set of such $\delta^{(r',r'')}(x, y)$ is a vector space basis of J, so the ideal that these generate is J itself. ∎

The generalization of Theorems 7.9.1 and 7.9.2 to the ring $F[x, y]$ will take some work. The result of this work will be the two forms of the Hilbert nullstellensatz. First, we will introduce some terminology.

The *radical* of the ideal I is given by

$$\sqrt{I} = \{s(x, y) \mid s(x, y)^m \in I, m \in N\}.$$

An ideal I is called a *radical ideal* if $\sqrt{I} = I$. A radical ideal is defined by the property that whenever $p(x, y)^m \in I$, then $p(x, y) \in I$. It is informative to contrast a radical ideal with a prime ideal. A *prime ideal* is defined by the property that if $p(x, y)q(x, y)$ is an element of the ideal I, then $p(x, y)$ or $q(x, y)$ must be an element of I. Every prime ideal is necessarily a radical ideal.

Every ideal with at least one zero is a proper ideal. Because every ideal with multiple zeros is contained in an ideal with only one zero, it is enough to consider only a maximal ideal I over the extension field with a single zero at (a, b). But this ideal has a minimal basis $\{x - a, y - b\}$ and a footprint with area 1, so it is not the ring $F[x, y]$.

The nullstellensatz asserts that $I(\mathcal{Z}(I)) = \sqrt{I}$ for any ideal I in the ring of bivariate polynomials over an algebraically closed field F. Thus $I(\mathcal{Z}(I))$ contains a polynomial $p(x, y)$ if and only if $p(x, y)$ itself, or a power of $p(x, y)$, is already in I. In particular, if $p(x, y)$ is an irreducible polynomial in an algebraically closed field, then $I(\mathcal{Z}(\langle p(x, y)\rangle)) = \langle p(x, y)\rangle$.

We shall first state the weak nullstellensatz. The method of proof used for the discrete case does not work here because there is an infinite number of points at which zeros must be voided. Therefore a much different proof is needed.

Theorem 7.9.3 (weak nullstellensatz) *The only ideal in the ring $F[x, y]$ over F that has no affine zeros in any extension of F is $F[x, y]$ itself.*

Proof: This follows from Theorem 7.9.4 below. ∎

Any ideal I has a minimal basis and a footprint Δ. In Section 7.4, we saw that if I is a locator ideal with no affine zeros in any extension field, then the footprint is empty. The only exterior corner is $(0, 0)$, so the minimal basis is $\{1\}$. Thus $I = \langle 1 \rangle = F[x, y]$. The upcoming proof of the nullstellensatz will make use of the trivariate version of this fact: the only ideal in the ring $F[x, y, z]$ that has no affine zeros in any extension of F is $F[x, y, z]$ itself.

In our applications, we must deal with fields that are not algebraically closed, so we want to impose the notion of algebraic closure as weakly as possible. For this, it is enough to embed the field F into its smallest extension field that contains all the zeros of I. Our definition of $\mathcal{Z}(I)$, as the set of zeros in the smallest extension field of F that contains all the zeros of I, actually makes unnecessary the condition that F be algebraically closed.

Theorem 7.9.4 (nullstellensatz) *If F is an algebraically closed field, then the ideal $I(\mathcal{Z}(J)) \subset F[x, y]$ is equal to the set of bivariate polynomials $p(x, y) \in F[x, y]$ such that some power $p^m(x, y)$ is an element of J.*

Proof: Let $\{g_\ell(x, y) \mid \ell = 1, \ldots, M\}$ be a minimal basis for J. We will embed $R = F[x, y]$ into the ring of trivariate polynomials $\widetilde{R} = F[x, y, t]$. Each basis polynomial $g_\ell(x, y) \in R$ will be regarded as the polynomial $g_\ell(x, y, t) \in \widetilde{R}$, which, in fact, does not depend on t. The ideal J becomes the ideal of $\widetilde{R}$ consisting of all trivariate polynomials of the form $\Sigma_{\ell=1}^{M} a(x, y, t) g_\ell(x, y)$.

Suppose that $p(x, y) \in I(\mathcal{Z}(J))$. We wish to show that $p^m(x, y) \in J$ for some m. First, we will form a new polynomial in three variables, given by $\tilde{p}(x, y, t) = 1 - tp(x, y)$. Then consider the ideal $\tilde{J} \subset F[x, y, t]$, defined by appending $\tilde{p}(x, y, t)$ as an additional generator.

$$\tilde{J} = \langle g_1(x, y), g_2(x, y), \dots, g_M(x, y), 1 - tp(x, y) \rangle.$$

Step (1) We shall first prove that the ideal $\tilde{J}$ has no affine zeros, so the footprint of $\tilde{J}$ has area 0. Then we can conclude that $\tilde{J} = \langle 1 \rangle$.

Consider any (α, β, γ) in any extension of F. If $(\alpha, \beta) \in \mathcal{Z}(J)$, then (α, β) is a zero of $p(x, y)$, so (α, β, γ) is not a zero of $1 - tp(x, y)$ for any γ. If $(\alpha, \beta) \notin \mathcal{Z}(J)$, then for some ℓ, $g_\ell(\alpha, \beta) \neq 0$. Because $g_\ell(x, y) \in \tilde{J}$, $g_\ell(\alpha, \beta, \gamma) \neq 0$, so (α, β, γ) is not a zero of $\tilde{J}$ for any γ. In either case, (α, β, γ) is not a zero of $\tilde{J}$. Because every proper ideal has a zero, we can conclude that $\tilde{J} = F[x, y, t]$.

Step (2) From step (1), we can write the following:

$$1 = \sum_{\ell=1}^{M} a_\ell(x, y, t) g_\ell(x, y) + a_{M+1}(x, y, t)(1 - tp(x, y)).$$

Let m be the largest exponent of t appearing in any term and set $t = z^{-1}$. Multiply through by z^m to clear the denominator. Then

$$z^m = \sum_{\ell=1}^{M} z^m a_\ell(x, y, z^{-1}) g_\ell(x, y) + z^m a_{M+1}(x, y, z^{-1}) \left(1 - \frac{p(x, y)}{z} \right).$$

Let $b_\ell(x, y, z) = z^m a_\ell(x, y, z^{-1})$ so that

$$z^m = \sum_{\ell=1}^{M} b_\ell(x, y, z) g_\ell(x, y) + z^m a_{M+1}(x, y, z^{-1}) \left(1 - \frac{p(x, y)}{z} \right).$$

Now replace z by $p(x, y)$. Then

$$p^m(x, y) = \sum_{\ell=1}^{M} b_\ell(x, y, p(x, y)) g_\ell(x, y).$$

But $b_\ell(x, y, p(x, y))$ when expanded is a polynomial in $F[x, y]$. Thus $p^m(x, y) \in J$, as was to be proved. ∎

7.10   Cyclic complexity of arrays

The one-dimensional cyclic complexity property is as follows: The weight of a one-dimensional sequence of length n is equal to the cyclic complexity of its Fourier transform. Is it possible to generalize this to a cyclic complexity property for the two-dimensional Fourier transform? The answer is "yes" if we provide appropriate definitions for the terms "weight" and "linear complexity" for arrays. The first term is immediate. The Hamming weight, denoted $\mathrm{wt}(v)$, of an n by n array v is the number of nonzero components of the array. The two-dimensional Fourier transform of v is the array V. We want to define the cyclic complexity of the array V. Then we will prove the statement that the weight of v is equal to the cyclic complexity of V.

To define the cyclic complexity of the array V, we resort to the language of polynomials, replacing the array V by the bivariate polynomial

$$V(x,y) = \sum_{j'=0}^{n-1} \sum_{j''=0}^{n-1} V_{j'j''} x^{j'} y^{j''},$$

which we regard as an element of the ring $F^{\circ}[x,y] = F[x,y]/\langle x^n - 1, y^n - 1\rangle$. The array V, represented by the polynomial $V(x,y)$, has the inverse two-dimensional Fourier transform v. The weight of v is the number of values of $v_{i'i''} = (1/n^2)V(\omega^{-i'}, \omega^{-i''})$ that are nonzero.

Every n by n array V is associated with a locator ideal, which we will now introduce. A *locator polynomial* for the array V (or the polynomial $V(x,y)$) is a nonzero polynomial, $\Lambda(x,y)$, that satisfies

$$\Lambda(x,y)V(x,y) = 0.$$

We are usually interested in the case in which $V(x,y)$ is doubly periodic, with period n in both directions. Then we regard $V(x,y)$ as an element of $F^{\circ}[x,y]$, and, for emphasis, we then sometimes write the locator polynomial as $\Lambda^{\circ}(x,y)$ as a reminder of this periodic property. Then,

$$\Lambda^{\circ}(x,y)V(x,y) = 0 \qquad (\mathrm{mod}\ \langle x^n - 1, y^n - 1\rangle).$$

This polynomial product is equivalent to a two-dimensional cyclic convolution, $\Lambda^{\circ} * {*}V$, on the arrays formed by the coefficients of these polynomials. The properties of the Fourier transform tell us that the polynomial product is equal to zero if and only if

$$\Lambda^{\circ}(\omega^{-i'}, \omega^{-i''})V(\omega^{-i'}, \omega^{-i''}) = 0 \qquad \begin{array}{l} i' = 0, \ldots, n-1 \\ i'' = 0, \ldots, n-1. \end{array}$$

Because the n by n array $\boldsymbol{v} = \left[(1/n^2)V(\omega^{-i'}, \omega^{-i''})\right]$ has finite weight, the locator polynomial (now called simply $\Lambda(x, y)$) needs only a finite number of zeros to satisfy this equation. Therefore locator polynomials for $\boldsymbol{V}$ do exist and can be specified by any array $\lambda_{i'i''} = (1/n^2)\Lambda(\omega^{-i'}, \omega^{-i''})$ that has zeros wherever $v_{i'i''}$ is nonzero. The name "locator polynomial" refers to the fact that the set of nonzeros of the polynomial $V(x, y)$ is contained in the set of zeros of $\Lambda(x, y)$. In this sense, $\Lambda(x, y)$ "locates" the nonzeros of $V(x, y)$. However, $\Lambda(x, y)$ may also have some additional zeros; these then are at the zeros of $V(x, y)$. Thus the zeros of $\Lambda(x, y)$ do not fully locate the nonzeros of $V(x, y)$. To fully specify the nonzeros, we will consider the set of all such $\Lambda(x, y)$. It is trivial to verify that the set of such $\Lambda(x, y)$ forms an ideal in the ring of bivariate polynomials.

Definition 7.10.1 *The locator ideal for the array $\boldsymbol{V}$ (or the polynomial $V(x, y)$) is given by*

$$\Lambda(V) = \{\Lambda(x, y) \mid \Lambda(x, y) \text{ is a locator polynomial for } \boldsymbol{V}\}.$$

To reconcile this definition with the earlier Definition 7.7.1, simply note that the polynomial $V(x, y)$ has a finite number of bicyclic nonzeros and that $\Lambda(V)$ is the ideal consisting of all polynomials with zeros at the nonzeros of $V(x, y)$.

This definition of the locator ideal $\Lambda(V)$ is very different from our earlier way of specifying ideals in terms of generators in the form

$$\Lambda = \langle \Lambda^{(\ell)}(x, y) \mid \ell = 1, \ldots, L \rangle.$$

To reconcile this definition with Definition 7.7.1, we remark that the locator ideal of $V(x, y)$ is the locator ideal for the set of nonzeros of $V(x, y)$. A major task of Chapter 8 will be the development of an efficient algorithm for computing, from the array $\boldsymbol{V}$, the minimal basis $\{\Lambda^{(\ell)}(x, y)\}$ for the locator ideal of $\boldsymbol{V}$.

Now we are ready to discuss linear complexity. We shall define the linear complexity of the (two-dimensional) array $\boldsymbol{V}$ in terms of the locator ideal of $\Lambda(V)$. To understand this, we first examine this version of the locator ideal in one dimension:

$$\Lambda(V) = \{\Lambda(x) \mid \Lambda(x)V(x) = 0 \pmod{x^n - 1}\}.$$

To state the linear complexity of the sequence $\boldsymbol{V}$ (or the polynomial $V(x)$), we could first observe that, because every ideal in one variable is a principal ideal, every polynomial in $\Lambda(V)$ is a polynomial multiple of a generator of the ideal (perhaps normalized so that $\Lambda_0 = 1$). Then we could define the linear complexity of $\boldsymbol{V}$ as the degree of this minimal degree univariate polynomial in $\Lambda(V)$. This generator polynomial is the smallest degree polynomial whose zeros annihilate the nonzeros of $V(x)$. This, in essence, is the definition of linear complexity given in Chapter 1. In the ring of bivariate polynomials, however, an ideal need not be generated by a single polynomial, so this form of the definition does not generalize to two dimensions. Accordingly, we must define linear complexity in a way that does not assume a principal ideal. Therefore we use the following, slightly more cumbersome, definition of linear complexity. The *linear complexity* $\mathcal{L}(V)$ of the one-dimensional periodic sequence V is the number of values of the nonnegative integer j, such that x^j is not the leading monomial of any $\Lambda(x)$ in $\Lambda(V)$. More simply, the linear complexity of the sequence V is the area of the footprint $\Delta(\Lambda(V))$.

Now this definition is in a form that carries over to two dimensions, provided a total order that respects monomial multiplication is defined on the set of bivariate monomials. We will usually use the graded order because, in this total order, the number of monomials proceeding a given monomial is finite. Some other total orders may also be suitable. The generalization to two-dimensional arrays is now immediate.

Definition 7.10.2   *The linear complexity $\mathcal{L}(V)$ of the array V is the area of the footprint of the locator ideal of V.*

We end the section with the linear complexity property of the two-dimensional Fourier transform. This is a generalization of the linear complexity property of sequences, which was discussed in Section 1.5.

Theorem 7.10.3   *Let F contain an element of order n. The weight of a two-dimensional n by n array over F is equal to the linear complexity of its two-dimensional Fourier transform.*

Proof: Because F contains an element of order n, an n-point Fourier transform exists in F. The weight of an array is equal to the number of zeros of its locator ideal, as was stated in the remark following Definition 7.10.1. The number of zeros of the locator ideal is equal to the area of its footprint, and this is the definition of linear complexity.   ∎

Definition 7.10.2 can be expressed concisely as $\mathcal{L}(V) = \|\Delta(\Lambda(V))\|$, while Theorem 7.10.3 can be expressed as $\mathrm{wt}(v) = \mathcal{L}(V)$.

7.11   Enlarging an ideal

We are now ready to return to the example started in Section 7.6. Recall that the task stated in that section was to enlarge the ideal $I = \langle x^3 + x^2y + xy + x + 1, y^2 + y \rangle$ of $GF(2)[x, y]$ – whose footprint was shown in Figure 7.9 – so that the point $(2, 1)$ is an exterior corner of the new footprint, as shown in Figure 7.10. To do this, the polynomial $g_3(x, y) = x^2y + ax^2 + bxy + cx + dy + e$, with the leading monomial in the new exterior corner, is appended to the basis, and the constants a, b, c, d, and e, with the leading monomial in the new exterior corner, are chosen so that the new basis

$$\mathcal{G} = \{x^3 + x^2 y + xy + x + 1, \quad y^2 + y, \quad g_3(x,y)\}$$

is a reduced basis. First, compute the first of the two conjunction polynomials that involve $g_3(x,y)$. Let

$$f(x,y) = yg_3(x,y) - x^2(y^2 + y)$$
$$= ax^2 y + bxy^2 + cxy + dy^2 + ey - x^2 y,$$

and choose the five constants such that

$$R_{\mathcal{G}}[f(x,y)] = 0.$$

This means that $f(x,y)$ is in the ideal generated by $\mathcal{G}$. Because $\mathcal{G}$ is a reduced basis, division by $\mathcal{G}$ is straightforward. Simply set $x^3 = x^2 y + xy + x + 1$, $y^2 = y$, and $x^2 y = ax^2 + bxy + cx + dy + e$ whenever possible in $f(x,y)$, performing the steps of this reduction in any convenient order. This yields the following:

$$R_{\mathcal{G}}[f(x,y)] = a(a+1)x^2 + (c + ba)xy + c(a+1)x + (e + da)y + e(a+1),$$

where a, b, c, d, and e are now to be chosen from extension fields of $GF(2)$ so that the right side is equal to zero. We conclude from the first term that we must choose either $a = 0$ or $a = 1$. The other coefficients, then, must be as follows. If $a = 0$, then $c = 0$ and $e = 0$. If $a = 1$, then $c = b$ and $e = d$.

Next, compute the other conjunction polynomial. Let

$$f(x,y) = xg_3(x,y) - y(x^3 + x^2 y + xy + x + 1)$$
$$= ax^3 + bx^2 y + cx^2 + dxy + ex - x^2 y^2 - xy^2 - xy - y,$$

and further restrict the free coefficients so that

$$R_{\mathcal{G}}[f(x,y)] = 0.$$

If $a = 0$, then this requires that b satisfies $b^3 + b + 1 = 0$ and that d satisfies $d = (b+1)^{-1}$. If $a = 1$, this requires that b satisfies $b^3 + b + 1 = 0$ and d satisfies $d = b^{-1}$. Thus with α, α^2, and α^4 as the three elements of $GF(8)$, satisfying $\alpha^3 + \alpha + 1 = 0$, we have exactly six possibilities for $g_3(x, y)$:

$$g_3(x, y) = x^2 y + \alpha xy + \alpha^4 y,$$
$$g_3(x, y) = x^2 y + \alpha^2 xy + \alpha y,$$
$$g_3(x, y) = x^2 y + \alpha^4 xy + \alpha^2 y,$$
$$g_3(x, y) = x^2 y + x^2 + \alpha xy + \alpha x + \alpha^6 y + \alpha^6,$$
$$g_3(x, y) = x^2 y + x^2 + \alpha^2 xy + \alpha^2 x + \alpha^5 y + \alpha^5,$$
$$g_3(x, y) = x^2 y + x^2 + \alpha^4 xy + \alpha^4 x + \alpha^3 y + \alpha^3.$$

Note that there are six points in the footprint and six ways of appending a new polynomial to the basis in order to reduce the footprint to five points. We shall see that this is not a coincidence.

To develop this example a little further, choose the first of the $g_3(x, y)$ computed above as the new generator polynomial. This gives the new ideal generated by the reduced basis:

$$\mathcal{G} = \{x^3 + x^2 y + xy^2 + x + 1, \ y^2 + y, \ x^2 y + \alpha xy + \alpha^4 y\}.$$

The new ideal is a larger ideal whose footprint has area 5. How can this new reduced basis be further expanded, in turn, so that the area of the footprint is decreased to 4? There are two ways in which a new exterior corner can be specified to reduce the footprint to area 4, and these are shown in Figure 7.17. Thus the new basis polynomial is either

$$g_4(x, y) = x^2 + axy + bx + cy + d$$

or

$$g_4(x, y) = xy + ax + by + c,$$

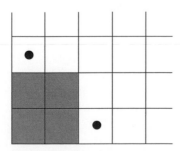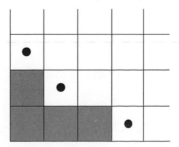

Figure 7.17. Possible new exterior corners.

where, as before, the constants a, b, c, and d are yet to be specified. We will consider each polynomial in turn. First, we append to $\mathcal{G}$ the polynomial

$$g_4(x, y) = x^2 + axy + bx + cy + d,$$

and choose a, b, c, and d such that the three conjunction polynomials including $g_4(x, y)$ are zero.

Let

$$f(x, y) = yg_4(x, y) - (x^2 y + \alpha xy + \alpha^4 y)$$
$$= axy^2 + (b - \alpha)xy + cy^2 + (d - \alpha^4)y.$$

This reduces to

$$R_{\mathcal{G}}[f(x, y)] = (a + b + \alpha)xy + (c + d + \alpha^4)y.$$

The coefficients are yet to be specified so that

$$R_{\mathcal{G}}[f(x, y)] = 0,$$

from which we conclude that $a + b = \alpha$ and $c + d = \alpha^4$.

To compute the second conjunction polynomial, let

$$f(x, y) = y^2 g_4(x, y) - x^2(y^2 + y)$$
$$= x^2 y + axy^3 + bxy^2 + cy^3 + dy^2.$$

From this, one can calculate

$$R_{\mathcal{G}}[f(x, y)] = (a + b + \alpha)xy + (c + d + \alpha^4)y$$
$$= 0,$$

which is the same condition encountered previously.

Finally, to compute the third conjunction polynomial, let

$$f(x, y) = xg_4(x, y) - (x^3 + x^2 y + xy + x + 1)$$
$$= (a + 1)x^2 y + bx^2 + (c + 1)xy + (d + 1)x + 1.$$

From this, one can calculate that

$$R_{\mathcal{G}}[f(x, y)] = [c + (a + 1)\alpha + 1 + ba]xy + [d + 1 + b^2]x$$
$$+ [bc + (a + 1)\alpha^4]y + 1 + bf$$
$$= 0.$$

From $1 + bd = 0$, we conclude that $d = b^{-1}$, and from $d + 1 + b^2 = 0$, we conclude that b satisfies $b^3 + b + 1 = 0$, so $b = \alpha, \alpha^2$, or α^4. We conclude that we have the following three possibilities:

$$
\begin{aligned}
a &= 0, & b &= \alpha, & c &= \alpha^3, & d &= \alpha^6; \\
a &= \alpha^4, & b &= \alpha^2, & c &= 1, & d &= \alpha^5; \\
a &= \alpha^2, & b &= \alpha^4, & c &= \alpha^6, & d &= \alpha^3.
\end{aligned}
$$

This specifies three possible polynomials for the new basis polynomial.

However, we have not yet found all the ways of appending a new polynomial to get a footprint of area 4. We can, instead, eliminate a different interior corner of the footprint. To do so, we append a polynomial of the form

$$g_4(x, y) = xy + ax + by + c,$$

choosing a, b, and c so that the three conjunction polynomials are all zero.

Let

$$
\begin{aligned}
f(x, y) &= xg_4(x, y) - (x^2 y + \alpha xy + \alpha^4 y) \\
&= ax^2 + (b + \alpha)xy + cx + \alpha^4 y.
\end{aligned}
$$

Then

$$R_G[f(x, y)] = ax^2 + (b + \alpha)(ax + by + c) + cx + \alpha^4 y = 0,$$

from which we conclude that $a = c = 0$ and b satisfies $b(b+\alpha) = \alpha^4$. This is solved by $b = \alpha^5$ or $b = \alpha^6$. This yields two more solutions that, together with the three earlier solutions, total five choices of a polynomial to append as a new generator polynomial. These are as follows:

$$
\begin{aligned}
g_4(x, y) &= x^2 + \alpha x + \alpha^3 y + \alpha^6, \\
g_4(x, y) &= x^2 + \alpha^4 xy + \alpha^2 x + y + \alpha^5, \\
g_4(x, y) &= x^2 + \alpha^2 xy + \alpha^4 x + \alpha^6 y + \alpha^3, \\
g_4(x, y) &= xy + \alpha^5 y, \\
g_4(x, y) &= xy + \alpha^6 y.
\end{aligned}
$$

For each choice of $g_4(x, y)$ to be appended to the basis, a new ideal is formed whose footprint has area 4. No other choice of $g_4(x, y)$, with its leading monomial in the current footprint, will give a new footprint of area 4; every other choice will give a footprint of an area smaller than 4, possibly area 0. The new footprint has either two or three exterior corners, so the reduced basis has only two or three generator polynomials, not four. This means that the old set of generator polynomials needs to be purged of superfluous polynomials to form a reduced basis.

Note that before appending polynomial $g_4(x, y)$, there were five points in the footprint and there are five choices for $g_4(x, y)$ that give a footprint of area 4. We made a similar observation earlier when there were six points in the footprint and six choices for $g_3(x, y)$ that give a footprint of area 5. Evidently, the new generator polynomial causes one of the zeros of the ideal to be eliminated. Each possible choice of a new generator polynomial corresponds to eliminating a different zero.

The lengthy calculations we have just finished are examples of a general procedure for enlarging a given ideal, I, in order to reduce the area of the footprint by 1, which we will develop in the remainder of this section. If the elements of the original ideal do not have a common nontrivial polynomial factor, then the footprint has finite area and the procedure systematically computes all ideals in which the given ideal is contained by removing points from the footprint, one by one, in various orders. Figure 7.18 shows the exterior corners of a typical footprint marked with dots and the interior corners marked with asterisks. Simply choose any interior corner of the footprint, then append to the reduced basis of the ideal a new polynomial whose leading monomial is in the chosen interior corner, and all of whose other monomials are in the footprint. Every other coefficient of the new polynomial is chosen in any way, provided only that all conjunction polynomials are zero.

The footprint Δ of $I = \langle g_1(x, y), \ldots, g_L(x, y) \rangle$ is a cascade set. We will require that the basis polynomials are arranged in staircase order. The new polynomial $p(x, y)$ that is appended will be a monic polynomial with all of its monomials in the footprint Δ and with its leading monomial corresponding to one of the interior corners of the footprint, as illustrated in Figure 7.18. This interior corner will have coordinates $(m_\ell - 1, n_{\ell+1} - 1)$ for some ℓ, as determined by two neighboring exterior corners with coordinates (m_ℓ, n_ℓ) and $(m_{\ell+1}, n_{\ell+1})$. These two neighboring exterior corners correspond to the two polynomials $g_\ell(x, y)$ and $g_{\ell+1}(x, y)$ of the reduced basis, which we will refer

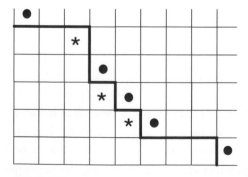

Figure 7.18. Exterior and interior corners of a footprint.

to more simply as $g(x, y)$ and $h(x, y)$, and write

$$g(x, y) = x^{m_\ell} y^{n_\ell} + g^-(x, y),$$
$$h(x, y) = x^{m_{\ell+1}} y^{n_{\ell+1}} + h^-(x, y).$$

To reduce the footprint, the single interior corner, corresponding to monomial $x^{m_\ell - 1} y^{n_{\ell+1} - 1}$, will be removed from the footprint Δ by appending to $\mathcal{G}$ the new polynomial

$$p(x, y) = x^{m_\ell - 1} y^{n_{\ell+1} - 1} + p^-(x, y),$$

where all coefficients of $p^-(x, y)$ have indices in the footprint Δ and occur prior to the leading monomial of $p(x, y)$ in the total order. This is written as follows:

$$p^-(x, y) = \sum_{\substack{(\ell', \ell'') \in \Delta \\ (\ell', \ell'') < (m_\ell - 1, n_{\ell+1} - 1)}} p_{\ell' \ell''} x^{\ell'} y^{\ell''}.$$

Our only remaining task is to show that the coefficients of this polynomial $p^-(x, y)$ can always be chosen so that the new polynomial $p(x, y)$ is a part of the reduced basis for a new ideal. This is a consequence of the following theorem. To earn the use of this theorem, we will need to go through a somewhat long and tedious, though elementary, algebraic proof.

Theorem 7.11.1 *A reduced basis for the ideal I in $F[x, y]$, with footprint of finite area, can be augmented by a single polynomial, perhaps with coefficients in an extension field F' of F, to produce a set of polynomials that contains a reduced basis for an ideal $I' \subset F'[x, y]$, with footprint of area smaller by 1.*

Proof: Choose any interior corner of the footprint of I. Let $p(x, y) = x^{m_\ell - 1} y^{n_{\ell+1} - 1} + p^-(x, y)$ be a polynomial with the leading monomial in that interior corner and with all other coefficients, yet to be determined, lying within the other cells of the footprint. The enlarged set of $L + 1$ generator polynomials is given by

$$\mathcal{G}' = \{g_1(x, y), \ldots, g(x, y), p(x, y), h(x, y), \ldots, g_L(x, y)\},$$

where $p(x, y)$ has been inserted between $g(x, y) = g_\ell(x, y)$ and $h(x, y) = g_{\ell+1}(x, y)$ to preserve the staircase order. This set generates the new ideal I' with footprint Δ'. To prove the theorem, we must show only that the coefficients of the polynomial $p(x, y)$ can be chosen so that all conjunction polynomials are equal to zero. By Corollary 7.6.2, for $\mathcal{G}'$ to be a reduced basis for I', it is enough for the coefficients $p_{\ell' \ell''}$ of $p(x, y)$ to be assigned so that the two conjunction polynomials of $p(x, y)$ with its two neighbors $g(x, y)$ and $h(x, y)$ are zero. This is because any conjunction polynomial not involving

$p(x, y)$ is surely zero, since appending $p(x, y)$ to the set of divisors cannot change a zero remainder to a nonzero remainder.   ∎

The conjunction polynomial of $p(x, y)$ and $g(x, y)$, set equal to zero, will impose relationships on the coefficients across rows of the array. Likewise, the conjunction polynomial of $p(x, y)$ and $h(x, y)$, set equal to zero, will impose relationships on the coefficients across columns of the array. We must show that these two sets of relationships are consistent and can be satisfied by at least one array of coefficients for $p(x, y)$.

By aligning the leading monomials, we can write the two conjunction polynomials of interest, set equal to zero, as follows:

$$R_{G'}[x^{m_\ell - m_{\ell+1} - 1} h(x, y) - yp(x, y)] = 0,$$
$$R_{G'}[y^{n_{\ell+1} - n_\ell - 1} g(x, y) - xp(x, y)] = 0.$$

How these two equations have been formed can be seen by reference to Figure 7.19, in which squares show some of the cells of the footprint Δ, those near the interior corner to be deleted. The exterior corners of Δ corresponding to $g(x, y)$ and $h(x, y)$ are marked by g and h. The interior corner of Δ that will be turned into an exterior corner of the new footprint is the square marked by p. The first equation comes from aligning $h(x, y)$ and $p(x, y)$ to the cell of Figure 7.19 marked by the upper asterisk. The second equation comes from aligning $g(x, y)$ and $p(x, y)$ to the cell of Figure 7.19 marked by the lower asterisk. The leading monomials cancel, so the two equations reduce to

$$R_{G'}[x^{m_\ell - m_{\ell+1} - 1} h^-(x, y)] = R_{G'}[yp^-(x, y)]$$

and

$$R_{G'}[y^{n_{\ell+1} - n_\ell - 1} g^-(x, y)] = R_{G'}[xp^-(x, y)].$$

Because $\mathcal{G}'$ is a reduced basis, these modulo $\mathcal{G}'$ reductions are computed simply by folding back terms into the quotient ring $F[x, y]/I'$. However, it may be necessary to enter the extension field F' to find the coefficients of $p(x, y)$.

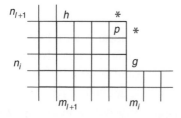

Figure 7.19. Computing the conjunction polynomials.

We will first restate the left sides of these two expressions, then the right sides. The left sides of the two expressions, with $m_\ell - m_{\ell+1} = m$ and $n_{\ell+1} - n_\ell = n$, can be written as follows:

$$R_{G'}[x^{m-1}h^-(x,y)] = R_G[x^{m-1}h^-(x,y)] - R_G[\tilde{a}p(x,y)],$$

$$R_{G'}[y^{n-1}g^-(x,y)] = R_G[y^{n-1}g^-(x,y)] - R_G[\tilde{b}p(x,y)],$$

where the coefficients $\tilde{a}$ and $\tilde{b}$ are given by the coefficient of $x^{m-1}y^{n-1}$ in the first term on the right. These equations can be abbreviated:

$$R_{G'}[x^{m-1}h^-(x,y)] = \tilde{h}(x,y) - \tilde{a}p(x,y),$$

$$R_{G'}[y^{n-1}g^-(x,y)] = \tilde{g}(x,y) - \tilde{b}p(x,y),$$

where $\tilde{h}(x,y)$ and $\tilde{g}(x,y)$ are polynomials all of whose coefficients are determined by $\mathcal{G}$ and so are also known. The expression says that to fold back $x^{m-1}h^-(x,y)$ onto the footprint of I', first fold back $x^{m-1}h^-(x,y)$ onto the footprint of I, using the polynomials of $\mathcal{G}$, then fold back the single exterior corner $(m-1, n-1)$ of that result onto the footprint of I', using the newly appended generator polynomial $p(x,y)$.

To simplify notation in the coming equations, we will write the first two coefficients of $p^-(x,y)$ as a and b, respectively:

$$p^-(x,y) = ax^{m-1}y^{n-2} + bx^{m-2}y^{n-1} + \cdots$$

These are the two coefficients of $p(x,y)$ that will hold our attention.

The right sides of the earlier two expressions, in the arithmetic of the quotient ring $F[x,y]/I$, are given by

$$R_{G'}[yp^-(x,y)] = yp^-(x,y) - ap(x,y),$$

$$R_{G'}[xp^-(x,y)] = xp^-(x,y) - bp(x,y),$$

where the coefficients a and b in the two equations are equal to the coefficients of $x^{m-1}y^{n-1}$ in the first terms of $yp^-(x,y)$ and $xp^-(x,y)$, respectively, on the right. The leading monomials of the terms $ap(x,y)$ and $bp(x,y)$ cancel the leading monomials of the terms $yp^-(x,y)$ and $xp^-(x,y)$, respectively. Combining the equations by pairs yields

$$(\tilde{a} - a)p(x,y) = \tilde{h}(x,y) - yp^-(x,y),$$

$$(\tilde{b} - b)p(x,y) = \tilde{g}(x,y) - xp^-(x,y).$$

These two equations are understood to be equations in the quotient ring $F[x,y]/I$, and the polynomials $\tilde{h}(x,y)$ and $\tilde{g}(x,y)$ are known. The monomials contained in the footprint of I form a vector-space basis of $F[x, y]/I$, which means that the polynomial equations can be rewritten as the following matrix equations:

$$(\tilde{a} - a)\boldsymbol{I}\boldsymbol{p} = \tilde{\boldsymbol{h}} - \boldsymbol{A}\boldsymbol{p},$$

$$(\tilde{b} - b)\boldsymbol{I}\boldsymbol{p} = \tilde{\boldsymbol{g}} - \boldsymbol{B}\boldsymbol{p},$$

from which we can solve the two equations for $\boldsymbol{p}$ to write

$$\boldsymbol{p} = [\boldsymbol{A} + (\tilde{a} - a)\boldsymbol{I}]^{-1}\tilde{\boldsymbol{h}};$$

$$\boldsymbol{p} = [\boldsymbol{B} + (\tilde{b} - b)\boldsymbol{I}]^{-1}\tilde{\boldsymbol{g}}.$$

The elements of the inverse matrices on the right are rational functions in a and b, respectively, of the form $r(a)/\det(a)$ and $r'(b)/\det'(b)$. The first rows are given by

$$\det(a)a - \sum_i r(a)h_i = 0,$$

$$\det(b)b - \sum_i r'(b)g_i = 0,$$

which yields a polynomial in a equal to zero and a polynomial in b equal to zero.

Furthermore, multiplying the earlier equation for the first conjunction polynomial by x and the earlier equation for the second conjunction polynomial by y yields

$$R_G[x^m h^-(x, y)] = R_G[y^n g^-(x, y)],$$

from which we obtain

$$x\tilde{h}(x, y) = y\tilde{g}(x, y)$$

in the ring $F[x, y]/I$. We must only show that there is at least one $p(x, y)$ satisfying this system of three equations in $F[x, y]/I$. It will suffice to work in the ring $F[x, y]$, deferring the modulo G reduction until later.

The terms on both sides in the monomial $x^{m-1}y^{n-1}$ cancel by design and need not be considered further. Equating coefficients of the other monomials yields

$$\tilde{h}_{i'i''} - Ap_{i'i''} = p_{i', i''-1} - ap_{i'i''}.$$

Setting $(i', i'') = (m - 1, n - 2)$, and recalling that $a = p_{m-1, n-2}$, yields

$$a(a - A) + \tilde{h}_{m-1, n-2} = p_{m-1, n-3},$$

which yields a polynomial in the unknown a that we can write as follows:

$$a(a - A) - p_{m-1,n-3} + \tilde{h}_{m-1,n-2} = 0.$$

The last term is the unknown $p_{m-1,m-3}$, which can be eliminated by using the equation obtained by setting the coefficient of the monomial $x^{m-1}y^{n-3}$ equal to zero,

$$\tilde{h}_{m-1,n-3} - Ap_{m-1,n-3} = p_{m-1,n-4} - ap_{m-1,n-3},$$

from which we obtain

$$(a - A)p_{m-1,n-3} = p_{m-1,n-4} - \tilde{h}_{m-1,n-3},$$

which can be used to eliminate $p_{n-1,n-3}$ from the earlier equation. Thus

$$a(a - A)^2 - (a - A)\tilde{h}_{m-1,n-2} = p_{m-1,n-4} - \tilde{h}_{m-1,n-3}.$$

Repeating this process, $p_{m-1,n-4}$ can be eliminated in terms of $p_{m-1,n-5}$, and so on. The process stops with $p_{m-1,0}$ because there is no $p_{m-1,-1}$. In this way, a polynomial in only the unknown a is obtained. All coefficients of this polynomial are known. Consequently, in some extension field F' of F, there is an element that satisfies this polynomial.

Once a is known, the sequence of equations just encountered can be used to solve for other unknown coefficients. Thus

$$(A - a)p_{m-1,0} = \tilde{h}_{m-1,0}$$

can then be solved for $p_{m-1,0}$, one by one;

$$(A - a)p_{m-1,i} = \tilde{h}_{m-1,i} - p_{m-1,i-1}$$

can be solved for $p_{m-1,i}$ for $i = 1, \ldots, n - 2$.

Now consider the second conjunction polynomial to be satisfied. Using the same arguments leads to

$$\tilde{g}_{i'i''} - Bp_{i'i''} = p_{i'-1,i''} - bp_{i'i''},$$

where B and b are defined analogously to A and a. In the same way as before, b can be found as a zero of a univariate polynomial. All remaining coefficients can then be determined.

Problems

7.1   Is the minimal basis for an ideal unique? Either prove that it is unique or give a counterexample.

7.2   How many monomial orders on N exist? How many monomial orders on N^2 exist?

7.3   Prove that $(j', j'') \preceq (k', k'')$ if $(j', j'') \leqq (k', k'')$.

7.4   Find two ideals, I_1 and I_2, in $F[x, y]$ such that $I_1 \bigcup I_2$ is not an ideal.

7.5   Show that every ideal in $F[x]$ is a principal ideal, but that principal ideals are atypical in $F[x, y]$.

7.6   Prove that an equivalent definition of a radical ideal is the following: an ideal I is a radical ideal if and only if no element of its reduced basis $\{g_\ell(x, y)\}$ can be expressed as a power of another element of I.

7.7   In the ring $GF(2)[x, y]$, how many ideals are there with footprint

$$\Delta = \{(0, 0), (0, 1), (0, 2), (1, 0), (1, 1)\}$$

in the graded order?

7.8   Show that every set of monomials in $F[x, y]$ contains a minimal basis for the ideal it generates. Is this minimal basis a reduced basis?

7.9   Prove that addition and multiplication in the quotient ring $F[x, y]/I$ are well defined; that is, prove that the sum and product in the quotient ring do not depend on the choice of representatives.

7.10 Generalize the treatment of the Hilbert basis theorem, given in this chapter, to polynomial rings in m variables. Specifically, prove that every ideal in the ring $F[x_1, \ldots, x_m]$ is finitely generated.

7.11 A *maximal ideal* of $F[x, y]$ is a proper ideal that is not properly contained in another proper ideal. Show that if F is algebraically closed, then every maximal ideal of $F[x, y]$ has the form $\langle x - a, y - b \rangle$.

7.12 Prove that if R is a noetherian ring, then $R[x]$ is a noetherian ring.

7.13 Let $\mathcal{G}$ be a reduced basis for an ideal under the graded order. Show that the staircase order on the leading monomials need not put the basis polynomials in the same order as does the graded order.

7.14 Prove that the area of the footprint of an ideal in $F[x, y]$ does not depend on the choice of monomial order.

7.15 Prove that if the number of affine zeros of an ideal is fewer than the area of the footprint of that ideal, then the ideal is not a locator ideal, and that there is a larger ideal with that same set of zeros.

7.16 (**Dickson's lemma**) Let S be any subset of N^n of the form $\bigcup_{\ell=1}^{\infty}(u_\ell + N^n)$, where the u_ℓ are elements of N^n. Prove that there exists a finite set of points of

N^n, denoted $v_1, v_2, \ldots, v_r$, such that

$$\mathcal{S} = \bigcup_{\ell=1}^{r} (v_\ell + N^n).$$

7.17 How many affine zeros does the ideal $I = \langle x^3 + xy^2 + x^2 + 1, y^2 + xy + y \rangle$ have? What is the area of the footprint of I? Compute the footprint for the graded order with $x \prec y$ and for the graded order with $y \prec x$.

7.18 Prove that every bivariate polynomial over the field F has a unique factorization in $F[x, y]$.

7.19 The Klein quartic polynomial is $x^3 y + y^3 + x$. Find a reduced basis for the ideal $\langle x^3 y + y^3 + x, x^7 - 1, y^7 - 1 \rangle$ in $GF(8)[x, y]$. What is the area of the footprint of the ideal? How many affine zeros does the Klein quartic polynomial have? How many monomials are in a monomial basis for this ideal? List the monomials.

7.20 If all polynomials of a reduced basis for an ideal are irreducible, is the ideal necessarily a prime ideal?

7.21 A footprint for a specific ideal in the ring $F[x, y, z]$ of trivariate polynomials has 11 exterior corners. Show that there is a set of 11 trivariate polynomials that generates this ideal.

7.22 Let I be a prime ideal of $F[x, y]$. Prove that the quotient ring $F[x, y]/I$ is a field.

Notes

The material of this chapter belongs to those branches of mathematics known as commutative algebra and computational algebraic geometry, but the presentation has been reshaped to conform with the traditional presentation of the subject of algebraic coding theory. Although most of the concepts hold in the ring of polynomials in m variables, I prefer to think through the development given in $F[x, y]$, partly to make it explicit and shaped for the application to coding theory, and partly because the development here is unconventional, and I want to be sure to get it right for the purpose at hand.

The Gröbner bases were introduced independently by Hironaka (under the name *standard basis*) and by Buchberger (1985). The later name, Gröbner basis, has come to be preferred in the literature. The definition of a Gröbner basis is not sufficiently restrictive for our needs, because the definition allows superfluous polynomials to be included within a Gröbner basis. The special cases of minimal bases and reduced bases are defined with more restrictions, and so are more useful. A treatment of Gröbner bases along the line of this book may be found in Lauritzen (2003).

I consider the notion of a footprint as a natural companion to the definition of a Gröbner basis, but I do not find this notion to be explicit in the literature. Because

I have used the footprint as a starting point for the discussion of ideals, and also for Bézout's theorem, I needed to introduce a name. My choice captures, at least for me, the correct intuition. I also consider the equality between the number of zeros of an ideal and the area of its footprint as a key theorem. This equality is implicit in the algorithm of Sakata, although he does not extract it as an independent fact. The notion of the footprint is explicit in that algorithm because it recursively computes an ideal by first finding the footprint, then fitting its exterior corners with basis polynomials. The proof of the affine form of Bézout's theorem, by using the area of the footprint, may be original. This approach also provides a generalization to a set of more than two mutually coprime polynomial by bounding the number of common affine zeros in terms of only the bidegrees of the generator polynomials.

The Buchberger theorem was used by Buchberger as a key step in the derivation of his algorithm. Because I find other uses for this theorem, I have given it a separate place in the development, and also a proof that I prefer.

The Hilbert basis theorem and the nullstellensatz were proved by Hilbert (1890, 1893). An insight into this theorem can be found in the lemma of Dickson (1913). Because we will deal with fields that are not algebraically closed, and only with points of the field that are nth roots of unity, we have also stated the nullstellensatz here in a restricted form that is suited to the tone of this book.

8 Computation of Minimal Bases

An ideal in the ring $F[x, y]$ is defined as any set of bivariate polynomials that satisfies a certain pair of closure conditions. Examples of ideals can arise in several ways. The most direct way to specify concretely an ideal in the ring $F[x, y]$ is by giving a set of generator polynomials. The ideal is then the set of all polynomial combinations of the generator polynomials. These generator polynomials need not necessarily form a minimal basis. We may wish to compute a minimal basis for an ideal by starting with a given set of generator polynomials. We shall describe an algorithm, known as the *Buchberger algorithm*, for this computation. Thus, given a set of generator polynomials for an ideal, the Buchberger algorithm computes another set of generator polynomials for that ideal that is a minimal basis.

A different way of specifying an ideal in the ring $F[x, y]$ is as a locator ideal for the nonzeros of a given bivariate polynomial. We then may wish to express this ideal in terms of a set of generator polynomials for it, preferably a set of minimal polynomials. Again, we need a way to compute a minimal basis, but starting now from a different specification of the ideal. We shall describe an algorithm, known as the *Sakata algorithm*, that performs this computation.

Both the Buchberger algorithm and the Sakata algorithm compute a minimal basis of an ideal, but they start from quite different specifications of the ideal. Consequently, the algorithms are necessarily very different in their structures. The Buchberger algorithm may be regarded as a generalization of the euclidean algorithm, and the Sakata algorithm may be regarded as a generalization of the Berlekamp–Massey algorithm.

8.1 The Buchberger algorithm

Every ideal in $F[x]$ is generated by a single polynomial. Thus, any ideal of $F[x]$ that is specified by two polynomials as $I = \langle f(x), g(x) \rangle$ can be re-expressed as $I = \langle h(x) \rangle$, where $h(x) = \text{GCD}\langle f(x), g(x) \rangle$. This follows from the well known relationship $\text{GCD}[f(x), g(x)] = a(x)f(x) + b(x)g(x)$, given as Corollary 3.4.2. If $f(x)$ and $g(x)$ are coprime, then the ideal I is $F[x]$ itself. The euclidean algorithm for polynomials finds the greatest common divisor of two polynomials, and so it may be regarded as a method for computing a single generator polynomial for the ideal $I = \langle h(x) \rangle \subset F[x]$ whenever I is specified by two polynomials as $I = \langle f(x), g(x) \rangle$. Thus the euclidean algorithm computes the (unique) minimal basis for an ideal specified by two polynomials in $F[x]$. From this point of view, the generalization of the euclidean algorithm to ideals in $F[x, y]$ (or $F[x_1, \ldots, x_m]$) is called the *Buchberger algorithm*. The bivariate ideal $I \subset F[x, y]$ may be specified by a set of generator polynomials, $\mathcal{G} = \{g_\ell(x, y) \mid \ell = 1, \ldots, L\}$, that need not form a minimal basis with respect to a given monomial order. Buchberger's algorithm is a method of computing a minimal basis for the ideal generated by $\mathcal{G}$. The Buchberger algorithm uses the bivariate division algorithm for polynomials.

We shall first describe a straightforward, though inefficient, form of Buchberger's algorithm. It consists of "core" iterations, followed by a cleanup step that discards unneeded generator polynomials. Each Buchberger core iteration begins with a set of monic polynomials, $\mathcal{G}_i = \{g_\ell(x, y) \mid \ell = 1, \ldots, L_i\}$, that generates the given ideal I and appends additional monic polynomials $g_\ell(x, y)$ for $\ell = L_i + 1, \ldots, L_{i+1}$ to produce the larger set $\mathcal{G}_{i+1}$, which also generates the same ideal. These additional polynomials are computed from the others as conjunction polynomials. Then the core iteration is repeated. The core iterations terminate at the first iteration during which no new polynomials are appended. At the termination of these core iterations, as we shall see from the Buchberger theorem, the enlarged set of polynomials contains a minimal basis and, in general, some extra polynomials that can then be discarded.

To extract a minimal basis from the final set of polynomials, simply compute the footprint, as defined by the set of leading monomials of these polynomials. This can be done by marking the leading monomials on a grid representing $\mathbb{N}^2$. Each leading monomial excludes from the footprint all points of the quarter plane above and to the right of it. This is shown in Figure 8.1. Each small circle corresponds to one of the

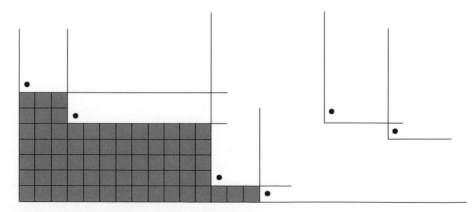

Figure 8.1. Output of the Buchberger algorithm.

polynomials of $\mathcal{G}^*$ by designating its leading monomial. The two rightmost circles designate leading monomials that do not play a role in defining the footprint; hence, those polynomials can be deleted from the basis. In this way, discard those polynomials that are not needed to fill all exterior corners of the footprint. The remaining polynomials (four in Figure 8.1) form a minimal basis for the ideal. Then it is straightforward to compute a reduced basis from the minimal basis.

Given the ordered set of polynomials $\mathcal{G}$, the core iteration of the Buchberger algorithm computes a conjunction polynomial for each pair of elements of $\mathcal{G}$. To compute a conjunction polynomial, as described in Section 7.3, take two bivariate monic polynomials from the set, align the polynomials as necessary, by multiplying each by a monomial so that both polynomials have the same leading monomial, then subtract the two aligned polynomials to cancel the leading monomials, and finally reduce modulo the ordered set $\mathcal{G}$ to produce a new polynomial in the ideal generated by the set $\mathcal{G}$.

The most straightforward (though inefficient) form of the Buchberger algorithm repeats this computation for each pair of polynomials to form the core iteration; this requires the computation of $\binom{L_i}{2}$ conjunction polynomials. If all $\binom{L_i}{2}$ conjunction polynomials are zero, the process halts. Otherwise, normalize each nonzero remainder polynomial to make it monic, and enlarge the set of generator polynomials by appending every distinct new monic polynomial to the set $\mathcal{G}$. Then repeat the core iteration for the new set $\mathcal{G}$.

Figure 8.2 shows an example in which an ideal is initially defined by three generator polynomials, represented by three small circles corresponding to the leading monomials of the three polynomials. Starting at each small dot is a quarter plane, consisting of all points above this point in the division order. No point in this quarter plane above a dot can be in the footprint of the ideal, nor can any point in this quarter plane be

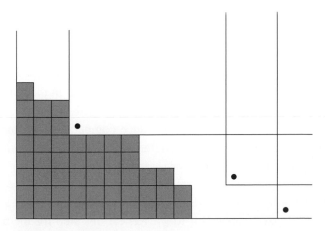

Figure 8.2. Footprint of an ideal.

the leading monomial of a polynomial generated in a subsequent Buchberger core
iteration.

Figure 8.2 also highlights the footprint of the ideal generated by these three polynomials. The Buchberger algorithm approaches the footprint of an ideal from above.
Each new conjunction polynomial, if not the zero polynomial, has a leading monomial
that is not dominated by the leading monomials of the polynomials already computed.
The algorithm halts with set $\mathcal{G}_i$ when each exterior corner of the footprint is occupied
by the leading monomial of a polynomial of $\mathcal{G}_i$. Other polynomials in the set $\mathcal{G}_i$ that do
not correspond to an exterior corner can be discarded.

The footprint of the set of leading monomials of the polynomials in $\mathcal{G}$ defines a
cascade set, which is nonincreasing under the core iteration. We must prove that the
algorithm eventually reduces the footprint of the increasing set of generator polynomials
to the footprint of the ideal.

The Buchberger algorithm is certain to halt, because at each step, unless all conjunction polynomials computed in that step are zero, new polynomials are produced
whose leading monomials are not divisible by the leading monomial of any previous
polynomial. This can be done only a finite number of times, because the footprint is
reduced at each step, either by deleting a finite number of squares, or by cropping one
or more infinite rows or columns along the horizontal or vertical axis; there are only a
finite number of such infinite rows or columns.

The following proposition says that any set which is unchanged by a core iteration
of the Buchberger algorithm must contain a minimal basis.

Proposition 8.1.1   *The Buchberger algorithm terminates with a set of polynomials
that contains a minimal basis for the ideal generated by* $\mathcal{G}$.

Proof: This is an immediate consequence of the Buchberger theorem, given as Theorem 7.6.1, which says that if $\mathcal{G}_i$ at any iteration does not contain a minimal basis, then
the computation does not yet terminate. ∎

This naïve form of the Buchberger algorithm is not acceptable, even for moderate
problems, because of its complexity, which is illustrated by recursively repeating the
assignment $L \to \binom{L}{2}$. Clearly, the number of computations is explosive (unless most
conjunction polynomials are zero). Indeed, L is doubly exponential in the number of
iterations n, being proportional to e^{ae^n}.

Embellishments to Buchberger's algorithm that reduce computational complexity
are easy to find. Prior to each Buchberger core iteration, reduce $\mathcal{G}_i$, if possible, as
follows. We may assume that all polynomials of $\mathcal{G}_i$ are monic. Let $\mathcal{G}_i^* \subseteq \mathcal{G}_i$ consist
of those polynomials that correspond to the exterior corners of $\Delta(\mathcal{G}_i)$. If $\mathcal{G}_i^* \neq \mathcal{G}_i$ for
each $g_\ell(x, y) \in \mathcal{G}_i$ that is not in $\mathcal{G}_i^*$, compute $r(x, y) = R_{\mathcal{G}_i^*}[g_\ell(x, y)]$. Delete $g_\ell(x, y)$
from $\mathcal{G}_i$, and, if $r(x, y)$ is nonzero, append it to $\mathcal{G}_i$. Repeat this step for the new $\mathcal{G}_i$
until $\mathcal{G}_i^* = \mathcal{G}_i$. At this point, every polynomial of $\mathcal{G}_i$ corresponds to an exterior corner

of $\Delta(\mathcal{G}_i)$. Then put the polynomials in staircase order and proceed with the next core iteration of the Buchberger algorithm. When the elements of $\mathcal{G}_i$ are arranged in staircase order, it is enough to compute the conjunction polynomials only for the $L - 1$ pairs of consecutive polynomials in $\mathcal{G}_i$, given by Corollary 7.6.2. The algorithm can stop when these conjunction polynomials are all zero.

8.2 Connection polynomials

The Sakata algorithm, to be developed in Section 8.4, computes a certain locator ideal – provided such an ideal is defined – by iteratively computing a set of polynomials that terminates with a set of minimal polynomials for that locator ideal. During the computations, the intermediate polynomials are called *connection polynomials* and their coefficients are called *connection coefficients*. These are named for the way they are used to connect the elements of a bivariate array.

For a given total order, which we usually take to be the graded order, the initial elements of the bivariate array V form a sequence. For each bi-index (j', j'') that occurs in the sequence V, $V_{j'j''}$ has a fixed place in the sequence. The jth element in the sequence is then denoted V_j, and each j of the sequence corresponds to one bi-index (j', j''). The bivariate recursion connects the given elements of the bivariate array $V = [V_{j'j''}]$. These are elements whose index j satisfies $j \preceq r$, where V_r is the last term of the sequence in the total order.

We define the polynomial $\Lambda(x, y)$ of bidegree s (corresponding to the bivariate sequence $\Lambda_0, \Lambda_1, \ldots, \Lambda_{s-1}, \Lambda_s$ in the total order) to be a connection polynomial for the bivariate sequence $V = V_0, V_1, \ldots, V_{r-1}, V_r$ if the bivariate recursion

$$\sum_{k'} \sum_{k''} \Lambda_{k'k''} V_{j'+k', j''+k''} = 0$$

is satisfied for every $j \preceq r$ for which all terms in the sum are contained in the sequences $\Lambda_0, \ldots, \Lambda_s$ and $V_0, \ldots, V_r$. This last condition is because the summation on the left can be executed only if all terms in the sum are given. The summations in the indices refer to ordinary addition on each component. Both the total order and the division order are in play, which leads to some interesting interactions between these two orders.

The definition of a connection polynomial is worded in such a way that if there is no $j = (j', j'')$ for which the defining sum can be executed, then $\Lambda(x, y)$ is a connection polynomial by default. It is easier to include this vacuous case as a connection polynomial than to exclude it. This means, however, that all $\Lambda(x, y)$ of sufficiently large degree are connection polynomials. Therefore the difference of two connection polynomials may fail to be a connection polynomial, so the set is not an ideal. Nevertheless, the set of all connection polynomials for a given V does have a footprint, and this footprint does have a set of exterior corners. The *minimal connection polynomials* are the monic polynomials with their leading monomials in these exterior corners.

Generally, the connection polynomials do not generate an ideal, so in general we do not refer to these polynomials as generator polynomials. However, we are ultimately interested in sets of connection polynomials that do form an ideal – specifically, a locator ideal. The Sakata algorithm in its normal use for decoding will terminate with a set of minimal connection polynomials, called a *minimal connection set*, that generates a locator ideal.

The bivariate recursion is rewritten more concisely by using a single index to refer to total order. Then we write the recurison as

$$\sum_{k=0}^{s} \Lambda_k V_{j+k} = 0,$$

where, as usual, $j + k$ always means $(j' + k', j'' + k'')$. We shall say that the bivariate polynomial $\Lambda(x, y)$ *produces* the element V_r of the array from the previous elements of the array $V_0, V_1, \ldots, V_{r-1}$ if we can find a bi-index $j = (j', j'')$, so this recursion involves the term V_r multiplied by a nonzero Λ_k and, otherwise, involves only earlier terms of the given sequence $V_0, V_1, \ldots, V_{r-1}$. Then the equation can be solved for V_r in terms of those V_j appearing earlier in the sequence. We shall say that $\Lambda(x, y)$ *reaches* the rth term of the array whenever all terms V_{j+k} on the left side of the above equation are from the given sequence, even if the sum does not equal zero. In general, the bivariate polynomial $\Lambda(x, y)$ need not reach a selected term of the sequence that precedes V_r simply because it reaches V_r. This situation is quite different from the univariate case.

The bivariate recursion has been defined with a plus sign in the subscript. Recall that a univariate recursion is written in the following form:

$$V_j = -\sum_{k=1}^{L} \Lambda_k V_{j-k}.$$

The univariate recursion is represented by the univariate connection polynomial $\Lambda(x)$, with coefficient Λ_0 equal to 1. It would be convenient to use the same form of recursion – with a minus sign – for the bivariate case, with j and k now representing bivariate indices (j', j'') and (k', k''). However, we would then run into problems. Whereas the univariate polynomial $\Lambda(x)$, after division by a suitable power of x, always has a nonzero Λ_0, the arbitrary bivariate polynomial $\Lambda(x, y)$ need not have a nonzero Λ_{00}, even after the monomial $x^a y^b$ of largest possible degree has been divided out.[1] The

---

[1] For consistency, the one-dimensional discussion could be reworked to use a plus sign, but we prefer to retain the conventional treatment in that case.

bivariate recursion given by

$$V_j = -\frac{1}{\Lambda_{00}} \sum_{k=1}^{L} \Lambda_k V_{j-k}$$

would fail to be meaningful whenever $\Lambda_{00} = 0$. Fortunately, this difficulty can be avoided because the leading monomial Λ_s of $\Lambda(x, y)$ is, by definition, always nonzero. To be able to reference the recursion to the nonzero leading monomial is why we simply define the recursion with a plus sign in the index of V instead of a minus sign.

By adopting the normalization convention that $\Lambda_s = 1$, the bivariate recursion $\sum_{k=0}^{s} \Lambda_k V_{j+k} = 0$ can be put into the following form:

$$V_{j+s} = -\sum_{k=0}^{s-1} \Lambda_k V_{j+k}.$$

The array Λ can be described by the monic bivariate polynomial $\Lambda(x, y)$ of bidegree s. We may also define $r = j + s$ to put the recursion into the following more convenient form:

$$V_r = -\sum_{k=0}^{s-1} \Lambda_k V_{k+r-s}.$$

The subtraction in the index is a bivariate subtraction. It is meaningful whenever $r \gtrsim s$, which is equivalent to the statement that $\Lambda(x, y)$ reaches r. Note that if $r \gtrsim s$ and $r' \succ r$, we cannot conclude that $r' \gtrsim s$. It is this interplay between the graded order and the division order that introduces a complicated and rich structure into the study of bivariate recursions that does not occur in the study of univariate recursions.

With the help of Figure 8.3, a geometric interpretation can be given to the notion of producing the element V_r of the sequence of bivariate elements. In the middle section of Figure 8.3 is shown the support of the sequence $V = V_0, \ldots, V_{31}$. The bi-index

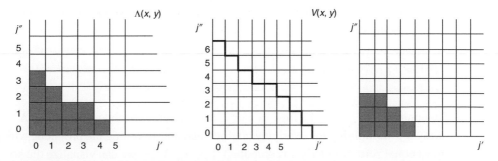

Figure 8.3. Illustrating the nature of a bivariate recursion.

associated with the term V_{31} is $(4, 3)$. If the sequence is the sequence of coefficients of the polynomial $V(x, y)$, then the leading monomial is $x^4 y^3$. The support of the polynomial $\Lambda(x, y)$ is shown on the left. The polynomial $\Lambda(x, y)$ reaches V_{31} because the leading monomial of $xy^2 \Lambda(x, y)$ agrees with the leading monomial of $V(x, y)$, and further, every monomial of $xy^2 \Lambda(x, y)$ corresponds to a monomial of $V(x, y)$. Therefore the linear recursion is defined. The right side of Figure 8.3 shows the support of the set of monomials $x^{j'} y^{j''}$, such that $x^{j'} y^{j''} \Lambda(x, y)$ has all its monomials in the set of monomials of $V(x, y)$.

For an example of a set of connection polynomials in the field $GF(8)$, constructed with the primitive element α satisfying $\alpha^3 = \alpha + 1$, consider the sequence V represented by the following polynomial:

$$V(x, y) = 0y^3 + \alpha^5 xy^2 + \alpha^6 x^2 y + \alpha^3 x^3 + \alpha^5 y^2 + 0xy + \alpha x^2 + \alpha^6 y + \alpha x + \alpha^3.$$

The polynomial $V(x, y)$ is specified up to terms with bidegree $(0, 3)$, but actually has bidegree $(1, 2)$ because the leading specified term of the sequence V is a zero. Therefore, in such cases, precision requires us to say that $V(x, y)$ has bidegree *at most* $r = (r', r'')$, but the sequence has a leading term V_r.

Among the many connection polynomials for $V(x, y)$ are the polynomials $\Lambda^{(1)}(x, y)$ and $\Lambda^{(2)}(x, y)$, given by

$$\Lambda^{(1)}(x, y) = x^2 + \alpha^6 x + 1,$$
$$\Lambda^{(2)}(x, y) = y + \alpha^6 x + \alpha^6.$$

To check that these are connection polynomials, it is convenient to write $V(x, y)$ as a two-dimensional array of coefficients as follows:

$$V = \begin{vmatrix} 0 & & & \\ \alpha^5 & \alpha^5 & & \\ \alpha^6 & 0 & \alpha^6 & \\ \alpha^3 & \alpha & \alpha & \alpha^3 \end{vmatrix}$$

Likewise, the two connection polynomials $\Lambda^{(1)}(x, y)$ and $\Lambda^{(2)}(x, y)$ are represented as the following arrays:

$$\Lambda^{(1)} = \begin{vmatrix} 1 & \alpha^6 & 1 \end{vmatrix} \quad \Lambda^{(2)} = \begin{vmatrix} 1 & \\ \alpha^6 & \alpha^6 \end{vmatrix}$$

To compute $\sum_{k=0}^{s} \Lambda_k^{(\ell)} V_k$ for $\ell = 1$ or 2, we visualize overlaying the array $\mathbf{\Lambda}^{(\ell)}$ on the array V, multiplying the overlying coefficients, and summing the products, which sum must equal zero. Similarly, to compute $\sum_{k=0}^{s} \Lambda_k^{(\ell)} V_{j+k}$, we can visualize the array $\mathbf{\Lambda}^{(\ell)}$

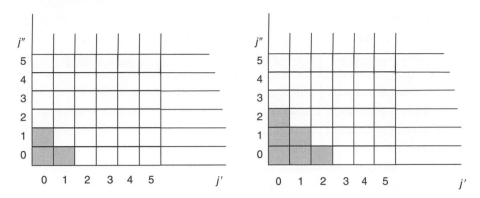

Figure 8.4. Points reached by two connection polynomials.

right-shifted and up-shifted by (j', j'') positions, and, provided that the shifted copy of $\Lambda^{(\ell)}$ lies within the support of V, laid on the array V in the new position. Again, the sum of products must equal zero. This view gives the algebraic structure a geometric interpretation. Indeed, this method of description might be called "geometric algebra."

Figure 8.4 shows the possible values of shifts, (j', j''), for the two polynomials $\Lambda^{(\ell)}(x, y)$ of the example. If point $j = (j', j'')$ is a shaded square in the appropriate half of Figure 8.4, then $x^{j'} y^{j''} \Lambda^{(\ell)}(x, y)$ lies within the support of $V(x, y)$. For each such point, it is simple to verify that the sum of the products equals zero.

With these as examples, we are now ready for the formal definition of a connection polynomial.

Definition 8.2.1 *The monic polynomial $\Lambda(x, y)$ of bidegree s is said to be a connection polynomial for the bi-index sequence $V = V_0, V_1, \ldots, V_r$ if*

$$\sum_{k=0}^{s} \Lambda_k V_{j+k} = 0$$

for all j satisfying $s \leq j \preceq r$.

The upper limit of $r - s$ is chosen so that all terms of the sum are from the given finite sequence V. The condition $s \leq j \preceq r$ mixes the division order and the graded order, as shown in Figure 8.5. The point j is smaller than r in the graded order, and the shaded points are smaller than (or equal to) j in the division order.

The set of all connection polynomials, called the connection set, has a footprint Δ, called the *connection footprint*, consisting of all (j', j'') such that $x^{j'} y^{j''}$ is not the leading monomial in the (graded) total order of any connection polynomial. Every exterior corner of Δ has a monic connection polynomial whose leading monomial lies in that exterior corner. These monic polynomials are called *minimal connection polynomials*.

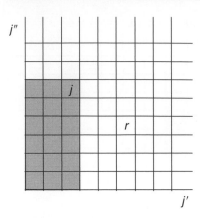

Figure 8.5. Division order and graded order.

Definition 8.2.2 *A minimal connection set for the finite bivariate sequence* $V = V_0, V_1, \ldots, V_r$ *is a set consisting of one monic connection polynomial for each exterior corner of the footprint of the connection set of that bivariate sequence.*

Any bivariate sequence that consists of a nonzero final term V_r preceded by all zeros has no connection polynomial that reaches V_r because then $\sum_k \Lambda_k V_{k+r-s}$ could not equal zero. In this case, the connection set only has polynomials that lie outside the support of V.

It may be that the connection set for $V_0, V_1, \ldots, V_{r-1}$ is not a connection set for $V_0, V_1, \ldots, V_{r-1}, V_r$ because it does not produce V_r. This means that for a bivariate sequence of length r, the connection set may lose its status as a connection set when the bivariate sequence is extended to length $r + 1$.

The minimal connection set has been introduced because it is a convenient form for the generalization of the agreement theorem and of Massey's theorem, and also because we have good algorithms for computing a minimal connection set. The minimal connection set differs in several ways from a minimal basis of the locator ideal for the nonzeros of the polynomial $V(x, y)$. The definition of a connection polynomial specifies that only a segment of length r of the sequence of coefficients is to be tested, and the equation of the test involves a sum of indices, rather than a difference of indices, as would be seen in a convolution. Also, the indices are not regarded as modulo n. In Section 8.3, we give a condition for equivalence of a minimal basis for a locator ideal for the nonzeros of $V(x, y)$ and for the set of minimal connection polynomials of the sequence of coefficients of $V(x, y)$. When this condition holds, as it does in many applications, the locator ideal can be determined by first computing the minimal connection set and then reciprocating the polynomials. Even more simply, when this condition holds (β, γ) is a zero of the locator ideal if and only if $(\beta^{-1}, \gamma^{-1})$ is a zero of the minimal connection set.

For a fairly large example of a minimal connection set, take the following polynomial over $GF(16)$, with primitive element α satisfying $\alpha^4 = \alpha + 1$:

$$V(x,y) = \alpha^6 y^5 + \alpha^7 y^4 x + \alpha^7 y^3 x^2 + \alpha^{12} y^2 x^3 + \alpha^5 yx^4 + \alpha^5 x^5$$
$$+ \ \alpha^5 y^4 + \alpha^4 y^3 x + 0 + \alpha^{12} yx^3 + \alpha^2 x^4$$
$$+ \ \alpha^6 y^3 + \alpha^{11} y^2 x + \alpha^{14} yx^2 + \alpha^7 x^3$$
$$+ \ \alpha^9 y^2 + \alpha^9 yx + \alpha^5 x^2$$
$$+ \ 0y + \alpha^{14} x$$
$$+ \ \alpha^9.$$

A minimal connection set for $V(x,y)$ is the set containing the following three polynomials:

$$\Lambda^{(1)}(x,y) = x^4 + \alpha^3 x^2 y + \alpha^5 x^3 + \alpha^{14} xy + \alpha^7 y + \alpha x + \alpha^{13};$$
$$\Lambda^{(2)}(x,y) = x^2 y + \alpha^{13} xy + \alpha^3 x^2 + \alpha^3 y + \alpha^6 x + \alpha^6;$$
$$\Lambda^{(3)}(x,y) = y^2 + \alpha^{10} xy + \alpha^{13} y + \alpha^{13} x + \alpha^{11}.$$

It is easiest to hand check that these are connection polynomials if the polynomial coefficients are arranged in the natural way as two-dimensional arrays:

$$\mathbf{\Lambda}^{(1)} = \begin{vmatrix} \alpha^7 & \alpha^{14} & \alpha^3 \\ \alpha^{13} & \alpha & 0 & \alpha^5 & 1 \end{vmatrix} \qquad \mathbf{\Lambda}^{(2)} = \begin{vmatrix} \alpha^3 & \alpha^{13} & 1 \\ \alpha^6 & \alpha^6 & \alpha^3 \end{vmatrix} \qquad \mathbf{\Lambda}^{(3)} = \begin{vmatrix} 1 \\ \alpha^{13} & \alpha^{10} \\ \alpha^{11} & \alpha^{13} \end{vmatrix}$$

Likewise, the coefficients of $V(x,y)$ can be arranged in the following array:

$$V = \begin{vmatrix} \alpha^6 \\ \alpha^5 & \alpha^7 \\ \alpha^6 & \alpha^4 & \alpha^7 \\ \alpha^9 & \alpha^{11} & 0 & \alpha^{12} \\ 0 & \alpha^9 & \alpha^{14} & \alpha^{12} & \alpha^5 \\ \alpha^9 & \alpha^{14} & \alpha^5 & \alpha^7 & \alpha^2 & \alpha^5 \end{vmatrix}$$

To verify that $\Lambda^{(1)}(x,y)$, for example, is a connection polynomial, simply overlay the array $\mathbf{\Lambda}^{(1)}$ on the array V in any compatible position and compute the sum of products. The sum must equal zero for every such compatible position.

We shall return to this example often. In Section 8.4, we give an algorithm that computes $\Lambda^{(1)}(x,y)$, $\Lambda^{(2)}(x,y)$, and $\Lambda^{(3)}(x,y)$ from $V(x,y)$. This is the *Sakata algorithm*. In Chapter 10, this example will be used in two examples of decoding. First, it will be used for decoding a code known as a *hyperbolic code*, then for decoding a code known as a *hermitian code*.

8.3   The Sakata–Massey theorem

Let $V = V_0, V_1, V_2, \ldots, V_{r-1}$ be a bivariate sequence arranged in the graded order, and let $\Lambda(x, y)$ be a monic bivariate polynomial, with leading index s in the graded order. This requires that Λ_s is nonzero. Suppose that

$$\sum_{k=0}^{s} \Lambda_k V_{j+k} = s \qquad s \succeq j \prec r,$$

provided all terms in the sum are defined. The last term of the bivariate sequence that appears in the sum corresponding to $j = r - s - 1$ and $k = s$ is V_{r-1}. If $r \leq s$, there is a unique way to extend the sequence by one symbol, V_r, so that

$$\sum_{k=0}^{s} \Lambda_k V_{r-s+k} = 0.$$

To find the required value of V_r, write

$$0 = \Lambda_s V_r + \sum_{k=0}^{s-1} \Lambda_k V_{r+k-s}.$$

Then, because $\Lambda_s = 1$,

$$V_r = -\sum_{k=0}^{s-1} \Lambda_k V_{k+r-s}$$

is the required next term in the sequence.

If each of two such polynomials, $\Lambda(x, y)$ and $\Lambda^*(x, y)$, produces the same sequence $V = V_0, V_1, V_2, \ldots, V_{r-1}$, will each of the two polynomials next produce the same symbol V_r? The answer, in general, is no. The following theorem gives a condition that ensures that the next term V_r will be the same. Because the theorem can be applied recursively, it says that if two bivariate recursions produce the same sequence for sufficiently many terms, then each will produce the same subsequent terms whenever those terms can be reached by both recursions. In particular, it says that the next terms are equal. Thus

$$\sum_{k=0}^{s-1} \Lambda_k V_{r+k-s} = \sum_{k=0}^{s^*-1} \Lambda_k^* V_{r+k-s^*}.$$

Theorem 8.3.1 (agreement theorem)   *Suppose that each of the two monic polynomials $\Lambda(x, y)$ and $\Lambda^*(x, y)$ produces the bi-index sequence $V_0, V_1, \ldots, V_{r-1}$. If $r \succeq s + s^*$,* *and if either produces the longer bi-index sequence V_0, V_1, ..., V_{r-1}, V_r, then so does the other.*

Proof: Because $r \gtrless s + s^*$, we can write

$$V_{r+k-s} = -\sum_{i=0}^{s^*-1} \Lambda_i^* V_{r+k+i-s-s^*}$$

and

$$V_{r+k-s^*} = -\sum_{k=0}^{s-1} \Lambda_k V_{r+k+i-s-s^*}.$$

Therefore,

$$\sum_{k=0}^{s-1} \Lambda_k V_{r+k-s} = -\sum_{k=0}^{s-1} \Lambda_k \sum_{i=0}^{s^*-1} \Lambda_i^* V_{r+k+i-s-s^*}$$

$$= -\sum_{i=0}^{s^*-1} \Lambda_i^* \sum_{k=0}^{s-1} \Lambda_k V_{r+k+i-s-s^*}$$

$$= \sum_{i=0}^{s^*-1} \Lambda_i^* V_{r+i-s^*},$$

as was to be proved.   ∎

Next, we generalize the Massey theorem to two dimensions. Recall that Massey's theorem in one dimension is an inequality relationship between the length of a minimum-length linear recursion that produces a sequence and the length of a minimum length linear recursion that produces a proper subsequence of that sequence. If $(\Lambda(x), L)$ is the shortest linear recursion that produces the univariate sequence $(V_0, V_1, \ldots, V_{r-1})$, and $(\Lambda(x), L)$ does not produce $V = (V_0, V_1, \ldots, V_{r-1}, V_r)$, then $\mathcal{L}(V) \geq \max[L, r + 1 - L]$.

The generalization to two dimensions follows.

Theorem 8.3.2 (Sakata–Massey theorem) *If $\Lambda(x, y)$, a polynomial of bidegree s, produces the bivariate sequence V_0, $V_1, \ldots, V_{r-1}$, but not the longer sequence $V_0, V_1, \ldots, V_{r-1}, V_r$, then, provided $r - s \gtrless 0$, the connection footprint for the sequence V_0, $V_1, \ldots, V_{r-1}$, V_r contains the point $r - s$.*

Proof: If $\Lambda(x, y)$ does not reach r, then the condition $r - s \gtrless 0$ does not hold, and there is nothing to prove, so we may suppose that $\Lambda(x, y)$ does reach r, but does not produce V_r. Suppose $\Lambda^*(x, y)$ exists with the leading monomial $x^{r'-s'} y^{r''-s''}$ that produces the sequence $V_0, V_1, \ldots, V_{r-1}, V_r$. Then it produces the sequence $V_0, V_1, \ldots, V_{r-1}$. But $\Lambda(x, y)$ produces the sequence $V_0, V_1, \ldots, V_{r-1}$. By the agreement theorem, because $r \gtrless s + (r - s)$, both $\Lambda(x, y)$ and $\Lambda^*(x, y)$ must produce the same value at V_r, contrary to the assumption of the theorem. Hence $x^{r'-s'} y^{r''-s''}$ cannot be the leading monomial of any such $\Lambda^*(x, y)$. Thus $r - s$ lies in the connection footprint. ∎

An alternative statement of the Sakata–Massey theorem is given in the following corollary. The notation $r - \Delta$ means the set that is obtained by inverting a copy of the set Δ, translating by r, and possibly truncating the set to avoid introducing negative coordinates. That is,

$$r - \Delta = \{(i', i'') \gtrless 0 \mid (r' - i', r'' - i'') \in \Delta\}.$$

Corollary 8.3.3 *Let Δ and Δ' be the connection footprints for the sequences V_0, $V_1, \ldots, V_r$ and $V_0, V_1, \ldots, V_r'$, respectively, where $V_r' \neq V_r$, and otherwise the two sequences are the same. Then*

$$\Delta \cup (r - \Delta') \quad \supset \quad \{(i', i'') \mid i' \leq r' \text{ and } i'' \leq r''\}.$$

Proof: Suppose $i' \leq r'$ and $i'' \leq r''$, and that $i \notin \Delta$. Then there is a minimal connection polynomial with the leading index i in the minimal connection set of $V_0, V_1, \ldots, V_r$. The Sakata–Massey theorem then says that the connection footprint for $V_0, V_1, \ldots, V_r'$ contains the bi-index $r - i$. That is, $i \in r - \Delta'$, as was to be proved. ∎

Corollary 8.3.3 implies that

$$\|\Delta \cup (r - \Delta')\| \quad \geq \quad (r' + 1)(r'' + 1),$$

where $\|\mathcal{S}\|$ denotes the cardinality of the set $\mathcal{S}$. Consequently,

$$\|\Delta\| + \|\Delta'\| \quad \geq \quad \|\Delta\| + \|(r - \Delta')\| \quad \geq \quad \|\Delta \cup (r - \Delta')\| \quad \geq \quad (r' + 1)(r'' + 1).$$

This is a deceptively powerful statement. Suppose one knows that both $\|\Delta\| \leq t$ and $\|\Delta'\| \leq t$. Then

$$\|\Delta\| + \|\Delta'\| \quad \leq \quad 2t.$$

If $(r' + 1)(r'' + 1) \geq 2t + 1$, the two statements are not compatible. The following proposition is the only way to reconcile this.

Proposition 8.3.4 *Let v have weight at most t and $v \Leftrightarrow V$. Let $(r' + 1)(r'' + 1) \geq 2t + 1$. Then in the graded order, there is exactly one way to extend the sequence $V_0, V_1, \ldots, V_{r-1}$ to the sequence $V_0, V_1, \ldots, V_{r-1}, \widehat{V}_r$ such that the connection footprint has an area not larger than t, and this is with $\widehat{V}_r = V_r$.*

This proposition provides a statement that will be important in decoding. It says when decoding two-dimensional codes whose error patterns have weight at most t, syndrome S_r is not needed if $(r' + 1)(r'' + 1) \geq 2t + 1$. Its value can be inferred from the other syndromes by the requirement that the footprint has an area not larger than t.

There is one last detail to cover in this section. Later, we shall need to compute the locator ideal of an array, but we will have powerful algorithms to compute the minimal connection set. The following theorem provides a condition, usually realized in our applications, that allows the algorithm for one problem to be used on another.

Theorem 8.3.5 *The (reciprocal) locator ideal and the minimal connection set of an n by n periodic array are equal if the linear complexity of the array is not larger than $n^2/4$.*

Proof: This is a simple consequence of the agreement theorem. The connection footprint is contained in the footprint of the locator ideal. Thus, in each case, the footprint has an area at most $n^2/4$. Two polynomials, one from each ideal, must agree at least until the midpoint of the n by n array. Hence, by the agreement theorem, they continue to agree thereafter.   ∎

8.4   The Sakata algorithm

The two-dimensional generalization of the Berlekamp–Massey algorithm is called the *Sakata algorithm*. It is a procedure for computing a minimal set of connection polynomials for a given bivariate sequence $V_0, V_1, \ldots, V_r$ in a given total order; we will use only the graded order in our examples. Recall that we have defined a minimal set of connection polynomials for the bivariate sequence as consisting of one monic connection polynomial for each exterior corner of the connection footprint of the sequence, one leading monomial in each exterior corner.

We may be given the sequence $V_0, V_1, \ldots, V_r$ as a low-order fragment of the sequence of coefficients of a bivariate polynomial, $V(x, y)$, representing the bispectrum V of a two-dimensional array v of Hamming weight at most t. If the sequence fragment is large enough, the set of connection polynomials generates a locator ideal for the set of nonzeros of $V(x, y)$.

Both the Buchberger algorithm and the Sakata algorithm are algorithms for computing a minimal basis for a bivariate ideal. However, they start with two quite different specifications of the ideal. The Buchberger algorithm starts with an ideal specified in terms of an arbitrary set of generator polynomials, not necessarily a minimal basis; the ideal is specified as follows:

$$I = \langle g_\ell(x, y) \mid \ell = 1, \ldots L \rangle.$$

The Buchberger algorithm re-expresses the ideal in terms of a new set of generator polynomials that forms a minimal basis. The Sakata algorithm starts only with an initial portion of the polynomial $V(x, y)$ and computes a set of minimal connection polynomials, which gives a minimal basis of the locator ideal for the set of nonzeros of the given polynomial $V(x, y)$, provided the number of nonzeros is not too large. The initial coefficients of the polynomial $V(x, y)$ are given in a specified total order, which we take to be the graded order, but no nonzero polynomial of the connection set is known at the start of the computation.

The Sakata algorithm is an iterative algorithm, which begins iteration r by knowing from iteration $r - 1$ a minimal connection set for $V_0, V_1, \ldots, V_{r-1}$, described by the set of M_{r-1} minimal polynomials $\{\Lambda^{(r-1,m)}(x, y) \mid m = 1, \ldots, M_{r-1}\}$, and it computes a minimal connection set for $V_0, \ldots, V_{r-1}, V_r$, described by the set of M_r minimal polynomials $\{\Lambda^{(r,m)}(x, y)\} \mid m = 1, \ldots, M_r\}$.

It is not necessary to specify the entire connection set; it is enough to specify a minimal connection set. To find a minimal connection set, it is enough first to find the connection footprint, then to find a monic connection polynomial for each exterior corner of the footprint. Of course, if the minimal connection set $\{\Lambda^{(r,m)}(x, y) \mid m = 1, \ldots, M_r\}$ were known, the footprint Δ_r at iteration r would be completely determined, but this minimal connection set is not known at the start of the iteration. The algorithm starts an iteration by computing the footprint Δ_r, then computing the set of monic polynomials to fit the exterior corners of the footprint. The computation of the footprint is an essential first step of each iteration of the algorithm.

The Sakata–Massey theorem has a central role in the algorithm. It describes how the footprint is updated from one iteration to the next. Figure 8.6 shows a hypothetical footprint after the iteration $r - 1$. The shaded region of the Figure 8.6 is the footprint. The footprint has *exterior corners* marked by filled circles and *interior corners* marked by open circles. In this example, the exterior corners tell us that a minimal connection set has four polynomials with bidegrees s equal to $(5, 0)$, $(3, 1)$, $(2, 3)$, and $(0, 4)$. To illustrate the role of the Sakata–Massey theorem, suppose that $r = (6, 3)$. The first three

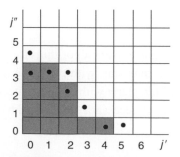

Figure 8.6. Footprint illustrating the Sakata–Massey theorem.

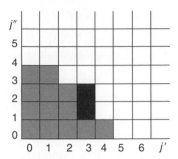

Figure 8.7. Footprint of a new connection set.

polynomials, those of bidegrees $(5, 0)$, $(3, 1)$, and $(2, 3)$, reach r. Computing $r - s$, we have $(6, 3) - (5, 0) = (1, 3)$, $(6, 3) - (3, 1) = (3, 2)$, and $(6, 3) - (2, 3) = (4, 0)$. This means that one or more of the points $(1, 3)$, $(3, 2)$, and $(4, 0)$ will need to be appended to the footprint if the connection polynomial corresponding to that exterior corner fails to produce V_r. Of these, only $(3, 2)$ is not already an element of the footprint. If the point $(3, 2)$ is appended, then all smaller points in the division order are appended also, which would lead to the new footprint, shown in Figure 8.7. The two new points of the new footprint have been highlighted in Figure 8.7. These two new points change the set of exterior corners. The point $(4, 1)$ is an exterior corner of this new footprint. Thus a new monic connection polynomial with bidegree $s = (4, 1)$ is needed, and the polynomial with bidegree $(3, 1)$ is no longer a connection polynomial for the longer sequence. It may also be necessary to change the other polynomials corresponding to other exterior corners, but their leading monomials will not change.

The algorithm updates three quantities at the rth iteration: the first is the footprint Δ_r; the second is the set of polynomials $\{\Lambda^{(r,m)}(x, y) \mid m = 1, \ldots, M_r\}$, which we call *minimal connection polynomials*, or *exterior polynomials*; and the third is the set of scratch polynomials $\{B^{(r,n)}(x, y) \mid n = 1, \ldots, N_r\}$, which we call *interior polynomials*. Each filled circle in Figure 8.6 designates one of the M_r minimal connection polynomials $\Lambda^{(r,m)}(x, y)$ by pointing to the exterior corner corresponding to its leading monomial. Each open circle represents one of the interior polynomials $B^{(r,n)}(x, y)$ by pointing to the box corresponding to its leading monomial. Each interior polynomial was a minimal connection polynomial of an earlier iteration, but multiplied by a suitable scalar. Although an interior polynomial was a connection polynomial at an earlier iteration, it is not a connection polynomials at the current (or any future) iteration.

During the rth iteration, all of these quantities are updated. If the connection footprint does not change, then, even if the minimal connection polynomials need to be modified, their leading monomials do not change. If the connection footprint grows larger, it will have one or more new exterior corners. In such a case, the corresponding minimal connection polynomials must be updated, and the number of minimal connection polynomials may change.

Now we will provide a brief outline of the algorithm.

- Test each minimal connection polynomial $\Lambda^{(r-1,m)}(x,y)$ that reaches r against the next term V_r.
- If one or more of the minimal connection polynomials that reaches r fails to produce the term V_r, use the Sakata–Massey theorem to compute the new footprint. The new footprint may be the same as the old footprint, or it may be larger.
- Form a new minimal connection polynomial to fit each exterior corner of the new footprint by taking a polynomial combination of the previous minimal connection polynomial and an interior polynomial so that the new polynomial produces the term V_r.
- If the footprint has become larger, update the set of interior polynomials for use in future iterations. Each interior polynomial is either unchanged from the previous iteration, or is a multiple of a discarded minimal connection polynomial from the previous iteration.

Next, we will fill in the details of the algorithm. The discussion, in part, mimics the discussion of the Berlekamp–Massey algorithm, given in Section 3.5. To understand the structure of the Sakata algorithm before continuing further into this section, it may be helpful first to study the lengthy example of a computation in Section 8.5.

Each minimal connection polynomial $\Lambda^{(r-1,m)}(x,y)$ for $m = 1, \ldots, M_{r-1}$, from the previous iteration, is tested against V_r as follows. Recall that the polynomial $\Lambda(x,y)$ of bidegree s reaches r if $r - s \geq 0$. Componentwise, this requires that $r' - s' \geq 0$ and $r'' - s'' \geq 0$. For each minimal connection polynomial $\Lambda^{(r-1,m)}(x,y)$ that reaches r, define the *discrepancy* as follows:

$$
\delta_r^{(r-1,m)} = \sum_{k=0}^{s} \Lambda_k^{(r-1,m)} V_{k+r-s}
$$

$$
= V_r - \left(-\sum_{k=0}^{s-1} \Lambda_k^{(r-1,m)} V_{k+r-s} \right)
$$

$$
= V_r - \widetilde{V}_r.
$$

The discrepancy $\delta_r^{(r-1,m)}$ will exist for some m. When it exists, it may be either zero or nonzero. For other m, $\Lambda^{(r-1,m)}(x,y)$ does not reach V_r and the discrepancy does not exist. When the discrepancy exists, we may also redefine the earlier discrepancies as follows:

$$
\delta_j^{(r-1,m)} = \sum_{k=0}^{s} \Lambda_k^{(r-1,m)} V_{k+j-s} \qquad j \preceq r.
$$

However, because of the definition of $\Lambda^{(r-1,m)}(x,y)$, this term $\delta_j^{(r-1,m)}$ must be equal to zero for all $j \prec r$ for which it exists.

If for $m = 1, \ldots, M_{r-1}$ every new discrepancy $\delta_r^{(r-1,m)}$ that exists is zero, the iteration is complete. Otherwise, the footprint and the set of the minimal connection polynomials must be updated. First, use the Sakata–Massey theorem to see whether the footprint is to be enlarged. This theorem says that if $\delta_r^{(r-1,m)}$ exists and is not zero, then the point $(r - s_m)$ is contained in the footprint Δ_r, where s_m is the bidegree of polynomial $\Lambda^{(r-1,m)}(x, y)$.

Let $\mathcal{M} = \{m_1, m_2, \ldots\}$ be the set of m for which the discrepancy $\delta_r^{(r-1,m)}$ exists and is nonzero. For each $m = 1, \ldots, M_{r-1}$, define the set

$$\mathcal{Q}_m = \{(j', j'') \mid (j', j'') \leqq (r' - s_m', r'' - s_m'')\}.$$

The set $\mathcal{Q}_m$ must be adjoined to the footprint whenever $m \in \mathcal{M}$. The new footprint is given by

$$\Delta_r = \Delta_{r-1} \bigcup \left(\bigcup_{m \in \mathcal{M}} \mathcal{Q}_m \right).$$

It may be that $\Delta_r = \Delta_{r-1}$, or it may be that Δ_r is strictly larger than Δ_{r-1}. If Δ_r is larger than Δ_{r-1}, then the number M_r of exterior corners of the new footprint may be equal to, or may be larger than, the number M_{r-1} of exterior corners of the old footprint.

We must compute one minimal connection polynomial for each exterior corner. To explain this, we will describe an experiment. Suppose that polynomial $\Lambda^{(r-1,m)}(x, y)$ of bidegree s gave a nonzero discrepancy at iteration r that led to the mth exterior corner. Consider the monic polynomial of bidegree r, given by

$$\Lambda(x, y) = x^{r'-s'} y^{r''-s''} \Lambda^{(r-1,m)}(x, y).$$

The polynomial $\Lambda(x, y)$ has the same coefficients as $\Lambda^{(r-1,m)}(x, y)$, but they are translated so that the leading monomial moves from s to r. Therefore $\Lambda^{(r-1,m)}(x, y)$ has the required leading monomial. Recall that the discrepancy is given by

$$\delta_j = \sum_{k=0}^{s} \Lambda_k^{(r-1,m)} V_{k+j-s}$$

for those $j \preceq r$ for which all terms of the sum are defined. Letting $\ell = k + s - t$, this can be expressed as follows:

$$\delta_j = \sum_{k=t-s}^{t} \Lambda_{k+s-t}^{(r-1,m)} V_{k+j-t}$$

$$= \sum_{\ell=0}^{s} \Lambda_{\ell}^{(r-1,m)} V_{j+\ell-s}$$

$$= \delta_{j}^{(r-1,m)} = \begin{cases} 0 & j \prec r \\ \delta_{r} \neq 0 & j = r \end{cases}$$

whenever δ_j is defined. Thus, although $\Lambda(x, y)$ has the desired bidegree of $r - s$, it fails to be a connection polynomial for the new sequence because $\delta_r \neq 0$. Recalling the development of the Berlekamp–Massey algorithm, we will try to modify this $\Lambda(x, y)$ so that, when computed with the new $\Lambda(x, y)$, δ_r becomes zero and all other δ_j remain zero.

Define a modified monic polynomial of bidegree $r - s$, given by

$$\Lambda(x, y) = x^{t'-s'} y^{t''-s''} \Lambda^{(r-1,m)}(x, y) - Ax^{b'} y^{b''} \Lambda^{(i-1,n)}(x, y),$$

where $\Lambda^{(i-1,n)}(x, y)$ is the nth of the minimal connection polynomials at a previous iteration, that is, at iteration $i - 1$.

The polynomial $\Lambda(x, y)$ is a connection polynomial for the longer sequence if $\delta_j = 0$ for all $j \preceq r$ for which δ_j is defined. We will ensure that this is so by careful specification of the parameters i, A, and $b = (b', b'')$. First, choose i and $\Lambda^{(i-1,n)}(x, y)$ so that $\delta_i^{(i-1,n)} \neq 0$. Let

$$A = \frac{\delta_r^{(r-1,m)}}{\delta_i^{(i-1,n)}},$$

and let $\bar{s}$ be the bidegree of $\Lambda^{(i-1,n)}(x, y)$. With this choice of $\Lambda(x, y)$, we repeat the computation of δ_j, given by

$$\delta_j = \sum_{k=0}^{t} \Lambda_k V_{j+k-t}$$

for all $j \preceq r$ for which all terms are defined. Therefore

$$\delta_j = \sum_{k=t-s}^{t} \Lambda_{k+s-t}^{(r-1,m)} V_{j+k-t} - \frac{\delta_r^{(r-1,m)}}{\delta_i^{(i-1,n)}} \sum_{k=0}^{t} \Lambda_{k-b}^{(i-1,n)} V_{j+k-t}.$$

We want δ_r to be zero, so we must choose b so that

$$\delta_r = \delta_r^{(r-1,m)} - \frac{\delta_r^{(r-1,m)}}{\delta_i^{(i-1,n)}} \delta_i^{(i-1,n)}$$

$$= 0,$$

and so the second sum is zero for $j \prec r$. In the second summation, recall that $\Lambda_{k-b}^{(i-1,n)} = 0$ for $k-b < 0$. Choose $b = (i-\bar{s}) - (r-s)$, and make the change of variables $\ell = k-b$ to obtain

$$\sum_{k=b}^{t} \Lambda_{k-b}^{(r-1,m)} V_{j+k-s} = \sum_{\ell=0}^{t-b} \Lambda_{\ell}^{(i-1,n)} V_{j+\ell+b-s}$$

$$= \begin{cases} \delta_i^{(i-1,n)} & j = r \\ 0 & j \preceq r. \end{cases}$$

The condition that $\delta^{(i-1,n)}$ is nonzero is satisfied by many $\Lambda^{(i-1,n)}(x,y)$. All that remains is to specify which of these polynomials should be chosen. The polynomials that need to be saved for this purpose, in fact, are the interior polynomials, expressed as

$$B^{(r-1,n)}(x,y) = \frac{1}{\delta_i^{(i-1,n)}} x^{b'} y^{b''} \Lambda^{(r-1,n)}(x,y),$$

one for each interior corner. The normalization by $\delta_i^{(i-1,n)}$ is for computational convenience, because it eliminates the need to store this term separately. Once these are saved, the previous minimal polynomials need not be saved; they can be discarded because it is certain they will not be used again.

8.5   An example

We now give an example of the Sakata algorithm in the field $GF(16)$, an example that is rather elaborate and will fill the entire section. The example will be further embellished in various ways in Chapter 12, where it becomes the basis for several examples of decoding two-dimensional codes; one example (in Section 12.3) is a code on the plane, and another (in Section 12.4) is a code on a curve.

We will continue with the example $V(x,y)$ in the field $GF(16)$ that appeared at the end of Section 8.2. The bivariate polynomial of this example, with terms arranged in the graded order, is as follows:

$$\begin{aligned} V(x,y) = \ & \alpha^6 y^5 + \alpha^7 y^4 x + \alpha^7 y^3 x^2 + \alpha^{12} y^2 x^3 + \alpha^5 yx^4 + \alpha^5 x^5 \\ & + \alpha^5 y^4 + \alpha^4 y^3 x + 0 + \alpha^{12} yx^3 + \alpha^2 x^4 \\ & + \alpha^6 y^3 + \alpha^{11} y^2 x + \alpha^{14} yx^2 + \alpha^7 x^3 \\ & + \alpha^9 y^2 + \alpha^9 yx + \alpha^5 x^2 \\ & + 0y + \alpha^{14} x \\ & + \alpha^9. \end{aligned}$$

The coefficients of this polynomial were obtained as the first 21 coefficients of the Fourier transform of a 15 by 15 array v. The full Fourier transform V is depicted in Chapter 12 (Figure 12.8) and the array v is depicted in Figure 12.7. The full arrays V and v play no role at the present time. However, in Chapter 12, we will refer back to this example; furthermore, additional components of the Fourier transform will be appended to $V(x, y)$, and the example will be continued.

In the graded order, the 21 coefficients of $V(x, y)$ form the following sequence:

$$V_0, \ldots, V_{20} = \alpha^9, \alpha^{14}, 0, \alpha^5, \alpha^9, \alpha^9, \alpha^7, \alpha^{14}, \alpha^{11}, \alpha^6, \alpha^2, \alpha^{12},$$
$$0, \alpha^4, \alpha^5, \alpha^5, \alpha^5, \alpha^{12}, \alpha^7, \alpha^7, \alpha^6.$$

The polynomial $V(x, y)$ will also be represented by arranging its coefficients in the following array:

$$V = \begin{vmatrix} \alpha^6 \\ \alpha^5 & \alpha^7 \\ \alpha^6 & \alpha^4 & \alpha^7 \\ \alpha^9 & \alpha^{11} & 0 & \alpha^{12} \\ 0 & \alpha^9 & \alpha^{14} & \alpha^{12} & \alpha^5 \\ \alpha^9 & \alpha^{14} & \alpha^5 & \alpha^7 & \alpha^2 & \alpha^5 \end{vmatrix}$$

As we shall see, this representation makes the recurring product $V(x, y)\Lambda(x, y)$ easy to compute.

The Sakata algorithm requires 21 iterations to work its way through the 21 terms of the sequence V. We will work through each of these 21 steps. The algorithm is initialized with the empty set as the footprint, and with the polynomials $\Lambda^{(-1,1)}(x, y) = 1$ as a single exterior polynomial and $B^{(-1,1)}(x, y) = 1$ as a single interior polynomial. As we proceed through the 21 steps, the Sakata algorithm will compute the footprint at each iteration. This will form a sequence of footprints shown in Figure 8.8. At each iteration, the algorithm will also compute a minimal connection polynomial for each exterior corner of the footprint, and an interior polynomial for each interior corner of the footprint.

Before proceeding with the interations, it may be helpful to examine Table 8.1, which summarizes the iterates computed by the Sakata algorithm during the first six iterations of the example. If at step r, $\Lambda^{(r-1,n)}(x, y)$ is not updated, then $\Lambda^{(r,n)}(x, y) = \Lambda^{(r-1,n)}(x, y)$, and if $B^{(r-1,n)}(x, y)$ is not updated, then $B^{(r,n)}(x, y) = B^{(r-1,n)}(x, y)$.

Step (0) Set $r = 0 = (0, 0)$. Using polynomial $\Lambda^{(-1,1)}(x, y)$, compute $\delta_0^{(-1,1)}$:

$$\delta_0^{(-1,1)} = \sum_{k=0}^{s} \Lambda_k^{(-1,1)} V_{k-s}$$
$$= \Lambda_0^{(-1,1)} V_0 \quad = \quad \alpha^9.$$

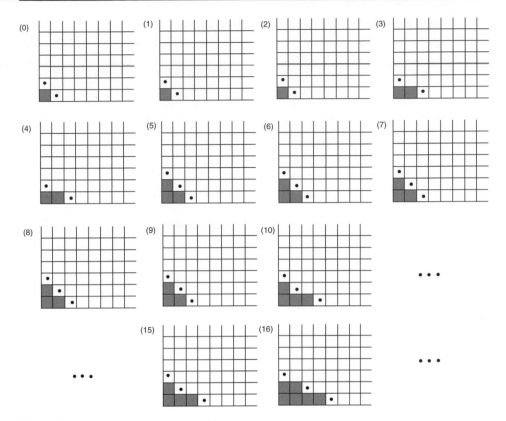

Figure 8.8. Illustrating the Sakata algorithm.

Because $\delta_0^{(-1,1)} \neq 0$, the point $r - s = (0,0)$ must be appended to the footprint. Then the new footprint is $\Delta = \{(0,0)\}$. Thus the new footprint has two exterior corners, $(1,0)$ and $(0,1)$. The two connection polynomials are given by

$$\Lambda^{(0,1)}(x,y) = x\Lambda^{(-1,1)}(x,y) + \alpha^9 B^{(-1,1)}(x,y)$$
$$= x + \alpha^9$$

and

$$\Lambda^{(0,2)}(x,y) = y\Lambda^{(-1,1)}(x,y) + \alpha^9 B^{(-1,1)}(x,y)$$
$$= y + \alpha^9,$$

which we abbreviate as follows:

$$\Lambda^{(0,1)} = \lfloor \alpha^9 \ 1 \qquad \Lambda^{(0,2)} = \begin{vmatrix} 1 \\ \alpha^9 \end{vmatrix}$$

Table 8.1. *The first six iterations of the example*

r	Footprint exterior corners	Polynomials over $GF(16)$ $\{\Lambda^{(r,m)}(x,y)\}$	$\{B^{(r,n)}(x,y)\}$
	(0,0)	$\{1\}$	$\{1\}$
$0 = (0,0)$	$\{(1,0),(0,1)\}$	$\{x + \alpha^9, y + \alpha^9\}$	$\{\alpha^6\}$
$1 = (1,0)$	$\{(1,0),(0,1)\}$	$\{x + \alpha^5, y + \alpha^9\}$	$\{\alpha^6\}$
$2 = (0,1)$	$\{(1,0),(0,1)\}$	$\{x + \alpha^5, y\}$	$\{\alpha^6\}$
$3 = (2,0)$	$\{(2,0),(0,1)\}$	$\{x^2 + \alpha^5 x + \alpha^{14}, y\}$	$\{\alpha^6, \alpha^7 x + \alpha^{12}\}$
$4 = (1,1)$	$\{(2,0),(0,1)\}$	$\{x^2 + \alpha^5 x + \alpha^{14}, y + \alpha x + \alpha^6\}$	$\{\alpha^6, \alpha^7 x + \alpha^{12}\}$
$5 = (0,2)$	$\{(2,0),(1,1),(0,2)\}$	$\{x^2 + \alpha^5 x + \alpha^{14}, \; xy + \alpha x^2 + \alpha^9 x + \alpha^{10}, \; y^2 + \alpha xy + \alpha^6 y + \alpha^4\}$	$\{\alpha^7 x + \alpha^{12}, \; \alpha^2 y + \alpha^3 x + \alpha^8\}$

(Note it is not possible to compute either $\delta_0^{(0,1)}$ or $\delta_0^{(0,2)}$. The two polynomials $\Lambda^{(0,1)}(x,y)$ and $\Lambda^{(0,2)}(x,y)$ vacuously satisfy the condition to be connection polynomials.)

Because the footprint is enlarged in this step, it is also necessary to change the set of interior polynomials. There is only one interior corner, so we define

$$B^{(0,1)}(x,y) = \Lambda^{(-1,1)}(x,y)/\delta_0^{(-1,1)}$$

$$= \alpha^6.$$

Step (1) Set $r = 1 = (1,0)$. Because bideg $\Lambda^{(0,1)}(x,y) = (1,0) \leqq (1,0)$, $\delta_1^{(0,1)}$ exists. Because bideg $\Lambda^{(0,2)}(x,y) = (0,1) \not\leqq (1,0)$, $\delta_1^{(0,2)}$ does not exist. Using polynomial $\Lambda^{(0,1)}(x,y)$ and $r - s = (0,0)$, we compute $\delta_1^{(0,1)}$,

$$\delta_1^{(0,1)} = \sum_{k=0}^{s} \Lambda_k^{(0,1)} V_{r-s+k}$$

$$= \alpha^9 \alpha^9 + \alpha^{14} = \alpha^3 + \alpha^{14} = 1 \neq 0.$$

Therefore

$$\Lambda^{(1,1)}(x,y) = \Lambda^{(0,1)}(x,y) + \delta_1^{(0,1)} B^{(0,1)}(x,y)$$

$$= x + \alpha^5.$$

(As a check, note that $\delta_1^{(1,1)}$, which is calculated by using $\Lambda^{(1,1)}(x,y)$, is zero). Because $\delta_1^{(0,2)}$ does not exist, the corresponding polynomial is not changed. Thus

$$\Lambda^{(1,2)}(x,y) = \Lambda^{(0,2)}(x,y)$$

$$= y + \alpha^9.$$

We abbreviate the current minimal connection polynomials as follows:

$$\Lambda^{(1,1)} = \lfloor\, \alpha^5 \quad 1 \qquad \Lambda^{(1,2)} = \left\lfloor \begin{matrix} 1 \\ \alpha^9 \end{matrix} \right.$$

Step (2) Set $r = 2 = (0,1)$. Polynomial $\Lambda^{(1,1)}(x,y)$ does not reach $(0,1)$. Using polynomial $\Lambda^{(1,2)}(x,y) = y + \alpha^9$ and $r - s = (0,0)$, we compute $\delta_2^{(1,2)}$:

$$\delta_2^{(1,2)} = \sum_{k=0}^{s} \Lambda_k^{(1,2)} V_{r+k-s}$$

$$= \alpha^9 \alpha^9 + 0 = \alpha^3 \neq 0.$$

Because $r - s = (0,0)$ is already in the footprint, the footprint is not enlarged. The new minimal connection polynomials are

$$\Lambda^{(2,1)}(x,y) = \Lambda^{(1,1)}(x,y)$$

$$= x + \alpha^5$$

and

$$\Lambda^{(2,2)}(x,y) = \Lambda^{(1,2)}(x,y) + \alpha^3 B^{(1,1)}(x,y)$$

$$= y + \alpha^9 + \alpha^9$$

$$= y.$$

We abbreviate these minimal connection polynomials as follows:

$$\Lambda^{(2,1)} = \lfloor\, \alpha^5 \quad 1 \qquad \Lambda^{(2,2)} = \left\lfloor \begin{matrix} 1 \\ 0 \end{matrix} \right.$$

Step (3) Set $r = 3 = (2,0)$. Polynomial $\Lambda^{(2,2)}(x,y)$ does not reach $(2,0)$. Using polynomial $\Lambda^{(2,1)}(x,y) = x + \alpha^5$ and $r - s = (1,0)$, we compute $\delta_3^{(2,1)}$:

$$\delta_3^{(2,1)} = \sum_{k=0}^{s} \Lambda_k^{(2,1)} V_{(1,0)+k}$$

$$= \alpha^5 \alpha^{14} + \alpha^5$$

$$= \alpha^8 \neq 0.$$

Because $r - s = (1,0)$ is not already in the footprint, the footprint must be enlarged to include this point. A new exterior corner, $(2,0)$, is formed. The new minimal connection polynomials are

$$\Lambda^{(3,1)}(x,y) = x\Lambda^{(2,1)}(x,y) + \alpha^8 B^{(2,1)}(x,y)$$
$$= x^2 + \alpha^5 x + \alpha^{14}$$

and

$$\Lambda^{(3,2)}(x,y) = \Lambda^{(2,2)}(x,y)$$
$$= y,$$

which we abbreviate as follows:

$$\Lambda^{(3,1)} = \boxed{\begin{array}{ccc} \alpha^{14} & \alpha^5 & 1 \end{array}} \qquad \Lambda^{(3,2)} = \boxed{\begin{array}{c} 1 \\ 0 \end{array}}$$

As a check, note that if the new polynomial $\Lambda^{(3,1)}(x,y)$ is now used to compute $\delta_3^{(3,1)}$, then

$$\delta_3^{(3,1)} = \alpha^{14}\alpha^9 + \alpha^5\alpha^{14} + \alpha^5$$
$$= 0.$$

Because the footprint was enlarged in this step, it is also necessary to update the set of interior polynomials. The interior corners are at $(0,0)$ and $(1,0)$. Accordingly, define

$$B^{(3,1)}(x,y) = B^{(2,1)}(x,y)$$
$$= \alpha^6$$

and

$$B^{(3,2)}(x,y) = \Lambda^{(2,1)}(x,y) / \delta_3^{(2,1)}$$
$$= \alpha^7 x + \alpha^{12}.$$

Step (4) Set $r = 4 = (1,1)$. Polynomial $\Lambda^{(3,1)}(x,y)$ does not reach $(1,1)$. Using polynomial $\Lambda^{(3,2)}(x,y) = y$ and $r - s = (1,0)$, we compute $\delta_4^{(3,2)}$:

$$\delta_4^{(3,2)} = \sum_{k=0}^{s} \Lambda_k^{(3,2)} V_{(1,0)+k}$$
$$= 0 + \alpha^9 \neq 0.$$

Because $r - s = (1, 0)$ is already in the footprint, the footprint is not enlarged. The new minimal connection polynomials are

$$\Lambda^{(4,1)}(x, y) = \Lambda^{(3,1)}(x, y)$$
$$= x^2 + \alpha^5 x + \alpha^{14}$$

and

$$\Lambda^{(4,2)}(x, y) = \Lambda^{(3,2)}(x, y) + \alpha^9 B^{(3,2)}(x, y)$$
$$= y + \alpha x + \alpha^6,$$

which we abbreviate as follows:

$$\Lambda^{(4,1)} = \left\lfloor \alpha^{14} \ \ \alpha^5 \ \ 1 \right. \qquad \Lambda^{(4,2)} = \left\lfloor \begin{matrix} 1 \\ \alpha^6 & \alpha \end{matrix} \right.$$

The interior polynomial $B^{(3,2)}(x, y)$ is chosen because it has its first nonzero discrepancy at $i = 3 = (2, 0)$, so that

$$i - \text{bideg } B^{(3,2)}(x, y) = r - \text{bideg } \Lambda^{(3,2)}(x, y) = (1, 0),$$

and the contributions to the new discrepancy produced by each term of $\Lambda^{(4,2)}(x, y)$ cancel.

Step (5) Set $r = 5 = (0, 2)$. Only polynomial $\Lambda^{(4,2)}(x, y)$ reaches the point $(0, 2)$. Using polynomial $\Lambda^{(4,2)}(x, y) = y + \alpha x + \alpha^6$ and $r - s = (0, 1)$, we compute $\delta_5^{(4,2)}$:

$$\delta_5^{(4,2)} = \sum_{k=0}^{s} \Lambda_k^{(4,2)} V_{(0,1)+k}$$
$$- \alpha^6 \cdot 0 + \alpha \cdot \alpha^9 + 1 \cdot \alpha^9$$
$$= \alpha^{13} \neq 0.$$

Because $r - s = (0, 1)$ is not already in the footprint, the footprint must be enlarged to include this point. Consequently, two new exterior corners, $(1, 1)$ and $(0, 2)$, are formed, and new minimal connection polynomials are needed to go with these two new corners: one polynomial with bidegree $(1, 1)$ and one polynomial with bidegree $(0, 2)$.

The polynomial $x\Lambda^{(4,2)}(x, y)$ is one minimal connection polynomial with bidegree $(1, 1)$, as required, that has zero discrepancy for every point that it can reach, though there may be others.

The polynomial $y\Lambda^{(4,2)}(x, y)$ has bidegree $(0, 2)$ and does reach r, but with the nonzero discrepancy at r. The interior polynomial $B^{(4,1)}(x, y)$ is chosen because it has its first nonzero discrepancy at $i = 0$ and bidegree $\bar{s} = (0,0)$. Thus $i - \bar{s} = (0,0) = r - \text{bideg } y\Lambda^{(4,2)}(x,y)$. The three minimal connection polynomials now are

$$\Lambda^{(5,1)}(x,y) = \Lambda^{(4,1)}(x,y)$$
$$= x^2 + \alpha^5 x + \alpha^{14},$$

$$\Lambda^{(5,2)}(x,y) = x\Lambda^{(4,2)}(x,y)$$
$$= xy + \alpha x^2 + \alpha^6 x,$$

and

$$\Lambda^{(5,3)}(x,y) = y\Lambda^{(4,2)}(x,y) + \alpha^{13}B^{(4,1)}(x,y)$$
$$= y^2 + \alpha xy + \alpha^6 y + \alpha^4,$$

which we abbreviate as follows:

$$\Lambda^{(5,1)} = \begin{array}{|ccc} \alpha^{14} & \alpha^5 & 1 \\ \hline \end{array} \qquad \Lambda^{(5,2)} = \begin{array}{|cc|} 0 & 1 \\ 0 & \alpha^6 & \alpha \\ \hline \end{array} \qquad \Lambda^{(5,3)} = \begin{array}{|cc} 1 & \\ \alpha^6 & \alpha \\ \alpha^4 & 0 \\ \hline \end{array}$$

Because the footprint has changed, the interior polynomials must be updated:

$$B^{(5,1)}(x,y) = B^{(4,2)}(x,y)$$
$$= \alpha^7 x + \alpha^{12};$$

$$B^{(5,2)}(x,y) = \Lambda^{(4,2)}(x,y)/\delta_5^{(4,2)}$$
$$= \alpha^2 y + \alpha^3 x + \alpha^8.$$

Step (6) Set $r = 6 = (3,0)$. Only polynomial $\Lambda^{(5,1)}(x,y) = x^2 + \alpha^5 x + \alpha^{14}$ reaches $(3,0)$. Using this polynomial and $r - s = (1,0)$, we compute $\delta_6^{(5,1)}$:

$$\delta_6^{(5,1)} = \sum_{k=0}^{s} \Lambda_k^{(5,1)} V_{(1,0)+k}$$
$$= \alpha^{14}\alpha^{14} + \alpha^5\alpha^5 + \alpha^7 = 1 \neq 0.$$

Because $r - s = (1,0)$ is already in the footprint, the footprint is not enlarged. Interior polynomial $B^{(5,1)}(x,y)$ is chosen to revise $\Lambda^{(5,1)}(x,y)$ because it has its first nonzero discrepancy at $i = 3 = (2,0)$, so that

$$i - \text{bideg } B^{(5,1)}(x,y) = r - \text{bideg } \Lambda^{(5,1)}(x,y).$$

The minimal connection polynomials are

$$\Lambda^{(6,1)}(x,y) = \Lambda^{(5,1)}(x,y) + B^{(5,1)}(x,y)$$
$$= x^2 + \alpha^{13}x + \alpha^5,$$

$$\Lambda^{(6,2)}(x,y) = \Lambda^{(5,2)}(x,y)$$
$$= xy + \alpha x^2 + \alpha^6 x,$$

and

$$\Lambda^{(6,3)}(x,y) = \Lambda^{(5,3)}(x,y)$$
$$= y^2 + \alpha xy + \alpha^6 y + \alpha^4,$$

which we abbreviate as follows:

$$\Lambda^{(6,1)} = \begin{array}{|ccc|} \alpha^5 & \alpha^{13} & 1 \end{array} \qquad \Lambda^{(6,2)} = \begin{array}{|cc|} 0 & 1 \\ 0 & \alpha^6 & \alpha \end{array} \qquad \Lambda^{(6,3)} = \begin{array}{|cc|} & 1 \\ \alpha^6 & \alpha \\ \alpha^4 & 0 \end{array}$$

Step (7) Set $r = 7 = (2,1)$. Two polynomials, $\Lambda^{(6,1)}(x,y)$ and $\Lambda^{(6,2)}(x,y)$, reach the point $(2,1)$. Using polynomial $\Lambda^{(6,1)}(x,y)$ and $r - s = (0,1)$, we compute $\delta_7^{(6,1)}$:

$$\delta_7^{(6,1)} = \sum_{k=0}^{s} \Lambda_k^{(6,1)} V_{(0,1)+k}$$
$$= \alpha^5 \cdot 0 + \alpha^{13}\alpha^9 + \alpha^{14}$$
$$= \alpha \neq 0.$$

Using polynomial $\Lambda^{(6,2)}(x,y)$ and $r - s = (1,0)$, we compute $\delta_7^{(6,2)}$:

$$\delta_7^{(6,2)} = \sum_{r=0}^{s} \Lambda_k^{(6,2)} V_{(1,0)+k}$$
$$= 0 \cdot \alpha^{14} + \alpha^6 \alpha^5 + \alpha \alpha^7 + \alpha^{14}$$
$$= \alpha \neq 0.$$

Because $(0,1)$ and $(1,0)$ are already in the footprint, the footprint is not enlarged. The new minimal connection polynomials are

$$\Lambda^{(7,1)}(x,y) = \Lambda^{(6,1)}(x,y) + \alpha B^{(6,2)}(x,y)$$
$$= x^2 + \alpha^3 y + \alpha^{11}x + \alpha^6,$$

$$\Lambda^{(7,2)}(x,y) = \Lambda^{(6,2)}(x,y) + \alpha B^{(6,1)}(x,y)$$
$$= xy + \alpha x^2 + \alpha^{14}x + \alpha^{13},$$

and

$$\Lambda^{(7,3)}(x,y) = \Lambda^{(6,3)}(x,y)$$
$$= y^2 + \alpha xy + \alpha^6 y + \alpha^4,$$

which we abbreviate as follows:

$$\Lambda^{(7,1)} = \begin{vmatrix} \alpha^3 & \\ \alpha^6 & \alpha^{11} & 1 \end{vmatrix} \qquad \Lambda^{(7,2)} = \begin{vmatrix} 0 & 1 \\ \alpha^{13} & \alpha^{14} & \alpha \end{vmatrix} \qquad \Lambda^{(7,3)} = \begin{vmatrix} 1 \\ \alpha^6 & \alpha \\ \alpha^4 & 0 \end{vmatrix}$$

Step (8) Set $r = 8 = (1,2)$. Two polynomials, $\Lambda^{(7,2)}(x,y)$ and $\Lambda^{(7,3)}(x,y)$, reach the point $(1,2)$. Using polynomial $\Lambda^{(7,2)}(x,y)$ and $r - s = (0,1)$, we compute $\delta_8^{(7,2)}$:

$$\delta_8^{(7,2)} = \alpha^{13} \cdot 0 + \alpha^{14}\alpha^9 + 0 \cdot \alpha^9 + \alpha\alpha^{14} + \alpha^{11}$$
$$= \alpha^9 \neq 0.$$

Using polynomial $\Lambda^{(7,3)}(x,y)$ and $r - s = (1,0)$, we compute $\delta_8^{(7,3)}$:

$$\delta_8^{(7,3)} = \alpha^4\alpha^{14} + 0 \cdot \alpha^5 + \alpha^6\alpha^9 + 0 \cdot \alpha^7 + \alpha\alpha^{14} + \alpha^{11}$$
$$= \alpha^5 \neq 0.$$

Because $(0,1)$ and $(1,0)$ are already in the footprint, the footprint is not enlarged. The new minimal connection polynomials are

$$\Lambda^{(8,1)}(x,y) = \Lambda^{(7,1)}(x,y)$$
$$= x^2 + \alpha^3 y + \alpha^{11}x + \alpha^6,$$
$$\Lambda^{(8,2)}(x,y) = \Lambda^{(7,2)}(x,y) + \alpha^9 B^{(7,2)}(x,y)$$
$$= xy + \alpha x^2 + \alpha^{11}y + \alpha^5 x + \alpha^{14},$$

and

$$\Lambda^{(8,3)}(x,y) = \Lambda^{(7,3)}(x,y) + \alpha^5 B^{(7,1)}(x,y)$$
$$= y^2 + \alpha xy + \alpha^6 y + \alpha^{12}x + \alpha^{10},$$

which we abbreviate as follows:

$$
\Lambda^{(8,1)} = \begin{vmatrix} \alpha^3 & & \\ \alpha^6 & \alpha^{11} & 1 \end{vmatrix} \qquad \Lambda^{(8,2)} = \begin{vmatrix} \alpha^{11} & 1 & \\ \alpha^{14} & \alpha^5 & \alpha \end{vmatrix} \qquad \Lambda^{(8,3)} = \begin{vmatrix} & & 1 \\ & \alpha^6 & \alpha \\ \alpha^{10} & & \alpha^{12} \end{vmatrix}
$$

Step (9) Set $r = 9 = (0,3)$. One polynomial, $\Lambda^{(8,3)}(x,y)$, reaches the point $(0,3)$. Using polynomial $\Lambda^{(8,3)}(x,y)$ and $r - s = (0,1)$, we compute $\delta_9^{(8,3)}$:

$$
\delta_9^{(8,3)} = \alpha^{10} \cdot 0 + \alpha^{12}\alpha^9 + \alpha^6\alpha^9 + 0 \cdot \alpha^{14} + \alpha\alpha^{11} + 1 \cdot \alpha^6
$$
$$
= \alpha^{11} \neq 0.
$$

Because $(0,1)$ is already in the footprint, the footprint is not enlarged. The new minimal connection polynomials are

$$
\Lambda^{(9,1)}(x,y) = \Lambda^{(8,1)}(x,y)
$$
$$
= x^2 + \alpha^3 y + \alpha^{11} x + \alpha^6,
$$
$$
\Lambda^{(9,2)}(x,y) = \Lambda^{(8,2)}(x,y)
$$
$$
= xy + \alpha x^2 + \alpha^{11} y + \alpha^5 x + \alpha^{14},
$$

and

$$
\Lambda^{(9,3)}(x,y) = \Lambda^{(8,3)}(x,y) + \alpha^{11} B^{(8,2)}(x,y)
$$
$$
= y^2 + \alpha xy + y + \alpha^5 x + \alpha^2,
$$

which we abbreviate as follows:

$$
\Lambda^{(9,1)} = \begin{vmatrix} \alpha^3 & & \\ \alpha^6 & \alpha^{11} & 1 \end{vmatrix} \qquad \Lambda^{(9,2)} = \begin{vmatrix} \alpha^{11} & 1 & \\ \alpha^{14} & \alpha^5 & \alpha \end{vmatrix} \qquad \Lambda^{(9,3)} = \begin{vmatrix} & & 1 \\ & 1 & \alpha \\ \alpha^2 & & \alpha^5 \end{vmatrix}
$$

Step (10) Set $r = 10 = (4,0)$. One polynomial, $\Lambda^{(9,1)}(x,y)$, reaches the point $(4,0)$. Using the polynomial $\Lambda^{(9,1)}(x,y)$ and $r - s = (2,0)$, we compute $\delta_{10}^{(9,1)} = \alpha^5 \neq 0$. Because $(2,0)$ is not in the footprint, the footprint is enlarged to include this point. A new exterior corner at $(3,0)$ is created. The new minimal connection polynomials are

$$
\Lambda^{(10,1)}(x,y) = x\Lambda^{(9,1)}(x,y) + \alpha^5 B^{(9,1)}(x,y)
$$
$$
= x^3 + \alpha^3 xy + \alpha^{11} x^2 + \alpha^4 x + \alpha^2,
$$
$$
\Lambda^{(10,2)}(x,y) = \Lambda^{(9,2)}(x,y)
$$
$$
= xy + \alpha x^2 + \alpha^{11} y + \alpha^5 x + \alpha^{14},
$$

and

$$\Lambda^{(10,3)}(x,y) = \Lambda^{(9,3)}(x,y)$$
$$= y^2 + \alpha xy + y + \alpha^5 x + \alpha^2,$$

which we abbreviate as follows:

$$\Lambda^{(10,1)} = \begin{array}{|cc}
0 & \alpha^3 \\
\hline
\alpha^2 & \alpha^4 & \alpha^{11} & 1
\end{array}
\qquad
\Lambda^{(10,2)} = \begin{array}{|cc}
\alpha^{11} & 1 \\
\hline
\alpha^{14} & \alpha^5 & \alpha
\end{array}$$

$$\Lambda^{(10,3)} = \begin{array}{|c}
1 \\
1 & \alpha \\
\hline
\alpha^2 & \alpha^5
\end{array}$$

Because the footprint has changed, the interior polynomials must be updated so that there is an interior polynomial corresponding to each interior corner. These are

$$B^{(10,1)}(x,y) = \Lambda^{(9,1)}(x,y)/\delta_{10}^{(9,1)}$$
$$= \alpha^{10}x^2 + \alpha^{13}y + \alpha^6 x + \alpha;$$
$$B^{(10,2)}(x,y) = B^{(9,2)}(x,y)$$
$$= \alpha^2 y + \alpha^3 x + \alpha^8.$$

Step (11) Set $r = 11 = (3,1)$. Two polynomials, $\Lambda^{(10,1)}(x,y)$ and $\Lambda^{(10,2)}(x,y)$, reach the point $(3,1)$. Using polynomial $\Lambda^{(10,1)}(x,y)$ and $r-s = (0,1)$, we compute $\delta_{11}^{(10,1)} = \alpha^6 \neq 0$. Using the polynomial $\Lambda^{(10,2)}(x,y)$ and $r - s = (2,0)$, we compute $\delta_{11}^{(10,2)} = \alpha^6 \neq 0$. Because $(0,1)$ and $(2,0)$ are already in the footprint, the footprint is not enlarged. The new minimal connection polynomials are

$$\Lambda^{(11,1)}(x,y) = \Lambda^{(10,1)}(x,y) + \alpha^6 B_{10}^{(10,2)}(x,y)$$
$$= x^3 + \alpha^3 xy + \alpha^{11}x^2 + \alpha^8 y + \alpha^{14}x + \alpha^{13},$$
$$\Lambda^{(11,2)}(x,y) = \Lambda^{(10,2)}(x,y) + \alpha^6 B_{10}^{(10,1)}(x,y)$$
$$= xy + \alpha^{13}y + \alpha^{14}x + \alpha,$$

and

$$\Lambda^{(11,3)}(x,y) = \Lambda^{(10,3)}(x,y)$$
$$= y^2 + \alpha xy + y + \alpha^5 x + \alpha^2,$$

which we abbreviate as follows:

$$\Lambda^{(11,1)} = \begin{array}{|cc} \alpha^8 & \alpha^3 \\ \alpha^{13} & \alpha^{14} & \alpha^{11} & 1 \end{array} \qquad \Lambda^{(11,2)} = \begin{array}{|cc} \alpha^{13} & 1 \\ \alpha & \alpha^{14} \end{array}$$

$$\Lambda^{(11,3)} = \begin{array}{|cc} 1 \\ 1 & \alpha \\ \alpha^2 & \alpha^5 \end{array}$$

Step (12) Set $r = 12 = (2, 2)$. Two polynomials, $\Lambda^{(11,2)}(x, y)$ and $\Lambda^{(11,3)}(x, y)$, reach the point $(2, 2)$. Using polynomial $\Lambda^{(11,2)}(x, y)$ and $r - s = (1, 1)$, we compute $\delta_2^{(11,2)} = 0$. Using polynomial $\Lambda^{(11,3)}(x, y)$ and $r - s = (2, 0)$, we compute $\delta_{12}^{(11,3)} = 0$. Thus the minimal connection polynomials are unchanged:

$$\Lambda^{(12,1)}(x, y) = \Lambda^{(11,1)}(x, y)$$
$$= x^3 + \alpha^3 xy + \alpha^{11}x^2 + \alpha^8 y + \alpha^{14}x + \alpha^{13};$$
$$\Lambda^{(12,2)}(x, y) = \Lambda^{(11,2)}(x, y)$$
$$= xy + \alpha^{13}y + \alpha^{14}x + \alpha;$$
$$\Lambda^{(12,3)}(x, y) = \Lambda^{(11,3)}(x, y)$$
$$= y^2 + \alpha xy + y + \alpha^5 x + \alpha^2.$$

Step (13) Set $r = 13 = (1, 3)$. Two polynomials, $\Lambda^{(12,2)}(x, y)$ and $\Lambda^{(12,3)}(x, y)$, reach the point $(1, 3)$. Using polynomial $\Lambda^{(12,2)}(x, y)$ and $r - s = (0, 2)$, we compute $\delta_{13}^{(12,2)} = 0$. Using polynomial $\Lambda^{(12,3)}(x, y)$ and $r - s = (1, 1)$, we compute $\delta_{13}^{(12,3)} = 0$. Thus the minimal connection polynomials are unchanged:

$$\Lambda^{(13,1)}(x, y) = \Lambda^{(12,1)}(x, y)$$
$$= x^3 + \alpha^3 xy + \alpha^{11}x^2 + \alpha^8 y + \alpha^{14}x + \alpha^{13};$$
$$\Lambda^{(13,2)}(x, y) = \Lambda^{(12,2)}(x, y)$$
$$= xy + \alpha^{13}y + \alpha^{14}x + \alpha;$$
$$\Lambda^{(13,3)}(x, y) = \Lambda^{(12,3)}(x, y)$$
$$= y^2 + \alpha xy + y + \alpha^5 x + \alpha^2.$$

Step (14) Set $r = 14 = (0, 4)$. One polynomial, $\Lambda^{(13,3)}(x, y)$, reaches the point $(0, 4)$. Using polynomial $\Lambda^{(13,3)}(x, y)$ and $r - s = (0, 2)$, we compute $\delta_{14}^{(13,3)} = 0$. Thus the minimal connection polynomials are unchanged:

$$\Lambda^{(14,1)}(x, y) = \Lambda^{(13,1)}(x, y)$$
$$= x^3 + \alpha^3 xy + \alpha^{11} x^2 + \alpha^8 y + \alpha^{14} x + \alpha^{13};$$
$$\Lambda^{(14,2)}(x, y) = \Lambda^{(13,2)}(x, y)$$
$$= xy + \alpha^{13} y + \alpha^{14} x + \alpha;$$
$$\Lambda^{(14,3)}(x, y) = \Lambda^{(13,3)}(x, y)$$
$$= y^2 + \alpha xy + y + \alpha^5 x + \alpha^2.$$

Step (15) Set $r = 15 = (5, 0)$. One polynomial, $\Lambda^{(14,1)}(x, y)$, reaches the point $(5, 0)$. Using polynomial $\Lambda^{(14,1)}(x, y)$ and $r - s = (2, 0)$, we compute $\delta_{15}^{(14,1)} = \alpha^8 \neq 0$. Because $(2, 0)$ is already in the footprint, the footprint is not enlarged. The new minimal connection polynomials are

$$\Lambda^{(15,1)}(x, y) = \Lambda^{(14,1)}(x, y) + \alpha^8 B^{(14,1)}(x, y)$$
$$= x^3 + \alpha^3 xy + \alpha^5 x^2 + \alpha^{14} y + \alpha^{10},$$
$$\Lambda^{(15,2)}(x, y) = \Lambda^{(14,2)}(x, y)$$
$$= xy + \alpha^{13} y + \alpha^{14} x + \alpha,$$

and

$$\Lambda^{(15,3)}(x, y) = \Lambda^{(14,3)}(x, y)$$
$$= y^2 + \alpha xy + y + \alpha^5 x + \alpha^2,$$

which we abbreviate as follows:

$$\Lambda^{(15,1)} = \begin{array}{|cc}
\alpha^{14} & \alpha^3 \\
\alpha^{10} & 0 \quad \alpha^5 \quad 1 \\
\hline
\end{array}
\qquad
\Lambda^{(15,2)} = \begin{array}{|cc}
\alpha^{13} & 1 \\
\alpha & \alpha^{14} \\
\hline
\end{array}
\qquad
\Lambda^{(15,3)} = \begin{array}{|cc}
 & 1 \\
1 & \alpha \\
\alpha^2 & \alpha^5 \\
\hline
\end{array}$$

Step (16) Set $r = 16 = (4, 1)$. Two polynomials, $\Lambda^{(15,1)}(x, y)$ and $\Lambda^{(15,2)}(x, y)$, reach the point $(4, 1)$. Using polynomial $\Lambda^{(15,1)}(x, y)$ and $r - s = (1, 1)$, we compute $\delta_{16}^{(15,1)} = \alpha^5 \neq 0$. Using polynomial $\Lambda^{(15,2)}(x, y)$ and $r - s = (3, 0)$, we compute $\delta_{16}^{(15,2)} = \alpha^5 \neq 0$. Because neither $(1, 1)$ nor $(3, 0)$ is already in the footprint, the footprint is enlarged to include these two points. New exterior corners are created at $(4, 0)$ and $(2, 1)$. The new minimal connection polynomials are

$$\Lambda^{(16,1)}(x,y) = x\Lambda^{(15,1)}(x,y) + \alpha^5 B^{(15,2)}(x,y)$$
$$= x^4 + \alpha^3 x^2 y + \alpha^5 x^3 + \alpha^{14} xy + \alpha^7 y + \alpha x + \alpha^{13},$$
$$\Lambda^{(16,2)}(x,y) = x\Lambda^{(15,2)}(x,y) + \alpha^5 B^{(15,1)}(x,y)$$
$$= x^2 y + \alpha^{13} xy + \alpha^3 x^2 + \alpha^3 y + \alpha^6 x + \alpha^6,$$

and

$$\Lambda^{(16,3)}(x,y) = \Lambda^{(15,3)}(x,y)$$
$$= y^2 + \alpha xy + y + \alpha^5 x + \alpha^2,$$

which we abbreviate as follows:

$$\Lambda^{(16,1)} = \begin{vmatrix} \alpha^7 & \alpha^{14} & \alpha^3 \\ \alpha^{13} & \alpha & 0 & \alpha^5 & 1 \end{vmatrix} \qquad \Lambda^{(16,2)} = \begin{vmatrix} \alpha^3 & \alpha^{13} & 1 \\ \alpha^6 & \alpha^6 & \alpha^3 \end{vmatrix}$$

$$\Lambda^{(16,3)} = \begin{vmatrix} 1 \\ 1 & \alpha \\ \alpha^2 & \alpha^5 \end{vmatrix}$$

Because the footprint has changed, the interior polynomials must be updated with one interior polynomial at each interior corner. These are

$$B^{(16,1)}(x,y) = \Lambda^{(15,1)}(x,y)/\delta_{16}^{(15,1)}$$
$$= \alpha^{10} x^3 + \alpha^{13} xy + x^2 + \alpha^9 y + \alpha^5;$$
$$B^{(16,2)}(x,y) = \Lambda^{(15,2)}(x,y)/\delta_{16}^{(15,2)}$$
$$= \alpha^{10} xy + \alpha^8 y + \alpha^9 x + \alpha^{11};$$
$$B^{(16,3)}(x,y) = B^{(15,2)}(x,y)$$
$$= \alpha^2 y + \alpha^3 x + \alpha^8.$$

Step (17) Set $r = 17 = (3,2)$. Two polynomials, $\Lambda^{(16,2)}(x,y)$ and $\Lambda^{(16,3)}(x,y)$, reach the point $(3,2)$. Using polynomial $\Lambda^{(16,2)}(x,y)$ and $r-s = (1,1)$, we compute $\delta_{17}^{(16,2)} = 0$. Using polynomial $\Lambda^{(16,3)}(x,y)$ and $r - s = (3,0)$, we compute $\delta_{17}^{(16,3)} = \alpha^{13} \neq 0$. Because $(3,0)$ is already in the footprint, the footprint is not enlarged. The new minimal connection polynomials are

$$\Lambda^{(17,1)}(x,y) = \Lambda^{(16,1)}(x,y)$$
$$= x^4 + \alpha^3 x^2 y + \alpha^5 x^3 + \alpha^{14} xy + \alpha^7 y + \alpha x + \alpha^{13},$$
$$\Lambda^{(17,2)}(x,y) = \Lambda^{(16,2)}(x,y)$$
$$= x^2 y + \alpha^{13} xy + \alpha^3 x^2 + \alpha^3 y + \alpha^6 x + \alpha^6,$$

and

$$\Lambda^{(17,3)}(x,y) = \Lambda^{(16,3)}(x,y) + \alpha^{13} B^{(16,2)}(x,y)$$
$$= y^2 + \alpha^{10} xy + \alpha^{13} y + \alpha^{13} x + \alpha^{11},$$

which we abbreviate as follows:

$$\Lambda^{(17,1)} = \begin{array}{|ccc} \alpha^7 & \alpha^{14} & \alpha^3 \\ \alpha^{13} & \alpha & 0 & \alpha^5 & 1 \\ \hline \end{array} \qquad \Lambda^{(17,2)} = \begin{array}{|ccc} \alpha^3 & \alpha^{13} & 1 \\ \alpha^6 & \alpha^6 & \alpha^3 \\ \hline \end{array}$$

$$\Lambda^{(17,3)} = \begin{array}{|cc} 1 & \\ \alpha^{13} & \alpha^{10} \\ \alpha^{11} & \alpha^{13} \\ \hline \end{array}$$

Step (18) Set $r = 18 = (2,3)$. Two polynomials, $\Lambda^{(17,2)}(x,y)$ and $\Lambda^{(17,3)}(x,y)$, reach the point $(2,3)$. Using polynomial $\Lambda^{(17,2)}(x,y)$ and $r - s = (0,2)$, we compute $\delta_{18}^{(17,2)} = 0$. Using polynomial $\Lambda^{(17,3)}(x,y)$ and $r - s = (2,1)$, we compute $\delta_{18}^{(17,3)} = 0$. The minimal connection polynomials are unchanged:

$$\Lambda^{(18,1)}(x,y) = \Lambda^{(17,1)}(x,y)$$
$$= x^4 + \alpha^3 x^2 y + \alpha^5 x^3 + \alpha^{14} xy + \alpha^7 y + \alpha x + \alpha^{13};$$
$$\Lambda^{(18,2)}(x,y) = \Lambda^{(17,2)}(x,y)$$
$$= x^2 y + \alpha^{13} xy + \alpha^3 x^2 + \alpha^3 y + \alpha^6 x + \alpha^6;$$
$$\Lambda^{(18,3)}(x,y) = \Lambda^{(17,3)}(x,y)$$
$$= y^2 + \alpha^{10} xy + \alpha^{13} y + \alpha^{13} x + \alpha^{11}.$$

Step (19) Set $r = 19 = (1,4)$. One polynomial, $\Lambda^{(18,3)}(x,y)$, reaches the point $(1,4)$. Using polynomial $\Lambda^{(18,3)}(x,y)$ and $r - s = (1,2)$, we compute $\delta_{19}^{(18,3)} = 0$. The minimal connection polynomials are unchanged:

$$\Lambda^{(19,1)}(x,y) = \Lambda^{(18,1)}(x,y)$$
$$= x^4 + \alpha^3 x^2 y + \alpha^5 x^3 + \alpha^{14} xy + \alpha^7 y + \alpha x + \alpha^{13};$$
$$\Lambda^{(19,2)}(x,y) = \Lambda^{(18,2)}(x,y)$$
$$= x^2 y + \alpha^{13} xy + \alpha^3 x^2 + \alpha^3 y + \alpha^6 x + \alpha^6;$$
$$\Lambda^{(19,3)}(x,y) = \Lambda^{(18,3)}(x,y)$$
$$= y^2 + \alpha^{10} xy + \alpha^{13} y + \alpha^{13} x + \alpha^{11}.$$

Step (20) Set $r = 20 = (0,5)$. One polynomial, $\Lambda^{(19,3)}(x,y)$, reaches the point $(0,5)$. Using polynomial $\Lambda^{(19,3)}(x,y)$ and $r-s = (0,3)$, we compute $\delta_{20}^{(19,3)} = 0$. The minimal connection polynomials are unchanged:

$$\Lambda^{(20,1)}(x,y) = \Lambda^{(19,1)}(x,y)$$
$$= x^4 + \alpha^3 x^2 y + \alpha^5 x^3 + \alpha^{14} xy + \alpha^7 y + \alpha x + \alpha^{13};$$
$$\Lambda^{(20,2)}(x,y) = \Lambda^{(19,2)}(x,y)$$
$$= x^2 y + \alpha^{13} xy + \alpha^3 x^2 + \alpha^3 y + \alpha^6 x + \alpha^6;$$
$$\Lambda^{(20,3)}(x,y) = \Lambda^{(19,3)}(x,y)$$
$$= y^2 + \alpha^{10} xy + \alpha^{13} y + \alpha^{13} x + \alpha^{11}.$$

This completes the 21 iterations of the Sakata algorithm. The result of these 21 iterations is a set of three minimal connection polynomials for $V(x,y)$. There is one minimal connection polynomial for each exterior corner of the minimal connection footprint for $V(x,y)$. These three minimal connection polynomials are abbreviated as follows:

$$\mathbf{\Lambda}^{(20,1)} = \begin{array}{|ccccc|} \alpha^7 & \alpha^{14} & \alpha^3 & & \\ \alpha^{13} & \alpha & 0 & \alpha^5 & 1 \end{array} \qquad \mathbf{\Lambda}^{(20,2)} = \begin{array}{|ccc|} \alpha^3 & \alpha^{13} & 1 \\ \alpha^6 & \alpha^6 & \alpha^3 \end{array}$$

$$\mathbf{\Lambda}^{(20,3)} = \begin{array}{|cc|} 1 & \\ \alpha^{13} & \alpha^{10} \\ \alpha^{11} & \alpha^{13} \end{array}$$

It may be instructive at this point to recompute the discrepancy for each coefficient of $V(x,y)$ that each minimal connection polynomial can reach. All discrepancies will be zero. The minimal connection polynomials computed in this section will become more interesting in Sections 12.3 and 12.4. In Section 12.3, we will append more terms to $V(x, y)$, and will continue the iterations to find a locator ideal for decoding a hyperbolic code. In Section 12.4, we will append more terms to $V(x, y)$, and will continue the iterations to find a locator ideal for decoding an hermitian code.

8.6   The Koetter algorithm

The Sakata algorithm updates a set of connection polynomials during each iteration, but the pattern of the computation is somewhat irregular. For one thing, the number of minimal connection polynomials does not remain constant, but can grow from iteration to iteration. The Sakata algorithm is initialized with a single minimal connection polynomial, and as it iterates it gradually increases the number of minimal connection polynomials as well as the degrees and coefficients of these polynomials. Furthermore, each connection polynomial is tested and updated on an irregular schedule. In some applications of the Sakata algorithm, the irregular pattern in the way that the polynomials are processed may be considered a shortcoming because it does not map neatly onto a systolic implementation.

The *Koetter algorithm* is an alternative structure of the Sakata algorithm that forces a strong uniformity of structure onto the computation. It consists of a fixed number of connection polynomials, equal to the maximum number that might be needed. Not all of these connection polynomials correspond to exterior corners of the footprint. Each minimal connection polynomial is processed and updated during every iteration. The Koetter algorithm offers the advantage of a very regular structure and is therefore suitable for a systolic implementation, shown in Figure 8.9. The penalty for this regular structure is that extra computations are performed. Many cells of the systolic array will start out idle or performing computations that are unnecessary. These computations could be eliminated, but only by destroying the systolic structure of the algorithm.

For a finite field, the polynomials $x^q - x$ and $y^q - y$ are elements of every locator ideal. Hence the point $(q, 0)$ is not in the footprint, which means that there are at most q exterior corners in the footprint. A minimal basis for any ideal of $GF(q)[x, y]$ cannot have two polynomials whose leading monomials have the same value of j''. Hence there are at most q polynomials in any minimal basis of an ideal, though there may be fewer. During the iterations of the Sakata algorithm, there is one polynomial in the set of connection polynomials for each exterior corner of the footprint. Because the footprint has at most q exterior corners, there are at most q minimal polynomials, though there may be fewer.

The Koetter algorithm instead introduces q polynomial iterates, one for each value of j'', even though some may be unneeded. In each row of the (j', j'') plane, regard the first square that is not in the footprint as a potential exterior corner. The true exterior corners are some of these. In each row of the (j', j'') plane, the last cell that is in the

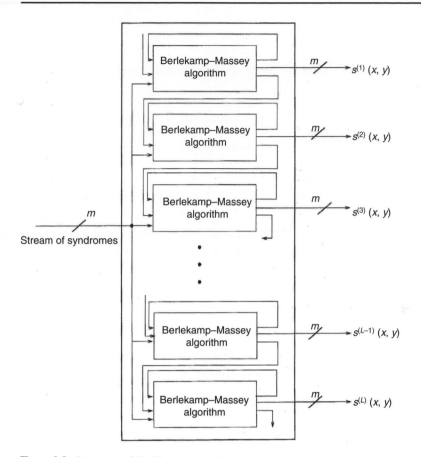

Figure 8.9. Structure of the Koetter algorithm.

footprint is regarded as a potential interior corner. The actual interior corners are some of the potential interior corners.

The Koetter algorithm always iterates q connection polynomials $\Lambda^{(\ell)}(x, y)$ for $\ell = 1, \ldots, q$, one for each row. It also uses q interior polynomials $B^{(\ell)}(x, y)$ for $\ell = 1, \ldots, q$, again one for each row. It is initialized with q interior polynomials and q exterior polynomials. Each minimal connection polynomial is initialized with a leading monomial and with a different value of j''. Each of q minimal connection polynomials is checked and possibly updated at every iteration.

The elegant structure of the Koetter algorithm is based on the observation that the essential structure of the Berlekamp–Massey algorithm can be made to appear multiple times as a module within the structure of the Sakata algorithm, one copy for each connection polynomial. The Koetter algorithm consists of q copies of the Berlekamp–Massey algorithm, all working in lock step on the same sequence of syndromes to embody the Sakata algorithm. The Koetter algorithm is also based on the observation

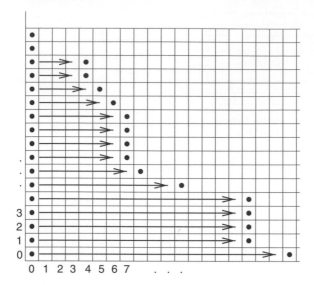

Figure 8.10. Initialization and growth of connection polynomials.

that the necessary interior polynomial for each Berlekamp–Massey module, implementing the Sakata algorithm, is always available within a neighboring Berlekamp–Massey module when needed. Thus, each Berlekamp–Massey module passes data to its cyclically adjacent neighbors. During each iteration, the interior polynomial is passed from module i to module $i + 1$, modulo q, where it is used if needed; otherwise, it is discarded.

Figure 8.10 provides an overview of how the minimal polynomials are developed by the Koetter algorithm. To start, the polynomial iterates, some or most of which may be superfluous, are initialized as $1, y, y^2, \ldots, y^{m-1}$. These are depicted in Figure 8.10 by the column of circles on the left. As the algorithm proceeds, iteration by iteration, the q polynomials are each updated in such a way that each polynomial $\Lambda^{(\ell)}(x, y)$ for $\ell = 1, \ldots q$ may be replaced by a new polynomial, whose leading term has a y degree that does not change, while the x degree may increase or may remain the same. Thus the leading monomial of each polynomial is regarded as moving across rows during the iterations, as depicted in Figure 8.10. Moreover, this is accomplished in such a way that the dots marking the leading monomials retain this staircase arrangement.

The Koetter algorithm was introduced to decode codes on curves, which will be defined in Chapter 10. Just as for the Sakata algorithm, the Koetter algorithm can be augmented in various ways to take advantage of other considerations in an application. Thus in Chapter 12, we shall see an example in which prior information about the ideal can be used to restrict the footprint further. In the example of Chapter 12, the polynomial $G(x, y) = x^{q+1} + y^q + y$, with leading monomial x^{q+1}, is known to be in the locator ideal. The monomials $x^{j'} y^{j''}$, $j' = 0, 1, \ldots$ and $j'' = 0, 1, \ldots, q - 1$, form a basis for $GF(q)/\langle G(x, y)\rangle$. We regard these monomials as arranged in the appropriate total order, such as the weighted order.

Problems

8.1 Given the first six elements under the graded order of the two-dimensional array V over $GF(16)$ as

$$
V = \begin{bmatrix} \vdots & & \\ \alpha^{12} & & \\ 0 & \alpha^{12} & \\ \alpha^{12} & \alpha^{2} & \alpha^{8} \ldots \end{bmatrix},
$$

calculate the locator polynomial $\Lambda(x, y)$ by executing the six steps of the Sakata algorithm.

8.2 An ideal $I \subseteq R[x, y, z]$ is given by

$$
I = \langle x + y + z - 13, x + y - z - 1, x + 5y + z - 17\rangle.
$$

Use the Buchberger algorithm to compute a Gröbner basis for I with respect to the lexicographic order $x \succeq y \succeq z$. Compare the steps of this computation to the steps of gaussian elimination used to solve the following linear systems of equations:

$$
x + y + z = 13,
$$
$$
x + y - z = 1,
$$
$$
x + 5y + z = 17.
$$

Is it appropriate to say that the Buchberger algorithm is a generalization of gaussian elimination to general polynomial equations?

8.3 Given the recursion $n^{(\ell)} = \binom{n^{(\ell-1)}}{2}$ for $\ell = 1, \ldots, t$, how does $n^{(t)}$ depend on t?

8.4 (a) Given the system of equations

$$
y^2 - x^3 + x = 0
$$

and

$$y^3 - x^2 = 0$$

to be solved, graph the equations and find the (three) solutions.

(b) Use the Buchberger algorithm to find a reduced basis for the ideal with these two polynomials as generators.

(c) From the reduced basis, find the zeros of the ideal.

8.5 Prove the division algorithm for bivariate polynomials.

8.6 (a) Using the Buchberger algorithm, find a reduced basis with respect to the lexicographic order for the ideal

$$I = \langle x^3 - y^2, x^2 - y^3 + y \rangle$$

over the field R.

(b) Is $I \cap R[x]$ a principal ideal? Give a reduced basis for $I \cap R[x]$.

8.7 Given an ideal $I \subset F[x, y]$ with two distinct term orders, does there exist a single set of generator polynomials that is a Gröbner basis with respect to both term orders?

8.8 Use the Sakata algorithm to find a set of connection polynomials for the following array:

$$V = \begin{vmatrix} \alpha^6 \\ \alpha^5 & \alpha^7 \\ \alpha^6 & \alpha^4 & \alpha^7 \\ \alpha^9 & \alpha^{11} & 0 & \alpha^{12} \\ 0 & \alpha^9 & \alpha^{14} & \alpha^{12} & \alpha^5 \\ \alpha^9 & \alpha^{14} & \alpha^5 & \alpha^7 & \alpha^2 & \alpha^5 \end{vmatrix}$$

8.9 Using the Buchberger algorithm, find a reduced basis with respect to the graded order for the ideal $\langle x^3 y + x^3 + y \rangle$ in the ring $GF(2)[x, y]/\langle x^7 - 1, y^7 - 1 \rangle$.

8.10 A particular epicyclic hermitian code with minimum distance 33 is specified. If this code is only required to decode to the packing radius, what is the largest number of exterior corners that the footprint of the error locator ideal can have?

8.11 Is it true that for an ideal $I \subset F[x, y]$, the monic polynomial of least bidegree in the total order must appear in the reduced basis of I?

8.12 Describe how to use the Sakata algorithm to compute a minimal basis for the ideal with zeros at t specified points in the bicyclic plane. Express the algorithm in both the "transform domain" and the "code domain." Is the algorithm valid in the (a, b)-weighted degree?

8.13 Prepare a detailed flow diagram for the Koetter algorithm.

Notes

The notion of a basis for an ideal is a natural one, though it seems that it played a somewhat subdued role in the early development of commutative algebra. It was Buchberger's definition of a Gröbner basis and the development of his algorithm to compute such a basis that led to the emergence of the minimal basis in a more prominent role in applications and, indeed, to much of the development of computational commutative algebra. The Buchberger algorithm (Buchberger, 1985) for computing a Gröbner basis may be regarded as a generalization of the euclidean algorithm.

Imai (1977) was drawn independently to the notion of a minimal basis of an ideal in his study of bicyclic codes in order to generalize the notion of the generator polynomial to two dimensions. Imai also suggested the need for a two-dimensional generalization of the Berlekamp–Massey algorithm. This suggestion was the stimulus that led Sakata (then visiting Buchberger) to the quest for the algorithm that now bears his name. Sakata (1990) developed his algorithm for the decoding of bicyclic codes. His algorithm computes a locator ideal that will locate the nonzeros of the polynomial $V(x, y)$ anywhere in the bicyclic plane. Side conditions, such as the knowledge that all nonzeros of $V(x, y)$ fall on a curve, are easily accommodated by the Sakata algorithm, but are not an essential part of it. Thus the Sakata algorithm was already waiting and would soon be recognized as applicable to the decoding of codes on curves. One of Sakata's insights was that the footprint of the locator ideal, which he called the delta set, is an essential property of an ideal.

An alternative to the Sakata algorithm for computing a locator ideal is the Porter algorithm (Porter, 1988), which is a slight reformulation of the Buchberger algorithm to suit this task. The Porter algorithm and the Sakata algorithm are in much the same relationship for the two-dimensional problem as are the Sugiyama algorithm and the Berlekamp–Massey algorithm are for the one-dimensional problem.

Koetter (1996) studied the various papers that discussed the structure of the Sakata algorithm and sought a more elegant formulation. The Koetter algorithm can be viewed as a systolic reformulation of Sakata's algorithm. The Koetter algorithm is useful primarily in those applications in which an upper bound on the number of minimal polynomials is known, as is the typical case for codes on curves. Leonard (1995) describes the Sakata algorithm as a systematic and highly structured procedure for row reduction of a matrix.

9 Curves, Surfaces, and Vector Spaces

Now that we have studied the ring of bivariate polynomials and its ideals in some detail, we are nearly ready to resume our study of codes. In Chapter 10, we shall construct linear codes as vector spaces on plane curves. This means that the components of the vector space are indexed by the points of the curve. Over a finite field, a curve can have only a finite number of points, so a vector space on a curve in a finite field always has a finite dimension.

Before we can study codes on curves, however, we must study the curves themselves. In this chapter, we shall study curves over a finite field, specifically curves lying in a plane. Such curves, called *planar curves* or *plane curves*, are defined by the zeros of a bivariate polynomial. We shall also study vectors defined on curves – that is, vectors whose components are indexed by the points of the curve – and the weights of such vectors. Bounds on the weight of a vector on a curve will be given in terms of the pattern of zeros of its two-dimensional Fourier transform. These bounds are companions to the bounds on the weight of a vector on a line, which were given in Chapter 1, and bounds on the weight of an array on a plane, which were given in Chapter 4.

9.1 Curves in the plane

Recall from Section 5.4 that a bivariate polynomial over the field F is given by

$$v(x, y) = \sum_{i'} \sum_{i''} v_{i'i''} x^{i'} y^{i''},$$

where each sum is a finite sum and the coefficients $v_{i'i''}$ are elements of the field F. The degree of the bivariate polynomial $v(x, y)$ is the largest value of the sum $i' + i''$ for any nonzero term of the polynomial.

The bivariate polynomial $v(x, y)$ is *evaluated* at the point (β, γ) in the affine plane F^2 by the following expression:

$$v(\beta, \gamma) = \sum_{i'} \sum_{i''} v_{i'i''} \beta^{i'} \gamma^{i''}.$$

Recall that a zero of the polynomial $v(x, y)$ is a point, (β, γ), at which $v(\beta, \gamma)$ is equal to zero, that a singular point of $v(x, y)$ is a point, (β, γ), at which $v(\beta, \gamma)$ is equal to zero, and that all partial derivatives of $v(x, y)$ are equal to zero. A nonsingular point of the bivariate polynomial $v(x, y)$ is called a *regular point* of $v(x, y)$. A nonsingular polynomial (or a regular polynomial) is a polynomial that has no singular points anywhere in the projective plane. Thus the polynomial $v(x, y, z)$ is nonsingular if

$$\frac{\partial v(x, y, z)}{\partial x} = 0,$$

$$\frac{\partial v(x, y, z)}{\partial y} = 0,$$

and

$$\frac{\partial v(x, y, z)}{\partial z} = 0$$

are not satisfied simultaneously at any point of the projective curve $v(x, y, z) = 0$ in any extension field of F.

Theorem 9.1.1 *A nonsingular bivariate polynomial is absolutely irreducible.*

Proof: Suppose the bivariate polynomial $v(x, y)$ is reducible in F or in an extension field of F. Then

$$v(x, y) = a(x, y)b(x, y),$$

where both $a(x, y)$ and $b(x, y)$ each have degree at least 1 and have coefficients in F or a finite extension of F. By the Bézout theorem, we know that the homogeneous forms of the polynomials $a(x, y, z)$ and $b(x, y, z)$ have at least one common projective zero. Therefore

$$\frac{\partial v(x, y, z)}{\partial x} = a(x, y, z)\frac{\partial b(x, y, z)}{\partial x} + b(x, y, z)\frac{\partial a(x, y, z)}{\partial x}$$

is equal to zero at a common projective zero of $a(x, y, z)$ and $b(x, y, z)$. By a similar argument, we see that the y and z partial derivatives of $v(x, y, z)$ are also zero at this same point, so the polynomial $v(x, y, z)$ is singular. ∎

It is essential to the validity of the theorem that a nonsingular polynomial be defined as one that has no singular points anywhere in the projective plane. For example, (over any field) the polynomial $x^2 y + x$ has no singular points in the affine plane, but it is not irreducible. The reason that Theorem 9.1.1 does not apply is that, in the projective plane, the polynomial $x^2 y + x$ has a singular point at $(0, 1, 0)$.

The converse of the theorem is not true. For example, the polynomial $x^3 + y^3 + y^2 + y$ over $GF(3)$ is absolutely irreducible, as can be verified by multiplying out each trial factorization over $GF(3)$ or extensions of $GF(3)$. However, this polynomial is singular because the point $(0, 1, 1)$ is not a regular point.

A polynomial of the form

$$v(x, y) = x^a + y^b + g(x, y),$$

where a and b are coprime and deg $g(x, y) < b < a$ is always regular at every point of the affine plane provided $g(x, y)$ is not divisible by x or y. Besides these affine points, the polynomial $v(x, y)$ has exactly one zero at infinity, at $(0, 1, 0)$, which is a regular point if and only if $a = b + 1$. Such polynomials are always absolutely irreducible.

A *curve* in the affine plane (or *affine curve*) is the set of all affine zeros of an irreducible bivariate polynomial over F. A curve will be denoted by $\mathcal{X}$. Thus for the irreducible polynomial $v(x, y)$, the curve is given by

$$\mathcal{X} = \{(\beta, \gamma) \in F^2 \mid v(\beta, \gamma) = 0\}.$$

The *zeros* of the bivariate polynomial $v(x, y)$ are the *points* of the curve $\mathcal{X}$. The zeros in the affine plane are also called *affine points* of the curve. A curve in the projective plane is the set of all projective zeros of an irreducible bivariate polynomial over F.

A *regular curve* (or a *smooth curve*) in the affine plane is the set of all affine zeros of a regular irreducible bivariate polynomial over F. A regular curve in the projective plane is the set of all projective zeros of such a nonsingular polynomial over F. For example, over the real field, the set of zeros of the polynomial $v(x, y) = (x^2 + y^2 - 1)(x^2 + y^2 - 4)$ does not define a regular curve because $v(x, y)$ is a reducible polynomial. Each irreducible factor, however, does define a regular curve.

We shall give several examples of curves stated as problems as follows.

Problem 1 In the plane Q^2, how many points are on the circle

$$x^2 + y^2 = 1?$$

Problem 2 In the plane Q^2, how many points are on the curve

$$x^m + y^m = 1$$

for $m > 2$?

Problem 3 In the plane $GF(8)^2$, how many points are on the curve

$$x^3 y + y^3 + x = 0?$$

This curve is called the *Klein curve*.

Problem 4 In the plane $GF(q^2)^2$, how many points are on the curve

$$x^{q+1} + y^{q+1} + 1 = 0?$$

This set of points in the plane $GF(q^2)^2$ is called an *hermitian curve*.

Problems of this sort are referred to as "finding the number of rational points on a curve" because of the intuitive notion in problems 1 and 2 that the curve exists in some extension of the rational field, say the real field, and the task is to find how many rational points the curve passes through. In problems 3 and 4, the term "rational" is used only in a suggestive sense. The curve is thought of in a larger, algebraically closed field, and the task is to count the number of "rational" points in the ground field (e.g., points in $GF(8)^2$ or points in $GF(q^2)^2$) through which the curve passes.

Problems of this kind can be very difficult. Until recently, one of these four problems had remained unsolved for more than three centuries, even though many excellent mathematicians devoted much of their lives toward finding its solution.

We are interested only in problems 3 and 4. Since the field is not too large, these two problems, if all else fails, can be solved through direct search by simply trying all possibilities; there is only a finite number.

9.2   The Hasse–Weil bound

The *genus* is an important parameter of a polynomial – or of a curve – that is difficult to define in full generality. We have defined the genus (in Definition 5.3.1) only for the case of a nonsingular, bivariate polynomial of degree d by the so-called Plücker formula: $g = \binom{d-1}{2}$. In Section 9.7, we will discuss another method of determining the genus of a polynomial by counting the so-called *Weierstrass gaps*.

The *Hasse–Weil bound* states that the number of rational points n on a curve of genus g in the projective plane over $GF(q)$ satisfies

$$n \le q + 1 + 2g\sqrt{q}$$

if the curve of genus g is defined by an absolutely irreducible bivariate polynomial. This is a deep theorem of algebraic geometry, which we will not prove. *Serre's improvement* of the Hasse–Weil bound is a slightly stronger statement. It states that the number of rational points n on a curve in the projective plane over $GF(q)$, defined by an absolutely irreducible polynomial of genus g, satisfies

$$n \le q + 1 + g\lfloor 2\sqrt{q}\rfloor.$$

When searching for bivariate polynomials of a specified degree over $GF(q)$ with a large number of rational points, if a polynomial is found that satisfies this inequality with equality, then the search need not continue for that value of q and that degree, because no bivariate polynomial of that degree can have more rational points. If we consider only irreducible nonsingular plane curves of degree d, this inequality can be rewritten as follows:

$$n \leq q+1+\binom{d-1}{2}\lfloor 2\sqrt{q}\rfloor.$$

Our statement of the Hasse–Weil bound suits our needs, but it is not the fullest statement of this bound. The full statement of the Hasse–Weil bound is given as an interval:

$$q+1-g\lfloor 2\sqrt{q}\rfloor \leq n \leq q+1+g\lfloor 2\sqrt{q}\rfloor.$$

However, for the polynomials we shall consider, the inequality on the left side is empty of information, so, for our purposes, there is no need to consider the lower inequality.

One can conclude from the Hasse–Weil bound that, for any sequence of absolutely irreducible polynomials of increasing genus g over the field $GF(q)$, the number of rational points $n(g)$ satisfies $\overline{\lim}_{g\to\infty}(n(g)/g) \leq 2\sqrt{q}$. This statement can be strengthened. The *Drinfeld–Vlăduț bound* is the statement that, in certain special cases, $\overline{\lim}_{g\to\infty}(n(g)/g) \leq \sqrt{q} - 1$. The Drinfeld–Vlăduț bound holds with equality if q is a square. The Drinfeld–Vlăduț bound states that the number of points on curves of large genus grows only as the square root of the field size, though with g as a proportionality factor.

9.3   The Klein quartic polynomial

Now we are ready to solve problems 3 and 4 of Section 9.1, which we will do in this section and in Section 9.4.

The *Klein polynomial*, over the field $GF(8)$, is given by

$$G(x,y) = x^3y + y^3 + x.$$

The Klein polynomial is absolutely irreducible. It follows from the Plücker formula that the Klein polynomial has genus given by

$$g = \frac{1}{2}(4-1)(4-2) = 3.$$

When written as a trivariate homogeneous polynomial, the Klein polynomial is given by

$$G(x,y,z) = x^3y + y^3z + z^3x.$$

The *Klein curve* $\mathcal{X}$ is the set of zeros of the Klein polynomial.

The Serre improvement to the Hasse–Weil bound says that the number of rational points of the Klein polynomial satisfies

$$n \leq 8 + 1 + 3\lfloor 2\sqrt{8} \rfloor = 24.$$

We shall show that $n = 24$ by finding all 24 rational points in the projective plane.

Clearly, three of the points are $(0,0,1)$, $(0,1,0)$, and $(1,0,0)$. We need to find 21 more rational points; they will be of the form $(\beta, \gamma, 1)$. Let β be any nonzero element of $GF(8)$. Then γ must be a solution, if there is one, of the equation

$$y^3 + \beta^3 y + \beta = 0.$$

Make the change of variables $y = \beta^5 w$ to get

$$\beta(w^3 + w + 1) = 0.$$

This equation has three zeros in $GF(8)$, namely $w = \alpha, \alpha^2$, and α^4. Therefore $y = \beta^5\alpha, \beta^5\alpha^2$, and $\beta^5\alpha^4$ are the three values of y that go with the value β for the variable x. Because there are seven nonzero values in $GF(8)$, and β can be any of these seven values, this yields 21 more zeros of the Klein polynomial. Altogether we have 24 zeros of the polynomial, namely $(0,0,1)$, $(0,1,0)$, $(1,0,0)$, and $(\beta, \gamma, 1)$, where β is any nonzero element of $GF(8)$ and $\gamma = \beta^5\alpha, \beta^5\alpha^2$, and $\beta^5\alpha^4$.

Figure 9.1 shows the projective curve over $GF(8)$ and the points of the Klein curve lying in the projective plane. The projective plane consists of the eight by eight affine plane, with coordinates labeled by the elements of $GF(8)$ and the nine points of the plane at infinity. The row labeled ∞, which is appended as the top row of the figure, denotes points of the form $(x, 1, 0)$, together with the single point $(1, 0, 0)$. These points of the projective plane at infinity form a copy of the projective line. The 24 dots in Figure 9.1 are the 24 points of the Klein quartic.

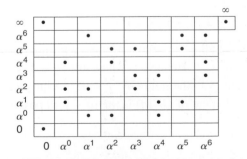

Figure 9.1. Klein quartic in the projective plane over $GF(8)$.

The Klein curve has two points at infinity, as shown in Figure 9.1. This is an intrinsic property of the Klein curve and not an accident of the coordinate system. Under any invertible transformation of the coordinate system, the curve will still have two points at infinity.

9.4   The hermitian polynomials

The *hermitian polynomial* over the field[1] $GF(q^2)$, in homogeneous trivariate form, is given by

$$G(x, y, z) = x^{q+1} + y^{q+1} - z^{q+1}.$$

We sometimes refer to this polynomial as the *Fermat version* of the hermitian polynomial. We will usually deal with the hermitian polynomials over fields of characteristic 2, so we will usually write the hermitian polynomial as

$$G(x, y, z) = x^{q+1} + y^{q+1} + z^{q+1},$$

with all plus signs.

The *hermitian curve* $\mathcal{X}$ is the set of zeros of the hermitian polynomial. It is straightforward to show that the hermitian polynomial is nonsingular, because the three partial derivatives set equal to zero reduce to $x = y = z = 0$, which is not a point of the projective plane. Therefore it follows from the Plücker formula that the genus of the hermitian polynomial is $g = (1/2)q(q-1)$.

The Hasse–Weil bound says that the number of rational points on the curve $\mathcal{X}$ satisfies

$$n \leq q^2 + 1 + q^2(q-1),$$

which reduces to

$$n \leq q^3 + 1.$$

We shall show that $n = q^3 + 1$ by finding all $q^3 + 1$ zeros. Choose a primitive element α in $GF(q^2)$; the element α has order $q^2 - 1$. Hence α^{q-1} has order $q + 1$, and α^{q+1} has order $q - 1$. This means that for any β in the field $GF(q^2)$, β^{q+1} is an element of $GF(q)$, as is $\beta^{q+1} + 1$. Each nonzero element of $GF(q)$ is the $(q+1)$th power of any of $q + 1$ elements of $GF(q^2)$.

---

[1] It is common practice here to write the field as $GF(q^2)$. If it were written $GF(q)$, then the exponents in $G(x, y, z)$ would be $\sqrt{q} + 1$, which is needlessly awkward.

First, we will find the zeros that lie in the projective plane among the points at infinity. These are the points with $z = 0$. They are not visible when looking only in the affine plane. If $z = 0$ in projective coordinates, then $y = 1$, and the polynomial reduces to

$$x^{q+1} + 1 = 0.$$

This means that there are $q + 1$ zeros of the form $(\alpha^{(q-1)i}, 1, 0)$ for $i = 0, \ldots, q$.

All other zeros lie in the affine plane. If $z = 1$ then

$$x^{q+1} + y^{q+1} + 1 = 0.$$

By the same argument used for $z = 0$, we conclude that there are $q + 1$ points on the curve of the form $(x, 0, 1)$ and $q + 1$ points of the form $(0, y, 1)$. All that remains is to count the number of points of the form $(x, y, 1)$, with both x and y nonzero. In such a case, x^{q+1} (or y^{q+1}) cannot equal one, because this would require that $y^{q+1} = 0$ (or $x^{q+1} = 0$). Otherwise, x^{q+1}, denoted γ, can be one of the $q - 2$ elements of $GF(q)$, excluding the elements zero and one. Thus we have the following pair of simultaneous equations:

$$x^{q+1} = \gamma;$$
$$y^{q+1} = \gamma + 1.$$

There are $(q - 2)(q + 1)^2$ solutions of this form, because, for each of $q - 2$ choices for γ, there are $q + 1$ solutions to the first equation and $q + 1$ solutions to the second equation.

Altogether, we have found

$$(q - 2)(q + 1)^2 + 3(q + 1) = q^3 + 1$$

zeros. Because the Hasse–Weil bound states that the number of zeros cannot be greater than $q^3 + 1$, we have found all of them. A hermitian polynomial has all the zeros that the Hasse–Weil bound allows for a polynomial of its degree. In part, this is why it is regarded as desirable for constructing codes.

The affine points of the projective curve have the form $(\beta, \gamma, 1)$, and are represented more simply as (β, γ). The bicyclic points of the affine plane have the form $(\omega^{i'}, \omega^{i''})$, and so exclude the affine points with zero in either coordinate. The curve has $(q - 2)(q + 1)^2$ points in the bicyclic plane. The $2(q + 1)$ points of the form $(\beta, 0)$ or $(0, \gamma)$ are in the affine plane, but not in the bicyclic plane. In addition to these are the $q + 1$ points at infinity, which have the form $(\beta, 1, 0)$.

Summarizing, the polynomial

$$G(x, y) = x^{q+1} + y^{q+1} + 1$$

has $q^3 + 1$ projective zeros: $q + 1$ zeros of the form $(\beta, 1, 0)$; $2(q + 1)$ zeros of the form $(\beta, 0, 1)$ or $(0, \gamma, 1)$; and $(q - 2)(q + 1)^2$ zeros of the form $(\beta, \gamma, 1)$, with β and γ both nonzero.

For the special case of $GF(4) = GF(2^2)$, the hermitian polynomial

$$G(x, y, z) = x^3 + y^3 + z^3$$

has $2^3 + 1 = 9$ zeros in the projective plane over $GF(4)$.

For the special case of $GF(16) = GF(4^2)$, the hermitian polynomial

$$G(x, y, z) = x^5 + y^5 + z^5$$

has $4^3 + 1 = 65$ zeros in the projective plane over $GF(16)$.

For the special case of $GF(64) = GF(8^2)$, the hermitian polynomial

$$G(x, y, z) = x^9 + y^9 + z^9$$

has $8^3 + 1 = 513$ zeros in the projective plane over $GF(64)$.

For the special case of $GF(256) = GF(16^2)$, the hermitian polynomial

$$G(x, y, z) = x^{17} + y^{17} + z^{17}$$

has $16^3 + 1 = 4097$ zeros in the projective plane over $GF(256)$.

Figure 9.2 shows the hermitian curve in the projective plane over $GF(4)$. Figure 9.3 shows the hermitian curve in the projective plane over $GF(16)$.

An alternative version of the hermitian curve is also used – and is often preferred – because it has fewer points at infinity. The *Stichtenoth version* of the hermitian curve is based on the polynomial

$$G(x, y) = y^q + y - x^{q+1}.$$

The reason this polynomial is also said to define the hermitian curve is that it is related by a coordinate transformation to the polynomial used previously. To obtain

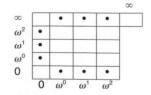

Figure 9.2. Hermitian curve in $GF(4)^2$.

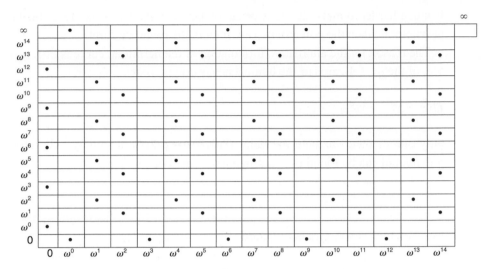

Figure 9.3. The hermitian curve in $GF(16)^2$.

this correspondence, start with the earlier homogeneous trivariate polynomial in the variables (u, v, w),

$$G(u, v, w) = u^{q+1} + v^{q+1} - w^{q+1},$$

and make the following change of variables:

$$\begin{bmatrix} u \\ v \\ w \end{bmatrix} = \begin{bmatrix} 1 & 0 & 1 \\ 1 & 1 & 0 \\ 1 & 1 & 1 \end{bmatrix} \begin{bmatrix} x \\ y \\ z \end{bmatrix}.$$

The hermitian polynomial then becomes

$$(x + z)^{q+1} + (x + y)^{q+1} - (x + y + z)^{q+1}$$

or

$$(x^q + z^q)(x + z) + (x^q + y^q)(x + y) - (x^q + y^q + z^q)(x + y + z).$$

Expanding the products and canceling like terms leads to the polynomial

$$x^{q+1} - y^q z - yz^q,$$

which is the Stichtenoth version of the polynomial. Incidently, this transformation provides convincing evidence for the proposition that it is often easier to treat a polynomial in the homogeneous trivariate form, even when only the affine plane is of interest. It is much harder to make the transformation between the two versions of the hermitian polynomial when the polynomial is written in the bivariate form.

The reason for the alternative formulation of the hermitian curve is that, as we shall see, of the $q^3 + 1$ zeros of the homogeneous polynomial

$$x^{q+1} - y^q z - yz^q,$$

only one of them, namely the point $(0, 1, 0)$, is a zero at infinity. All zeros of this polynomial can be found by rearranging the zeros of

$$u^{q+1} + v^{q+1} - w^{q+1},$$

by using the follwing transformation:

$$
\begin{bmatrix} x \\ y \\ z \end{bmatrix} = \begin{bmatrix} 1 & 0 & 1 \\ 1 & 1 & 0 \\ 1 & 1 & 1 \end{bmatrix}^{-1} \begin{bmatrix} u \\ v \\ w \end{bmatrix} = \begin{bmatrix} 1 & 1 & 1 \\ 1 & 0 & 1 \\ 0 & 1 & 1 \end{bmatrix} \begin{bmatrix} u \\ v \\ w \end{bmatrix}.
$$

Instead, we will find the zeros directly. The affine zeros are the solutions of

$$x^{q+1} - y^q - y = 0,$$

which are obtained by taking, in turn, each nonzero element γ of $GF(q^2)$ for x and finding the zeros of

$$y^q + y = \gamma^{q+1}.$$

To see that there are q such zeros for each γ, let $y = \gamma^{q+1} w$ and note that $(\gamma^{q+1})^q = \gamma^{q+1}$ in $GF(q^2)$, so if $\gamma \neq 0$, the equation becomes

$$\gamma^{q+1}(w^q + w + 1) = 0.$$

In general, this polynomial is not irreducible. In particular, we have the following factorizations into irreducible polynomials:

$$x^4 + x + 1 = x^4 + x + 1;$$
$$x^8 + x + 1 = (x^2 + x + 1)(x^6 + x^5 + x^3 + x^2 + 1);$$
$$x^{16} + x + 1 = (x^8 + x^6 + x^5 + x^3 + 1)(x^8 + x^6 + x^5 + x^4 + x^3 + x + 1).$$

Next, we will show that $w^q + w + 1$ has q zeros in the field $GF(q^2)$ when q is a power of 2.

Proposition 9.4.1 *Let $q = 2^m$. The polynomial $x^q + x + 1$ over $GF(2)$ has all its zeros in $GF(q^2)$.*

Proof: It is enough to show that $x^q + x + 1$ divides $x^{q^2-1} + 1$, because all the zeros of the latter polynomial are in $GF(q^2)$. But q is a power of 2, so

$$(x^q + x + 1)(x^q + x + 1)^{q-1} = (x^q + x + 1)^q$$

$$= x^{q^2} + x^q + 1.$$

Therefore

$$x(x^{q^2-1} - 1) = x^{q^2} - x$$

$$= (x^q + x + 1)[(x^q + x + 1)^{q-1} - 1].$$

Thus $x^q + x + 1$ divides $x^{q^2-1} - 1$. But $x^{q^2-1} - 1$ completely factors over $GF(q^2)$ into $q^2 - 1$ distinct linear factors. Therefore $x^q + x + 1$ completely factors over $GF(q^2)$ into q distinct linear factors, and the proof is complete. ∎

From the proposition, we can conclude that $x^{2^m} + x + 1$ is always reducible over $GF(2)$ if m is larger than 2 by the following argument. Because $2^{2^m} = (2^m)^2$, only if m is equal to 2, $GF((2^m)^2)$ is a proper subfield of $GF(2^{2^m})$ whenever m is greater than 2. Therefore if α generates $GF(2^{2^m})$, α is not an element of $GF((2^m)^2)$. However, if $x^{2^m} + x + 1$ were irreducible, it could be used to extend $GF(2)$ to $GF(2^{2^m})$ by appending α to $GF(2)$ and setting $\alpha^{2^m} = \alpha + 1$, in which case α would be a zero of $x^{2^m} + x + 1$. However, the proposition says that the zeros are in $GF((2^m)^2)$, and α is not in $GF((2^m)^2)$. Thus $x^{2^m} + x + 1$ is not irreducibile if m is larger than 2.

In addition to the affine zeros of $G(x, y, z)$, we can find the zeros at infinity from

$$G(x, y, 0) = x^{q+1} = 0.$$

We conclude that only the point $(0, 1, 0)$ is a zero at infinity.

Summarizing, the polynomial over $GF(q^2)$

$$G(x, y) = x^{q+1} - y^q - y$$

has $q^3 + 1$ projective zeros: one zero of the form $(0, 1, 0)$, q zeros of the form $(0, \gamma, 1)$, and $q^3 - q$ zeros of the form $(\beta, \gamma, 1)$, with β and γ both nonzero. In particular, the curve has $q^3 + 1$ points in the projective plane, q^3 points in the affine plane, and $q^3 - q$ points in the bicyclic plane.

Figure 9.4 shows the Stichtenoth version of the hermitian curve in the projective plane over $GF(4)$. There are eight points in the affine plane and one point at infinity. Figure 9.5 shows the Stichtenoth version of the hermitian curve in the projective plane

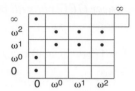

Figure 9.4. Alternative hermitian curve in $GF(4)^2$.

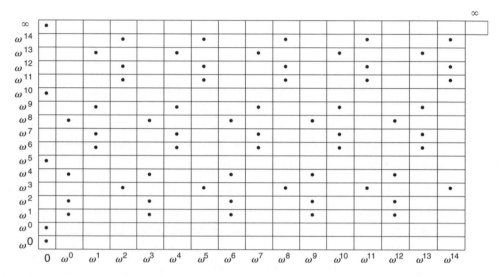

Figure 9.5. Alternative hermitian curve in $GF(16)^2$.

over $GF(16)$. There are 64 points in the affine plane that are zeros of the Stichtenoth polynomial, and one point at infinity.

9.5   Plane curves and the two-dimensional Fourier transform

In a finite field, the operation of evaluating a polynomial at all nonzero points of the field and the operation of computing a one-dimensional Fourier transform are essentially the same. Likewise, the operation of evaluating a bivariate polynomial at the points of the bicyclic plane and the operation of computing a two-dimensional Fourier transform are essentially the same. For the same reason, the computation of the points of a curve in a finite field is closely related to the two-dimensional Fourier transform. This relationship is most natural if the curve is restricted to the bicyclic plane because this is where the Fourier transform is defined. The curve is obtained simply by computing the two-dimensional Fourier transform and noting where the zeros occur.

Using the Fourier transform to evaluate the curve does not evaluate $G(x, y)$ at points of the form $(0, \gamma)$ or $(\beta, 0)$, nor at points at infinity. Because the two-dimensional Fourier transform has many strong and useful properties, it is often useful to shorten a curve to the bicyclic plane by deleting those points. If points other than the points of the bicyclic plane are to be used, they must then be regarded as special points, and they must be handled separately.

We shall regard the curve $\mathcal{X}$ as lying in the code domain and the polynomial $G(x, y)$ as lying in the transform domain. The inverse Fourier transform $g_{i'i''} = (1/n^2) G(\omega^{-i'}, \omega^{-i''})$ can be used to compute the curve from the polynomial. In the language of the Fourier transform, the bicyclic curve, defined by the n by n array G, is the set of points of the bicyclic plane at which the inverse Fourier transform g is equal to zero:

$$\mathcal{X} = \left\{ (\omega^{-i'}, \omega^{-i''}) \mid g_{i'i''} = G(\omega^{-i'}, \omega^{-i''}) = 0 \right\}.$$

The zero elements of g comprise the curve $\mathcal{X}$ within the bicyclic plane. The bicyclic plane over a finite field can be regarded as a torus, so this segment of the curve lies on the torus. Accordingly, we shall call this segment the *epicyclic* segment of the curve.

Given any plane curve $\mathcal{X}$, a polynomial $C(x, y)$ can be evaluated on the bicyclic points of $\mathcal{X}$ by computing all the components of the inverse two-dimensional Fourier transform c, then down-selecting to only those components of c lying on the curve. This forms the vector $c(\mathcal{X})$, whose components are indexed by the points of $\mathcal{X}$. Indeed, any vector that is defined on the points of the curve $\mathcal{X}$ can be regarded as so computed by the evaluation of a polynomial, which evaluation can be found by using the two-dimensional Fourier transform.

The polynomial $C(x, y)$, evaluated along the curve $\mathcal{X}$, becomes the vector $c(\mathcal{X}) = [C(P_\ell) \mid \ell = 0, \ldots, n - 1]$, where P_ℓ for $\ell = 1, \ldots, n$ are the points of the curve $\mathcal{X}$. Thus, to evaluate $C(x, y)$ along the curve $\mathcal{X}$, one may compute the array c with the components

$$c_{i'i''} = \frac{1}{n^2} C(\omega^{-i'}, \omega^{-i''}),$$

then restrict c to the curve $\mathcal{X}$:

$$c(\mathcal{X}) = [c_{i'i''} \mid (\omega^{-i'}, \omega^{-i''}) \in \mathcal{X}]$$
$$= [C(P_\ell) \mid P_\ell \in \mathcal{X}].$$

Any vector v of blocklength n on the curve $\mathcal{X}$ is a vector whose components are indexed by the n points of the curve. The ℓth component of v, indexed by the point P_ℓ, is denoted v_ℓ for $\ell = 0, \ldots, n - 1$. Although v is a vector of blocklength n, it is sometimes convenient to think of its n components embedded into an N by N array with the ℓth component of $\boldsymbol{v}$ placed at the position along the curve $\mathcal{X}$ corresponding to the point P_ℓ. In such a case, we will refer to the vector as $\boldsymbol{v}(\mathcal{X})$ and the two-dimensional array as $\boldsymbol{v}$. To form this representation, simply zero-pad $\boldsymbol{v}(\mathcal{X})$ to form the N by N array, called $\boldsymbol{v}$, indexed by i' and i'', with $v_{i'i''}$ appearing at the (i', i'') location of the array if $(i', i'') \in \mathcal{X}$, and a zero appearing at the (i', i'') location if $(i', i'') \notin \mathcal{X}$. Then we may take the two-dimensional Fourier transform of the array $\boldsymbol{v}$, with components given by

$$
\begin{aligned}
V_{j'j''} &= \sum_{i'=0}^{n-1} \sum_{i''=0}^{n-1} v_{i'i''} \omega^{i'j'} \omega^{i''j''} \\
&= \sum_{(i',i'')\in\mathcal{X}} v_{i'i''} \omega^{i'j'} \omega^{i''j''}.
\end{aligned}
$$

Using the polynomial

$$
v(x, y) = \sum_{i'=0}^{n-1} \sum_{i''=0}^{n-1} v_{i'i''} x^{i'} y^{i''},
$$

this can be expressed as

$$
\begin{aligned}
V_{j'j''} &= v(\omega^{j'}, \omega^{j''}) \\
&= v(P_\ell^{-1}).
\end{aligned}
$$

What has been accomplished here? Our collection of one-dimensional Fourier transforms of vectors $\boldsymbol{v}$ of various blocklengths n has been enriched by establishing a kind of modified Fourier transform that lies between a one-dimensional Fourier transform and a two-dimensional Fourier transform. The components of vector $\boldsymbol{v}$ are assigned to the points of a curve in a zero-padded two-dimensional array which is transformed by an N by N two-dimensional Fourier transform. We shall see that many properties of the vector $\boldsymbol{v}$ can be deduced from the properties of the two-dimensional Fourier transform.

9.6   Monomial bases on the plane and on curves

The ring of bivariate polynomials $F[x, y]$ is closed under addition and scalar multiplication, so it can be regarded as a vector space. Therefore, as a vector space, it can be expressed in terms of any of its vector-space bases. We shall use the set of bivariate monomials as a basis. Using the graded order to order the monomials, the vector-space basis is $\{1, x, y, x^2, xy, y^2, x^3, x^2y, xy^2, y^3, x^4, x^3y, \ldots\}$. The number of monomials is infinite, so the number of elements in the basis is infinite. Thus, regarded as a vector space, the ring $F[x, y]$ is an infinite-dimensional vector space over F.

Just as $F[x, y]$ can be regarded as a vector space, so too the quotient ring $F[x, y]/\langle G(x, y)\rangle$ can be regarded as a vector space. It is a vector subspace of $F[x, y]$. Does this vector subspace have a basis that is contained in the basis for $F[x, y]$? In general, it is not true that a basis for a vector subspace can be found as a subset of a given basis for the entire vector space. However, if we choose a monomial basis for $F[x, y]$, then there is a subset of the basis that forms a basis for $F[x, y]/\langle G(x, y)\rangle$. The ring $F[x, y]/\langle G(x, y)\rangle$, when regarded as a vector space, always has a vector-space basis consisting of monomials.

The ring $F°[x, y] = F[x, y]/\langle x^n - 1, y^n - 1\rangle$ of bivariate polynomials modulo $\langle x^n - 1, y^n - 1\rangle$ mimics the ring $F[x, y]$, but it is much simpler because all polynomials have a componentwise degree smaller than (n, n). The ring $F°[x, y]$ also can be regarded as a vector space. It, too, has a basis consisting of bivariate monomials; now the only monomials in the basis are those of componentwise degree smaller than (n, n). There are n^2 such monomials, and they are linearly independent, so $F°[x, y]$ is a vector space of dimension n^2. We will often be interested in the case in which $F = GF(q)$ and n is equal to $q - 1$ (or perhaps a divisor of $q - 1$). Then the vector space over $GF(q)$ has dimension $(q - 1)^2$.

The quotient ring $F°[x, y]/\langle G(x, y)\rangle = F[x, y]/\langle G(x, y)\rangle°$, where

$$\langle G(x, y)\rangle° = \langle G(x, y), x^n - 1, y^n - 1\rangle,$$

can be regarded as a vector subspace of $F°[x, y]$. We are interested in the relationship between the dimension of this subspace and the number of bicyclic zeros of $G(x, y)$. Equivalently, we want to find the relationship between the number of bicyclic zeros of $G(x, y)$ and the number of monomials in a monomial basis for the quotient ring $F°[x, y]/\langle G(x, y)\rangle$. The discrete nullstellensatz tells us that if $G(x, y)$ has no bicyclic zeros, then $\langle G(x, y)\rangle° = F°[x, y]$, so the quotient ring is the trivial ring $\{0\}$.

Theorem 9.6.1 *Let $F = GF(q)$ and let $F°[x, y] = F[x, y]/\langle x^{q-1} - 1, y^{q-1} - 1\rangle$. The number of monomials in a monomial basis for $F°[x, y]/\langle G(x, y)\rangle$ is equal to the number of bicyclic rational zeros of $G(x, y)$.*

Proof: Let $p_0, \ldots, p_{N-1}$ be the N points of the bicyclic plane over $GF(q)$, where $N = (q - 1)^2$. Let $P_0, \ldots, P_{n-1}$ be the n bicyclic rational zeros of $G(x, y)$. Then $\{P_0, \ldots, P_{n-1}\} \subset \{p_0, \ldots, p_{N-1}\}$. Let $\varphi_0, \ldots, \varphi_{N-1}$ be the N monomials of componentwise degree at most $(q - 2, q - 2)$, where, again, $N = (q - 1)^2$. Consider the N by N matrix $\boldsymbol{M}$ whose elements are the N monomials evaluated at the N points of the plane. This matrix (which is actually the matrix corresponding to the two-dimensional

Fourier transform) has full rank:

$$
M = \begin{bmatrix}
\varphi_0(p_0) & \varphi_0(p_1) & \cdots & \varphi_0(p_{N-1}) \\
\varphi_1(p_0) & \varphi_1(p_1) & \cdots & \varphi_1(p_{N-1}) \\
\vdots & & & \vdots \\
\varphi_{N-1}(p_0) & \varphi_{N-1}(p_1) & \cdots & \varphi_{N-1}(p_{N-1})
\end{bmatrix}.
$$

Now delete each column that does not correspond to one of the n rational zeros of $G(x,y)$, as given by the set $\{P_0, \ldots, P_{n-1}\}$. Because the original matrix has full rank, the surviving n columns remain linearly independent, so there must be a linearly independent set of n rows. The rows in any linearly independent set of rows correspond to n monomials, and these monomials form a basis. ∎

For an example of this theorem, let $G(x,y)$ be the Klein quartic polynomial given by

$$G(x,y) = x^3 y + y^3 + x,$$

and let $F^\circ[x,y] = GF(8)[x,y]/\langle x^7 - 1, y^7 - 1\rangle$. The reduced basis $\mathcal{G}^*$ for $\langle G(x,y)\rangle^\circ \subset F^\circ[x,y]$ is $\{x^7 - 1, y^7 - 1, x^3 y + y^3 + x, xy^5 + x^2 y^2 + x^5 + y\}$, which can be verified by using the Buchberger theorem to check that all conjunction polynomials are zero. The footprint of this ideal is shown in Figure 9.6. The area of the footprint is 21, and $G(x,y)$ has 21 bicyclic zeros. Moreover, the ring $F^\circ[x,y]/\langle G(x,y)\rangle$ has dimension 21. Multiplication in this ring is polynomial multiplication modulo $\mathcal{G}^*$.

Theorem 9.6.1 says that if $G(x,y)$ has n bicyclic rational zeros, then there are n monomials in any monomial basis for $F^\circ[x,y]/\langle G(x,y)\rangle$. This is also equal to the area of the footprint of the ideal, and the monomials corresponding to the points of the footprint are independent, so those monomials must form a basis.

Let $G(x,y)$ be a bivariate polynomial of the form

$$G(x,y) = x^a + y^b + g(x,y),$$

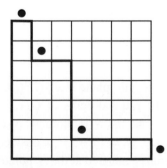

Figure 9.6. Footprint corresponding to the Klein polynomial.

where a and b are coprime and $a > b > \deg g(x, y)$. A polynomial of this form has exactly one zero at infinity, which is a regular point if and only if $a = b + 1$. The Stichtenoth version of an hermitian polynomial is an example of a polynomial of this form.

For this form of polynomial $G(x, y)$, we have

$$x^a = -y^b - g(x, y)$$

in the quotient ring $F[x, y]/\langle G(x, y)\rangle$. Therefore the monomial x^a is linearly dependent on other monomials. To form a basis, either x^a or some other suitable monomial must be deleted from the set of monomials. We choose to delete x^a. Likewise, monomials such as x^{a+1} or $x^a y$ are also linearly dependent on other monomials and can be deleted to form a basis. Thus a monomial basis for the ring $F[x, y]/\langle G(x, y)\rangle$ is $\{x^{i'} y^{i''} | i' = 0, \ldots, a - 1;\ i'' = 0, \ldots\}$. Similarly, a monomial basis for the ring $F^\circ[x, y]/\langle G(x, y)\rangle$ is $\{x^{i'} y^{i''} | i' = 0, \ldots, a - 1;\ i'' = 0 \ldots, n - 1\}$.

In general, there will be many ways to choose n monomials to form a basis. We want to determine if there is a preferred choice. In particular, we want to define a monomial basis that supports a total order on the basis monomials and has a certain desirable form. This desired total order will be the weighted order – if there is one – that is implied by the polynomial $G(x, y)$.

Definition 9.6.2   *The weight function ρ on a quotient ring of polynomials of the form*

$$R = F[x, y]/\langle G(x, y)\rangle$$

is the function[2]

$$\rho : R \to N \cup \{-\infty\},$$

satisfying, for any polynomials $f(x, y)$, $g(x, y)$, and $h(x, y)$ of R, the following properties:

(1) $\rho(\lambda f) = \rho(f)$ for every nonzero scalar λ;
(2) $\rho(f + g) \leq \max[\rho(f), \rho(g)]$ with equality if $\rho(f) < \rho(g)$;
(3) if $\rho(f) < \rho(g)$ and h is nonzero, then $\rho(hf) < \rho(hg)$;
(4) if $\rho(f) = \rho(g)$, then for some nonzero scalar λ, $\rho(f - \lambda g) < \rho(f)$;
(5) $\rho(fg) = \rho(f) + \rho(g)$.

A weight function need not exist for every such quotient ring. A function that satisfies properties (1)–(4) is called an *order function*. Note that it follows from property (5) that $\rho(0)$ is $\pm\infty$. By convention, we choose $\rho(0) = -\infty$. Thus it follows from property (4) that $\rho(g) = -\infty$ only if $g = 0$. It follows from property (5) that $\rho(1) = 0$. Without

---

[2] A *weight function* should not be confused with the *weight* of a vector.

loss of generality, we will require that for any weight function, the set of weights has no common integer factor.

A weight function, if it exists, assigns a unique weight to each monomial of that ring, and thus orders the monomials. The *weighted bidegree* of a polynomial is the largest weight of any of its monomials. The weight function ρ, applied to monomials, satisfies $\rho(\varphi_i \varphi_j) = \rho(\varphi_i) + \rho(\varphi_j)$. Because each monomial φ_i has a unique weight, $\rho(\varphi_i) = \rho_i$, the weight can be used as an alternative index on the monomials, now writing a monomial with an indirect index as φ_{ρ_i}. This indirect indexing scheme allows us to write

$$\varphi_{\rho_i + \rho_j} = \varphi_{\rho_i} \varphi_{\rho_j},$$

now referring to the monomial by its weight.

As an example of a weight function, consider the ring of bivariate polynomials over the field $GF(16)$ modulo the hermitian polynomial $x^5 + y^4 + y$. The properties defining a weight function require that, because $\rho(x^5) = \rho(y^4 + y) = \rho(y^4)$, we must have $5\rho(x) = 4\rho(y)$. Let $\rho(x) = 4$ and $\rho(y) = 5$. Then the weight function[3] on monomials is defined by

$$\rho(x^{i'} y^{i''}) = 4i' + 5i''.$$

This weight function extends linearly to all polynomials in the ring. It is now straightforward to verify that all properties of a weight function are satisfied. The weight function implies a weighted order as a total order on the monomials of the ring. Thus the monomials and weights are given by

$\varphi_i =$	1	x	y	x^2	xy	y^2	x^3	$x^2 y$	xy^2	$\ldots$
$\rho(\varphi_i) =$	0	4	5	8	9	10	12	13	14	$\ldots$
	y^3	x^4	$x^3 y$	$x^2 y^2$	xy^3	y^4	$x^4 y$	$x^3 y^2$		$\ldots$
	15	16	17	18	19	20	21	22		$\ldots$

Note that x^5 is the first monomial not in the ring, so the repetition $\rho(x^5) = \rho(y^4)$, which threatened to be a violation of uniqueness, does not occur. With indirect indexing, this list is reversed to write

$\rho_i =$	0	4	5	8	9	10	12	13	14	$\ldots$
$\varphi_{\rho_i} =$	1	x	y	x^2	xy	y^2	x^3	$x^2 y$	xy^2	$\ldots$
	y^3	x^4	$x^3 y$	$x^2 y^2$	xy^3	y^4	$x^4 y$	$x^3 y^2$		$\ldots$
	15	16	17	18	19	20	21	22		$\ldots$

[3] If a fixed polynomial $p(x, y)$ admits a weight function on the set of monomials spanning $F[x, y]/\langle p(x, y)\rangle$, then the weight function of a monomial is called the *pole order* of the monomial. If any straight line tangent to the curve defined by the polynomial does not intersect the curve elsewhere, then that polynomial will induce a weight function on the set of monomials modulo that polynomial. For this reason, codes defined on such a curve (admitting a weight function on monomials) are called *one-point codes*.

As before, the weights are all different. Again, although $\rho(x^5) = 20$ would be a repetition, the monomial x^5 is not in the basis. Note that the integers $1, 2, 3, 6, 7$, and 11 do not appear as weights. We shall discuss these missing integers, called *Weierstrass gaps*, in some detail in the following section. The number of missing integers is six, which equals the genus g of the hermitian polynomial. We shall see that this is not a coincidence.

Our second example illustrates the fact that a weight function need not exist for every ring of the form $F[x, y]/\langle G(x, y)\rangle$. For some purposes, such a ring is flawed and is not useful. Nevertheless, to illuminate further the notion of a weight function, we shall show that there is a subring of such a ring on which a weight function does exist. Consider the ring of bivariate polynomials over the field $GF(8)$ modulo the Klein polynomial $x^3 y + y^3 + x$. In this ring, $x^3 y = y^3 + x$, so the monomial $x^3 y$ is not in the basis. The monomial basis for $F[x, y]/\langle x^3 y + y^3 + x\rangle$ is easiest to see when arranged in the array shown in Figure 9.7.

In this ring, $x^3 y = y^3 + x$, so $\rho(x^3 y) = \rho(y^3 + x)$. The properties of a weight function require that $\rho(x^3 y) = \max[\rho(y^3), \rho(x)]$. Thus

$$3\rho(x) + \rho(y) = \max[3\rho(y), \rho(x)]$$
$$= 3\rho(y),$$

because $\rho(x)$ and $\rho(y)$ are both nonnegative. Thus, one concludes that $3\rho(x) = 2\rho(y)$ so that $\rho(x) = 2$ and $\rho(y) = 3$. This implies that the weight function on monomials must be defined so that $\rho(x^{i'} y^{i''}) = 2i' + 3i''$. Finally, we can write $\rho(x^3 + y^2) + \rho(y) = \rho(x^3 y + y^3) = \rho(x)$. Thus $\rho(x^3 + y^2) = -1$. But a weight cannot be negative, so we conclude that a weight function does not exist.

To continue this example, we will find the subring $R' \subset F[x, y]/\langle G(x, y)\rangle$ on which a weight function can be defined. We do this by eliminating certain troublesome monomials. The requirement that $\rho(x) = 2$ and $\rho(y) = 3$ implies the following assignment of weights to monomials:

$$
\begin{array}{cccccccccccccc}
\varphi_i = & 1 & x & y & x^2 & xy & y^2 & x^3 & x^2 y & xy^2 & y^3 & x^4 & x^2 y^2 & \cdots \\
\rho(\varphi_i) = & 0 & 2 & 3 & 4 & 5 & 6 & 6 & 7 & 8 & 9 & 8 & 10 & \cdots
\end{array}
$$

y^6	xy^6	$x^2 y^6$	—	—	—	—
y^5	xy^5	$x^2 y^5$	—	—	—	—
y^4	xy^4	$x^2 y^4$	—	—	—	—
y^3	xy^3	$x^2 y^3$	—	—	—	—
y^2	xy^2	$x^2 y^2$	—	—	—	—
y	xy	$x^2 y$	—	—	—	—
1	x	x^2	x^3	x^4	x^5	x^6

Figure 9.7. Monomial basis for $F[x, y]/\langle x^3 y + y^3 + x\rangle$.

y^6	xy^6	x^2y^6	–	–
y^5	xy^5	x^2y^5	–	–
y^4	xy^4	x^2y^4	–	–
y^3	xy^3	x^2y^3	–	–
y^2	xy^2	x^2y^2	–	–
y	xy	x^2y	–	–
1	–	–	–	–

Figure 9.8. New monomial basis for $F[x,y]/\langle x^3y + y^3 + x \rangle$.

The monomial x^3y does not appear because $x^3y = y^3 + x$. By inspection, we see once again that this choice of $\rho(\varphi_i)$ is not a weight function, because it does not assign unique weights to the monomials. It is possible, however, to salvage something that does have a weight function. By eliminating all powers of x except x^0, this becomes

$$\begin{array}{rccccccccccc}
\varphi_i & = 1 & y & xy & y^2 & x^2y & xy^2 & y^3 & x^2y^2 & xy^3 & \cdots \\
\rho(\varphi_i) & = 0 & 3 & 5 & 6 & 7 & 8 & 9 & 10 & 11 & \cdots
\end{array}$$

Although the monomials x and x^2 have unique weights, they were eliminated as well to ensure that the new set of monomials is closed under multiplication of monomials. Now the set of monomials has a total order defined by the polynomial weights. The new monomial basis may be easier to see when it is arranged as in the two-dimensional array in Figure (9.8).

The ring R' generated by these monomials is a subring of R. The set of monomials can also be regarded as a basis for a vector space, so R' is also a vector space. Within the ring R', multiplication is reduced modulo the Klein polynomial, just as for the ring $GF(8)[x,y]/\langle x^3y + y^3 + x \rangle$. We end this exercise with the conclusion that the ring R' does have a weight function even though the larger ring does not.

9.7   Semigroups and the Feng–Rao distance

The set of integer values taken by a weight function has a simple algebraic structure. This structure has a formal name.

Definition 9.7.1 *A semigroup S is a set with an associative operation, denoted $+$, such that $x + y \in S$ whenever $x, y \in S$.*

We will consider only semigroups with an identity element, denoted 0. A semigroup with an identity element is also called a *monoid*. Note that the definition of a semigroup does not require that the addition operation has an inverse. Thus a semigroup is a weaker structure than a group. We will be interested only in those semigroups that are subsets of the natural integers N, with integer addition as the semigroup operation. Given the semigroup $S \subset N$, with the elements listed in their natural order as integers, we shall denote the rth element of the semigroup by ρ_r. Any set of integers generates such a semigroup by taking all possible sums. The smallest set of integers that generates a semigroup of integers is called the set of *generators* for that semigroup.

Definition 9.7.2   *The gaps of the integer semigroup S are the elements of S^c.*

The set of gaps of S is called the *gap sequence*. The elements of the integer semigroup S are called *nongaps*.

For example, the semigroup generated by 3, 5, and 7 is $\{\rho_1, \rho_2, \ldots\} = \{0, 3, 5, 6, 7, 8, 9, \ldots\}$. There are three gaps in this semigroup. The gap sequence is $\{1, 2, 4\}$. The semigroup generated by 4 and 5 is $\{\rho_1, \rho_2, \ldots\} = \{0, 4, 5, 8, 9, 10, 12, 13, \ldots\}$. There are six gaps in the semigroup generated by 4 and 5. The gap sequence is $\{1, 2, 3, 6, 7, 11\}$. A semigroup of this kind, formed by at least two coprime integers, always has only a finite number of gaps.

The reason we have introduced the notion of a semigroup is because the set of weights defined by a weight function is always a semigroup. If the ring $F[x, y]/\langle G(x, y)\rangle$ has a weight function, then the set of weights forms a semigroup. The number of gaps of the semigroup is equal to the number of missing weights in the weight function for $F[x, y]/\langle G(x, y)\rangle$. More formally, in algebraic geometry, gaps of a semigroup that arise in this way with reference to a polynomial $G(x, y)$ are called *Weierstrass gaps*. It is well known in algebraic geometry that if $F[x, y]/\langle G(x, y)\rangle$ has a weight function, then the number of Weierstrass gaps of the resulting integer semigroup is equal to the genus g of $G(x, y)$. For our purpose, we could take this to be the definition of the genus of a polynomial whenever it can be applied. If this method of finding the genus and the Plücker formula both apply, then they will agree. Both methods apply if, on the one hand, the quotient ring $F[x, y]/\langle G(x, y)\rangle$ has a weight function, and, on the other hand, if the polynomial $G(x, y)$ is nonsingular.

Lemma 9.7.3   *Let $s \neq 0$ be any element of the semigroup $S \subset N$. Then*

$$\|S - (s + S)\| = s.$$

Proof: By definition, $s + S = \{s + a \mid a \in S\}$. Because S is a semigroup, $s + S \subset S$. Therefore in the set subtraction of the lemma, each element of $s + S$ deletes one element of S. If the last gap of S is at integer t, then the last gap of $s + S$ is at integer $t + s$. Set S has $t + s - g$ elements smaller than $t + s$. Set $s + S$ has $t - g$ elements smaller than $t + s$. Because $s + S \subset S$, we conclude that

$$\|S - (s + S)\| = (t + s - g) - (t - g) = s,$$

as was to be proved. ∎

By using the lemma on the weight function of a ring, the following theorem is obtained.

Theorem 9.7.4   *If the ring $R \subset F[x, y]$ has the weight function ρ and $f(x, y) \in R$, then*

$$\dim[R/\langle f(x, y)\rangle] = \rho(f).$$

Proof: Every monomial has a unique weight, and these weights form a semigroup. Let wt $f(x, y) = s$. Then the weights of the ideal $\langle f \rangle$ are the elements of the set $s + S$. The elements of $R/\langle f \rangle$ have weights in the set $S - \langle s + S \rangle$. The lemma says that the cardinality of this set is s.   ∎

Every element ρ_{r+1} of the semigroup $S \subset N$ is an integer. Most elements of S can be written as the sum of two smaller elements of S. Thus for most elements, ρ_{r+1}, we can write

$$\rho_{r+1} = \rho_i + \rho_j.$$

In fact, most elements of S can be written as the sum of two smaller elements in more than one way. Define

$$\mathcal{N}_r = \{(i, j) \mid \rho_{r+1} = \rho_i + \rho_j\}.$$

We will be interested in $\|\mathcal{N}_r\|$, the cardinality of the set $\mathcal{N}$. One way to compute $\mathcal{N}_r$ is to form an array with $\rho_i + \rho_j$ listed in location ij. For our example generated by 3, 5, and 7, this array has the form shown in Figure (9.9).

Then $\|\mathcal{N}_r\|$ is equal to the number of times that ρ_{r+1} appears in this array.

It may be easier to understand this array if a space is inserted for each of the g gaps. Then the array is as shown in Figure (9.10).

With these spaces inserted, the back-diagonal connecting any integer in the first row with that same integer in the first column crosses every appearance of this same integer

	ρ_1	ρ_2	ρ_3	ρ_4	ρ_5	ρ_6	ρ_7	ρ_8	ρ_9	$\cdots$
ρ_1	0	3	5	6	7	8	9	10	11	$\cdots$
ρ_2	3	6	8	9	10	11	12	13	14	
ρ_3	5	8	10	11	12	13	14	15		
ρ_4	6	9	11	12	13	14	15			
ρ_5	7	10	12	13	14	15				
ρ_6	8	11	13	14	15					
ρ_7	9	12	14	15						
ρ_8	10	13	15							
ρ_9	11	14								
$\vdots$	$\vdots$									

Figure 9.9. Array of semigroup elements.

	ρ_1	–	–	ρ_2		ρ_3	ρ_4	ρ_5	ρ_6	ρ_7	ρ_8	ρ_9	$\cdots$
ρ_1	0	–	–	3	–	5	6	7	8	9	10	11	$\cdots$
–	–	–	–	–	–	–	–	–	–	–	–	–	
–	–	–	–	–	–	–	–	–	–	–	–	–	
ρ_2	3	–	–	6	–	8	9	10	11	12	13	14	
–	–	–	–	–	–	–	–	–	–	–	–	–	
ρ_3	5	–	–	8	–	10	11	12	13	14	15		
ρ_4	6	–	–	9	–	11	12	13	14	15			
ρ_5	7	–	–	10	–	12	13	14	15				
ρ_6	8	–	–	11	–	13	14	15					
ρ_7	9	–	–	12	–	14	15						
ρ_8	10	–	–	13	–	15							
ρ_9	11	–	–	14	–								
$\vdots$	$\vdots$												

Figure 9.10. Array of semigroup elements augmented with gaps.

and no other integer. Now it is easy to see that $\|\mathcal{N}_r\| = r + 1 + g - f(r)$, where $f(r)$ is a function of r that is eventually equal to $2g$. The term $f(r)$ can be understood by observing that each back-diagonal can cross each gap value at most twice, once for that gap in the horizontal direction and once for that gap in the vertical direction. Eventually, a back-diagonal crosses each gap exactly twice. The remaining term $r + 1 + g$ can be seen, because ρ_r is the counting sequence with the g gaps deleted.

Definition 9.7.5   *The Feng–Rao distance profile is given by*

$$d_{\mathrm{FR}}(r) = \min_{s \ge r} \|\mathcal{N}_s\|,$$

where $\|\mathcal{N}_r\|$ is the cardinality of the set

$$\mathcal{N}_r = \{(i,j) \mid \rho_{r+1} = \rho_i + \rho_j\}.$$

To compute $\|\mathcal{N}_r\|$, we must examine ρ_{r+1}.

As an example, we will compute $d_{\mathrm{FR}}(r)$ for the integer semigroup generated by 3, 5, and 7. This is $\rho_i = 0, 3, 5, 6, 7, 8, 9, 10, 11, \ldots$ If $r = 1$, then $\rho_2 = 3$, which (because $\rho_1 = 0$) can be written as either $\rho_1 + \rho_2 = \rho_2$, or $\rho_2 + \rho_1 = \rho_2$. Thus $\mathcal{N}_1 = \{(1,2), (2,1)\}$ and $\|\mathcal{N}_1\| = 2$. If $r = 2$, then $\rho_3 = 5$, and $\mathcal{N}_2 = \{(1,3),(3,1)\}$ and $\|\mathcal{N}_2\| = 2$. If $r = 3$, then $\rho_4 = 6$, which can be written in three ways: as $\rho_1 + \rho_4 = \rho_4$, as $\rho_4 + \rho_1 = \rho_4$, or as $\rho_2 + \rho_2 = \rho_4$. Thus $\mathcal{N}_3 = \{(1,4),(2,2),(4,1)\}$, and $\|\mathcal{N}_3\| = 3$. It is not true that $\|\mathcal{N}_r\|$ is nondecreasing, as can be seen by noting that $\mathcal{N}_4 = \{(1,5),(5,1)\}$, so $\|\mathcal{N}_4\| = 2$. Continuing in this way, we obtain in the following sequence:

$$\|\mathcal{N}_r\| = 2, 2, 3, 2, 4, 4, 5, 6, 7, \ldots$$

The gap sequence for this semigroup is $\{1, 2, 4\}$. Because the number of gaps is finite, eventually the sequence $\|\mathcal{N}_r\|$ will become simply the counting sequence.[4]

The Feng–Rao distance profile $d_{FR}(r)$ is obtained by taking the minimum of all terms of the sequence that do not precede the rth term. Because eventually the sequence is monotonically increasing, this minimum is straightforward to evaluate. The sequence $d_{FR}(r)$ for $r = 1, 2, 3, \ldots$ is given by

$$d_{FR}(r) = \min_{s \geq r} \|\mathcal{N}_s\|$$

$$= 2, 2, 2, 2, 4, 4, 5, 6, 7, \ldots$$

As a second example, we compute $d_{FR}(r)$ for the integer semigroup generated by 4 and 5. This semigroup corresponds to the hermitian polynomial $x^5 + y^4 + y$. The sequence of integers forming the integer semigroup is given by

$$\rho_r = 0, 4, 5, 8, 9, 10, 12, 13, 14, 15, 16, 17, 18, 19, 20, 21, \ldots,$$

and the gap sequence is $\{1, 2, 3, 6, 7, 11\}$. Then

$$\|\mathcal{N}_r\| = 2, 2, 3, 4, 3, 4, 6, 6, 4, 5, 8, 9, 8, 9, 10, \ldots$$

and

$$d_{FR}(r) = 2, 2, 3, 3, 3, 4, 4, 4, 4, 5, 8, 8, 8, 9, 10, \ldots$$

Eventually, this becomes the counting sequence.

The Feng–Rao distance profile cannot be expressed by a simple formula, so the following weaker, but simpler, distance profile is often useful. In many applications, it is entirely adequate.

Definition 9.7.6   *The Goppa distance profile for an integer semigroup with g gaps is given by*

$$d_\Gamma(r) = r + 1 - g.$$

This definition applies, indirectly, to the polynomial $G(x, y)$ only if $G(x, y)$ has a weight function. Alternatively, we can define the Goppa distance profile directly in terms of the polynomial $G(x, y)$.

Definition 9.7.7   *The Goppa distance profile for the polynomial $G(x, y)$ with genus g is*

$$d_\Gamma(r) = r + 1 - g.$$

---

[4] In passing, we point out the curiosity that $\|\mathcal{N}_r\|$ is essentially the autocorrelation function of the indicator function of the gap sequence.

Whenever the ring $F[x,y]/\langle G(x,y)\rangle$ has a weight function, three are two definitions of the Goppa distance profile for the polynomial $G(x,y)$ and these are consistent. This is because the number of gaps g in the weight sequence of polynomial $G(x,y)$ is equal to the genus g of $G(x,y)$, whenever $G(x,y)$ has a weight function.

For example, for a polynomial of genus 3 with weight function generated by 3, 5, and 7, the number of gaps is three. Then, for $r = 1, 2, 3, \ldots$,

$$d_\Gamma(r) = -1, 0, 1, 2, 3, 4, 5, 6, 7, \ldots,$$

as compared with the Feng–Rao distance profile

$$d_{FR}(r) = 2, 2, 2, 2, 4, 4, 5, 6, 7, \ldots$$

Note that the two sequences eventually agree.

This example illustrates the following theorem.

Theorem 9.7.8

$$d_\Gamma(r) \le d_{FR}(r)$$

with equality if r is larger than the largest gap.

Proof: The Goppa distance profile is defined as $d_\Gamma(r) = r + 1 - g$. As we have seen, eventually the Feng–Rao distance profile $d_{FR}(r)$ is the counting sequence with g terms deleted. ∎

The following theorem, stated only for a semigroup with two generators a' and a'', gives an alternative and simple graphical method of computing the Feng–Rao distance profile. Let

$$\mathcal{L}'_r = \{(r', r'') \mid a'r' + a''r'' = \rho_{r+1}\}.$$

Let $\mathcal{L}_r = \text{hull}\{(r', r'') \in \mathcal{L}'_r \mid r' < a'\}$, where "hull" denotes the cascade hull of the indicated set.

Theorem 9.7.9 *The Feng–Rao distance profile can be stated as follows:*

$$d_{FR}(r) = \min_{s \ge r} \|\mathcal{L}_s\|.$$

Proof: Every ρ_ℓ is a linear combination of a' and a'', so ρ_{r+1} can be decomposed as the following:

$$\rho_{r+1} = \rho_\ell + \rho_m$$
$$= \ell'a' + \ell''a'' + m'a' + m''a''$$
$$= (\ell' + m')a' + (\ell'' + m'')a''.$$

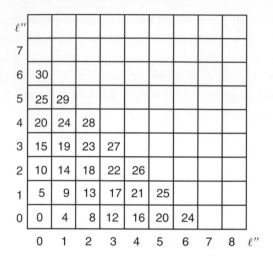

Figure 9.11. One construction of a Feng–Rao distance profile.

Every way in which (r', r'') can be decomposed as $(\ell' + m', \ell'' + m'')$ yields one way of writing ρ_{r+1} as the sum of ρ_ℓ and ρ_m. Because we must not count the same decomposition twice, r' is restricted to be less than a'. ∎

A graphical method, illustrated in Figure 9.7, of computing the Feng–Rao distance profile is developed with the aid of Theorem 9.7.9 for the case in which the semigroup has two generators. In this example, the generators are 4 and 5. Each square in Figure 9.7 is given a score, consisting of four counts for each step on the horizontal axis and five counts for each step on the vertical axis. The vertical axis is restricted to be less than 4, which ensures that no score appears twice in the array. Each individual square and the square with zero score define a rectangle. The number of unit squares in the rectangle is its area, which is $\|\mathcal{N}_r\|$. As the individual squares are visited in order as determined by the scores, the sequence of areas give the Feng–Rao distance profile $\|\mathcal{N}_r\|$.

Definition 9.7.10 *The hyperbolic distance profile for a semigroup sequence with two generators a' and a'' is given by*

$$\|\mathcal{H}_r\| = (r' + 1)(r'' + 1);$$
$$d_{\mathrm{H}}(r) = \min_{s \geq r} \|\mathcal{H}_s\|,$$

where r' and r'' are the integers satisfying

$$\rho_{r+1} = a'r' + a''r''.$$

For an example of the hyperbolic distance profile, consider the semigroup generated by the integers 4 and 5. Then

$$\rho_r = 0, 4, 5, 8, 9, 10, 12, 13, 14, 15, 16, 17, 18, 19, 20, 21, 22, 23, 24, 25, \ldots,$$

$$\|\mathcal{N}_r\| = 2, 2, 3, 4, 3, 4, 6, 6, 4, 5, 8, 9, 8, 9, 10, 12, 12, 13, 14, 15, \ldots,$$

$$\|\mathcal{H}_r\| = 2, 2, 3, 4, 3, 4, 6, 6, 4, 5, 8, 9, 8, 6, 10, 12, 12, 7, 12, \ldots$$

Therefore

$$d_{FR}(r) = 2, 2, 3, 3, 3, 4, 4, 4, 4, 5, 8, 8, 8, 9, 10, 12, 12, 13, 14, 15, 16, \ldots,$$

$$d_H(r) = 2, 2, 3, 3, 3, 4, 4, 4, 4, 5, 6, 6, 6, 6, 7, 7, 7, 7, \ldots,$$

$$d_\Gamma(r) = -, -, -, -, 0, 1, 2, 3, 4, 5, 6, 7, 8, 9, 10, 11, 12, 13, 14, 15, 16, \ldots.$$

For this example, $d_{FR}(r)$ is at least as large as $d_H(r)$. In fact, the maximum of $d_H(r)$ and $d_\Gamma(r)$ gives a good underbound to $d_{FR}(r)$.

9.8   Bounds on the weights of vectors on curves

We now return to our continuing task of giving bounds on the weights of vectors on lines, planes, and curves, completing the task here by giving bounds on the weights of vectors on curves.

Recall that a one-dimensional vector of blocklength n over the field F is given by

$$v = [v_i \mid i = 0, \ldots, n - 1].$$

In Section 1.8, we developed bounds on the weight of the one-dimensional vector v in terms of the pattern of zeros in its one-dimensional Fourier transform V. Recall further that a two-dimensional n by n array over the field F is given by

$$v = [v_{i'i''} \mid i' = 0, \ldots, n - 1; \ \ i'' = 0, \ldots, n - 1].$$

In Section 5.7, we developed bounds on the weight of the two-dimensional array v in terms of the two-dimensional pattern of zeros in its two-dimensional Fourier transform V. Now we will develop bounds on the weights of vectors defined on plane curves, regarding such a curve as embedded in a two-dimensional array.

The vector v of blocklength n over F on the curve $\mathcal{X}$ is given by

$$v = [v_\ell \mid P_\ell \in \mathcal{X}, \ell = 0, \ldots, n - 1],$$

where the curve is the set of points

$$\mathcal{X} = \{P_\ell = (\beta_\ell, \gamma_\ell) \mid G(\beta_\ell, \gamma_\ell) = 0\}$$

and $G(x, y)$ is the irreducible bivariate polynomial defining the curve $\mathcal{X}$. The vector v will also be referred to as $v(\mathcal{X})$. In this section, we shall develop bounds on the weight of the vector $v(\mathcal{X})$ on the curve $\mathcal{X}$ in terms of the two-dimensional pattern of zeros in its two-dimensional Fourier transform V. For simplicity, we will usually require that $G(x, y)$ be a nonsingular, irreducible bivariate polynomial.

We shall give two bounds: the Goppa bound and the Feng–Rao bound. In general, the Goppa bound is weaker than the Feng–Rao bound, although in most situations the two are equivalent. The Goppa bound, which was discovered earlier, is often preferred because its much simpler form makes it more convenient to use, and it is usually just as good. The proof of the Feng–Rao bound uses rank arguments and gaussian elimination, and it requires that a weight function exists for $G(x, y)$. Our indirect proof of the Goppa bound, by reference to the Feng–Rao bound, does not apply to those $G(x, y)$ for which a weight function does not exist. Then a more direct proof must be used. We will not provide this direct proof for such $G(x, y)$, leaving the proof as an exercise. After we give the proof of the Feng–Rao bound, we will give yet another bound as an alternative that is proved by using the Sakata–Massey theorem to bound the area of a connection footprint.

Goppa bound   Let $\mathcal{X}$ be a curve of length n, defined by a regular polynomial of genus g. The only vector $v(\mathcal{X})$ of weight $d_\Gamma(r) - 1$, or less, on the curve $\mathcal{X}$, that has two-dimensional Fourier transform components $V_{j'j''}$ equal to zero for $j' + j'' \leq r$, is the all-zero vector.

Proof: Provided a weight function exists for $G(x, y)$, the Goppa bound is a special case of the Feng–Rao bound. Then the Goppa bound can be inferred from the Feng–Rao bound, which is given next, using the properties of weight functions. We do not provide a proof of the Goppa bound for curves without a weight function.   ∎

The proof of the Feng–Rao bound requires careful attention to indexing, because sometimes we index components by i and sometimes by the weight ρ_i. Because a weight is unique, specifying the weight of a component designates a unique component. This is a form of indirect indexing. Let V_{ρ_i} denote the component of V corresponding to a monomial with weight ρ_i. Spectral component V_{ρ_i} is defined as a Fourier transform component expressed as follows:

$$V_{\rho_i} = \sum_{\ell=1}^{n} v_\ell \varphi_{\rho_i}(P_\ell).$$

To understand the meaning of this expression, it may be useful to first write

$$V_{j'j''} = \sum_{i'=0}^{n-1}\sum_{i''=0}^{n-1} v_{i'i''}\omega^{i'j'}\omega^{i''j''}$$

$$= \sum_{i'=0}^{n-1}\sum_{i''=0}^{n-1} v_{i'i''}x^{j'}y^{j''} \, |_{x=\omega^{i'},y=\omega^{i''}} \; .$$

The term $v_{i'i''}$ is zero for all (i',i'') that are not on the curve $\mathcal{X}$. Now replace the bi-index (i',i'') by ℓ, which indexes only the points of $\mathcal{X}$, and sum only over such points. Replace the bi-index (j',j'') by ρ_i, which indexes the sequence of monomials $\varphi_{\rho_i}(x,y)$. The expression of the Fourier transform becomes

$$V_{\rho_i} = \sum_{\ell} v_\ell \varphi_{\rho_i}(x,y) \, |_{(x,y)=P_\ell}$$

$$= \sum_{\ell=1}^{n} v_\ell \varphi_{\rho_i}(P_\ell),$$

as was stated earlier.

We will now introduce a new array, W, with elements

$$W_{ij} = V_{\rho_i+\rho_j}.$$

Again, we caution against possible confusion regarding indices. The indices $\imath$ and $\jmath$ in this matrix are indices in the total order, which we will distinguish by font in the remainder of this section. Thus the pair $(\imath, \jmath)$ refers to a pair of points in the total order. Each individual index in the total order itself corresponds to a pair of componentwise indices. Thus the term $V_{\rho_\imath}$, written in terms of its componentwise indices, is $V_{j'(\imath),j''(\imath)}$. The index $\imath$ can be thought of as an indirect address pointing to the bi-index (j',j''), pointing to $V_{j'j''}$ in its defining two-dimensional array. Thus the pair $(\imath, \jmath)$ can be interpreted as $((j'(\imath),j''(\imath)), (j'(\jmath),j''(\jmath)))$.

To understand the matrix W, it is best to inspect an example. The example we give is the semigroup generated by 3, 5, and 7. This semigroup is $\{0, 3, 5, 6, 7, 8, 9, \ldots\}$. The matrix W, written explicitly, is as follows:

$$W = \begin{bmatrix} V_0 & V_3 & V_5 & V_6 & V_7 & V_8 & V_9 & \cdots \\ V_3 & V_6 & V_8 & V_9 & V_{10} & V_{11} & V_{12} & \cdots \\ V_5 & V_8 & V_{10} & V_{11} & V_{12} & V_{13} & V_{14} & \cdots \\ V_6 & V_9 & V_{11} & V_{12} & V_{13} & V_{14} & V_{15} & \cdots \\ V_7 & V_{10} & V_{12} & V_{13} & V_{14} & V_{15} & V_{16} & \cdots \\ V_8 & V_{11} & V_{13} & V_{14} & V_{15} & V_{16} & V_{17} & \cdots \\ \vdots & \vdots & & & \vdots & \vdots & & \end{bmatrix}.$$

The indices in the first row run through all elements of the semigroup in their natural order, as do the indices in the first column. The indices of other elements of W are found by adding the indices of the elements in the corresponding first row and first column. If false columns and false rows, corresponding to the gaps of the semigroup, are inserted into the matrix with false entries represented by hyphens, it is easy to fill in the matrix. In row j are the components of V indexed by the semigroup, starting with ρ_j, and skipping an element at each gap,

$$
W = \begin{bmatrix}
V_0 & - & - & V_3 & - & V_5 & V_6 & V_7 & V_8 & V_9 & \cdots \\
- & - & - & - & - & - & - & - & - & - \\
- & - & - & - & - & - & - & - & - & - \\
V_3 & - & - & V_6 & - & V_8 & V_9 & V_{10} & V_{11} & V_{12} \\
- & - & - & - & - & - & - & - & - & - \\
V_5 & - & - & V_8 & - & V_{10} & V_{11} & V_{12} & V_{13} & V_{14} \\
V_6 & - & - & V_9 & - & V_{11} & V_{12} & V_{13} & V_{14} & V_{15} \\
V_7 & - & - & V_{10} & - & V_{12} & V_{13} & V_{14} & V_{15} & V_{16} \\
V_8 & - & - & V_{11} & - & V_{13} & V_{14} & V_{15} & V_{16} & V_{17} & \cdots \\
\vdots & & & & & & & & \vdots &
\end{bmatrix}.
$$

The false columns and false rows make it easy to see the pattern by which the elements are arranged in W. The actual matrix W, however, does not have these false columns and false rows.

We are now ready to derive the Feng–Rao bound in terms of the matrix W. The Feng–Rao distance profile $d_{\mathrm{FR}}(r)$ was defined in Definition 9.7.5.

Feng–Rao bound Suppose that the bivariate polynomial $G(x, y)$ has a weight function. The only vector v of length n on the curve $G(x, y) = 0$ having weight $d_{\mathrm{FR}}(r) - 1$ or less, whose two-dimensional Fourier transform components are equal to zero for all indices smaller than $r + 1$ in the weighted order, is the all-zero vector.

Proof: Because the weight of the vector v is nonzero, there must be a smallest integer, r, such that $V_{r+1} \neq 0$. We will prove the bound for that value of r. Because the Feng–Rao distance $d_{\mathrm{FR}}(r)$ is nondecreasing, the statement is also true for smaller r.

Consider the bivariate spectral components arranged in the weighted order

$$
V_{\rho_0}, V_{\rho_1}, V_{\rho_2}, V_{\rho_3}, \ldots, V_{\rho_i}, \ldots, V_{\rho_r}, V_{\rho_{r+1}}.
$$

Recall that the matrix W is defined with the elements

$$
W_{ij} = V_{\rho_i + \rho_j}.
$$

The proof consists of proving two expressions for the rank of W; one is that

$$\text{wt } v = \text{rank } W,$$

and the other is that

$$\text{rank } W \geq \|\mathcal{N}_r\|,$$

where $\|\mathcal{N}_r\|$ is defined in Definition 9.7.5 as the cardinality of the set $\{(\imath, \jmath) \mid \rho_{r+1} = \rho_\imath + \rho_\jmath\}$. Putting these last two expressions together proves the theorem.

To relate the rank of W to the weight of v, write

$$W_{\imath\jmath} = \sum_{\ell=1}^{n} v_\ell \varphi_{\rho_\imath + \rho_\jmath}(P_\ell) = \sum_{\ell=1}^{n} v_\ell \varphi_{\rho_\imath}(P_\ell) \varphi_{\rho_\jmath}(P_\ell).$$

We will write W as a matrix factorization. Let V be the diagonal matrix given by

$$V = \begin{bmatrix} v_1 & 0 & \cdots & 0 \\ 0 & v_2 & & 0 \\ \vdots & & \ddots & \\ 0 & & & v_n \end{bmatrix},$$

and let M be the matrix

$$M = [M_{\imath\ell}] = [\varphi_{\rho_\imath}(P_\ell)].$$

The factorization of W can be written

$$[W_{\imath\jmath}] = [\varphi_{\rho_\imath}(P_\ell)] \begin{bmatrix} v_1 & 0 & \cdots & 0 \\ 0 & v_2 & & 0 \\ \vdots & & \ddots & \\ 0 & & & v_n \end{bmatrix} [\varphi_{\rho_\jmath}(P_\ell)]^{\mathrm{T}},$$

and now has the form

$$W = MVM^{\mathrm{T}}.$$

The two outer matrices on the right side, M and M^{T}, have full rank. Hence the rank of W is equal to the number of nonzero elements on the diagonal of the diagonal matrix. Consequently,

$$\text{wt } v = \text{rank } W,$$

so the weight of v is equal to the rank of W.

It remains to bound the rank of W in terms of the Feng–Rao distance. The array W has the form

$$W = \begin{bmatrix} V_{\rho_0} & V_{\rho_1} & V_{\rho_2} & V_{\rho_3} & V_{\rho_4} & V_{\rho_5} & \cdots \\ V_{\rho_1} & V_{\rho_1+\rho_1} & V_{\rho_1+\rho_2} & V_{\rho_1+\rho_3} & V_{\rho_1+\rho_4} & V_{\rho_1+\rho_5} & \\ V_{\rho_2} & V_{\rho_2+\rho_1} & V_{\rho_2+\rho_2} & V_{\rho_2+\rho_3} & & & \\ V_{\rho_3} & V_{\rho_3+\rho_1} & V_{\rho_3+\rho_2} & & & & \\ \vdots & & & & & & \end{bmatrix}.$$

How many times does V_{ρ_k} appear in W? Each distinct V_{ρ_i} corresponds to a unique monomial, and each monomial has a unique weight. The number of times that $V_{\rho_{r+1}}$ appears is the same as the number of times that $\rho_{r+1} = \rho_i + \rho_j$. This means that $V_{\rho_{r+1}}$ appears $\|\mathcal{N}_r\|$ times in the array W. The indices are elements of a semigroup.

To find the rank of W, observe that it has the form

$$W = \begin{bmatrix} 0 & & & & & & * \\ & & & & & * & \\ & & & & * & & \\ & & & * & & & \\ & & * & & & & \\ & * & & & & & \\ * & & & & & & \end{bmatrix},$$

where each element denoted by an asterisk is in a different row and a different column. All elements denoted by an asterisk are equal, and each is equal to $V_{\rho_{r+1}}$. We have chosen r such that $V_{\rho_{r+1}}$ is the first nonzero term in the sequence.

Each element of W above or to the left of an appearance of an asterisk is zero. Therefore the number of linearly independent columns of W is at least as large as the number of times $V_{\rho_{r+1}}$ appears in W. Thus we have

$$\text{rank } W \geq \|\mathcal{N}_r\|.$$

Now we collect all the pieces,

$$\text{wt } v = \text{rank } W \geq \|\mathcal{N}_r\| \geq \min_{s \geq r} \|\mathcal{N}_s\| = d_{\text{FR}}(r),$$

to conclude the proof of the Feng–Rao bound. ∎

This proof of the Feng–Rao bound is somewhat subtle, but the method is direct, and, despite the indirect indexing, is essentially straightforward. An alternative proof, of a slightly different statement, that uses the Sakata–Massey theorem is much different and much simpler, but it does require that the Sakata–Massey theorem has been proved previously. This alternative proof uses the Sakata–Massey theorem to bound the area of the footprint of the locator ideal.

The Sakata–Massey theorem says that if $\Lambda(x,y)$ has bidegree s and produces $V_0, V_1, \ldots, V_{r-1}$, but not $V_0, V_1, \ldots, V_{r-1}, V_r$, then the footprint contains the point $r - s$. But the trivial polynomial $\Lambda(x,y) = 1$, for which $s = 0$, produces the sequence $V_0, \ldots, V_{r-1}$ if all terms of this sequence are zero, but $\Lambda(x,y)$ fails to produce the next term V_r if V_r is nonzero. Therefore the footprint of the locator ideal for the sequence $0, 0, \ldots, 0, V_r$ contains the point $r = (r', r'')$. This means that the area of the footprint is at least $(r' + 1)(r'' + 1)$. Thus

$$\text{wt } \boldsymbol{v} \geq (r' + 1)(r'' + 1)$$

if V_r is the first nonzero spectral component. This argument has shown that the weight of $\boldsymbol{v}$ is bounded by the hyperbolic distance profile, which is not as strong as the Feng–Rao distance profile.

As an example, consider the hermitian polynomial $x^5 + y^4 + y$. The weight function for this polynomial is given by

$$\rho(x^{i'} y^{i''}) = 4i' + 5i''.$$

The basis monomials are

y^3	xy^3	x^2y^3	x^3y^3	x^4y^3	x^5y^3	
y^2	xy^2	x^2y^2	x^3y^2	x^4y^2	x^5y^2	
y	xy	x^2y	x^3y	x^4y	x^5y	$\cdots$
1	x	x^2	x^3	x^4	x^5	

The weighted order based on this weight function assigns the following order to the monomials

9	13	17					
5	8	12	16				
2	4	7	11	15			
0	1	3	6	10	14	18	$\cdots$

The Sakata–Massey theorem says that if V_r is the first nonzero term in the sequence $V_0, V_1, V_2, \ldots, V_j, \ldots$, then the footprint Δ_r contains the square marked r. Because the footprint must be a cascade set, it contains the rectangle defined by the square marked 0 and the square marked r. It is easy to write down the sequence consisting of the areas of the locator footprints Δ_r. Thus

$$\|\Delta_r\| = 1, 2, 2, 3, 4, 3, 4, 6, 6, 4, 5, 8, 9, 8, 6, 10, \ldots$$

Now recall that the Feng–Rao distance profile can be written as follows:

$$d_{FR}(r) = \min_{s \geq r} \|\Delta_s\|$$

$$= 1, 2, 2, 3, 3, 3, 4, 4, 4, 4, 5, 6, 6, 6, 6, \ldots$$

We conclude, as before, that if V_r is the first nonzero spectral component in the ordered sequence, then wt $v \geq d_{FR}(r)$.

Problems

9.1   In the plane Q^2, how many points are on the circle $x^2 + y^2 = 1$?

9.2   Let $p(x, y, z) = x^3 + y^3 + z^3$. How many points does the polynomial $p(x, y, z)$ have in the projective plane over $GF(2)$, over $GF(4)$, and over $GF(8)$?

9.3   Show that the Klein polynomial is absolutely irreducible.

9.4   Let $v = (x, y, z)$ be a vector of elements of $GF(q^2)$, where q is a power of 2, and let $v^\dagger = (x^q, y^q, z^q)$ be the vector of q-ary conjugates of the elements of v. Show that the hermitian polynomial can be written $v \cdot v^\dagger = 0$. Use this representation to derive the Stichtenoth form of the polynomial.

9.5   The polynomial $p(x, y) = x^5 + y^2 + y + 1$ over $GF(2)$ has genus 2. Show that this polynomial has $2^{12} + 2^8 + 1$ zeros over $GF(2^{12})$. Compare this with the Hasse–Weil bound. Show that $p(x, y)$ has $2^8 - 2^6 + 1$ zeros over $GF(2^8)$. Compare this with the Hasse–Weil bound.

9.6   Prove that the semigroup S, formed under integer addition by two coprime integers, has at most a finite number of gaps.

9.7   Suppose the semigroup S is generated by integers b and $b + 1$ under integer addition.

(a) Prove that the number g of gaps is $\binom{b}{2}$.

(b) Prove that the largest gap is $2g - 1$.

9.8   Prove the following. A polynomial of the form

$$G(x, y) = x^a + y^b + g(x, y),$$

for which a and b are coprime and $a > b > \deg g(x, y)$, is regular at every affine point. A polynomial of this form has exactly one point at infinity, which is a regular point if and only if $a = b + 1$.

9.9   Is the polynomial $p(x, y) = x^2 + y^2$ irreducible over $GF(3)$? Is it absolutely irreducible? Are these statements true over all finite fields?

9.10 Prove the following. A polynomial of the form

$$G(x, y) = x^a + y^b + g(x, y),$$

where $a = b + 1$ and $b > \deg g(x, y)$, is absolutely irreducible.

9.11 Prove that a weight function does not exist for the Klein polynomial.

9.12 Prove that a weight function for a ring, if it exists, assigns a unique weight to each monomial of that ring and so orders the monomial.

9.13 Does a transformation of coordinates exist that will represent the Klein curve with only one point at infinity?

9.14 (a) Show that the curve defined by the hermitian polynomial intersects any straight line tangent to the curve only once. What is the multiplicity of this intersection? (A code defined on such a curve is sometimes called a one-point code.)

(b) Show that the curve defined by the Klein quartic polynomial intersects any straight line tangent to the curve at a second point.

9.15 Prove that the projective curve defined by

$$x^m y + y^m z + z^m x = 0$$

has no singular points over $GF(q)$ if $\mathrm{GCD}(m^2 - m + 1, q) = 1$.

9.16 An *elliptic polynomial* over the field F is a polynomial of the form

$$y^2 + a_1 xy + a_3 y = x^3 + a_2 x^2 + a_4 x + a_6,$$

where the coefficients are the elements of F. Suppose that $F = GF(q)$. Estimate how many points in the bicyclic plane over $GF(q)$ are zeros of this polynomial. The set of rational points of an elliptic polynomial is called an *elliptic curve*.

9.17 The polynomial

$$p(x, y) = x^{10} + y^8 + x^3 + y$$

has genus 14. Show that it has 65 (rational) zeros in the projective plane over $GF(8)$, only one of them at infinity. (This is first of a series of polynomials associated with the *Suzuki group*.)

9.18 Let $\{\rho_r\}$ be an integer semigroup generated by a finite set of integers, and let

$$\mathcal{N}_r = \{(i, j) \mid \rho_i + \rho_j = \rho_{r+1}\}.$$

Define the indicator function of the gap sequence as a sequence of zeros and ones according to the presence or absence of a gap. Let $\phi(r)$ be the autocorrelation function of the indicator function. Describe $\|\mathcal{N}_r\|$ in terms of $\phi(r)$ for the integer semigroup generated by the integers 3, 5, and 7. Prove the appropriate statement for the general case.

9.19 (a) Show that the reduced basis for the ideal in the ring $GF(8)[x, y]/\langle x^7 - 1, y^7 - 1\rangle$, generated by $x^3y + x^3 + y$, is given by

$$\{x^3 + x^2 + x + y + 1, x^2y + x^2 + y^2 + xy + x + y + 1, y^3 + x^2 + 1\}.$$

(b) What is the area of the footprint of this ideal?

(c) How many bicyclic zeros does $x^3y + x^3 + y$ have?

9.20 Let $F°[x, y] = GF(2)[x, y]/\langle x^{15} - 1, y^{15} - 1\rangle$. Find a reduced basis for the ideal $F°[x, y]/\langle x^{17} + y^{16} + y\rangle$. Sketch the footprint for this ideal. What is the area of the footprint?

9.21 Prove that the graphical construction of the Feng–Rao distance is valid. Can this graphical construction be generalized to a semigroup with three generators?

9.22 Prove the Goppa bound.

Notes

I have tried to emphasize a parallel structure in the treatment of the bounds on the weights of vectors in Chapter 1, bounds on the weights of arrays in Chapter 4, and bounds on the weights of vectors on curves in Chapter 9. In each of these chapters, the bounds are presented as bounds on individual vectors. Then, to define linear codes, the bounds are applied to sets of vectors in Chapter 2, Chapter 5, and Chapter 10.

The Feng–Rao bound is implicit in the work of Feng and Rao (1994). It was made explicit by Pellikaan (1992), who first realized that the Feng–Rao distance is sometimes larger than the Goppa distance. Consequently, the Feng–Rao bound is stronger than the Goppa bound. The role of semigroups in the analysis of codes on curves was recognized by Garcia, Kim, and Lax (1993). The practice of writing the matrices with blanks at the gap locations apparently first appeared in Feng, Wei, Rao, and Tzeng (1994). This practice makes the structure of the matrix more apparent. The roles of the weight function and the order function were developed by Høholdt, van Lint, and Pellikaan (1998).

The dissertation by Dabiri (1996) discussed monomial bases of quotient rings. The discussion of monomial bases of quotient rings is closely related to the celebrated Riemann–Roch theorem of algebraic geometry. The Hasse–Weil bound is discussed in Stichtenoth (1993). These theorems are examples of the powerful methods of algebraic geometry that were put to good use in the development of codes on curves. The genus and gonality are properties of a polynomial whose full definition requires more algebraic geometry than this book has required. The Drinfeld–Vlădut bound (Drinfeld and

Vlădut, 1993) is an example of a contribution to algebraic geometry that came from the study of the theory of codes.

The alternative form of the hermitian polynomials was suggested by Stichtenoth (1988) as a way to move some points of the hermitian curve from the line at infinity into the affine plane. His form of the hermitian curve has only one point at infinity, and it is the preferred form for hermitian codes.

10 Codes on Curves and Surfaces

Codes on curves, along with their decoding algorithms, have been developed in recent years by using rather advanced topics of mathematics from the subject of algebraic geometry, which is a difficult and specialized branch of mathematics. The applications discussed in this book may be one of the few times that the somewhat inaccessible topics of algebraic geometry, such as the Riemann–Roch theorem, have entered the engineering literature. With the benefit of hindsight, we shall describe the codes in a more elementary way, without much algebraic geometry, emphasizing connections with bicyclic codes and the two-dimensional Fourier transform.

We shall discuss the hermitian codes as our primary example and the Klein codes as our secondary example. The class of hermitian codes, in its fullest form, is probably large enough to satisfy whatever needs may arise in communication systems of the near future. Moreover, this class of codes can be used to illustrate general methods that apply to other classes of codes. The Klein codes comprise a small class of codes over $GF(8)$ with a rather rich and interesting structure, though probably not of practical interest.

An hermitian code is usually defined on a projective plane curve or on an affine plane curve. These choices for the definition are most analogous to the definitions of a doubly extended or singly extended Reed–Solomon code. In studying some code properties – especially in connection with decoding – it is also natural, however, to use the term "hermitian code" to refer to the code defined on the bicyclic plane. This consists of all points of the affine plane that do not have a zero in either coordinate. The distinction is also significant in the discussion of dual codes. Accordingly, we shall define "codes on projective curves," "codes on affine curves," and "codes on epicyclic curves." These are analogous to Reed–Solomon codes on the projective line (doubly extended Reed–Solomon codes), Reed–Solomon codes on the affine line (singly extended Reed–Solomon codes), and Reed–Solomon codes on the cyclic line (cyclic Reed–Solomon codes).

In this chapter, we shall define our codes from the point of view of the encoder. More specifically, the codes on curves studied in this chapter are obtained by puncturing codes on the plane. This leaves us with an issue later when we deal with the task of decoding, because decoders are immediately suitable for shortened codes, not for punctured codes. Fortunately, for many codes of interest, a punctured code can also be viewed as the shortened form of a different code. Then the encoder can treat the code as a punctured code, while the decoder can treat it as a shortened code. This comment will become more fully explained in Chapter 11 when we discuss shortened codes.

10.1 Beyond Reed–Solomon codes

The Reed–Solomon codes (and other BCH codes) are very successful in practical applications, and they continue to satisfy most needs for designers of digital communication systems and digital storage systems. However, in the field $GF(2^m)$, a cyclic Reed–Solomon code cannot have a blocklength larger than $2^m - 1$ (or $2^m + 1$ on the projective line). For example, over $GF(256)$, the longest *cyclic* Reed–Solomon code has blocklength $n = 255$. The longest *projective* Reed–Solomon code has blocklength $n = 257$.

The Reed–Solomon code chosen for the *Voyager* and *Galileo* spacecraft is the $(255, 223, 33)$ cyclic code over $GF(256)$. We will compare the hermitian codes to this code. For the purpose of our comparison, it is convenient to replace this code with the $(256, 224, 33)$ affine Reed–Solomon code over $GF(256)$. This code can correct sixteen symbol errors; each code symbol is an eight-bit byte.

In practice, very long records of data must be transmitted, much longer than 224 bytes. To use the $(256, 224, 33)$ code, these data records are broken into blocks of 224 bytes each, and each data block is individually encoded into a 256-byte codeword. The codewords are then transmitted sequentially, possibly interleaved, to form a long codestream. The codewords are uncoupled, and the structure of one codeword provides no help in correcting errors in a different codeword.

Consider a message of 3584, eight-bit bytes broken into sixteen blocks of 224 bytes each, and encoded into sixteen extended Reed–Solomon codewords of 256 bytes each. Concatenating or interleaving these sixteen codewords to form a single codeword of 4096 bytes gives a $(4096, 3584, 33)$ code. The minimum distance is still 33 because, given two distinct codewords, the underlying Reed–Solomon codewords might be identical in fifteen out of the sixteen blocks. Thus the $(4096, 3584, 33)$ code is only guaranteed to correct sixteen byte errors. There will be patterns of seventeen errors – all occurring in the same Reed–Solomon codeword – that cannot be corrected. However, the code will correct other patterns with more than sixteen errors, provided that not more than sixteen errors occur in any single codeword. It may even be able to correct 256 error bytes, but only if they are properly distributed, sixteen byte errors to every 256-byte codeword. Of course, random errors need not be so cooperative. If seventeen byte errors fall into the same 256-byte codeword, that error pattern cannot be corrected.

Other ways to obtain a code over $GF(256)$ with length larger than $q + 1$ are to use a larger locator field or to use a larger symbol field. If the larger symbol field is equal to the larger locator field, then the code is still a Reed–Solomon code. For example, one may use a Reed–Solomon code over $GF(2^{12})$ instead of a Reed–Solomon code over $GF(2^8)$. This means that the arithmetic of the encoder and decoder will be in the larger field, and so may be more expensive or slower. There might also be the minor inconvenience of transcribing the code alphabet to the channel alphabet. Finally, the twelve-bit error symbols might not be well matched to a given channel error mechanism. For example, errors may arise on eight-bit boundaries because of the structure of a communication system.

If only the locator field is enlarged, but not the symbol field, the appropriate code is a BCH code in $GF(q)$. A BCH code may be unsatisfactory, because the number of check symbols may be excessive. For example, a BCH code of blocklength $q^2 - 1$ over $GF(q)$ typically requires about four check symbols per error to be corrected. A Reed–Solomon code of blocklength $q - 1$ over $GF(q)$ requires only two check symbols per error to be corrected.

Yet another approach, and the main subject of this chapter, is to use the points of an hermitian curve to index the components of the codeword.

The hermitian codes that we will describe are more attractive than the other approaches, as judged by the error correction. We will describe a $(4096, 3586)$ hermitian code that has a minimum distance not less than 391. This code can correct 195 byte errors no matter how they are distributed.

The hermitian code can also be compared to a shortened BCH code. A summary comparison is as follows.

(1) A $(4096, 3584, 33)$ interleaved Reed–Solomon code over $GF(256)$ that corrects any pattern of sixteen byte errors, a small fraction of patterns of 256 byte errors, and a number of intermediate cases, including all burst errors of length 256 bytes.
(2) A $(4096, 3586, 391)$ hermitian code over $GF(256)$ that corrects any pattern of 195 byte errors.
(3) A $(4096, 3585, 264)$ shortened BCH code over $GF(256)$ that corrects any pattern of 131 byte errors.
(4) A $(4096, 3596, 501)$ Reed–Solomon code over $GF(4096)$ that corrects any pattern of 250 symbol errors in a codeword of twelve-bit symbols.

The code in example (1) will correct any pattern of 16 byte errors as contrasted with the hermitian code which will correct any pattern of 195 byte errors. However, this may be an over-simplified comparison. Both codes can correct many error patterns beyond the packing radius, though it can be difficult to compare two codes with respect to this property. In particular, the interleaved Reed–Solomon code can correct burst errors or length 256 bytes. Thus it may be preferred in an application that makes primarily burst errors. The code of example (2) is clearly better than the code of example (3), which is shortened from a $(4369, 3858, 264)$ BCH code. Moreover, the decoder for the BCH code requires computations in $GF(2^{16})$, while the decoder for the hermitian code requires computations only in $GF(2^8)$. A more detailed comparison of the codes in examples (1) and (2) requires a specification of the statistics of the error model of a particular application, and also a detailed specification of the decoders behavior for patterns of error beyond the packing radius. We will not provide such a comparison.

10.2   Epicyclic codes

The affine plane over $GF(q)$ consists of those points of the projective plane over $GF(q)$ whose z coordinate is not equal to zero. The bicyclic plane over $GF(q)$ consists of those points of the affine plane over $GF(q)$ whose x and y coordinates are both not equal to zero. Recall that a bicyclic code is defined in Section 6.1 as the set of two-dimensional arrays c with elements indexed by the points of the bicyclic plane, and whose bispectra satisfy

$$C_{j'j''} = 0 \text{ for } (j', j'') \in \mathcal{A},$$

where the defining set $\mathcal{A}$ is a subset of $\{0, \ldots, q-2\}^2$. No compelling rule is known for choosing the defining set $\mathcal{A}$ to obtain exceptional bicyclic codes. Accordingly we will describe some bicyclic codes that can be punctured or shortened to give noteworthy codes.

The notions of puncturing and shortening were introduced in Section 2.1. Choose any subset $\mathcal{B}$ of the bicyclic plane $\{0, \ldots, q-2\}^2$ with points indexed as $\{i', i''\}$. Let $P_0, P_1, P_2, \ldots, P_{n-1}$, indexed by ℓ, denote the n elements of $\mathcal{B}$. Each such P_ℓ is a point of the bicyclic plane. The point P_ℓ can be written as $(\alpha^{i'}, \alpha^{i''})$ or[1] $(\omega^{-i'}, \omega^{-i''})$, and sometimes we may refer to P_ℓ as $P_{(i', i'')}$, and to ℓ as (i', i''), for $(i', i'') \in \mathcal{B}$. To *puncture* the two-dimensional code $\mathcal{C}$, delete all components of the code not indexed by the elements of $\mathcal{B}$. Then the *punctured code*, denoted $\mathcal{C}(\mathcal{B})$, and whose codewords are denoted $c(\mathcal{B})$, has blocklength n equal to $\|\mathcal{B}\|$. The codeword components are given by

$$c_\ell = C(P_\ell) \quad \ell = 0, \ldots, n-1,$$

where

$$C(x, y) = \sum_{j'=0}^{N-1} \sum_{j''=0}^{N-1} C_{j'j''} x^{j'} y^{j''}$$

is a bispectrum polynomial, satisfying $C_{j'j''} = 0$ if (j', j'') is an element of the two-dimensional defining set $\mathcal{A}$. The punctured code $\mathcal{C}(\mathcal{B})$ consists of codewords $c(\mathcal{B})$ that

---

[1] The negative signs can be used when we want the polynomial evaluation to have the form of the inverse two-dimensional Fourier transform.

are obtained by discarding all components of c that are indexed by elements of $\mathcal{B}^c$, the complement of set $\mathcal{B}$. When there is no possibility of confusion with the underlying two-dimensional code $\mathcal{C}$, we may refer to $\mathcal{C}(\mathcal{B})$ simply as $\mathcal{C}$ and to codeword $c(\mathcal{B})$ simply as c.

In a variation of this construction, we may instead *shorten* the two-dimensional code $\mathcal{C}$. The shortened code $\mathcal{C}'(\mathcal{B})$, whose codewords are denoted $c'(\mathcal{B})$, consists of only the codewords of $\mathcal{C}$ for which every component indexed by an element of $\mathcal{B}^c$ is equal to zero, such components are then deleted. Thus a codeword of the subcode is any array c that satisfies the following two constraints:

$$C_{j'j''} = 0 \text{ for } (j',j'') \in \mathcal{A};$$
$$c_{i'i''} = 0 \text{ for } (i',i'') \in \mathcal{B}^c,$$

where both $\mathcal{A}$ and $\mathcal{B}$ are subsets of $\{0, \ldots, q-2\}^2$. The components of the codewords that are not indexed by elements of $\mathcal{B}$ are dropped from the codewords to form the codewords $c(\mathcal{B})$ of blocklength n'.

Thus, from the set of bispectrum polynomials $C(x, y)$ satisfying $C_{j'j''} = 0$ if $(j',j'') \in \mathcal{A}$, we form the punctured code,

$$\mathcal{C}(\mathcal{B}) = \{c(\mathcal{B}) = (c_0, \ldots, c_{n-1}) \mid c_\ell = C(P_\ell) \text{ for } \ell \in \mathcal{B}; C_{j'j''} = 0 \text{ for } (j',j'') \in \mathcal{A}\},$$

and the shortened code,

$$\mathcal{C}'(\mathcal{B}) = \{(c'(\mathcal{B}) = (c_0, \ldots, c_{n-1}) \mid c \in \mathcal{C}(\mathcal{B}) \text{ and } c_{i'i''} = 0 \text{ for } (i',i'') \in \mathcal{B}^c\},$$

as two alternatives. For a fixed $\mathcal{C}$, the punctured code has more codewords than the shortened code. This apparent disadvantage of shortening, however, is not real. Because a shortened code, in general, has a smaller dimension but a larger minimum distance than the corresponding punctured code, the apparent disadvantage goes away if one chooses a different code to shorten than the code that was chosen to puncture. This will be more evident in Chapter 11. Because the dropped components of a shortened code are always equal to zero, it is trivial for the receiver to reinsert those zeros to recover a noisy codeword of the original bicyclic code in the form of a $q - 1$ by $q - 1$ array. The dropped components of a noisy punctured codeword, however, are not as easy to recover. These components must be inferred from the received noisy components, a potentially difficult task. Thus we see that we prefer to encode a punctured code and we prefer to decode a shortened code. It is not trivial to reconcile this conflict.

It remains to specify the sets $\mathcal{A}$ and $\mathcal{B}$ so that the punctured code, or the shortened code, has good properties, and this is far from a simple task. Algebraic geometry now enters as part of the definition of the set $\mathcal{B}$, which is defined as the curve $\mathcal{X}$ in $GF(q)^2$. This curve $\mathcal{X}$ is the set of zeros of the bivariate polynomial $G(x, y)$ and the set $\mathcal{B}$

is chosen to be the curve $\mathcal{X}$. For this reason, we refer to such codes as "codes on curves."

In this section, codes on curves are restricted to only those points of the curve that lie in the bicyclic plane. Because the bicyclic plane over a finite field can be regarded as a torus, a code on the bicyclic plane can be regarded as defined on a curve on a torus. Certain automorphisms of the code are then seen in terms of corresponding translations on the torus. Bicyclic shifts on the bicyclic plane, or torus, that leave the curve invariant, map codewords to codewords. This is a consequence of the two-dimensional convolution theorem, which says that bicyclic translations of codewords correspond to multiplying components of the bispectrum by powers of ω, thereby leaving the defining set unaffected. We refer to codes with translation invariants on the bicyclic plane as *epicyclic codes*. Epicyclic codes are not themselves cyclic.

The underlying bicyclic code over $GF(q)$ is the $q-1$ by $q-1$ two-dimensional code whose defining set is $\mathcal{A} = \{(j',j'') \mid j' + j'' > J\}$ or its complement $\mathcal{A}^c = \{(j',j'') \mid j' + j'' \leq J\}$. In this chapter, we will puncture the bicyclic code with defining set $\mathcal{A}$. In the next chapter, we will shorten the bicyclic code with $\mathcal{A}^c$ as the defining set. The reason for choosing $\mathcal{A}$ as the defining set of the punctured code and $\mathcal{A}^c$ as the defining set of the shortened code is to respect the dual relationship between the two types of codes, to anticipate the role of the Feng–Rao bound, and to facilitate the following discussion of the dual relationship. However, either defining set could be complemented, and we sometimes do so to get an equivalent code.

We shall see in this chapter that the punctured code on a curve defined by a polynomial of degree m has dimension and designed distance satisfying

$$k = mJ - g + 1 \quad d = n - mJ.$$

We shall see in Chapter 11 that the shortened code on a curve defined by a polynomial of degree m has dimension and minimum distance satisfying

$$k = n - mJ + g - 1 \quad d = mJ - 2g + 2,$$

where n is the common blocklength of the two codes.

Although the performance formulas seem very different, they are actually equivalent. To see this, let $mJ' = n - mJ + 2g - 2$, and consider the performance of the shortened code with mJ' in place of mJ. Then

$$k = n - mJ + g - 1$$
$$= n - (n - mJ + 2g - 2) + g - 1$$
$$= mJ - g + 1.$$

Furthermore,

$$d = mJ' - 2g + 2$$
$$= (n - mJ + 2g - 2) - 2g + 2$$
$$= n - mJ.$$

Thus the punctured code with design parameter mJ has the same performance parameters as the shortened code with design parameter $n - mJ + 2g - 2$.

The punctured code is defined as follows. Let $G(x, y)$ be a nonsingular bivariate polynomial of degree m with coefficients in the field $GF(q)$. Let $P_0, P_1, P_2, \ldots, P_{n-1}$ be the rational bicyclic points of $G(x, y)$. These are the zeros of $G(x, y)$ in the bicyclic plane over the base field $GF(q)$.

For the defining set, let $\mathcal{A} = \{(j', j'') \mid j' + j'' > J\}$. Every bispectrum polynomial $C(x, y)$ has the coefficient $C_{j'j''} = 0$ if $j' + j'' > J$. Let $\mathcal{S}_J$ denote the set of bispectrum polynomials that consists of the zero polynomial and all bivariate polynomials of degree at most J and with coefficients in $GF(q)$. Thus

$$\mathcal{S}_J = \{C(x, y) \mid \deg C(x, y) \le J\} \cup \{0\}.$$

The epicyclic code $\mathcal{C}(\mathcal{X})$, lying on the curve $\mathcal{X}$ in the bicyclic plane, is the punctured code defined as follows:

$$\mathcal{C}(\mathcal{X}) = \{c(\mathcal{X}) \mid c_\ell = C(P_\ell) \text{ for } \ell = 0, \ldots, n - 1; \ C(x, y) \in \mathcal{S}_J\}.$$

The number of codewords in $\mathcal{C}(\mathcal{X})$ need not be the same as the number of polynomials in $\mathcal{S}_J$, because the same codeword might be generated by several polynomials in $\mathcal{S}_J$. Indeed, if $J \ge r$, then $G(x, y)$ itself will be in $\mathcal{S}_J$, and it will map into the all-zero codeword as will any polynomial multiple of $G(x, y)$.

It is evident that this construction gives a linear code. In Section 10.3, we show that the linear code has q^k codewords, where the dimension k is given by

$$k = \begin{cases} \dfrac{1}{2}(J + 1)(J + 2) & \text{if } J < m \\ \dfrac{1}{2}(J + 1)(J + 2) - \dfrac{1}{2}(J - m + 1)(J - m + 2) & \text{if } J \le m, \end{cases}$$

and with $d_{\min} \ge n - mJ$.

We can identify the coefficients of the bispectrum polynomial $C(x, y)$ with the components of the two-dimensional Fourier transform C of the codeword c. Thus the bispectrum C of codeword c is given by

$$C = \{C_{j'j''} \mid j' = 0, \ldots, q - 2; \ j'' = 0, \ldots, q - 2\},$$

satisfying $C_{j'j''} = 0$ for $j' + j'' > J$.

Encoder

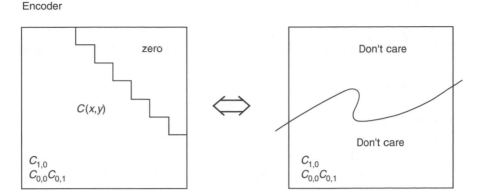

Figure 10.1. Computing a punctured codeword from its spectrum.

The relationship between the two-dimensional punctured codeword $c(\mathcal{X})$ and its two-dimensional spectrum $C(\mathcal{X})$, depicted symbolically in Figure 10.1, is immediately compatible with the notions of encoding. The left side depicts the coefficients of the bispectrum polynomial $C(x, y)$, arranged as an array $C = [C_{j'j''}]$. Because $C(x, y)$ has degree J, the array is such that $C_{j'j''} = 0$ if $j' + j'' > J$. The defining set $\mathcal{A}$ is labeled "zero." The doubly shafted arrow in Figure 10.1 denotes the inverse Fourier transform

$$c_{i'i''} = \frac{1}{n^2} \sum_{j'=0}^{n-1} \sum_{j''=0}^{n-1} C_{j'j''} \omega^{-i'j'} \omega^{-i''j''},$$

where, here, $n = q - 1$. The right side of Figure 10.1 depicts the codeword as a set of values lying along a planar curve. The curve denotes the set $\mathcal{B}$. The complement $\mathcal{B}^c$ is labeled "don't care." The codeword, when restricted to the bicyclic plane, is the vector that consists of those components of the two dimensional inverse Fourier transform lying on the curve, defined by $G(x, y) = 0$.

The convolution property of the Fourier transform shows that the bispectrum of a codeword retains zeros in the same components if the bicyclic plane, or torus, is cyclically shifted in the row direction or the column direction. This is because a bicyclic shift of the bicyclic plane corresponds to multiplication of the bispectral components by powers of ω, which means that a bispectral component that is zero remains zero. Thus a cyclic shift that preserves the curve also preserves the code.

The code definition is not immediately compatible with the notion of decoding, however. This is because the codeword bispectrum polynomial $C(x, y)$ is not immediately recoverable by computing the two-dimensional Fourier transform of the codeword. All values of $c_{i'i''} = C(\omega^{-i'}, \omega^{-i''})$, not on the curve, have been discarded. All that remains known of the structure is depicted in Figure 10.2. Those components of the array c that

Decoder

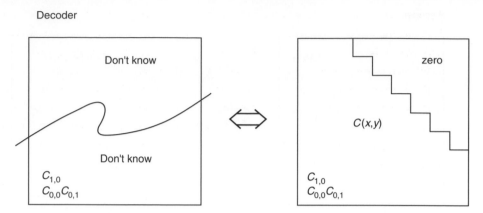

Figure 10.2. Computing a spectrum from its shortened codeword.

are not on the curve are unknown. The decoder must infer these unknown components to compute the Fourier transform, and it is far from obvious how this can be done.

The definition of the shortened form of the code reverses the situation. Now the definition is immediately compatible with the notions of decoding, because components of the array c, not on the curve, are known to be zero, but the definition is not compatible with the notions of encoding. Not every polynomial $C(x, y)$ can be used to encode a shortened code. Only those that evaluate to zero off the curve can be used. It is not immediately obvious how to select a bispectrum polynomial, $C(x, y)$, that will produce an array c with the required zero elements. Later, we will give a simple rectification of this difficulty for the case of bicyclic hermitian codes.

10.3   Codes on affine curves and projective curves

Let $G(x, y, z)$ be a regular, homogeneous, trivariate polynomial of degree m, with coefficients in the field $GF(q)$. Let $P_0, P_1, P_2, \ldots, P_{n-1}$ be the rational projective points of $G(x, y, z)$. These are the zeros of $G(x, y, z)$ in the projective plane of the base field $GF(q)$.

Let $\mathcal{S}_J$ denote the set that consists of the zero polynomial and all homogeneous trivariate polynomials $C(x, y, z)$ of degree at most J, and with coefficients in $GF(q)$. The punctured code $\mathcal{C}(\mathcal{X})$ in the projective plane is defined as follows:

$$\mathcal{C}(\mathcal{X}) = \{c(\mathcal{X}) \mid c_\ell = C(P_\ell) \text{ for } \ell = 0, \ldots, n-1; \ C(x, y, z) \in \mathcal{S}_J\}.$$

It is immediately evident that $\mathcal{C}(\mathcal{X})$ is a linear code, because the sum of two elements of $\mathcal{S}_J$ is an element of $\mathcal{S}_J$. Just as for codes on curves in the bicyclic plane, the number of codewords in $\mathcal{C}(\mathcal{X})$ need not be the same as the number of polynomials in $\mathcal{S}_J$, because the same codeword might be generated by several polynomials in $\mathcal{S}_J$. Possibly, if $J \geq m$, then $G(x, y, z)$ will be in $\mathcal{S}_J$, and it will map into the all-zero codeword as will any polynomial multiple of $G(x, y, z)$. Likewise, any two polynomials, whose difference is a multiple of $G(x, y, z)$, will map into the same codeword.

By working in the projective plane, one may obtain additional points of the curve at infinity, thereby increasing the blocklength of the code. Many popular curves have only a single point at infinity, which means that the blocklength of the projective code will be larger by one. This single additional component might not be considered worth the trouble of using projective coordinates. Possibly, a representation of a curve with a single point at infinity, if it exists, may be attractive precisely because the affine code is nearly as good as the projective code, and little is lost by choosing the convenience of the affine code.

The punctured code $\mathcal{C}(\mathcal{X})$ in the affine plane is defined in the same way. Let $G(x, y)$ be a nonsingular bivariate polynomial of degree m, with coefficients in the field $GF(q)$. Let $P_0, P_1, P_2, \ldots, P_{n-1}$ be the rational affine points of $G(x, y)$. These are the zeros of $G(x, y)$ in the affine plane over the base field $GF(q)$. Let $\mathcal{S}_J$ denote the set that consists of the zero polynomial and all bivariate polynomials $C(x, y)$ of degree at most J, and with coefficients in $GF(q)$. Then

$$\mathcal{C}(\mathcal{X}) = \{c(\mathcal{X}) \mid c_\ell = C(P_\ell) \text{ for } \ell = 0, \ldots, n - 1; \quad C(x, y) \in \mathcal{S}_J\}.$$

The code in the affine plane is the same as the code in the projective plane but with all points at infinity deleted. The code in the bicyclic plane is the same as the code in the affine plane but with all points with a zero coordinate deleted.

A lower bound on the minimum distance of code $\mathcal{C}$ in the affine plane or the bicyclic plane can be computed easily by using Bézout's theorem in the affine plane. The identical proof can be given in the projective plane by using Bézout's theorem in the projective plane.

Theorem 10.3.1   *The minimum distance of the code $\mathcal{C}$ on the smooth plane curve $\mathcal{X}$ satisfies*

$$d_{\min} \geq n - mJ,$$

where m is the degree of the polynomial defining $\mathcal{X}$.

Proof: Because $G(x, y)$ was chosen to be irreducible, $C(x, y)$ and $G(x, y)$ can have no common factor unless $C(x, y)$ is a multiple of $G(x, y)$. If $C(x, y)$ is a multiple of $G(x, y)$, it maps to the all-zero codeword. Therefore, by Bézout's theorem, either $C(x, y)$ maps to the all-zero codeword, or $C(x, y)$ has at most mJ zeros in common with $G(x, y)$ in the base field $GF(q)$. This means that the codeword has at least $n - mJ$ nonzero components.   ∎

Henceforth we shall assume that $J < n/m$. Otherwise, the bound of the theorem would be uninformative.

Next, we will determine the dimension k of the code $\mathcal{C}$. First, consider the dimension of the space $\mathcal{S}_J$. This is the number of different terms $x^{j'} y^{j''} z^{j'''}$, where $j' + j'' + j''' = J$. To count the number of such terms, write a string of j' zeros followed by a one, then a string of j'' zeros followed by a one, then a string of j''' zeros. This is a binary number of length $J + 2$ with J zeros and two ones. The number of such binary numbers is equal to the number of monomials of the required form. Thus

$$\dim \mathcal{S}_J = \binom{J+2}{2} = \frac{1}{2}(J+2)(J+1).$$

The code $\mathcal{C}$ is obtained by a linear map from $\mathcal{S}_J$ onto the space of vectors on n points. Therefore

$$k = \dim \mathcal{C} = \dim \mathcal{S}_J - \dim(\text{null space}).$$

If $J < m$, then no polynomial in $\mathcal{S}_J$ is a multiple of $G(x, y, z)$, so the dimension of the null space is zero. If $J \geq m$, then the null space is the space of all homogeneous polynomials of the form

$$C(x, y, z) = G(x, y, z)A(x, y, z),$$

where $A(x, y, z)$ is a homogeneous polynomial of degree $J - m$. Hence, reasoning as before, the null space has dimension $\binom{J-m+2}{2}$. We conclude that

$$k = \begin{cases} \dfrac{1}{2}(J+1)(J+2) & \text{if } J < m \\ \dfrac{1}{2}(J+1)(J+2) - \dfrac{1}{2}(J-m+1)(J-m+2) & \text{if } J \geq m. \end{cases}$$

The second case can be multiplied out as follows:

$$\frac{1}{2}(J+1)(J+2) - \frac{1}{2}(J-m+1)(J-m+2) = mJ - \frac{1}{2}(m-1)(m-2) + 1$$
$$= mJ - g + 1,$$

where $g = \binom{m-1}{2}$ is the genus of the polynomial $G(x, y, z)$. This is summarized in the following corollary, which applies equally to codes on the bicyclic, affine, or projective plane.

Corollary 10.3.2 *A code of blocklength n on a smooth plane curve of degree m has parameters satisfying the following conditions.*

(1) *If J < m:*

$$k = \frac{1}{2}(J + 2)(J + 1) \quad d_{\min} \geq n - mJ.$$

(2) *If J ≥ m:*

$$k = mJ - g + 1 \quad d_{\min} \geq n - k - g + 1.$$

If the code is punctured by dropping those rational points of $G(x, y, z)$ that have $z = 0$, then we need only deal with the affine points $(x, y, 1)$. Then we can think of S_J as containing all polynomials $C(x, y) = C(x, y, 1)$ whose degree is at most J. If the code is further punctured by dropping these rational points of $G(x, y)$, with x or y equal to zero, then the epicyclic form of the code is obtained. Evaluating $C(x, y)$ at those rational points of $G(x, y)$, with both x and y nonzero, is the same as computing the inverse Fourier transform

$$c_{i'i''} = \frac{1}{n^2} \sum_{i'=0}^{n-1} \sum_{i''=0}^{n-1} \omega^{-i'j'} \omega^{-i''j''} C_{j'j''}$$

and keeping $c_{i'i''}$ only if $(\omega^{-i'}, \omega^{-i''})$ is a zero of $G(x, y) = G(x, y, 1)$. These n values of $c_{i'i''}$ form the codeword.

As a simple example of these ideas, we will describe an unconventional construction of the doubly extended Reed–Solomon code, this time as a code in the projective plane. The polynomial of degree m equal to one given by

$$G(x, y, z) = x + y + z$$

has genus g equal to zero over any finite field $GF(q)$. It has $n - q + 1$ rational points, namely $(-1, 1, 0)$ and $(\alpha, -1 - \alpha, 1)$. We can choose any $J < n/m = q + 1$. Because $k = J + 1$, any $k \leq q + 1$ is possible, and

$$d_{\min} \geq n - k + 1.$$

Using the Singleton bound, we conclude that

$$d_{\min} = n - k + 1,$$

so this is a maximum-distance code. This amounts to yet another description of the doubly extended Reed–Solomon codes over $GF(q)$, this time as codes on a diagonal line in the projective plane over $GF(q)$.

10.4   Projective hermitian codes

Codes defined on hermitian curves, either punctured codes or shortened codes, are called *hermitian codes*. We shall examine those codes obtained by puncturing to the projective hermitian curve. The *Fermat version* of the homogeneous hermitian polynomial of degree $q + 1$,

$$G(x, y, z) = x^{q+1} + y^{q+1} - z^{q+1},$$

has genus $g = (1/2)q(q-1)$ and has $q^3 + 1$ zeros in the projective plane over the field $GF(q^2)$. The *Stichtenoth version* of the homogeneous hermitian polynomial of degree $q + 1$ over $GF(q^2)$ is

$$G(x, y, z) = x^{q+1} - y^q z - y z^q.$$

It also has genus $g = (1/2)q(q-1)$ and $q^3 + 1$ zeros in the projective plane over the field $GF(q^2)$. The two polynomials will form equivalent codes.

We can choose any integer, $J < n/m$, where $m = q + 1$ is the degree of $G(x, y, z)$. Because $n = q^3 + 1$, this becomes $J < q^2 - q + 1$. Then, by Corollary 10.3.2, for $J < q + 1$ the codes have performance described by

$$n = q^3 + 1;$$

$$k = \frac{1}{2}(J + 2)(J + 1);$$

$$d_{\min} \geq n - (q + 1)J.$$

For $J \geq q + 1$, the codes have performance described by

$$n = q^3 + 1;$$

$$k = (q + 1)J - g + 1;$$

$$d_{\min} \geq n - k - g + 1.$$

We will calculate these performance parameters for the fields $GF(4)$, $GF(16)$, $GF(64)$, and $GF(256)$.

For the field $GF(4)$, $q = 2$ and $g = 1$. Thus $m = 3$ and $J \leq 2$. Because J cannot be larger than 2 in this field, projective hermitian codes over $GF(4)$ can only have $J = 1$ or 2, and performance parameters given by

$$n = 9; \quad k = \frac{1}{2}(J + 2)(J + 1); \quad d_{\min} \geq 9 - 3J.$$

Thus there are only two codes: the $(9, 3, 6)$ code and the $(9, 6, 3)$ code over $GF(4)$, respectively.

For the field $GF(16)$, $q = 4$ and $g = 6$. Thus, $m = 5$ and $J \leq 12$. For $J = 1, \ldots, 4$, the performance parameters of the projective hermitian codes over $GF(16)$ are given by

$$n = 65; \quad k = \frac{1}{2}(J + 2)(J + 1); \quad d_{min} \geq 65 - 5J,$$

while for $J = 5, \ldots, 12$, the performance parameters are given by

$$n = 65; \quad k = 5J - 5; \quad d_{min} \geq 65 - 5J.$$

Thus, these hermitian codes in the projective plane over $GF(16)$, for $J = 1, \ldots, 4, 5, 6, \ldots, 11, 12$, have performance parameters given by $(65, 3, 60)$, $(65, 6, 55)$, $(65, 10, 50)$, $(65, 15, 45)$, $(65, 20, 40)$, $(65, 25, 35)$, $(65, 30, 30)$, $(65, 35, 25)$, $(65, 40, 20)$, $(65, 45, 15)$, $(65, 50, 10)$, and $(65, 55, 5)$.

For the field $GF(64)$, $q = 8$ and $g = 28$. Thus $m = 9$ and $J \leq 56$. For $J = 1, \ldots, 8$, the performance parameters are given by

$$n = 513; \quad k = \frac{1}{2}(J + 2)(J + 1); \quad d_{min} \geq 513 - 9J.$$

For $J = 9, 10, \ldots, 56$, the performance parameters of the projective hermitian codes over $GF(64)$ are given by

$$n = 513; \quad k = 9J - 27; \quad d_{min} \geq 513 - 9J.$$

Thus, these hermitian codes in the projective plane over $GF(64)$ have performance parameters, for $J = 1, \ldots, 8, 9, 10, 11, \ldots, 55, 56$, given by $(513, 3, 504)$, $(513, 6, 495)$, $\ldots$, $(513, 45, 441)$, $(513, 54, 432)$, $(513, 63, 423)$, $\ldots$, $(513, 477, 9)$.

For the field $GF(256)$, $q = 16$ and $g = 120$. Thus, $m = 17$ and $J \leq 240$. For $J = 1, \ldots, 16$, the performance parameters are given by

$$n = 4097; \quad k = \frac{1}{2}(J + 2)(J + 1); \quad d_{min} \geq 4097 - 17J,$$

while for $J = 17, \ldots, 240$, the performance parameters of the projective hermitian codes over $GF(256)$ are given by

$$n = 4097; \quad k = 17J - 119; \quad d_{min} \geq 4097 - 17J.$$

Thus, these hermitian codes in the projective plane over GF(256) have performance parameters, for $J = 1, \ldots, 16, 17, 18, 19, \ldots, 239, 240$, given by $(4097, 3, 4080)$, $(4097, 6, 4063)$, $\ldots$, $(4097, 153, 3825)$, $(4097, 170, 3808)$, $(4097, 187, 3791)$, $(4097, 204, 3774)$, $\ldots$, $(4097, 3944, 34)$, $(4097, 3961, 17)$.

10.5   Affine hermitian codes

An hermitian code can be further punctured to the affine plane. It is then called an *affine hermitian code*. An affine hermitian code is a code of a smaller block-length and with a simpler structure than a projective hermitian code. Accordingly, the encoders and decoders are simpler, both conceptually and in implementation. The parameters of the code depend on which form of the hermitian polynomial is used to define the curve. This is because the number of points of the curve that lie at infinity depend on which form of the hermitian polynomial is used. We will discuss the parameters of the affine hermitian codes, constructed first with the Fermat version of the hermitian polynomial, then with the Stichtenoth version of the hermitian polynomial.

When constructed from the Fermat version of the hermitian polynomial,

$$G(x, y) = x^{q+1} + y^{q+1} - 1,$$

the affine hermitian code has blocklength $n = q^3 - q$. Consequently, the blocklength and (the bound on) minimum distance of the shortened code are both reduced by $q + 1$ compared with the projective code. Therefore, the affine hermitian codes, constructed from the Fermat version of the hermitian polynomial, for $J < q+1$, have performance described by

$$n = q^3 - q;$$
$$k = \frac{1}{2}(J + 2)(J + 1);$$
$$d_{\min} \geq n - (q + 1)J,$$

while for $J \geq q + 1$, the codes have performance described by

$$n = q^3 - q;$$
$$k = (q + 1)J - g + 1;$$
$$d_{\min} \geq n - k - g + 1.$$

We will calculate these performance parameters for the fields $GF(4)$, $GF(16)$, $GF(64)$, and $GF(256)$.

For the field $GF(4)$, $q = 2$ and $g = 1$. The only hermitian codes are for $J = 1$ and 2. The only code worth mentioning is a $(7, 3, 4)$ code over $GF(4)$.

For the field $GF(16)$, $q = 4$ and $g = 6$. Thus $m = 5$ and $J \leq 11$. For $J = 1, \ldots, 4$, the performance parameters of the affine hermitian codes over $GF(16)$ based on the

Fermat version of the hermitian polynomial are given by

$$n = 60; \quad k = \frac{1}{2}(J+2)(J+1); \quad d_{\min} \geq 60 - 5J,$$

while for $J = 5, \ldots, 11$, the performance parameters of the codes are given by

$$n = 60; \quad k = 5J - 5; \quad d_{\min} \geq 60 - 5J.$$

Thus, these affine hermitian codes over $GF(16)$, for $J = 1, \ldots, 4, 5, 6, \ldots, 11$, have performance parameters given by $(60, 3, 55)$, $(60, 6, 50)$, $(60, 10, 45)$, $\ldots$, $(60, 30, 25)$, $(60, 35, 20)$, $\ldots$, and $(60, 50, 5)$.

For the field $GF(64)$, $q = 8$ and $g = 28$. Thus $m = 9$ and $J \leq 55$. For $J = 1, \ldots, 8$, the performance parameters of the affine hermitian codes over $GF(64)$ based on the Fermat version of the hermitian polynomial are given by

$$n = 504; \quad k = \frac{1}{2}(J+2)(J+1); \quad d_{\min} \geq 504 - 9J,$$

while for $J = 9, 10, \ldots, 55$, the performance parameters of the codes are

$$n = 504; \quad k = 9J - 27; \quad d_{\min} \geq 504 - 9J.$$

Thus, these affine hermitian codes over $GF(64)$, for $J = 1, \ldots, 8, 9, 10, \ldots, 55$, have performance parameters given by $(504, 3, 495)$, $(504, 6, 486)$, $\ldots$, $(504, 45, 432)$, $(504, 54, 443)$, $\ldots$, $(504, 468, 9)$.

For the field $GF(256)$, $q = 16$ and $g = 120$. Thus $m = 17$ and $J \leq 239$. For $J = 1, \ldots, 16$, the performance parameters of the affine hermitian codes over $GF(256)$ based on the Fermat version of the hermitian polynomial are given by

$$n = 4080; \quad k = \frac{1}{2}(J+2)(J+1); \quad d_{\min} \geq 4080 - 17J,$$

while for $J = 17, \ldots, 239$, the performance parameters of the codes are given by

$$n = 4080; \quad k = 17J - 119; \quad d_{\min} \geq 4080 - 17J.$$

Thus, these affine hermitian codes over $GF(256)$, for $J = 1, \ldots, 239$, have performance parameters given by $(4080, 3, 4063)$, $(4080, 6, 4046)$, $\ldots$, $(4080, 153, 3808)$, $(4080, 170, 3791)$, $\ldots$, $(4080, 3944, 17)$.

This completes our brief inventory of codes on the affine plane constructed from the Fermat version of the hermitian polynomial.

We now turn to the second variation on this topic. This is the topic of codes on the affine plane constructed from the Stichtenoth version of the hermitian polynomial. With the polynomial

$$G(x, y) = x^{q+1} + y^q + y,$$

the affine hermitian code has blocklength $n = q^3$. Consequently, the Stichtenoth version of the hermitian polynomial will produce codes of larger blocklength when evaluated in the affine plane. The affine hermitian codes, constructed with this polynomial, are nearly the same as the projective hermitian codes, except that the blocklength and the (bound on) minimum distance are both reduced by one.

For the field $GF(16)$, $q = 4$ and $g = 6$. Thus $m = 5$ and $J \leq 12$. For $J = 1, \ldots, 4$, the performance parameters of these affine hermitian codes over $GF(16)$ based on the Stichtenoth version of the hermitian polynomial are given by

$$n = 64; \quad k = \frac{1}{2}(J + 2)(J + 1); \quad d_{\min} \geq 64 - 5J,$$

while, for $J = 5, \ldots, 12$, the performance parameters of the codes are given by

$$n = 64; \quad k = 5J - 5; \quad d_{\min} \geq 64 - 5J.$$

Thus, these affine hermitian codes over $GF(16)$, for $J = 1, \ldots, 4, 5, 6, \ldots, 12$, have performance parameters given by $(64, 3, 59)$, $(64, 6, 54)$, $(64, 10, 49)$, $(64, 15, 44), \ldots, (64, 45, 14), (64, 50, 9), (64, 55, 4)$.

For the field $GF(64)$, $q = 8$ and $g = 28$. Thus $m = 9$ and $J \leq 56$. For $J = 1, \ldots, 8$, the performance parameters of these affine hermitian codes over $GF(64)$ based on the Stichtenoth version of the hermitian polynomial are given by

$$n = 512; \quad k = \frac{1}{2}(J + 2)(J + 1); \quad d_{\min} \geq 512 - 9J,$$

while, for $J = 9, 10, \ldots, 56$, the performance parameters of these affine hermitian codes are given by

$$n = 512; \quad k = 9J - 27; \quad d_{\min} \geq 512 - 9J.$$

Thus, these affine hermitian codes over $GF(64)$, for $J = 9, 10, \ldots, 56$, have performance parameters given by $(512, 3, 503)$, $(512, 6, 494)$, $\ldots$, $(512, 45, 440)$, $(512, 54, 431), (512, 63, 422), \ldots, (512, 477, 8)$.

For the field $GF(256)$, $q = 16$ and $g = 120$. Thus $m = 17$ and $J \leq 240$. For $J = 1, \ldots, 16$, the performance parameters of the affine hermitian codes over $GF(256)$ based on the Stichtenoth version of the hermitian polynomial are given by

$$n = 4096; \quad k = \frac{1}{2}(J + 2)(J + 1); \quad d_{\min} \geq 4096 - 17J,$$

while, for $J = 17, \ldots, 240$, the performance parameters of the affine hermitian codes are given by

$$n = 4096; \quad k = 17J - 119; \quad d_{\min} \geq 4096 - 17J.$$

Thus, the affine hermitian codes over $GF(256)$, for $J = 1, 2, \ldots, 239, 240$, have performance parameters given by $(4096, 3, 4079)$, $(4096, 6, 4062)$, $\ldots$, $(4096, 3944, 33)$, $(4096, 3961, 16)$.

10.6 Epicyclic hermitian codes

An hermitian code can be further punctured to the bicyclic plane. Much of the underlying structure of the hermitian codes stands out quite clearly when the code is restricted to the bicyclic plane, thereby defining an *epicyclic hermitian code*. Because of the simpler structure, some might even take the view that the epicyclic code is the more fundamental form of the hermitian code, just as some might take the view that the cyclic code is the more fundamental form of the Reed–Solomon code.

The bicyclic plane over a finite field can be regarded as a torus. The epicyclic form of the hermitian code, then, lies on a torus, and many of its automorphisms are shifts on the torus that leave the code invariant. There is also a simple characterization of the dual of an epicyclic hermitian code, which will be given in Section 10.7.

The epicyclic hermitian code over $GF(q^2)$, when using the Fermat form $x^{q+1} + y^{q+1} + 1$, has blocklength $n = (q-2)(q+1)^2 = q^3 - 3q - 2$, in contrast to the corresponding affine hermitian code, which has blocklength $n = q^3 - q$.

For the field $GF(16)$, $q = 4$ and $g = 6$. Thus $m = 5$ and $J \leq 9$. For $J = 1, \ldots, 4$, the performance parameters of these epicyclic hermitian codes over $GF(16)$ based on the Fermat form of the hermitian polynomial are given by

$$n = 50; \quad k = \frac{1}{2}(J+2)(J+1); \quad d_{\min} \geq 50 - 5J,$$

while, for $J = 5, \ldots, 9$, the performance parameters of the codes are given by

$$n = 50; \quad k = 5J - 5; \quad d_{\min} \geq 50 - 5J.$$

Thus, these epicyclic hermitian codes over $GF(16)$, for $J = 1, \ldots, 4, 5, 6, \ldots, 9$ have performance parameters given by $(50, 3, 45)$, $(50, 6, 40)$, $(50, 10, 35)$, $\ldots$, $(50, 30, 15)$, $(50, 35, 10)$, $(50, 40, 5)$.

For the field $GF(64)$, $q = 8$ and $g = 28$. Thus $m = 9$ and $J \leq 53$. For $J = 1, \ldots, 8$, the performance parameters of these epicyclic hermitian codes over $GF(64)$ based on the Fermat form of the hermitian polynomial are given by

$$n = 486; \quad k = \frac{1}{2}(J+2)(J+1); \quad d_{\min} \geq 486 - 9J,$$

while, for $J = 9, \ldots, 53$, the performance parameters of the codes are given by

$$n = 486; \quad k = 9J - 27; \quad d_{\min} \geq 486 - 9J.$$

Thus, these epicyclic codes over $GF(64)$, for $J = 1, \ldots, 53$, have performance parameters given by $(486, 3, 477)$, $(486, 6, 468)$, $\ldots$, $(486, 45, 414)$, $(486, 54, 405)$, $\ldots$, $(486, 450, 9)$.

For the field $GF(256)$, $q = 16$ and $g = 120$. Thus $m = 17$ and $J \leq 237$. For $J = 1, \ldots, 16$, the performance parameters of these epicyclic codes over $GF(256)$ based on the Fermat form of the hermitian polynomial are given by

$$n = 4046; \quad k = \frac{1}{2}(J + 2)(J + 1); \quad d_{\min} \geq 4046 - 17J,$$

while, for $J = 54, \ldots, 237$, the performance parameters of the codes are given by

$$n = 4046; \quad k = 17J - 119, \quad d_{\min} \geq 4046 - 17J.$$

Thus, these affine hermitian codes over $GF(256)$, for $J = 1, \ldots, 237$, have performance parameters given by $(4046, 3, 4029)$, $(4046, 6, 4012)$, $\ldots$, $(4046, 153, 3774)$, $(4046, 170, 3757)$, $\ldots$, $(4046, 3910, 17)$.

This completes our brief inventory of codes on the bicyclic plane constructed from the Fermat version of the hermitian polynomial.

We now turn to the second variation on the topic of epicyclic hermitian codes. This is the topic of epicyclic codes constructed from the Stichtenoth version of the hermitian polynomial. The epicyclic hermitian code over $GF(q^2)$, using the Stichtenoth form $x^{q+1} + y^q + y$, has blocklength $n = q^3 - q$, in contrast to the corresponding affine hermitian code, which has blocklength q^3.

For the field $GF(16)$, $q = 4$ and $g = 6$. Thus $m = 5$ and $J \leq 11$. For $J = 1, \ldots, 4$, the performance parameters of these epicyclic hermitian codes over $GF(16)$ based on the Stichtenoth version of the hermitian polynomial are given by

$$n = 60; \quad k = \frac{1}{2}(J + 2)(J + 1); \quad d_{\min} \geq 60 - 5J,$$

while, for $J = 5, \ldots, 11$, the performance parameters of the codes are given by

$$n = 60; \quad k = 5J - 5; \quad d_{\min} \geq 60 - 5J.$$

Thus, these epicyclic hermitian codes over $GF(16)$, for $J = 1, \ldots, 4, 5, 6, \ldots, 11$, have performance parameters given by $(60, 3, 55)$, $(60, 6, 50)$, $(60, 10, 45)$, $(60, 15, 40)$, $(60, 20, 35)$, $(60, 25, 30)$, $(60, 30, 25)$, $(60, 35, 20)$, $(60, 40, 15)$, $(60, 45, 10)$, and $(60, 50, 5)$.

For the field $GF(64)$, $q = 8$ and $g = 28$. Thus $m = 9$ and $J \leq 55$. For $J = 1, \ldots, 8$, the performance parameters of these epicyclic hermitian codes over $GF(64)$ based on the Stichtenoth version of the hermitian polynomial are given by

$$n = 504; \quad k = \frac{1}{2}(J + 1)(J + 1); \quad d_{\min} \geq 504 - 9J,$$

while, for $J = 9, \ldots, 55$, the performance parameters of the codes are given by

$$n = 504; \quad k = 9J - 27; \quad d_{min} \geq 504 - 9J.$$

Thus, these epicyclic hermitian codes over $GF(64)$, for $J = 1, \ldots, 55$, have performance parameters given by $(504, 3, 495)$, $(504, 6, 486)$, $\ldots$, $(504, 45, 432)$, $(504, 54, 423)$, $\ldots$, $(504, 459, 18)$, $(504, 468, 9)$.

For the field $GF(256)$, $q = 16$ and $g = 120$. Thus $m = 17$ and $J \leq 239$. For $J = 1, \ldots, 16$, the performance parameters of these epicyclic hermitian codes over $GF(256)$ based on the Stichtenoth polynomial are given by

$$n = 4080; \quad k = \frac{1}{2}(J + 2)(J + 1); \quad d_{min} \geq 4080 - 17J,$$

while, for $J = 17, \ldots, 239$, the performance parameters of the codes are given by

$$n = 4080; \quad k = 17J - 119; \quad d_{min} \geq 4080 - 17J.$$

Thus, these epicyclic hermitian codes over $GF(256)$, for $J = 1, \ldots, 239$, have performance parameters given by $(4080, 3, 4063)$, $(4080, 6, 4046)$, $\ldots$, $(4080, 153, 3808)$, $(4080, 170, 3791)$, $\ldots$, $(4080, 3944, 17)$.

10.7 Codes shorter than hermitian codes

An affine Reed–Solomon code over $GF(q^2)$ has blocklength q^2. An affine hermitian code over $GF(q^2)$ based on the polynomial $x^{q+1} + y^q + y$ has blocklength q^3. Thus the hermitian code is q times as long as the Reed–Solomon code. For example, the hermitian code over the field $GF(256)$ is sixteen times as long as the Reed–Solomon code over $GF(256)$. Although the notion of codes on curves was introduced in order to find long codes, for some applications the hermitian code may actually be too long. Are there good classes of codes whose blocklengths lie between q^2 and q^3? We shall give a sequence of codes for the field $GF(256)$ having blocklengths 256, 512, 1024, 2048, and 4096. A code from this sequence of blocklength 256 is a Reed–Solomon code over $GF(256)$. A code from this sequence of blocklength 4096 is an hermitian code over $GF(256)$.

Recall that the hermitian polynomial $x^{17} + y^{16} + y$ has 4096 affine zeros, so it can be used to form a code of blocklength 4096. A polynomial that is similar to the hermitian polynomial is $x^{17} + y^2 + y$. We shall see that this alternative polynomial has 512 affine zeros. It can used to form a code of blocklength 512 over $GF(256)$. This is twice as long as a Reed–Solomon code over this field. In this sense, the code that is based on polynomial $x^{17} + y^2 + y$ is the simplest generalization of a Reed–Solomon code in a family that includes the hermitian code.

The bivariate polynomial $x^{17} + y^2 + y$ has genus 8, and it has all the zeros allowed by the Hasse–Weil bound. Such a polynomial is called a *maximal polynomial*. The polynomial $x^{17} + y^2 + y$ is singular, having a singular point at infinity, and so its genus cannot be determined from the Plücker formula. The genus can be determined from the cardinality of the gap sequence[2] of the integer semigroup generated by 2 and 17. The semigroup consists of the sequence $\rho_r = 0, 2, 4, 6, 8, 10, 12, 14, 16, 17, 18, 19, 20, \ldots$ The gaps in this sequence form the set $\{1, 3, 5, 7, 9, 11, 13, 15\}$. Because there are eight gaps, the genus of the polynomial $x^{17} + y^2 + y$ is 8.

The Hasse–Weil bound

$$n \leq q + 1 + g \lfloor 2\sqrt{q} \rfloor$$

then says that the number of rational points of the polynomial $x^{17} + y^2 + y$ is at most 513. We shall see that this polynomial has as many zeros as the Hasse–Weil bound allows. First, note that there is one projective zero at infinity. This is the point $(0, 1, 0)$. Then, note that at any value of x, say $x = \gamma$, there can be at most two values of y that satisfy $y^2 + y + \gamma^{17} = 0$. To see that this polynomial in y always has two solutions, note that γ^{17} is always an element of $GF(16)$. Let $\beta = \gamma^{17}$. Then $y^2 + y + \beta$ either factors in $GF(16)$, and so has two zeros in $GF(16)$, or is irreducible in $GF(16)$, and so has two zeros in $GF(256)$. Either way, the polynomial $y^2 + y + \gamma^{17}$ has two zeros for each of 256 values of γ. This gives 512 zeros in the affine plane $GF(256)^2$ and a total of 513 zeros in the projective plane.

We are nearly ready to describe all of the promised collection of polynomials that give good codes of various lengths. First, we will return to the hermitian polynomial over $GF(256)$ to analyze its structure in more detail. This polynomial, which has 4080 bicyclic zeros, can be written in the form

$$x^{17} + y^{16} + y = x^{17} + 1 + p(y),$$

where

$$
\begin{aligned}
p(y) &= y^{16} + y + 1 \\
&= (y^8 + y^6 + y^5 + y^3 + 1)(y^8 + y^6 + y^5 + y^4 + y^3 + y + 1).
\end{aligned}
$$

Because each of its two (irreducible) factors has degree 8, we conclude that the univariate polynomial $p(y)$ has all its zeros in $GF(256)$. Now we can see that some, but not all, of the zeros of $x^{17} + y^{16} + y$ occur where both $x^{17} + 1 = 0$ and $p(y) = 0$. Because the first equation has seventeen solutions and the second has sixteen solutions, the pair

---

[2] This could have been given as a more general definition of the genus since it applies to more polynomials than does the Plücker formula.

has 272 solutions. More generally, we can separate the polynomial as

$$x^{17} + y^{16} + y = (x^{17} + \beta) + (y^{16} + y + \beta),$$

where β is a nonzero element of $GF(16)$. The polynomial $x^{17} + \beta$ has seventeen zeros. The polynomial $y^{16} + y + \beta$ has sixteen zeros, which can be found by making the substitution $y = \beta w$. This yields

$$y^{16} + y + \beta = \beta(w^{16} + w + 1),$$

and the zeros of $y^{16} + y + \beta$ are easily found from the zeros of $w^{16} + w + 1$. There are fifteen nonzero values of β. Finally, the zeros of the original bivariate polynomial are simply computed by pairing, for each value of β, the zeros of two univariate polynomials. This gives $15 \cdot 16 \cdot 17 = 4080$ bicyclic zeros.

The observation that the hermitian polynomial can be constructed by finding a polynomial $p(y)$ that has sixteen zeros leads us to define the following list of polynomials:

$$G(x,y) = x^{17} + y^2 + y;$$
$$G(x,y) = x^{17} + y^4 + y;$$
$$G(x,y) = x^{17} + y^8 + y^4 + y^2 + y;$$
$$G(x,y) = x^{17} + y^{16} + y.$$

In each case, the polynomial has the form

$$G(x,y) = x^{17} + 1 + p(y),$$

where $p(y)$ is a univariate polynomial over $GF(2)$ that has all its zeros in $GF(256)$. The four univariate polynomials $p(y)$ are given by

$$p(y) = y^2 + y + 1;$$
$$p(y) = y^4 + y + 1;$$
$$p(y) = y^8 + y^4 + y^2 + y + 1;$$
$$p(y) = y^{16} + y + 1;$$
$$= (y^8 + y^6 + y^5 + y^3 + 1)(y^8 + y^6 + y^5 + y^4 + y^3 + y + 1).$$

Those four polynomials have two, four, eight, and sixteen zeros, respectively, in $GF(256)$. By repeating the previous analysis of the hermitian polynomial, we can conclude that the four bivariate polynomials $G(x,y)$ have $4080, 2040, 1020$, and 510 bicyclic zeros, respectively. These polynomials can be used to construct codes whose blocklengths are multiples of 255. In particular, the codes have blocklengths equal to 255 times two, four, eight, or sixteen.

Problems

10.1   Prepare a table comparing the following codes:
  (a) the hermitian codes over $GF(64)$ of blocklength 512;
  (b) the composite Reed–Solomon codes over $GF(64)$ of blocklength 512;
  (c) the BCH codes over $GF(64)$ of blocklength 512;
  (d) the Reed–Solomon codes over $GF(64^2)$ of blocklength 256.

10.2   Prepare a list of hermitian codes over $GF(1024)$.

10.3   Factor $p(y) = y^{16} + y + 1$ over $GF(2)$. Show that all zeros of $p(y)$ are in $GF(256)$. Repeat for the polynomials $y^8 + y^4 + y^2 + y + 1$, $y^4 + y + 1$, and $y^2 + y + 1$.

10.4   (a) What is the gap sequence for the polynomial

$$p(x, y) = x^{17} + y^4 + y?$$

  (b) What is the genus of this polynomial? Why?
  (c) How many rational zeros does the Hasse–Weil bound allow for this polynomial?
  (d) Find the rational zeros.

10.5   (a) How many mod-q conjugates are there of an element of $GF(q^2)$?
  (b) Define the norm of the vector v over $GF(q^2)$ as $\langle v \cdot v^* \rangle$, where v^* denotes the vector whose components are the conjugates of the components of v (with $\beta^* = \beta$ if β is an element of $GF(q)$). Let $v = (x, y, z)$. What is the norm of v?

10.6   Show that the polynomial

$$p(x) = (x^{q+1} + y^{q+1} + 1)^{q-1} + 1$$

  is irreducible over $GF(q^2)$.

10.7   Consider the polynomial

$$p(x, y, z) = x^{15}y^3 + y^{15}z^3 + z^{15}x^3$$

  over $GF(64)$.
  (a) What does Serre's improvement of the Hasse–Weil bound say about the number of rational points?
  (b) Show that the polynomial has 948 rational points.
  (c) What are the parameters of the codes obtained by evaluating polynomials along this curve?

10.8   Prove that the dual of a shortened linear code is a puncture of the dual code.

10.9   Using Problem 10.8, show that a shortened epicyclic hermitian code is equivalent to a punctured epicyclic hermitian code.

10.10 A *hyperelliptic curve* is a curve formed from a nonsingular polynomial of genus g of the form

$$p(x, y) = y^2 + h(x)y - f(x),$$

where $\deg f(x) = 2g + 1$ and $\deg h(x) < g$. Show that the zeros of the polynomial

$$p(x, y) = x^{17} + y^2 + y$$

form a hyperelliptic curve. Graph this polynomial over the real field.

10.11 Consider the polynomial

$$p(x, y, z) = x^r y + y^r z + z^r x$$

over the field $GF(q)$. Show that the polynomial is irreducible if q and $r^2 - r + 1$ are coprime.

10.12 Extend the $(49, 35, 7)$ bicyclic hyperbolic code over $GF(8)$ to the projective plane. What are the performance parameters?

10.13 Let $GF(2^7) = GF(2)[x]/\langle x^7 + x + 1 \rangle$. For $y = x^4 + x^3 + 1 \in GF(2^7)^*$, determine $a \in \{0, 1, \ldots, 126\}$ such that $y = x^a$. (Hint: Use the euclidean algorithm to write y as a quotient $y = g(x)/h(x)$ of two polynomials $g(x)$ and $h(x)$, each of degree at most 3 in x.)

10.14 Construct a $(91, 24, 25)$ binary code by puncturing first a $(92, 24, 26)$ binary code, which is obtained by starting with a $(23, 8, 13)$ Klein code over $GF(8)$ and replacing each octal symbol by four bits, one of which is a simple parity check on the other three. Is there a cyclic $(91, 24, 25)$ binary code?

10.15 Prepare a table of code parameters for the codes based on the polynomials of Section 10.7.

10.16 Find the change of variables that will change the Stichtenoth version of the hermitian polynomial into the Fermat version of the hermitian polynomial. Verify correctness by working through the change of variables. Is there any other change of variables with binary coefficients that will give another form of the hermitian polynomial? Why?

Notes

The Reed–Solomon codes, although introduced on the affine line, were originally popularized in cyclic form. These cyclic codes were later lengthened to the affine line and the projective line. The hermitian codes were introduced in the projective plane and studied there, and only later were they shortened to the affine plane or the bicyclic plane. Perhaps this difference in history is because Reed–Solomon codes were popularized and applied by engineers, whereas hermitian codes were first discussed by mathematicians.

The idea of using the points of an algebraic curve to index the components of a code is due to Goppa (1977, 1981). The very powerful and elegant theorems of algebraic geometry, notably the Reimann-Roch theorem and the Hasse–Weil theorem, were immediately available to investigate the nature of such codes. These powerful theorems led the early research away from the practical issues of encoder and decoder design for codes of moderate blocklength and toward the study of asymptotic statements about the performance of very large codes. A major milestone in this direction is Tsfasman, Vlădut, and Zink (1982), which proved that in fields at least as large as $GF(49)$, there exist codes on curves whose performance is not only better than known codes, but better than known asymptotic bounds on the performance of codes of very large blocklength. The paper by Justesen *et al.* (1989) developed the notions of codes on curves more directly, using the theorems of algebraic geometry only with a light touch to determine the performance of the codes, thereby making the codes more accessible to those with little or no algebraic geometry, and opening the door to many later developments. Hermitian codes comprise an elegant family of codes on curves that form a rather compelling generalization of Reed–Solomon codes, and so have been widely studied. In this book, the family of hermitian codes is our preferred instance of a family of codes on curves. In this chapter, we view the hermitian codes as punctured codes; in the next chapter, we view them as shortened codes. Indeed, these codes are the same when restricted to the bicyclic plane. The punctured codes are also called evaluation codes. The true minimum distances of hermitian codes were determined by Yang and Kumar (1988).

The term "epicyclic code" was introduced in this chapter because I found it desirable to speak of codes restricted to the bicyclic plane as a class, and no suitable standard term seemed to exist. Though the codes themselves are not cyclic, the codes of interest do have several internal cyclic properties, so the term "epicyclic" seems to fit. With this term, moreover, we can complete our classification, begun in Chapters 2 and 5, for codes on lines and codes on planes, by introducing the names "codes on epicyclic curves," "codes on affine curves," and "codes on projective curves."

One may even choose to take the view that, just as the cyclic form is the more elementary form of the Reed–Solomon code, so too the epicyclic form is the more elementary form of the hermitian code. Certainly, the points outside the bicyclic plane have a different character and must be treated much differently within the usual locator decoding and encoding algorithms.

11 Other Representations of Codes on Curves

In contrast to the class of Reed–Solomon codes, which was introduced by engineers, the class of hermitian codes was introduced by mathematicians as an example of an important class of algebraic geometry codes. In this chapter, we shall reintroduce hermitian codes as they might have appeared had they been discovered by the engineering community. Some additional insights will be exposed by this alternative formulation. In particular, we will shift our emphasis from the notion of punctured codes on curves to the notion of shortened codes on curves. We then give constructions of hermitian codes as quasi-cyclic codes and as linear combinations of Reed–Solomon codes akin to the Turyn construction. Much of the structure of hermitian codes stands out quite clearly when a code is restricted to the bicyclic plane (or torus), thereby forming an epicyclic hermitian code. If one takes the view that the cyclic form is the more fundamental form of the Reed–Solomon code, then perhaps one should take the parallel view that the epicyclic form is the more fundamental form of the hermitian code. In particular, we shall see that, for the epicyclic form of an hermitian code, there is no difference between a punctured code and a shortened code. This is important because the punctured code is compatible with encoding and the shortened code is compatible with decoding. In Section 11.2, we shall provide a method for the direct construction of shortened epicyclic hermitian codes.

The epicyclic hermitian code inherits certain automorphisms from the underlying curve. An epicyclic hermitian code can be converted into a quasi-cyclic code. For example, the fifty components of the epicyclic hermitian codeword c over $GF(16)$ can be serialized in any way to form a one-dimensional codeword. We shall see in Section 11.3 that, under one such serialization, this fifty-point one-dimensional codeword has the quasi-cyclic form of ten concatenated segments: $c = |\ c_0\ |\ c_1\ |\ \cdots\ |\ c_9\ |$, where each of the ten segments consists of five components taking values in $GF(16)$. A cyclic shift of c by one segment (or five components) produces another codeword. Also, each of the ten segments can be individually cyclically shifted by one component to produce another codeword of the same code.

We shall also see in this chapter how some codes on curves can be constructed from Reed–Solomon codes in the same field. To this end, in Section 11.6 we will give a Turyn representation of hermitian codes in terms of Reed–Solomon codes.

11.1   Shortened codes from punctured codes

There are several close relationships between the punctured version and the shortened version of an epicyclic code which we will explore in this section. For one thing, the punctured codes and the shortened codes have the simple relationship of duality. Just as the dual of a cyclic Reed–Solomon code is a cyclic Reed–Solomon code, so, too, the dual of a punctured epicyclic hermitian code is a shortened epicyclic hermitian code. Indeed, the dual of a punctured epicyclic code on any curve is a shortened epicyclic code on that same curve. For another thing, when restricted to the bicyclic plane, the punctured version and the shortened version of an hermitian code have equivalent performance, and indeed are the same code.

We saw in Chapter 10 that the dimension and the minimum distance of a punctured code $C_J(\mathcal{X})$ on a curve $\mathcal{X}$ are given by

$$k = mJ - g + 1 \quad d_{\min} \geq n - mJ.$$

In this chapter, we shall see that the dimension and the minimum distance of a shortened code $C'_J(\mathcal{X})$ on a curve $\mathcal{X}$ are given by

$$k = n - mJ + g - 1 \quad d_{\min} \geq mJ - 2g + 2.$$

These performance descriptions appear to be quite different, but it is only a matter of the choice of J. If mJ in the second pair of formulas is replaced by $n - mJ + 2g - 2$, the second pair of formulas reduces to the first pair of formulas.

More strikingly, for epicyclic hermitian codes, we will make a stronger statement. Not only is the performance of a punctured epicyclic hermitian code *equivalent to* a shortened epicyclic hermitian code, a punctured epicyclic hermitian codes *is* a shortened epicyclic hermitian code. The same hermitian code can be described either way.

Recalling the notions of puncturing and shortening, the punctured form of a code on the plane curve $\mathcal{X}$ is given by

$$C(\mathcal{X}) = \left\{ c(\mathcal{X}) \mid c \Leftrightarrow C; C_{j'j''} = 0 \text{ if } (j',j'') \in \mathcal{A} \right\},$$

and the shortened form of the hermitian code is given by

$$C'_J(\mathcal{X}) = \{ c(\mathcal{X}) \mid c \Leftrightarrow C; C_{j'j''} = 0 \text{ if } (j',j'') \in \mathcal{A}', \ c(\mathcal{X}^c) = 0 \},$$

where $\mathcal{A}$ and $\mathcal{A}'$ are the defining sets of the two codes. If $\mathcal{A} = \mathcal{A}' = \{ (j',j'') \mid j' + j'' \leq J \}$, then both the code $C(\mathcal{X})$ and the code $C'(\mathcal{X})$ are obtained from the same primitive bicyclic code $\mathcal{C} = \{c \mid c \Leftrightarrow C; \deg C(x,y) \leq J\}$; the first code is obtained by puncturing $\mathcal{C}$; the second, by shortening $\mathcal{C}$.

Instead of choosing the same defining set for the two codes, it is more common to use complementary sets as the two defining sets, and this is the form that we used to state the performance equations at the start of the section. For this purpose, set $\mathcal{A} = \{(j',j'') \mid j' + j'' \leq J\}$ and set $\mathcal{A}' = \mathcal{A}^c$. Then, in the language of polynomials, the codes are defined as

$$\mathcal{C}_J(\mathcal{X}) = \{c(\mathcal{X}) \mid c \Leftrightarrow C; \deg C(x,y) \leq J\}$$

and

$$\mathcal{C}'_J(\mathcal{X}) = \{c(\mathcal{X}) \mid c \Leftrightarrow C; \deg \widetilde{C}(x,y) \leq J, c(\mathcal{X}^c) = 0\},$$

where a polynomial $\widetilde{C}(x,y)$ is the reciprocal of a polynomial $C(x,y)$. While these constructions of a punctured code $\mathcal{C}_J(\mathcal{X})$ and a shortened code $\mathcal{C}'_J(\mathcal{X})$ appear to give different codes, for epicyclic hermitian codes, they actually are equivalent constructions in that they define the same set of codes. Specifically, we will prove that, for epicyclic hermitian codes, $\mathcal{C}_J(\mathcal{X})$ is equivalent to $\mathcal{C}'_{J+q^2-2}(\mathcal{X})$ and is the dual of $\mathcal{C}'_{2q^2-5-J}$. This first statement is important because the punctured form is more suitable for encoding, while the shortened form is more suitable for decoding.

The simplest demonstration that the two constructions give dual codes is to start from the general fact that the dual of any shortened linear code is a puncture of the dual of the linear code (see Problem 10.8). The punctured form of the hermitian code $\mathcal{C}_J(\mathcal{X})$ arises by puncturing the bicyclic code defined by the set of bispectrum polynomials $C(x,y)$ that satisfy $\deg C(x,y) \leq J$. This means that the (unpunctured) bicyclic code has defining set $\mathcal{A} = \{(j',j'') \mid j' + j'' > J\}$ (or $j' + j'' \geq J + 1$). The dual $\mathcal{C}^\perp$ of this bicyclic code has defining set $\mathcal{A}^c = \{(j',j'') \mid j' + j'' \leq J\}$. But this defining set is not in the standard form of our definition of a shortened code; it must be reciprocated. The two-dimensional spectrum is defined on a $q - 1$ by $q - 1$ array with indices running from zero to $q - 2$. The reciprocal of j' and j'' are $q - 2 - j'$ and $q - 2 - j''$. This means that the reciprocal of $\mathcal{A}^c$ is given by

$$\widetilde{\mathcal{A}}^c = \{(j',j'') \mid (q - 2 - j') + (q - 2 - j'') \leq J\}$$
$$= \{(j',j'') \mid j' + j'' \geq 2q - 4 - J\}.$$

With q replaced by q^2, we conclude that the reciprocal of the dual hermitian code has spectral components that satisfy $\widetilde{C}^\perp_{j'j''} = 0$ for $j' + j'' > 2q^2 - 5 - J$.

Next we will show that the punctured form of the hermitian code with parameter J is the reciprocal of the shortened form of the hermitian code with parameter $J + q^2 - 1$. This is an important observation because it says that we may encode by viewing the code as a punctured code and decode by viewing the same code as a shortened code.

Define the hermitian *mask polynomial* over $GF(q^2)$, with $q = 2^m$, as follows:

$$H(x, y) = G(x, y)^{q-1} + 1$$

$$= (x^{q+1} + y^{q+1} + 1)^{q-1} + 1.$$

For any nonzero β and γ, β^{q+1} and γ^{q+1} have order $q - 1$, and so β^{q+1} and γ^{q+1} are elements of $GF(q)$. Thus $G(\beta, \gamma) = \beta^{q+1} + \gamma^{q+1} + 1$ is always an element of $GF(q)$. This means that $G(\omega^{-i'}, \omega^{-i''})^{q-1}$ can only be zero or one. Therefore $h_{i'i''} = H(\omega^{-i'}, \omega^{-i''})$ equals zero if $g_{i'i''}$ is not zero, and equals one if $g_{i'i''}$ is zero. Then because $h_{i'i''}^2 = h_{i'i''}$ in the bicyclic plane, the convolution theorem tells us that $H(x, y)^2 = H(x, y)$ modulo $\langle x^{q^2-1} - 1, y^{q^2-1} - 1 \rangle$.

This conclusion can also be reached directly. Recall that q is a power of 2, and using $(\beta + 1)^2 = \beta^2 + 1$, write the following:

$$H(x, y) = \left(\left(x^{q+1} + y^{q+1} + 1 \right)^{q-1} + 1 \right)^2$$

$$= \left(x^{q+1} + y^{q+1} + 1 \right)^{2q-2} + 1$$

$$= \left(x^{q+1} + y^{q+1} + 1 \right)^q \left(x^{q+1} + y^{q+1} + 1 \right)^{q-2} + 1.$$

Because q is a power of 2, and $x^{q^2} = x \pmod{x^{q^2-1} - 1}$ and $y^{q^2} = y \pmod{y^{q^2-1} - 1}$, the first term becomes

$$(x^{q+1} + y^{q+1} + 1)^q = x^{q^2} x^q + y^{q^2} y^q + 1$$

$$= x^{q+1} + y^{q+1} + 1.$$

Therefore

$$H(x, y)^2 = H(x, y) \quad (\text{mod } \langle x^{q^2-1} - 1, y^{q^2-1} - 1 \rangle),$$

from which we conclude, as before, that $H(x, y)$ is a bivariate idempotent polynomial.

The mask polynomial $H(x, y)$ can be used to redefine the epicyclic hermitian code as follows. Instead of evaluating the polynomial $C(x, y) \in S_J$ on the bicyclic plane, first multiply $C(x, y)$ by $H(x, y)$, then evaluate $D(x, y) = H(x, y)C(x, y)$ on the bicyclic plane. Let

$$\mathcal{D}(\mathcal{X}) = \{ d \mid d \Leftrightarrow D; \ D(x, y) = H(x, y)C(x, y); \ C(x, y) \in S_J \}.$$

Then

$$d_{i'i''} = \begin{cases} c_{i'i''} & \text{if } (\omega^{-i'}, \omega^{-i''}) \in \mathcal{X} \\ 0 & \text{if } (\omega^{-i'}, \omega^{-i''}) \notin \mathcal{X}. \end{cases}$$

Because only those n points along the curve $\mathcal{X}$ are used to form the codeword, this actually changes nothing about the codeword. Thus

$$\mathcal{D}(\mathcal{X}) = \mathcal{C}(\mathcal{X}).$$

Nevertheless, this reformulation makes the task of the decoder much more accessible. Whereas $c_{i'i''}$ is not given to the decoder at points not on the curve, $d_{i'i''}$ is known by the decoder to be zero at those points. This means that the decoder can proceed as if $\mathcal{D}(\mathcal{X})$ were the code. The following theorem says, moreover, that $\mathcal{D}(\mathcal{X})$ is the shortened code.

Theorem 11.1.1 *The punctured epicyclic hermitian code $\mathcal{C}_J(\mathcal{X})$ and the shortened epicyclic hermitian code $\mathcal{C}'_{q^2-1-J}(\mathcal{X})$ are equivalent.*

Proof: Let c be the codeword of the punctured code $\mathcal{C}_J(\mathcal{X})$ corresponding to the spectrum polynomial $C(x,y) \in \mathcal{S}_J$. This polynomial satisfies $\deg C(x,y) \leq J$, and the polynomial $H(x,y)$ satisfies $\deg H(x,y) = q^2 - 1$. Therefore $D(x,y) = H(x,y)C(x,y) \in \mathcal{S}_{J+q^2-1}$.

Evaluating $D(x,y)$ on the curve $\mathcal{X}$ gives the same result as evaluating $C(x,y)$ on the curve $\mathcal{X}$, so c is a codeword of $\mathcal{C}_{J+q^2-1}$. But evaluating $D(x,y)$ at all points P of the bicyclic plane gives $D(P) = 0$ for all $P \notin \mathcal{X}$, so we conclude that d is also a codeword of the shortened code $\mathcal{C}'_{J+q^2-1}(\mathcal{X})$. Hence $\mathcal{C}'_{J+q^2-1}(\chi) \supseteq \mathcal{C}_j(\chi)$.

To show that every codeword of the shortened code $\mathcal{C}'_{J+q^2-1}(\mathcal{X})$ can be formed in this way, suppose that c' corresponds to the spectrum polynomial $C'(x,y) \in \mathcal{S}_{J+q^2-1}$. Because c' is a codeword of the shortened code, $C'(P)$ must be zero for any point P not on the curve. That is, $C'(P) = 0$ whenever $H(P) = 0$. A variation of the nullstellensatz, given in Theorem 7.9.2 as the weak discrete nullstellensatz, states that the ideal $I(\mathcal{Z}(J))$ is equal to J. Let $I = \langle H(x,y) \rangle$. This means that $C'(x,y)$ is a multiple of $H(x,y)$, say $C(x,y)H(x,y)$. Hence $\mathcal{C}'_{J+q^2-1}(\mathcal{X}) \subseteq \mathcal{C}_J(\mathcal{X})$. ∎

Theorem 11.1.1 asserts that a punctured epicyclic hermitian code can also be regarded as a shortened epicyclic hermitian code. This means, of course, that the dimension and minimum distance of a code agree for the two descriptions. To verify this, recall that the punctured epicyclic hermitian code $\mathcal{C}_J$ has the following performance parameters:

$$k = mJ - g + 1 \quad d_{\min} \geq n - mJ.$$

The shortened epicyclic hermitian code, on the other hand, denoted $\mathcal{C}'_{q^2-2-J} = \mathcal{C}'_{J'}$, with $J' = q^2 - 2 - J$, has dimension

$$k = n - mJ' + g - 1.$$

But, because $n = q^3 - q$, this becomes

$$k = q^3 - q - (q+1)(q^2 - 2 - J) + \frac{1}{2}q(q-1) - 1$$

$$= mJ + 1 - g,$$

which is the same as the dimension of C_J. In a similar way, the shortened code C'_{q^2-2-J}, with $J' = q^2 - 2 - J$, has minimum distance

$$d_{\min} \geq mJ' - 2g + 2$$

$$= n - mJ.$$

Thus, as promised by Theorem 11.1.1, the code viewed as a shortened code has the same performance as the code viewed as a punctured code. This is in accordance with the earlier discussion that said that for every punctured epicyclic code there is a shortened epicyclic code on the same curve with the same performance. Here, however, we have gone even further for the special case of an hermitian code. In this case, the code not only has the same performance, it is the same code but for reciprocation.

For example, the hermitian curve over the field $GF(16)$ is based on the polynomial

$$G(x,y) = x^5 + y^5 + 1,$$

which has the inverse Fourier transform

$$g_{i'i''} = G(\omega^{-i'}, \omega^{-i''}).$$

The punctured epicyclic codeword $c(\mathcal{X})$, corresponding to spectrum polynomial $C(x,y)$, consists only of those components of the inverse Fourier transform c, with components $c_{i'i''} = C(\omega^{-i'}, \omega^{-i''})$, for which $(\omega^{-i'}, \omega^{-i''})$ is on the curve $\mathcal{X}$. The components of c not on the curve are discarded. Therefore an inverse Fourier transform cannot recover $C(x,y)$ directly from the punctured codeword $c(\mathcal{X})$ because the missing components of c are not available. Instead, by setting the missing components to zero, and taking an inverse Fourier transform, the product $H(x,y)C(x,y)$ is recovered instead of just $C(x,y)$.

The mask polynomial is given by

$$H(x,y) = G(x,y)G(x,y)^2 + 1$$

$$= x^5y^{10} + x^{10}y^5 + y^{10} + x^{10} + y^5 + x^5 \quad (\text{mod } \langle x^{15} - 1, y^{15} - 1 \rangle).$$

It has the inverse Fourier transform

$$h_{i'i''} = H(\omega^{-i'}, \omega^{-i''})$$

$$= g_{i'i''}^3 + 1.$$

Because $G(x, y)$ has sixty bicyclic zeros, and the bicyclic plane has 225 points, we see that $H(x, y)$ has exactly 165 bicyclic zeros. It is easy to check further that $H(x, y)$ has a total of 183 projective zeros. The two curves $\mathcal{X}$ and $\mathcal{Y}$, defined by $G(x, y)$ and $H(x, y)$, are disjoint in the bicyclic plane and together completely fill the bicyclic plane over $GF(16)$. Of course, from Bézout's theorem, we know that $G(x, y)$ and $H(x, y)$ have seventy-five common zeros somewhere, though none of them are in the bicyclic plane over $GF(16)$.

As a final illustration, note that the hermitian curve over $GF(16)^2$, based on the polynomial

$$G(x, y) = x^5 + y^4 + y,$$

has a mask polynomial over $GF(16)^2$, given by

$$H(x, y) = G(x, y)^3 + 1 \quad (\mathrm{mod}\langle x^{15} - 1, y^{15} - 1\rangle)$$
$$= y^3 + y^6 + y^9 + y^{12} + x^5(y^2 + y^8) + x^{10}(y + y^4),$$

which is equal to its own square, so it can take only the values zero and one. This polynomial has the property that

$$H(\beta, \gamma) = \begin{cases} 1 & \text{if } G(\beta, \gamma) = 0 \\ 0 & \text{if } G(\beta, \gamma) \neq 0. \end{cases}$$

To verify this, one performs the following polynomial multiplication:

$$G(x, y)H(x, y) = (y + y^4)(x^{15} - 1) + y(y^{15} - 1),$$

which says that every point (β, γ) is a zero of $G(x, y)H(x, y)$. Finally, evaluate $H(x, y)$ on the curve $x^5 + y^4 + y = 0$ to get

$$H(x, y)|_{x^5 + y^4 + y = 0} = y^3 + y^6 + y^9 + y^{12}.$$

The right side equals one for every nonzero value of y in $GF(16)$. Therefore $G(x, y)$ and $H(x, y)$ have no common zeros in the bicyclic plane.

11.2   Shortened codes on hermitian curves

The family of punctured codes on the curve corresponding to the polynomial $G(x, y)$, as defined in Section 10.3 by the defining set $A_J = \{(j', j'') \mid j' + j'' > J\}$, contains codes of dimension k for $k = mJ - g + 1$ and for $J = m, m+1, \ldots$, where m

is the degree and g is the genus of the polynomial $G(x, y)$. Because the designed distance of a punctured code was given by $d^* = n - mJ$ as a consequence of Bézout's theorem, the maximum degree J of the set of bispectrum polynomials $C(x, y)$ plays an important role and J appears in the performance formulas multiplied by m, the degree of $G(x, y)$. Accordingly, within this family of punctured codes, as J increases, the dimension k increases, by multiples of m, and the designed distance d^* decreases by multiples of m. This is a somewhat sparse family of codes. However, there are many other defining sets between $\mathcal{A}_J$ and $\mathcal{A}_{J+1}$, and also many codes between $\mathcal{C}_J$ and $\mathcal{C}_{J+1}$. Instead of evaluating polynomials in a set whose total degree is constrained, to enlarge the class of codes defined on $G(x, y)$, we will evaluate polynomials in a set whose weighted degree is constrained. For example, most of the hermitian codes over $GF(256)$, as described in Section 10.5, have dimensions that are spaced by multiples of 17. We might want to have a code whose dimension lies between two of these available dimensions. Moreover, as already mentioned, these codes are punctured codes, and so are not immediately compatible with the task of decoding. For these reasons, one may want to give an alternative definition of the codes. In this section, we shall define the hermitian codes in a more deliberate way, as shortened codes, that enables us to enlarge the family of codes. To do so, we will replace the degree of the bispectrum polynomial by the weighted degree, which is a more delicate notion of degree that gives each relevant monomial a unique weight whenever the polynomial admits a weight function. Accordingly, we will restrict the choice of the polynomial $G(x, y)$ to those polynomials that admit a weight function. Then, instead of using Bézout's theorem to bound the minimum distance of a punctured code, we shall use the Feng–Rao bound to bound the minimum distance of a shortened code. Among the codes constructed in this way are the same codes as before, as well as many new codes.

A linear code on the curve $\mathcal{X}$ over $GF(q)$ is any vector space over $GF(q)$ whose components are indexed by the n points of the curve $\mathcal{X}$. The codewords are the elements of this vector space. One defines a code by specifying the vector space on the curve. The method that we have used in Chapter 10 to specify the vector space is by constraints on the two-dimensional Fourier transform, setting to zero certain components of the bispectrum. The defining set $\mathcal{A}$ of the code $\mathcal{C}$ specifies those components of the two-dimensional Fourier transform $C(x, y)$ in which every codeword $\boldsymbol{c}$ is constrained to be zero. In that chapter, the polynomial $C(x, y)$ is constrained only by its degree, which does not require the introduction of a monomial order. In this section, we will constrain $C(x, y)$ using the weighted graded order on monomials.

The introduction of a monomial order is not the only change to be found in this section. In addition to introducing the weighted graded order as the monomial order, we will also change the codes from punctured codes on curves to shortened codes on curves. These two changes go together well, and so we introduce them at the same time. In addition, we will use the Feng–Rao bound instead of the Bézout theorem. The

Feng–Rao bound applies directly to the weighted graded order and it applies directly to the shortened codes.

The Feng–Rao bound, given in Section 9.8, states that the only vector v of length n on the curve $G(x, y) = 0$ having weight $d_{\mathrm{FR}}(r) - 1$ or less, whose two-dimensional Fourier transform components are equal to zero for all indices smaller than $r + 1$ in the weighted order, is the all-zero vector. Because the Feng–Rao distance cannot be stated analytically, we will usually use the weaker Goppa distance instead. The Goppa distance profile is $d_\Gamma(r) = r + 1 - g$, where r is the number of monomials in the defining set. Thus, for a code on the curve defined by $G(x, y)$ and with the first r monomials in the weighted order as the defining set, the minimum distance satisfies

$$d_{\min} \geq d_\Gamma(r)$$
$$= r + 1 - g,$$

as asserted by the Goppa bound.

To restate this expression in terms of a defining set $\mathcal{A}$, recall that the only monomials that need to be counted are those with $j' < m$, where m is the degree of $G(x, y)$. To count these monomials, observe that there are m such monomials with $j' + j'' = j$ for large enough j, fewer than m monomials for small j, and that j takes on J values. Thus there are fewer than mJ monomials. The defining set has the following form:

$$\mathcal{A} = \{(j', j'') \mid j' + j'' \leq J; j' < m\},$$

and the number of monomials is the area of this trapezoidal set. By a straightforward calculation, we will conclude that the area of this set is $\|\mathcal{A}\| = mJ - g + 1$. One way to organize this calculation is to observe that there can be up to m monomials for each value of $j = j' + j''$, and there are $J + 1$ values of j. Because some values of j have fewer than m monomials, we can write the area as follows:

$$r = \|\mathcal{A}\|$$
$$= m(J + 1) - \sum_{j=0}^{m-1} j$$
$$= m(J + 1) - \frac{1}{2}m(m - 1)$$
$$= mJ - \frac{1}{2}(m - 1)(m - 2) + 1.$$

Therefore, for such a defining set, the designed distance d^* is given by

$$d^* = r + 1 - g$$
$$= mJ - 2g + 2,$$

as asserted earlier. Finally, the dimension of the code is given by

$$k = n - r$$
$$= n - mJ + g - 1.$$

For example, the hermitian curve over $GF(16)$ can be defined by using the polynomial $G(x, y) = x^5 + y^4 + y$. We have seen that for this polynomial, the weights of the monomials can be defined by setting $\rho(x) = 4$ and $\rho(y) = 5$. Then $\rho(x^{j'} y^{j''}) = 4j' + 5j''$. These monomials weights are shown in Figure 11.1.

To construct a code, we will select all j' and j'' such that $4j' + 5j'' < m$ as the indices of the defining set $\mathcal{A}$. For each choice of m, one obtains a hermitian code over $GF(16)$. In particular, if $m = 31$, then the code consists of all arrays on the affine plane $GF(16)^2$ such that $C_{j'j''}$ is zero for all j', j'', corresponding to $4j' + 5j'' \leq 31$. The blocklength of the affine code is 64 because there are sixty-four points on the affine curve using the polynomial $x^9 + y^8 + y$. The dimension of the code is 38 because there are twenty-six monomials for which $4j' + 5j'' \leq 31$. These twenty-six monomials are $1, x, y, x^2, \ldots, x^4 y^3$. The minimum distance of the code is at least 21 according to the Feng–Rao bound. Thus, as asserted, one obtains a $(64, 38, 21)$ code. Clearly, it is a simple matter to increase or decrease the defining set by one element to make the dimension of the code smaller or larger by one. The minimum distance of the code is then determined by the Feng–Rao bound.

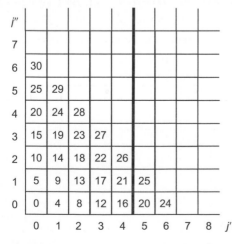

Figure 11.1. Weights of monomials for $x^5 + y^4 + y$.

11.3   Quasi-cyclic hermitian codes

In this section, we shall see that the hermitian codes over $GF(q^2)$ can be viewed as quasi-cyclic codes over that same field.

Our first example is the Fermat version of the hermitian curve over the bicyclic plane, as shown in Figure 11.2, for $GF(16)$ (for which $q = 4$). This curve has fifty points restricted to the torus $GF(16)^{*2}$. The fifty components of an epicyclic hermitian codeword lie on the fifty points of the shortened hermitian curve. It is clear from Figure 11.2 that the bicyclic portion of the hermitian curve is mapped onto itself if the bicyclic plane is cyclically shifted by three places in the row direction (or by three places in the column direction). To see that under such a shift a codeword of an epicyclic hermitian code is mapped to another codeword of that same code, we refer to the translation property of the Fourier transform. This property says that a cyclic translate of the two-dimensional codeword in the bicyclic plane by three places in the row direction (or in the column direction) is equivalent to multiplying the bispectrum componentwise by $\omega^{3j'}$ (or by $\omega^{3j''}$). The codeword c of C corresponds to a bispectrum C, which, in turn, is represented by the polynomial $C(x, y)$. If the polynomial $C(x, y)$ with coefficient $C_{j'j''}$ is replaced by a new polynomial, $B(x, y)$, with coefficients $B_{j'j''} = C_{j'j''}\omega^{3j'}$ (or $B_{j'j''} = C_{j'j''}\omega^{3j''}$), then the degree of the polynomial is unchanged. This means that $B(x, y)$ is also in the set $\mathcal{S}_J$. Consequently, the cyclic translation of a codeword by three places, either rowwise or columnwise, in the bicyclic plane is another codeword.

The fifty components of codeword c, lying on the fifty points of the hermitian curve, can be serialized in any way to form a one-dimensional codeword. For example, the

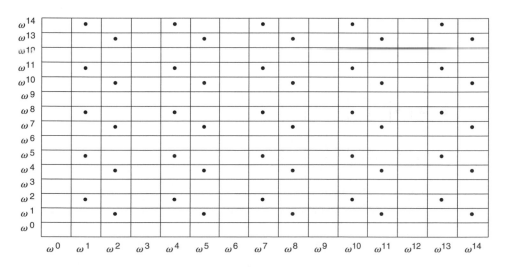

Figure 11.2. Hermitian curve over $GF(16)$ in the bicyclic plane.

fifty components can be serially ordered by reading the points of the curve across rows. As a one-dimensional vector, this codeword has the form of ten concatenated segments,

$$c =| c_0 | c_1 | \cdots | c_9 |,$$

where each segment consists of the five components in one of the nonzero rows of the two-dimensional array.

Certain automorphisms of the epicyclic hermitian code are obvious consequences of the underlying curve, as shown for $GF(16)$ in Figure 11.2. A fifty-point codeword, written by rows as a fifty-point vector, produces another codeword when cyclically shifted by five places. Hence the code is a *quasi-cyclic code*, which is the term given to a code that is not cyclic but is invariant under cyclic shifts of b places, $b \neq 1$. Moreover, the code is composed of ten segments, each of which has length 5. If each of the ten segments is individually cyclically shifted by one place, then another codeword of the same hermitian code is obtained.

For a second example, the intersection of the Stichtenoth version of the hermitian curve over $GF(16)$ with the bicyclic plane, as shown in Figure 11.3, has sixty points. The sixty components of the corresponding epicyclic hermitian codeword lie on the sixty points of the hermitian curve restricted to the torus. It is clear from Figure 11.3 that a cyclic shift by one place in the row direction, followed by a cyclic shift by five places in the column direction, will leave the curve unchanged. The code is invariant under this bicyclic shift because, by the convolution theorem, the degree of $C(x, y)$ is not changed by cyclic shifts of the array c. Now it is easy to see several ways to serialize the hermitian code in a way that forms a one-dimensional, quasi-cyclic code with $b = 4$. Thus, for example, a one-dimensional vector can be written in the form of

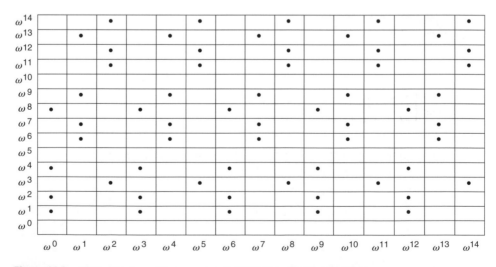

Figure 11.3. Alternative hermitian curve over $GF(16)$ in the bicyclic plane.

15 concatenated segments,

$$c =| \; c_0 \mid c_1 \mid \cdots \mid c_{13} \mid c_{14} \; |,$$

where each segment consists of the four components of a column written in order, starting from the component with three zeros below it. Clearly the cyclic shift of c by four places gives another codeword, and the cyclic shift of each segment by one place also gives another codeword.

11.4 The Klein codes

A small family of codes over $GF(8)$ of blocklength 24 on the projective plane, or of blocklength 21 on the bicyclic plane, can be constructed by using the Klein quartic polynomial. We call these *Klein codes*.

The Klein polynomial,

$$G(x, y) = x^3 y + y^2 + x,$$

has degree $r = 4$ and genus $g = 3$. The homogeneous form of the Klein polynomial,

$$G(x, y, z) = x^3 y + y^3 z + z^3 x,$$

has 24 zeros in the projective plane over $GF(8)$. Thus the codes have blocklength $n = 24$ in the projective plane. To define a code, we can choose any $J < n/r = 6$, thereby obtaining a code whose minimum distance d_{min} satisfies

$$d_{min} \geq n - rJ,$$

and whose dimension k satisfies

$$k = \begin{cases} \frac{1}{2}(J+2)(J+1) & J < 4 \\ rJ - g + 1 & J \geq 4. \end{cases}$$

Thus we have the following Klein codes over the field $GF(8)$:

$$
\begin{array}{llll}
J = 1 & (24,3) & d_{min} \geq 20; \\
J = 2 & (24,6) & d_{min} \geq 16; \\
J = 3 & (24,10) & d_{min} \geq 12; \\
J = 4 & (24,14) & d_{min} \geq 8; \\
J = 5 & (24,18) & d_{min} \geq 4.
\end{array}
$$

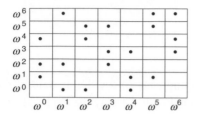

Figure 11.4. Klein curve in the bicyclic plane.

	ω^0	ω^1	ω^2	ω^3	ω^4	ω^5	ω^6
ω^6			c_5				c_{15}	c_{16}
ω^5				c_6	c_7		c_{17}	
ω^4		c_{19}		c_8				c_{18}
ω^3					c_9	c_{10}		c_{20}
ω^2		c_0	c_1		c_{11}			
ω^1		c_2				c_{12}	c_{13}	
ω^0			c_3	c_4		c_{14}		

Figure 11.5. Quasi-cyclic serialization of the Klein code.

The Klein curve restricted to the bicyclic plane is shown in Figure 11.4. When so restricted, the Klein curve has twenty-one points. Therefore the epicyclic form of a Klein code, which lies on this set of twenty-one points, has blocklength 21.

By restricting the Klein curve to the bicyclic plane, which can be regarded as a torus, several automorphisms of the Klein code become more evident as automorphisms of the epicyclic Klein code. If the bicyclic plane is cyclically shifted by one place along the row direction, then cyclically shifted by two places along the column direction, the Klein curve is mapped onto itself. This means that a codeword will map onto another codeword under this bicyclic shift, provided the new spectrum polynomial also has a degree at most J. But by the convolution theorem, under this bicyclic shift the spectrum coefficient $C_{j'j''}$ is replaced by $C_{j'j''}\alpha^{j'}\alpha^{2j''}$, which coefficient is still zero if $j' + j'' \leq J$. Therefore this particular bicyclic shift takes a codeword onto a codeword. This bicyclic shift property is similar to the cyclic shift property of a cyclic code. It can be used to put the Klein code in the form of a one-dimensional, quasi-cyclic code.

The twenty-one components of codeword c, lying on the twenty-one points of the Klein curve, can be serialized in any way to form a one-dimensional codeword. For example, the twenty-one components can be serially ordered by reading across by rows. To arrange the Klein code in the form of a quasi-cyclic code, it is enough to arrange the components sequentially in an order that respects the bicyclic shift described above. Figure 11.5 labels the twenty-one components of the Klein code to give a serialization that forms a quasi-cyclic code. Other serializations with this property are readily apparent.

11.5   Klein codes constructed from Reed–Solomon codes

In Section 6.9, we saw that the Turyn representation of the binary Golay code is a concatenation of three binary codewords in the form $c = |\ c_0\ |\ c_1\ |\ c_2\ |$. The individual codewords are given by

$$
\begin{bmatrix} c_0 \\ c_1 \\ c_2 \end{bmatrix} = \begin{bmatrix} 1 & 0 & 1 \\ 1 & 1 & 0 \\ 1 & 1 & 1 \end{bmatrix} \begin{bmatrix} b_0 \\ b_1 \\ b_2 \end{bmatrix},
$$

where b_1 and b_2 are any codewords of the $(8,4,3)$ extended Hamming code over $GF(2)$ and b_0 is any codeword of the $(8,4,4)$ reciprocal extended Hamming code over $GF(2)$. We shall see in this section that an epicyclic Klein code over $GF(8)$ has a similar representation as a linear combination of three Reed–Solomon codes over $GF(8)$. Each codeword c of the $(21, k, d)$ Klein code is represented as a concatenation of the form $c = |\ c_0\ |\ c_1\ |\ c_2\ |$. The individual codewords are given by

$$
\begin{bmatrix} c_0 \\ c_1 \\ c_2 \end{bmatrix} = \begin{bmatrix} 1 & \alpha^2 & \alpha \\ 1 & \alpha^4 & \alpha^2 \\ 1 & \alpha & \alpha^4 \end{bmatrix} \begin{bmatrix} b_0 \\ b_1 \\ b_2 \end{bmatrix},
$$

where α is the primitive element used to construct $GF(8)$, and b_0, b_1, and b_2 are codewords from three different Reed Solomon codes over $GF(8)$.

This representation is interesting because of its similarity to the Turyn representation. It also provides a convenient method of encoding the Klein code; first encode the data into three Reed–Solomon codewords, then perform the indicated linear transformation.

For example, we shall see that, to express the $(21, 7, 12)$ epicyclic Klein code with this representation, b_0 is a codeword of a $(7, 3, 5)$ Reed–Solomon code over $GF(8)$ with defining set $\{3, 4, 5, 6\}$; b_1 is a codeword of a $(7, 2, 6)$ Reed–Solomon code over $GF(8)$ with defining set $\{0, 1, 2, 3, 4\}$; and b_2 is a $(7, 2, 6)$ Reed–Solomon code over $GF(8)$ with defining set $\{5, 6, 0, 1, 2\}$. Together, these three Reed–Solomon codes encode seven data symbols, and the dimension of the underlying Klein code equals 7. The concatenation $b = |\ b_0\ |\ b_1\ |\ b_2\ |$ has minimum distance 5 because b_0 has minimum distance 5. We can conclude that the matrix operation ensures that $c = |\ c_0\ |\ c_1\ |\ c_2\ |$ has minimum distance 12 by showing that this gives a representation of a Klein code with minimum distance 12.

Our first step in developing this representation of the Klein code is to study the two-dimensional Fourier transform of a sparse array over $GF(8)$ of the following form:

$$c = \begin{bmatrix} 0 & 0 & 0 & 0 & 0 & 0 & 0 \\ c_{10} & c_{11} & c_{12} & c_{13} & c_{14} & c_{15} & c_{16} \\ c_{20} & c_{21} & c_{22} & c_{23} & c_{24} & c_{25} & c_{26} \\ 0 & 0 & 0 & 0 & 0 & 0 & 0 \\ c_{40} & c_{41} & c_{42} & c_{43} & c_{44} & c_{45} & c_{46} \\ 0 & 0 & 0 & 0 & 0 & 0 & 0 \\ 0 & 0 & 0 & 0 & 0 & 0 & 0 \end{bmatrix}.$$

Because the indices of the three rows that are allowed to be nonzero form a conjugacy class, the structure of $GF(8)$ and the structure of the Fourier transform interact and thus simplify the relationship between this c and its bispectrum. This is the same interaction that was used to derive a semifast Fourier transform algorithm in Section 1.10. In that section, we saw that the seven-point Fourier transform in $GF(8)$,

$$\begin{bmatrix} V_0 \\ V_1 \\ V_2 \\ V_3 \\ V_4 \\ V_5 \\ V_6 \end{bmatrix} = \begin{bmatrix} 1 & 1 & 1 & 1 & 1 & 1 & 1 \\ 1 & \alpha^1 & \alpha^2 & \alpha^3 & \alpha^4 & \alpha^5 & \alpha^6 \\ 1 & \alpha^2 & \alpha^4 & \alpha^6 & \alpha^1 & \alpha^3 & \alpha^5 \\ 1 & \alpha^3 & \alpha^6 & \alpha^2 & \alpha^5 & \alpha & \alpha^4 \\ 1 & \alpha^4 & \alpha^1 & \alpha^5 & \alpha^2 & \alpha^6 & \alpha^3 \\ 1 & \alpha^5 & \alpha^3 & \alpha & \alpha^6 & \alpha^4 & \alpha^2 \\ 1 & \alpha^6 & \alpha^5 & \alpha^4 & \alpha^3 & \alpha^2 & \alpha^1 \end{bmatrix} \begin{bmatrix} 0 \\ v_1 \\ v_2 \\ 0 \\ v_4 \\ 0 \\ 0 \end{bmatrix},$$

can be reduced to

$$\begin{bmatrix} V_0 \\ V_1 \\ V_2 \\ V_3 \\ V_4 \\ V_5 \\ V_6 \end{bmatrix} = \begin{bmatrix} 1 & 0 & 0 \\ 0 & 1 & 0 \\ 0 & 0 & 1 \\ 1 & 1 & 0 \\ 0 & 1 & 1 \\ 1 & 1 & 1 \\ 1 & 0 & 1 \end{bmatrix} \begin{bmatrix} 1 & 1 & 1 \\ \alpha^1 & \alpha^2 & \alpha^4 \\ \alpha^2 & \alpha^4 & \alpha^1 \end{bmatrix} \begin{bmatrix} v_1 \\ v_2 \\ v_4 \end{bmatrix},$$

from which we can extract the inverse relationship,

$$\begin{bmatrix} v_1 \\ v_2 \\ v_4 \end{bmatrix} = \begin{bmatrix} 1 & \alpha^2 & \alpha^1 \\ 1 & \alpha^4 & \alpha^2 \\ 1 & \alpha^1 & \alpha^4 \end{bmatrix} \begin{bmatrix} V_0 \\ V_1 \\ V_2 \end{bmatrix}.$$

To apply this to our problem, recall that the two-dimensional Fourier transform

$$C_{j'j''} = \sum_{i'=0}^{n-1} \sum_{i''=0}^{n-1} c_{i'i''} \omega^{i'j'} \omega^{i''j''}$$

can be represented by the following diagram:

$$c \leftrightarrow B$$

$$\updownarrow \qquad \updownarrow$$

$$b \leftrightarrow C.$$

For the seven by seven, two-dimensional Fourier transform we are studying, a horizontal arrow denotes a one-dimensional, seven-point Fourier transform relationship along every row of the array, and a vertical arrow denotes a one-dimensional, seven-point Fourier transform relationship along every column of the array. The rows of the array b are the spectra of the rows of c (viewed as row codewords). The columns of B are the spectra of the columns of c (viewed as column codewords).

Thus, four rows of B are zero rows, namely all rows other than rows numbered 1, 2, and 4. Retaining only the three nonzero rows, we can write

$$\begin{bmatrix} B_1 \\ B_2 \\ B_4 \end{bmatrix} = \begin{bmatrix} 1 & \alpha^2 & \alpha \\ 1 & \alpha^4 & \alpha^2 \\ 1 & \alpha & \alpha^4 \end{bmatrix} \begin{bmatrix} C_0 \\ C_1 \\ C_2 \end{bmatrix}.$$

Now refer to the earlier diagram and take the seven-point inverse Fourier transform of each of the three rows of this equation. The inverse Fourier transform of row $c_{j''}$ is row $b_{i''}$, and the inverse Fourier transform of row $B_{j''}$ is row $c_{i''}$. Thus

$$\begin{bmatrix} c_1 \\ c_2 \\ c_4 \end{bmatrix} = \begin{bmatrix} 1 & \alpha^2 & \alpha \\ 1 & \alpha^4 & \alpha^2 \\ 1 & \alpha & \alpha^4 \end{bmatrix} \begin{bmatrix} b_0 \\ b_1 \\ b_2 \end{bmatrix},$$

and all other rows of c are zero. This simplified expression, which we have derived for a special case of the Fourier transform, will be especially useful.

Now we return to the study of the Klein code, which we recast to fit the above form of the Fourier transform. The method can be motivated by examining Figure 11.4. If the i'th column of Figure 11.4 is cyclically downshifted by $5i'$ places, the curve is "twisted" into the simple form shown in Figure 11.6.

Thus our reformulation of the Klein codes uses the twist property of the two-dimensional Fourier transform, but formulated in the language of polynomials. Because we want to reserve the notation $G(x, y)$ and $C(x, y)$ for these polynomials after the twist

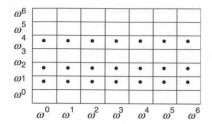

Figure 11.6. Twisted Klein curve in the bicyclic plane.

operation, in this section the Klein polynomial before the twist operation is denoted $G'(x, y)$ and the codeword spectrum polynomial before the twist operation is denoted $C'(x, y)$. Replace the variable y by $x^5 y^3$ so that the polynomial

$$G'(x, y) = x^3 y + y^3 + x$$

becomes the twisted polynomial

$$
\begin{aligned}
G(x, y) &= G'(x, x^5 y^3) \\
&= x^8 y^3 + x^{15} y^9 + x \\
&= x(y^3 + y^2 + 1),
\end{aligned}
$$

by using the fact that $x^8 = x$ in the ring of polynomials over $GF(8)$ modulo the ideal $\langle x^7 - 1, y^7 - 1 \rangle$. Under this transformation, the Klein curve takes the simple form shown in Figure 11.6. To show this, let $g_{i'i''} = G(\omega^{i'}, \omega^{i''})$, choosing $\omega = \alpha$. Then $g_{i'i''} = G(\alpha^{i'}, \alpha^{i''})$, and

$$
g = \begin{bmatrix}
g_{00} & g_{01} & g_{02} & g_{03} & g_{04} & g_{05} & g_{06} \\
0 & 0 & 0 & 0 & 0 & 0 & 0 \\
0 & 0 & 0 & 0 & 0 & 0 & 0 \\
g_{30} & g_{31} & g_{32} & g_{33} & g_{34} & g_{35} & g_{36} \\
0 & 0 & 0 & 0 & 0 & 0 & 0 \\
g_{50} & g_{51} & g_{52} & g_{53} & g_{54} & g_{55} & g_{56} \\
g_{60} & g_{61} & g_{62} & g_{63} & g_{64} & g_{65} & g_{66}
\end{bmatrix}.
$$

Now the twenty-one bicyclic zeros of the twisted Klein polynomial $G(x, y)$ have become very orderly. Indeed, the twenty-one zeros in the bicyclic plane form three "lines." (This pattern of zeros is a consequence of the fact that the *gonality* of the Klein polynomial is 3, a term that we will not define.) Because the codeword components must be zero everywhere that g is nonzero, the codeword c has the special form for which we have given the Fourier transform at the start of this section.

There is one further comment that is needed here. The evaluation $G(\alpha^{i'}, \alpha^{i''})$ is not the inverse Fourier transform that we have been using. The inverse Fourier transform instead corresponds to $G(\alpha^{-i'}, \alpha^{-i''})$. Thus we are actually forming a reciprocal of the Klein curve that we have defined previously. This is a convenient modification because it allows us to consider $C(x, y)$ as a reciprocal of the bispectrum polynomial instead of the bispectrum polynomial itself. This means that instead of the condition that $\deg C(x, y) \leq J$, we have $C_{j'j''} = 0$ if $j' + j'' < J$. This is consistent with our convention for shortened codes.

There is no longer any reason to retain the original indices on the $\mathbf{c}$ vectors. Accordingly, we now redefine $\mathbf{c}_4$ as $\mathbf{c}_0$, and write

$$
\begin{bmatrix} \mathbf{c}_0 \\ \mathbf{c}_1 \\ \mathbf{c}_2 \end{bmatrix} = \begin{bmatrix} 1 & \alpha & \alpha^4 \\ 1 & \alpha^2 & \alpha \\ 1 & \alpha^4 & \alpha^2 \end{bmatrix} \begin{bmatrix} \mathbf{b}_0 \\ \mathbf{b}_1 \\ \mathbf{b}_2 \end{bmatrix},
$$

where $\mathbf{b}_0 \leftrightarrow \mathbf{C}_0$, $\mathbf{b}_1 \leftrightarrow \mathbf{C}_1$, and $\mathbf{b}_2 \leftrightarrow \mathbf{C}_2$.

All that remains to do is to describe $\mathbf{C}_0$, $\mathbf{C}_1$, and $\mathbf{C}_2$. These come from three rows of the bivariate polynomial $C(x, y)$ as a consequence of twisting $C'(x, y)$. Evidently, $\mathbf{C}_0, \mathbf{C}_1$, and $\mathbf{C}_2$ are vectors of blocklength 7 over $GF(8)$, except that certain of their components are equal to zero. This observation follows from the fact that polynomial $C'(x, y)$ is arbitrary, except that $C'_{j'j''} = 0$ if $j' + j'' < J$. Thus, to describe $\mathbf{C}_0$, $\mathbf{C}_1$, and $\mathbf{C}_2$, we need to observe what happens to the zero coefficients of $C'(x, y)$ under the twist operation. Replace y by $x^5 y$ to obtain

$$
C(x, y) = C'(x, x^5 y) = \sum_{j'=0}^{6} \sum_{j''=0}^{6} x^{j'} x^{5j''} y^{j''} C'_{j'j''}
$$

so that $C_{j'j''} = C'_{j'-5j'', j''}$. Thus $C_{j'j''} = 0$ if $((j' - 5j'')) + j'' < J$. This means: if $J = 2$, $\mathbf{C}_0$ has its components equal to zero for $j' = 0, 1, 2$; $\mathbf{C}_1$ has its components equal to zero for $j' = 5, 6$; and $\mathbf{C}_2$ has its components equal to zero for $j' = 3$. In addition, the constraints

$$
\mathbf{C}_3 = \mathbf{C}_1 + \mathbf{C}_0,
$$
$$
\mathbf{C}_4 = \mathbf{C}_2 + \mathbf{C}_1,
$$
$$
\mathbf{C}_5 = \mathbf{C}_2 + \mathbf{C}_1 + \mathbf{C}_0,
$$
$$
\mathbf{C}_6 = \mathbf{C}_2 + \mathbf{C}_0,
$$

must be satisfied. Because $\mathbf{C}_3$, $\mathbf{C}_4$, $\mathbf{C}_5$, and $\mathbf{C}_6$ are not constrained by the required bispectral zeros, they are completely defined by the constraint equations.

If $J = 3$, the situation is more complicated, because the equation of the curve creates other constraining relationships among the spectral components. The constraint $((j' - 5j'')) + j'' < 3$ means that $\mathbf{C}_0$ has its components equal to zero for $j' = 0, 1, 2, 3$;

Table 11.1. *Preliminary defining sets*

	$J = 2$	$J = 3$	$J = 4$	$J = 5$
A_0'	$\{0,1,2\}$	$\{0,1,2,3\}$	$\{0,1,2,3,4\}$	$\{0,1,2,3,4,5\}$
A_1'	$\{5,6\}$	$\{5,6,0\}$	$\{5,6,0,1\}$	$\{5,6,0,1,2\}$
A_2'	$\{3\}$	$\{3,4\}$	$\{3,4,5\}$	$\{3,4,5,6\}$
A_3'	$\{-\}$	$\{1\}$	$\{1,2\}$	$\{1,2,3\}$
A_4'	$\{-\}$	$\{-\}$	$\{6\}$	$\{6,0\}$
A_5'	$\{-\}$	$\{-\}$	$\{-\}$	$\{4\}$
A_6'	$\{-\}$	$\{-\}$	$\{-\}$	$\{-\}$

Table 11.2. *Actual defining sets*

	$J = 2$	$J = 3$	$J = 4$	$J = 5$
A_0	$\{0,1,2\}$	$\{0,1,2,3\}$	$\{0,1,2,3,4\}$	$\{0,1,2,3,4,5\}$
A_1	$\{5,6\}$	$\{5,6,0,1\}$	$\{5,6,0,1,2\}$	$\{5,6,0,1,2,3,4\}$
A_2	$\{3\}$	$\{3,4\}$	$\{3,4,5,6\}$	$\{3,4,5,6,0\}$

C_1 has its components equal to zero for $j' = 5,6,0$; C_2 has its components equal to zero for $j' = 3,4$; and C_3 has its components equal to zero for $j' = 1$. Then because $C_3 = C_1 + C_0$, this last condition also requires that $C_{1j'} = 0$ for $j' = 1$.

To find the defining sets, in general, from the constraint $((j' - 5j'')) + j'' < J$, we will first form the preliminary table (Table 11.1).

Then, to accommodate the constraints relating the spectral components, the actual defining sets of the three Reed–Solomon codes are found to be as given in Table 11.2.

Thus we see that C_0, C_1, and C_2 each has a cyclically sequential set of terms in its defining set, so each is the spectrum of a Reed–Solomon code with a defining set as tabulated. (These codes are actually defined with the primitive element α^{-1}.) Thus the twenty-one-point Klein codeword can be expressed as the concatenation $c = |\ c_0\ |\ c_1\ |\ c_2\ |$ of three sections, given by

$$
\begin{bmatrix} c_0 \\ c_1 \\ c_2 \end{bmatrix} = \begin{bmatrix} \alpha & \alpha^2 & 1 \\ \alpha^2 & \alpha^4 & 1 \\ \alpha^4 & \alpha & 1 \end{bmatrix} \begin{bmatrix} b_0 \\ b_1 \\ b_2 \end{bmatrix},
$$

and $b_0 \in C_1$, $b_1 \in C_2$, and $b_2 \in C_3$, where C_0, C_1, and C_2 are the appropriate Reed–Solomon codes.

For $J = 2$, the three Reed–Solomon codes have spectra with defining sets $\{0, 1, 2\}$, $\{5, 6\}$, and $\{3\}$, respectively. Altogether, there are six check symbols and fifteen data symbols.

For $J = 3$, the three Reed–Solomon codes have spectra with defining sets $\{0, 1, 2, 3\}$, $\{5, 6, 0, 1\}$, and $\{3, 4\}$. Altogether, there are ten check symbols and eleven data symbols.

For $J = 4$, the three Reed–Solomon codes have spectra with defining sets $\{0, 1, 2, 3, 4\}$, $\{5, 6, 0, 1, 2\}$, and $\{3, 4, 5, 6\}$. Altogether, there are fourteen check symbols and seven data symbols.

For $J = 5$, the three Reed–Solomon codes have spectra with defining sets $\{0, 1, 2, 3, 4, 5\}$, $\{5, 6, 0, 1, 2, 3, 4\}$, and $\{3, 4, 5, 6, 0\}$. Altogether there are eighteen check symbols and three data symbols.

By restricting the code to the bicyclic plane, three codeword components at infinity have been dropped. We may want to reinsert these components. If the bicyclic (x, y) plane is extended to the projective (x, y, z) plane, there are three more zeros of the Klein polynomial $G'(x, y, z)$ at $(0, 0, 1)$, $(0, 1, 0)$, and $(1, 0, 0)$. This means that the three components at infinity are C'_{J0}, C'_{0J}, and C'_{00}, and they have a simple correspondence to components extending the three Reed–Solomon codes.

11.6 Hermitian codes constructed from Reed–Solomon codes

The hermitian codes over $GF(q^2)$ have been defined in two ways: first, using the Fermat version of the hermitian polynomial,

$$x^{q+1} + y^{q+1} + 1,$$

and second using the Stichtenoth version of the hermitian polynomial,

$$x^{q+1} + y^q + y.$$

In the projective plane, the codes defined by using these two polynomials are equivalent. When restricted to the bicyclic plane, however, these two forms of the epicyclic hermitian code are quite different. The first has blocklength $n = (q-2)(q+1)^2 = q^3 - 3q - 2$; the second has blocklength $n = q(q^2 - 1)$. We have also seen in Section 10.7 that either of the two cases can be viewed as a quasi-cyclic code, though with two different blocklengths. In this section, we show that the shortened hermitian codes can be represented in a manner similar to the Turyn representation. The first case can be represented as a linear combination of $q + 1$ shortened Reed–Solomon codes over $GF(q^2)$, each of blocklength $(q - 2)(q + 1)$. The second case can be represented as a linear combination of q cyclic Reed–Solomon codes over $GF(q^2)$, each of blocklength $q^2 - 1$. We shall give these two constructions only in the field $GF(16)$. First, we will describe one hermitian code over $GF(16)$ as a linear combination of four Reed–Solomon codes of blocklength 16 by starting with the Stichtenoth form of the hermitian polynomial. Then we will describe an hermitian code as a linear combination of five shortened Reed–Solomon codewords, each of blocklength 10, by starting with the Fermat form of the hermitian polynomial. The hermitian codes in any other field of characteristic 2 of the form $GF(q^2)$ can be treated in similar ways.

The formulation of the shortened hermitian code as a linear combination of four Reed–Solomon codes is obtained by appropriately twisting the plane so that the hermitian curve becomes four straight lines. The twist property of the Fourier transform explains what happens to the codeword spectrum. We will first apply the twist operation to the Stichtenoth version of the hermitian polynomial. Because we want to reserve the notation $G(x, y)$ and $C(x, y)$ for these polynomials after the twist operation, in this section the hermitian polynomial prior to the twist operation is denoted $G'(x, y)$ and the codeword spectrum polynomial prior to the twist operation is denoted $C'(x, y)$. With w replacing y, this polynomial is $G'(x, w) = x^{q+1} + w^q + w$. Now replace w by $x^{q+1}y \pmod{x^{q^2-1} - 1}$; then the polynomial becomes

$$G(x, y) = G'(x, x^{q+1}y)$$
$$= x^{q+1}(y^q + y + 1).$$

The curve shown in Figure 11.3, with $q = 4$, now takes the simple form portrayed in Figure 11.7. The zeros of $G(x, y)$ now are only along the four rows of the (x, y) plane at which $y^q + y + 1 = 0$; these are the four rows indexed by the set $\{\alpha, \alpha^2, \alpha^4, \alpha^8\}$. Thus, under the transformation of coordinates, the hermitian curve $\mathcal{X}$ has become four straight lines. In the general case, the hermitian curve of blocklength $n = q(q^2 - 1)$ is twisted into q straight lines, each with $q^2 - 1$ points.

To find the bispectrum C of codeword c, compute the two-dimensional Fourier transform. Because c is in the shortened hermitian code $\mathcal{C}(\mathcal{X})$, only the components of c on the curve $\mathcal{X}$ can have nonzero values, so there is no need to compute the

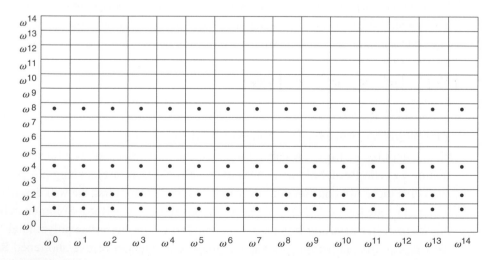

Figure 11.7. Twisted hermitian curve in the bicyclic plane.

Fourier transforms of the other rows. Thus the four Fourier transforms along the four rows can be computed first; then the Fourier transforms along the columns can be computed.

Recall that the two-dimensional Fourier transform given by

$$C_{j'j''} = \sum_{i'=0}^{n-1} \sum_{i''=0}^{n-1} c_{i'i''} \alpha^{i'j'} \alpha^{i''j''}$$

can be represented by the following diagram:

$$c \leftrightarrow B$$

$$\updownarrow \quad \updownarrow$$

$$b \leftrightarrow C,$$

where now a horizontal arrow denotes a one-dimensional, fifteen-point Fourier transform relationship along every row of the array, and a vertical arrow denotes a one-dimensional, fifteen-point Fourier transform relationship along every column of the array. The columns of the array b are the spectra of the columns of c (viewed as column codewords). The rows of B are the spectra of the rows of c (viewed as row codewords).

We must compute the one-dimensional Fourier transform of each column. Accordingly, it is appropriate to study the one-dimensional Fourier transform of a vector of blocklength 15 that is nonzero only in components indexed by 1, 2, 4, and 8. This instance of the Fourier transform was studied in Section 1.10 for another reason. There we observed that if only these four components of v are nonzero, then the first four components of the Fourier transform, given by

$$\begin{bmatrix} V_0 \\ V_1 \\ V_2 \\ V_3 \end{bmatrix} = \begin{bmatrix} 1 & 1 & 1 & 1 \\ \omega^1 & \omega^2 & \omega^4 & \omega^8 \\ \alpha^2 & \alpha^4 & \alpha^8 & \alpha^{16} \\ \alpha^3 & \alpha^6 & \alpha^{12} & \alpha^{24} \end{bmatrix} \begin{bmatrix} v_1 \\ v_2 \\ v_4 \\ v_8 \end{bmatrix},$$

are sufficient to determine all fifteen components of the Fourier transform. It is straightforward to compute the inverse of the matrix to write

$$\begin{bmatrix} v_1 \\ v_2 \\ v_4 \\ v_8 \end{bmatrix} = \begin{bmatrix} \alpha^{14} & \alpha^2 & \alpha & 1 \\ \alpha^{13} & \alpha^4 & \alpha^2 & 1 \\ \alpha^{11} & \alpha^8 & \alpha^4 & 1 \\ \alpha^7 & \alpha & \alpha^8 & 1 \end{bmatrix} \begin{bmatrix} V_0 \\ V_1 \\ V_2 \\ V_3 \end{bmatrix}.$$

This is the inverse Fourier transform augmented by the side information that v_i is equal to zero for all values of i, except $i = 1, 2, 4,$ and 8. Thus the four components V_0,

V_1, V_2, and V_3 are sufficient to recover $\boldsymbol{v}$, and hence are equivalent to the full Fourier transform. Thus all fifteen components of V can be computed from V_0, V_1, V_2, and V_3. To make this explicit, recall that x^2 is a linear function in a field of characteristic 2. This means that the terms $(\alpha^{j''})^2$, $(\alpha^{j''})^4$, and $(\alpha^{j''})^8$ are actually linear functions of $\alpha^{j''}$. This implies that

$$V_j + V_k = (\alpha^j + \alpha^k)v_1 + (\alpha^j + \alpha^k)^2 v_2 + (\alpha^j + \alpha^k)^4 v_4 + (\alpha^j + \alpha^k)^8 v_8$$
$$= V_\ell,$$

where ℓ is determined by $\alpha^\ell = \alpha^j + \alpha^k$. This relationship constrains the spectrum polynomial $V(x)$. Accordingly, the four values of V_j for $j = 0, 1, 2$, and 3 determine all other values of V_j.

This constraint, applied to the fifteen components of the vector V, yields the following relationships:

$V_4 = V_1 + V_0,$

$V_5 = V_2 + V_1,$

$V_6 = V_3 + V_2,$

$V_7 = V_3 + V_1 + V_0,$

$V_8 = V_2 + V_0,$

$V_9 = V_3 + V_1,$

$V_{10} = V_2 + V_1 + V_0,$

$V_{11} = V_3 + V_2 + V_1,$

$V_{12} = V_3 + V_2 + V_1 + V_0,$

$V_{13} = V_3 + V_2 + V_0,$

$V_{14} = V_3 + V_0.$

In this way, all fifteen components of the Fourier transform V can be computed from V_0, V_1, V_2, and V_3.

We return now to the study of the twisted hermitian codes. The twisted codeword array $\boldsymbol{c}$ is nonzero only on the four lines illustrated in Figure 11.7, and the two-dimensional Fourier transform $\boldsymbol{c}$ must satisfy the constraint $\deg C(x, y) \leq J$. This gives two constraints, one in the code domain and one in the transform domain, both of which must be satisfied. We will work backwards from the transform domain. According to the earlier discussion, it is enough to work with the first four components of C if the others are constrained by the equations given earlier. Because only four rows of $\boldsymbol{c}$ are nonzero, only four rows of $\boldsymbol{B}$ are nonzero. Accordingly, we can collapse the inverse

Fourier transform on columns as follows:

$$
\begin{bmatrix} B_1 \\ B_2 \\ B_4 \\ B_8 \end{bmatrix} = \begin{bmatrix} \alpha^{14} & \alpha^2 & \alpha & 1 \\ \alpha^{13} & \alpha^4 & \alpha^2 & 1 \\ \alpha^{11} & \alpha^8 & \alpha^4 & 1 \\ \alpha^7 & \alpha & \alpha^8 & 1 \end{bmatrix} \begin{bmatrix} C_0 \\ C_1 \\ C_2 \\ C_3 \end{bmatrix}.
$$

Note that the rows of this matrix are related to each other by a squaring operation.

Now take the fifteen-point Fourier transform of the four vectors on the right $(C_0, C_1, C_2,$ and $C_3)$ and the vectors on the left $(B_1, B_2, B_4,$ and $B_8)$. The Fourier transform can be interchanged with the matrix multiplication in this equation to give our desired representation:

$$
\begin{bmatrix} c_1 \\ c_2 \\ c_4 \\ c_8 \end{bmatrix} = \begin{bmatrix} \alpha^{14} & \alpha^2 & \alpha & 1 \\ \alpha^{13} & \alpha^4 & \alpha^2 & 1 \\ \alpha^{11} & \alpha^8 & \alpha^4 & 1 \\ \alpha^7 & \alpha & \alpha^8 & 1 \end{bmatrix} \begin{bmatrix} b_0 \\ b_1 \\ b_2 \\ b_3 \end{bmatrix}.
$$

The four codewords b_0, b_1, b_2, and b_3 are Reed–Solomon codewords given by the inverse Fourier transforms of C_0, C_1, C_2, and C_3. This representation in the manner of the Turyn representation gives the hermitian codeword c as the concatenation $c = |c_1|c_2|c_4|c_8|$ in terms of the four Reed–Solomon codewords b_0, b_1, b_2, and b_3. All of this, of course, is merely a property of the Fourier transform over the finite field $GF(16)$, as described in Section 1.10. This is our desired representation of the Stichtenoth version of the hermitian code over $GF(16)$: as a matrix combination of the four Reed–Solomon codes, denoted b_0, b_1, b_2, and b_3. To complete the description, we must specify the four Reed–Solomon codes from which the four codewords b_0, b_1, b_2, and b_3 are taken. These codes are completely defined by their spectral zeros, which can be found by examining the movement of the spectral zeros of the hermitian codes under the twist operation.

Recall that deg $C'(x, w) \leq J$ and $C(x, y) = C'(x, x^{q+1}y)(\bmod x^{q^2-1} - 1)$. By writing $C(x, y) = \sum_{j'=0}^{n-1} C_{j'}(y)x^{j'}$, we can conclude that $C_{j'j''} = 0$ if $((j' - (q+1)j'')) + j'' > J$. This constrains various components of C_0, C_1, C_2, and C_3 to be zero either directly or indirectly, because these components are related to other components of C_j that are constrained to be zero.

For an hermitian code over $GF(16)$, $q = 4$ and $g = 6$. For a straightforward example, let $J = 4$. This gives a $(50, 15, 30)$ epicyclic hermitian code, which can be expressed by a Turyn representation. Then, we have that C_0 has components equal to zero for $j' = 0, 1, 2, 3$; that C_1 has components equal to zero for $j' = 2, 3, 4$; that C_2 has components equal to zero for $j' = 4, 5$; and that C_3 has components equal to zero for $j' = 6$.

For a more complicated example, let $J = 6$. Then, we have that C_0 has components equal to zero for $j' = 0, 1, 2, 3, 4, 5$; that C_1 has components equal to zero for $j' = 2, 3, 4, 5, 6$; that C_2 has components equal to zero for $j' = 4, 5, 6, 7$; and that C_3 has components equal to zero for $j' = 6, 7, 8$. However, this does not complete the enumeration of the spectral zeros. We also know that C_4 has components equal to zero for $j' = 8, 9$, and C_5 has components equal to zero for $j' = 10$. Because $C_4 = C_1 + C_0$, we obtain the additional constraint that $C_{0,8} = C_{1,8}$ and $C_{0,9} = C_{1,9}$. Similarly, because $C_5 = C_2 + C_1$, we have the additional constraint $C_{1,10} = C_{2,10}$. Thus, in this example, the Reed–Solomon codewords cannot be specified independently. Some spectral components must be constrained to take the same values.

To conclude this section, we will develop a Turyn representation for the alternative formulation of the shortened hermitian code, based on the Fermat version of the hermitian polynomial. For this purpose, we will apply the twist operation to the bicyclic plane, observing its effect on the Fermat version of the hermitian polynomial. To do so, replace y by xy, as follows:

$$G(x, y) = G'(x, xy)$$
$$= x^{q+1}(y^{q+1} + 1) + 1.$$

The twisted hermitian curve in the bicyclic plane is the set of zeros of this polynomial, shown in Figure 11.8. These zeros lie on five lines in the bicyclic plane into which the hermitian curve has been twisted. These zeros mark the coordinates of the hermitian code. However, not every point of these lines is a zero. Accordingly, we will describe the hermitian code as a linear combination of five shortened Reed–Solomon codes of blocklength 10. Because the points of the twisted curve do not fill out full rows of the matrix, the Fermat version of the hermitian polynomial does not work quite as neatly under the twist operation. As shown in Figure 11.8, some columns of the matrix contain no points of the curve. This is why Reed–Solomon codes that underlie this form of the hermitian code are shortened Reed–Solomon codes. Moreover, the twisted curve lies on the rows indexed by elements of two conjugacy classes. These are the conjugacy classes of α^0 and α^3, based on the primitive element α.

Recall that the two-dimensional Fourier transform

$$C_{j'j''} = \sum_{i'=0}^{n-1} \sum_{i''=0}^{n-1} c_{i'i''} \alpha^{i'j'} \alpha^{i''j''}$$

can be represented by the following diagram:

$$c \leftrightarrow B$$

$$\updownarrow \quad \updownarrow$$

$$b \leftrightarrow C,$$

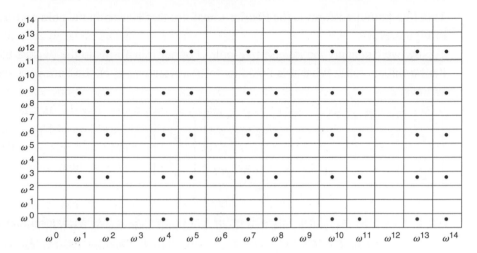

Figure 11.8. Another twisted hermitian curve in the bicyclic plane.

where now a horizontal arrow denotes a one-dimensional, fifteen-point Fourier transform relationship along every row of the array, and a vertical arrow denotes a one-dimensional, fifteen-point Fourier transform relationship along every column of the array. The rows of the array b are the spectra of the rows of c (viewed as row codewords). The columns of B are the spectra of the columns of c (viewed as column codewords).

We will need a fifteen-point Fourier transform in $GF(16)$ of a vector that is zero in all components except those indexed by $0, 3, 6, 9,$ or 12. It is enough to write out only the columns of the Fourier transform matrix corresponding to these five unconstrained components. It is also enough to compute only the first five components of V. The other components can be computed from these five, using the fact that the spectrum is periodic with period 5. The five-point Fourier transform written with $\omega = \alpha^3$ and vector $v = (v_0, v_3, v_6, v_9, v_{12})$ takes the following form:

$$
\begin{bmatrix} V_0 \\ V_1 \\ V_2 \\ V_3 \\ V_4 \end{bmatrix} = \begin{bmatrix} 1 & 1 & 1 & 1 & 1 \\ 1 & \alpha^3 & \alpha^6 & \alpha^9 & \alpha^{12} \\ 1 & \alpha^6 & \alpha^{12} & \alpha^3 & \alpha^9 \\ 1 & \alpha^9 & \alpha^3 & \alpha^{12} & \alpha^6 \\ 1 & \alpha^{12} & \alpha^9 & \alpha^6 & \alpha^3 \end{bmatrix} \begin{bmatrix} v_0 \\ v_3 \\ v_6 \\ v_9 \\ v_{12} \end{bmatrix}.
$$

The inverse of this Fourier transform is given by

$$
\begin{bmatrix} v_0 \\ v_3 \\ v_6 \\ v_9 \\ v_{12} \end{bmatrix} = \begin{bmatrix} 1 & 1 & 1 & 1 & 1 \\ 1 & \alpha^{12} & \alpha^9 & \alpha^6 & \alpha^3 \\ 1 & \alpha^9 & \alpha^3 & \alpha^{12} & \alpha^6 \\ 1 & \alpha^6 & \alpha^{12} & \alpha^3 & \alpha^9 \\ 1 & \alpha^3 & \alpha^6 & \alpha^9 & \alpha^{12} \end{bmatrix} \begin{bmatrix} V_0 \\ V_1 \\ V_2 \\ V_3 \\ V_4 \end{bmatrix}.
$$

This five-point transform is the Fourier transform that is used in the column direction in Figure 11.8.

Now we can apply the inverse five-point Fourier transforms to the twisted form of the hermitian code. What this amounts to is a restatement of the hermitian code as follows:

$$
\begin{bmatrix} c_0 \\ c_3 \\ c_6 \\ c_9 \\ c_{12} \end{bmatrix} = \begin{bmatrix} 1 & 1 & 1 & 1 & 1 \\ 1 & \alpha^{12} & \alpha^9 & \alpha^6 & \alpha^3 \\ 1 & \alpha^9 & \alpha^3 & \alpha^{12} & \alpha^6 \\ 1 & \alpha^6 & \alpha^{12} & \alpha^3 & \alpha^9 \\ 1 & \alpha^3 & \alpha^6 & \alpha^9 & \alpha^{12} \end{bmatrix} \begin{bmatrix} b_0 \\ b_1 \\ b_2 \\ b_3 \\ b_4 \end{bmatrix},
$$

where b_0, b_1, b_2, b_3, and b_4 are five Reed–Solomon codewords, from different codes, with spectra C_0, C_1, C_2, C_3, and C_4, respectively. Because c_0, c_3, c_6, c_9, and c_{12} are to be shortened to the same ten components, b_0, b_1, b_2, b_3, and b_4 can be shortened to those ten components as well. The concatenation $c = |c_0|c_3|c_6|c_9|c_{12}|$ then results in a representation of the hermitian code in the manner of the Turyn representation.

To conclude, we must describe the spectral zeros of the five individual Reed–Solomon codewords b_0, b_1, b_2, b_3, and b_4. Recall that deg $C'(x, w) \le J$, and that $C(x, y) = C'(x, xy)(\bmod x^{q^2-1} - 1)$. Furthermore, $C(x, y) = \sum_{j'=0}^{n-1} C_{j'}(y)x^{j'}$. From these we conclude that $C_{j'j''} = 0$ if $((j' - j'')) + j' > J$. For each j'', the Reed–Solomon spectrum $C_{j''}$ has its j'th component equal to zero for all j', satisfying $((j' - j'')) > J - j'$.

For a straightforward example, let $J = 4$. Then we have that C_0 has components equal to zero for $j' = 0, 1, 2, 3$; that C_1 is equal to zero for $j' = 1, 2, 3$; that C_2 is equal to zero for $j' = 2, 3$; that C_3 is equal to zero for $j' = 3$; and that C_4 has no spectral components constrained to zero. There are ten spectral zeros set equal to zero, so $n - k = 10$.

Problems

11.1   The Stichtenoth version of the hermitian polynomial over $GF(16)$ is as follows:

$$
p(x, y) = x^{17} + y^{16} + y.
$$

Sketch the epicyclic form of this hermitian curve over $GF(16)$ on a 15 by 15 grid. Then permute the elements of the horizontal axis and the vertical axis to rearrange the points of the hermitian curve into blocks. What automorphisms of the curve are evident in this description?

11.2   What are the dimension and minimum distance of the epicyclic form of the hermitian code over $GF(16)$, based on the Stichtenoth form of the hermitian polynomial? What are the dimension and minimum distance of the binary subfield-subcode of this code?

11.3   Let $\mathcal{A} = \{(j',j'') \mid j',j'' \geq 0; j'+j'' \leq J; j' < m\}$. Compute $\|\mathcal{A}\|$ in two ways. One way is by summing the number of terms with j fixed, where $j = j' + j''$. A second way is by first writing

$$\|\mathcal{A}\| = \frac{1}{2}(J+1)(J+2) - \frac{1}{2}(J+1-m)(J+2-m)$$

then expanding the product terms and simplifying the expression. How is this approach explained?

11.4   Let $p(x,y) = x^5 y + y^5 + x$ be a polynomial over $GF(64)$. How many bicyclic zeros does this polynomial have?

11.5   Sketch the Klein curve in the bicyclic plane. What are the bicyclic automorphisms of the Klein curve?

11.6   Let $c = |c_0|c_1|c_2|$, where

$$\begin{bmatrix} c_0 \\ c_1 \\ c_2 \end{bmatrix} = \begin{bmatrix} 1 & \alpha^2 & \alpha^1 \\ 1 & \alpha^4 & \alpha^2 \\ 1 & \alpha^1 & \alpha^4 \end{bmatrix} \begin{bmatrix} b_0 \\ b_1 \\ b_2 \end{bmatrix}.$$

Note that b_0 is a Reed–Solomon code with spectral zeros at $B_{0,1}$ and $B_{0,2}$; b_1 is a $(7,5,3)$ Reed–Solomon code with spectral zeros at $B_{1,5}$ and $B_{1,6}$; and b_2 is a $(7,6,2)$ Reed–Solomon code with a spectral zero at $B_{2,0}$. Show that the octal code has minimum distance 6. Show that the binary subfield-subcode is the binary Golay code.

11.7   State the Turyn representation of the Klein code of blocklength 24. Explain the relationship of the three Reed–Solomon extension components to the three components of the Klein code at infinity.

11.8   The epicyclic codes over $GF(16)$ based on the polynomial

$$p(x,y) = x^{17} + y^2 + y$$

have dimension 510. Describe these codes as the combination of Reed–Solomon codes. Also describe the bicyclic automorphisms of the codes.

11.9   The polynomial

$$p(w) = w^5 + w^2 + 1$$

is primitive over $GF(32)$. Using the substitution $w = y^{11}z$, $p(w)$ can be used to form the homogeneous polynomial

$$p(x,y,z) = x^2 y^2 z^5 + x^7 z^2 + y^9$$

over $GF(32)$. Verify that this polynomial has two singular points. Does this polynomial have a weight function? Determine the genus of the polynomial

$$p(x, y) = y^9 + x^7 + x^2 y^2$$

by examining the gap sequence.

11.10 Construct a table of the family of codes on the hermitian curve over $GF(16)$ by using the Feng–Rao bound.

11.11 The shortened epicyclic hermitian codes are based on the Feng–Rao bound, which allows a code to be defined with defining set equal to the first r monomials in the graded order for any value of r. Can a *punctured* epicyclic hermitian code be constructed as well by evaluating polynomials consisting of only the first r monomials in the graded order for any value of r?

11.12 Show that an epicyclic code is a self-dual code if $J = q^2 - 2$. What are the parameters of the self-dual binary subfield-subcode if $q = 16$?

Notes

A code on an algebraic plane curve can be constructed from a bicyclic code either by puncturing or shortening to the curve. These methods have different attributes and the resulting codes are analyzed by different methods. The punctured codes are best analyzed by using Bézout's theorem. The shortened codes were originally analyzed using the Riemann–Roch theorem, but this theorem is not easily accessible and it is so powerful that it may hide some of the basic structure. For this purpose, the Riemann–Roch theorem has been superseded by the Feng–Rao bound.

The use of the weighted degree and the Feng–Rao bound to construct a large family of shortened codes on curves was introduced by Feng and Rao (1994), and was formally developed in the Ph.D. thesis of Duursma (1993). Høholdt, van Lint, and Pellikaan (1998) further refined this approach by introducing the use of an order function.

The Klein codes were introduced by Hansen (1987) and were later studied by many others, including Pellikaan (1998). The representation of a Klein code as a linear combination of Reed–Solomon codes was discussed by Blahut (1992). A similar representation for an hermitian code in terms of Reed–Solomon codes was discovered independently by Yaghoobian and Blake (1992). The representation for the Stichtenoth form of the hermitian code in terms of Reed–Solomon codes seems to be original here. The similarity of these representations to the Turyn representation is striking. Feng (1999) studied the relationship between the Turyn representation of the Golay code and codes over $GF(8)$ expressed on the Klein quartic. The automorphism group of an hermitian code, which is independent of representation, was studied by Xing (1995).

At the present time, large surveys of plane curves over finite fields with many rational points do not exist. Justesen *et al.* (1989) provided some computer-generated curves, which have been incorporated into the exercises of this book.

12 The Many Decoding Algorithms for Codes on Curves

Codes based on the two-dimensional Fourier transform can be decoded by methods analogous to those methods discussed in Chapter 3 for decoding codes that are based on the one-dimensional Fourier transform, such as the Reed–Solomon codes and other BCH codes. Just as for decoding one-dimensional cyclic codes, the task of finding the errors may be divided into two subtasks: finding the locations of the errors, then finding the magnitudes of the errors. In particular, the family of locator decoding algorithms introduces the notion of a bivariate error-locator polynomial, $\Lambda(x, y)$, into one step of the decoding. However, we no longer have the neat equality that we had in one dimension between the degree of the locator polynomial $\Lambda(x)$ and the number of its zeros. It now takes several polynomials to specify a finite number of bivariate zeros, and so, in two dimensions, we use the locator ideal instead of a locator polynomial as was used in one dimension. Now we have a neat equality between the number of zeros of the locator ideal $\{\Lambda_\ell(x, y) \mid \ell = 1, \ldots, L\}$ and the area of the locator footprint.

The methods for decoding two-dimensional bicyclic codes can also be applied to the decoding of codes on curves. However, restricting a code to a curve in general increases the minimum distance. This means that the decoding algorithm must then be strengthened to reach the minimum distance of the code.

In this chapter, we study decoding algorithms for both bicyclic codes and codes on curves. We give two examples of decoding, both for codes over $GF(16)$. One is an example of decoding a hyperbolic code, and the other is an example of decoding an hermitian code. In each of the two examples, the code has a defining set that provides the set of syndromes. The defining sets in the two examples are not the same, but these two sets do have a large intersection. We shall choose exactly the same error pattern for the two codes. Because the error patterns are the same, the syndromes in the intersection of the two defining sets are equal. Therefore the decoding steps corresponding to these common syndromes are the same. This allows an instructive comparison of the same decoding algorithm when used for two different codes against the same error pattern. The two cases deal with the missing syndromes similarly, though not in exactly the same way.

12.1   Two-dimensional syndromes and locator ideals

We shall study the decoding of the two-dimensional noisy codeword $v = c + e$ for both the case in which c is a codeword on the full bicyclic plane and the case in which c is a codeword restricted to a curve of the bicyclic plane. In either case, the codeword is transmitted and the channel makes errors. The senseword is $v = c + e$. In the first case, the components of the senseword will cover the full bicyclic plane. In the second case, the components of the senseword will be restricted to a curve in the plane corresponding to the definition of the code. Accordingly, in that case, we will regard the senseword as padded with zeros and arranged in the form of a full two-dimensional array, with zeros filling all the elements of the array that are not part of the curve. In this way, the decoding of codes on a plane and the decoding of codes on a curve are unified. Of course, in the computations one can suppress those components of the shortened codeword that are known to be zero, but conceptually we consider those components to be there. (If the code had been punctured to lie on a curve, rather than shortened, then there is an additional difficulty because it is not evident to the decoder how to restore the components that have been dropped. This is why we study only the decoding of shortened codes, rather than punctured codes.)

The senseword, which is the codeword c corrupted by an additive error vector e, has the following components:

$$v_{i'i''} = c_{i'i''} + e_{i'i''} \quad i' = 0, \ldots, n-1$$
$$i'' = 0, \ldots, n-1.$$

If the error vector e is nonzero in at most t places with $t \leq (d_{\min} - 1)/2$, then the decoder should be able to recover the codeword (or the data symbols defining the codeword). The senseword v has the Fourier transform V, given by

$$V_{j'j''} = \sum_{i'=0}^{n-1} \sum_{i''=0}^{n-1} \omega^{i'j'} \omega^{i''j''} v_{i'i''} \quad \begin{array}{l} j' = 0, \ldots, n-1 \\ j'' = 0, \ldots, n-1, \end{array}$$

which is easily computed from the senseword. It immediately follows from the linearity of the Fourier transform that

$$V_{j'j''} = C_{j'j''} + E_{j'j''} \quad \begin{array}{l} j' = 0, \ldots, n-1 \\ j'' = 0, \ldots, n-1. \end{array}$$

But, by construction of the code,

$$C_{j'j''} = 0 \quad (j', j'') \in \mathcal{A}.$$

Hence

$$V_{j'j''} = E_{j'j''} \quad (j',j'') \in \mathcal{A}.$$

Whenever we need a reminder that we know $E_{j'j''}$ only for $(j',j'') \in \mathcal{A}$, we will introduce the alternative notation

$$S_{j'j''} = V_{j'j''} = E_{j'j''} \quad (j',j'') \in \mathcal{A},$$

and refer to $S_{j'j''}$ for $(j',j'') \in \mathcal{A}$ as a *two-dimensional syndrome*. Sometimes it is convenient to overreach the original intent of the terminology and refer to the components $E_{j'j''}$ for $(j',j'') \notin \mathcal{A}$ as *missing syndromes*, again referring to them for this purpose by $S_{j'j''}$.

The error array E and the syndrome array S may be represented as bivariate polynomials $E(x,y)$ and $S(x,y)$, defined by

$$E(x,y) = \sum_{j'=0}^{n-1} \sum_{j''=0}^{n-1} E_{j'j''} x^{j'} y^{j''} \quad \text{and} \quad S(x,y) = \sum_{(j',j'') \in \mathcal{A}} S_{j'j''} x^{j'} y^{j''}.$$

Thus the syndrome polynomial is the error spectrum polynomial "cropped" to the complete defining set $\mathcal{A}$.

The error-locator polynomial that was used in the one-dimensional case is replaced in the two-dimensional case by the *error-locator ideal*. The error-locator ideal is given by

$$\Lambda = \{\Lambda(x,y) \mid \Lambda(x,y)E(x,y) = 0\},$$

where the polynomial product is interpreted in the cyclic form, meaning modulo $\langle x^n - 1, y^n - 1 \rangle$. However, the error-locator ideal must be computed from the expression

$$\Lambda = \{\Lambda(x,y) \mid \Lambda(x,y)S(x,y) = 0\},$$

with the understanding that only terms of the polynomial product involving known coefficients of $S(x,y)$ can be used. We express the locator ideal in terms of a minimal basis as follows:

$$\Lambda = \langle \Lambda^{(\ell)}(x,y) \mid \ell = 1, \ldots, L \rangle.$$

The first task of decoding is to compute the minimal basis $\{\Lambda^{(\ell)}(x,y) \mid \ell = 1, \ldots, L\}$ from the syndrome polynomial $S(x,y)$.

After the minimal polynomials $\Lambda^{(\ell)}(x,y)$ of the locator ideal are known, the full error spectrum can be recovered by any of several methods. One method is to use the set of recursions

$$\sum_{k'}\sum_{k''}\Lambda^{(\ell)}_{k'k''}E_{j'-k',j''-k''} = 0$$

for $\ell = 1,\ldots,L$, which follows from the definition of the locator ideal. To use this set of recursions, choose j',j'', and ℓ so that only one unknown component of E appears in the equation; the equation then can be solved for that component, which then becomes a known component. This process is repeated to obtain, one by one, the other components of E, stopping when all are known. From the full error spectrum E, the error pattern e is computed as an inverse Fourier transform.

An alternative procedure to find e is first to find the zeros of the ideal Λ. This gives the location of the errors. The magnitudes can be computed by setting up a system of linear equations in the unknown error magnitudes, or by a two-dimensional generalization of the Forney formula.

Not all coefficients of the polynomial $E(x,y)$ are known initially; only the syndrome polynomial $S(x,y)$ is known at first. We will start out by computing the connection set for $S(x,y)$ because algorithms are available for computing the connection set for a sequence of syndromes. If the error pattern is correctible, it is trivial to convert the connection set for $S(x,y)$ to a basis for the locator ideal for $E(x,y)$.

We studied the two-dimensional generalization of the Berlekamp–Massey algorithm, known as the Sakata algorithm, in Chapter 8. The Sakata algorithm computes, for each r, a set of minimal connection polynomials $\{\Lambda^{(\ell,r)}(x,y)\}$ for the sequence $S_0, S_1, \ldots, S_{r-1}$, and the footprint Δ of this connection set. The pair $(\{\Lambda^{(\ell,r)}(x,y)\}, \Delta)$ has the same role for the Sakata algorithm that $(\Lambda^{(\ell)}(x), L)$ has for the Berlekamp–Massey algorithm. In this chapter, the Sakata algorithm will be regarded as a computational module that can be called upon as needed. For the purposes of this chapter, it is not necessary to understand why the algorithm works, only to understand how to use it.

12.2　The illusion of missing syndromes

The decoding of a code on the plane $GF(q)^2$ using the Sakata algorithm, or a code on a curve (such as an hermitian code), is based on finding the locator ideal by first computing the connection set and its footprint. If the error pattern is correctable, then the set of minimal connection polynomials generates the locator ideal. This is similar to the one-dimensional case, but with a more elaborate structure. In the two-dimensional case, however, there may be additional considerations. In particular, even though the number of errors t satisfies $2t + 1 \le d_{\min}$, there may not be enough syndromes to set up the number of linear equations in the coefficients of $\Lambda(x,y)$ needed to ensure that the connection set can be computed by inverting the set of linear equations. Indeed, for the hermitian codes, a careful analysis would show that enough linear equations to find the coefficients of $\Lambda(x, y)$ are available only if t satisfies the inequality $2t + 1 \leq d - g$, where g is the genus of the hermitian curve and d is the designed distance of the code.

The decoding situation is similar to the situation for a few BCH codes, such as the $(23, 12, 7)$ binary Golay code and the $(127, 43, 31)$ binary BCH code. For these codes, locator decoding, as by using the Berlekamp–Massey algorithm, unembellished, decodes only to the BCH radius, given as the largest integer not larger than $(d - 1)/2$, where d is the designed distance of the BCH code. To mimic this, define the false decoding radius of an hermitian code as $(d - g - 1)/2$. The Sakata algorithm, unembellished, can decode only to the false decoding radius. There are not enough syndromes for the Sakata algorithm to reach the actual packing radius or even the Goppa radius. We will refer to the additional needed syndromes as *missing syndromes*.

This limitation, however, is due to a deficiency of the locator decoding algorithm in its elementary form, not a limitation of the code. Any code with the minimum distance $d_{\min}$ can correct up to $(d_{\min} - 1)/2$ errors, and the set of syndromes contains all the information that the code provides about the error pattern. Therefore the syndromes must uniquely determine the error if the error weight is not greater than the packing radius $(d_{\min} - 1)/2$. Thus it must be possible to determine the missing syndromes from the given syndromes. There must be a way to extract full value from the known syndromes. We will show that, from a certain point of view, it is only an illusion that needed syndromes are missing. Every needed syndrome is either known, or is implicit in the other syndromes. Moreover, and more subtly, a missing syndrome is completely determined only by those syndromes appearing earlier in the total order. It follows from the Sakata–Massey theorem that each missing syndrome can be determined just at the time it is needed in the Sakata recursion. A simple procedure determines this missing syndrome.

In the coming sections, we illustrate a method, called *syndrome filling*, that appends the missing syndromes to the given syndromes as they are needed. The connection set is computed recursively by Sakata's algorithm. Whenever the next needed syndrome is unknown, that iteration is altered so that the missing syndrome is found before the procedure is continued. Later, in Section 12.7, we will prove that the method of syndrome filling is sound.

There are two ways to fill the missing syndromes for codes on curves, but only one way to fill missing syndromes for codes on the full plane. For a code on a curve, some of the syndromes are implied by the equation of the curve and can be inferred from that equation. Otherwise, the missing syndromes can be filled because, as implied by the Sakata–Massey theorem, there is only one value of the missing syndrome for which the recursion can continue under the condition that $2\|\Delta\| < d_{\min}$. By finding this unique value, the missing syndrome is filled and the recursion continues.

Syndrome filling is easiest to understand for unshortened bicyclic codes, in particular hyperbolic codes, because there is only one mechanism for syndrome filling. This makes it easy to prove that syndrome filling works for these codes. The decoding of hyperbolic codes will be studied in Section 12.3, when the equation of the curve comes into play.

For a code on a curve, it is more difficult to give a formal proof that syndrome filling always works. This is because the method of syndrome filling is interconnected with the method of using the equation of the curve to estimate implied syndromes that are related to the given syndrome. The decoding of hermitian codes will be studied in Section 12.4. The proof that syndrome filling works for codes on curves is more complicated, and hence will be deferred until Section 12.7.

12.3 Decoding of hyperbolic codes

A hyperbolic code, with the designed distance d, is defined in Section 6.4 as a two-dimensional cyclic code on the bicyclic plane $GF(q)^2$, with the defining set given by

$$\mathcal{A} = \{(j', j'') \mid (j' + 1)(j'' + 1) < d\}.$$

The defining set $\mathcal{A}$ is described as the set of bi-indices bounded by a hyperbola. The two-dimensional syndromes are computed from the noisy senseword,

$$S_{j'j''} = \sum_{i'=0}^{n-1} \sum_{i''=0}^{n-1} \omega^{i'j'} \omega^{i''j''} v_{i'i''},$$

for all $(j', j'') \in \mathcal{A}$.

We will choose our example of a hyperbolic code so that we can build on the example of the Sakata algorithm that was given in Section 8.5. That example applied the Sakata algorithm to the array of syndromes that is repeated here in Figure 12.1. A larger pattern of syndromes is shown in Figure 12.2. These are the syndromes for a senseword corresponding to a $(225, 190, 13)$ hyperbolic code over $GF(16)$. The syndromes are the spectral components of the senseword lying below the defining hyperbola. We will use the additional syndromes, given in Figure 12.2, to continue the example begun in Section 8.5. By using the support of the complete defining set $\mathcal{A}$ of the code, the syndromes in Figure 12.2 are cropped from the bispectrum of the error pattern shown in Figure 12.9.

Because this code has designed distance 13, we are assured that if the senseword is within distance 6 of any codeword, then it can be uniquely decoded. Perhaps surprisingly, the Sakata algorithm, together with the filling of missing syndromes, will also uniquely decode many sensewords that are at a distance larger than 6 from the nearest codeword. We shall see that the pattern of syndromes in Figure 12.1 actually corresponds to a pattern of seven errors, and this pattern will be correctly decoded.

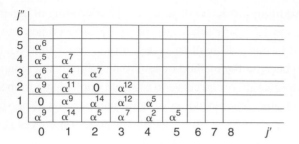

Figure 12.1. Initial set of syndromes.

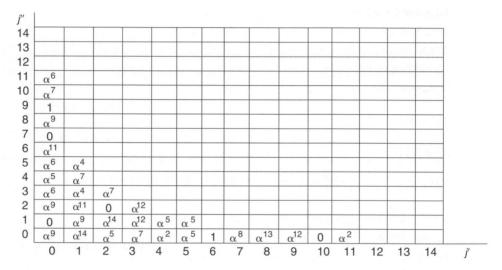

Figure 12.2. Syndromes for decoding a $(225, 190, 13)$ hyperbolic code.

It may be helpful to the understanding of syndrome filling to recall the development of the Sakata–Massey theorem. This development begins with the two-dimensional agreement theorem, given as Theorem 8.3.1 in Chapter 8.

Recall that the agreement theorem states the following. "Suppose $\Lambda(x, y)$ and $\Lambda^*(x, y)$ each produces the bi-index sequence V_0, V_1, ..., V_{r-1} in the graded order. If $r \gtrsim s + s^*$, and if either produces the longer bi-index sequence $V_0, V_1, \ldots, V_{r-1}, V_r$, then so does the other."

The Sakata–Massey theorem then follows as the statement that if $\Lambda(x, y)$, of bidegree s, produces the sequence $V_0, V_1, \ldots, V_{r-1}$ but does not produce the next term V_r then the connection footprint Δ' for the longer sequence $V_0, V_1, \ldots, V_r$ contains the point $r - s = (r' - s', r'' - s'')$. This statement, in turn, leads to the statement that $\Delta \cup (r - \Delta')$ has an area at least $(r' + 1)(r'' + 1)$, where Δ is the connection footprint for the original sequence.

The Sakata algorithm processes the syndrome sequence $S_0, S_1, \ldots$ (in the graded order) to produce the set of minimal connection polynomials $\{\Lambda^{(\ell)}(x, y)\}$. In the example of Figure 12.2, the first 23 syndromes in the graded order are known, but the 24th syndrome is missing. We argue that, whenever a syndrome of a senseword within (nearly) the packing radius is missing, the Sakata–Massey algorithm tells us that it can always (often) be filled in the unique way that minimizes the growth of the footprint of the connection set.

We employ the recursive structure of the Sakata algorithm for the computation. At the start of the rth step, the decoder has previously computed the connection set and the connection footprint Δ_{r-1} for the partial sequence $S_0, S_1, \ldots, S_{r-1}$ in the graded order. Then the connection set for the longer sequence $S_0, S_1, \ldots, S_{r-1}, S_r$ must be computed. If S_r is a missing syndrome, the decoder must first find S_r in order to continue. We first show that only one choice of the missing syndrome will give a locator ideal with a footprint of area t or less. This means that if S_r is the correct value of the missing syndrome, then the locator footprint Δ for that sequence satisfies

$$\Delta(S_0, S_1, \ldots, S_{r-1}, S_r) \leq t,$$

whereas if S'_r is any incorrect value of the missing syndrome, then

$$\Delta(S_0, S_1, \ldots, S_{r-1}, S'_r) > t.$$

Let Δ be the footprint of the connection set for the correct sequence, and let Δ' be the connection set for the incorrect sequence. By the corollary to the Sakata–Massey theorem, Corollary 8.3.3, for $r = (r', r'')$, the set $\Delta \cup (r - \Delta')$ must have area at least $(r' + 1)(r'' + 1)$, which is at least $2t + 1$ if r is not in the defining set. Because Δ has an area at most t, this means that Δ' has area at least $t + 1$, which cannot be, so S'_r is identified as an incorrect value of the rth syndrome. This statement is true for any incorrect value S'_r of the syndrome. Thus S_r is identified as the correct value of that syndrome, as was claimed.

We will describe the first 45 iterations of the Sakata algorithm that are needed to decode the pattern of seven errors underlying the syndromes of Figure 12.2. The first 21 iterations of the Sakata algorithm (from step (0) to step (20)) are the same as the 21 iterations of the long example given in Section 8.5, which applied the Sakata algorithm to the array of syndromes repeated in Figure 12.1.

That example in Section 8.5 terminated after step (20) with the three minimal connection polynomials:

$$\Lambda^{(20,1)}(x, y) = x^4 + \alpha^3 x^2 y + \alpha^5 x^3 + \alpha^{14} xy + \alpha^7 y + \alpha x + \alpha^{13};$$
$$\Lambda^{(20,2)}(x, y) = x^2 y + \alpha^{13} xy + \alpha^3 x^2 + \alpha^3 y + \alpha^6 x + \alpha^6;$$
$$\Lambda^{(20,3)}(x, y) = y^2 + \alpha^{10} xy + \alpha^{13} y + \alpha^{13} x + \alpha^{11},$$

which we abbreviate as follows:

$$\Lambda^{(20,1)} = \begin{vmatrix} \alpha^7 & \alpha^{14} & \alpha^3 \\ \alpha^{13} & \alpha & 0 & \alpha^5 & 1 \end{vmatrix} \qquad \Lambda^{(20,2)} = \begin{vmatrix} \alpha^3 & \alpha^{13} & 1 \\ \alpha^6 & \alpha^6 & \alpha^3 \end{vmatrix}$$

$$\Lambda^{(20,3)} = \begin{vmatrix} 1 \\ \alpha^{13} & \alpha^{10} \\ \alpha^{11} & \alpha^{13} \end{vmatrix}$$

Turning to Figure 12.2, we can immediately continue the example with two more iterations because we are given the next two syndromes.

Step (21) Set $r = 21 = (6,0)$. Syndrome $S_{6,0}$ is known and is equal to 1. One polynomial, $\Lambda^{(20,1)}(x,y)$, reaches the point $(6,0)$. Using polynomial $\Lambda^{(20,1)}(x,y)$ and $r - s = (2,0)$, we compute the discrepancy $\delta_{21}^{(20,1)}$ to be $\alpha^6 \neq 0$. Because $(2,0)$ is already in the footprint, the footprint is not enlarged. The new minimal connection polynomials corresponding to the exterior corners are

$$\begin{aligned} \Lambda^{(21,1)}(x,y) &= \Lambda^{(20,1)}(x,y) + \alpha^6 x B^{(20,2)}(x,y) \\ &= x^4 + \alpha^9 x^2 y + \alpha^5 x^3 + x^2 + \alpha^7 y + \alpha^5 x + \alpha^{13}, \end{aligned}$$

$$\begin{aligned} \Lambda^{(21,2)}(x,y) &= \Lambda^{(20,2)}(x,y) \\ &= x^2 y + \alpha^{13} xy + \alpha^3 x^2 + \alpha^3 y + \alpha^6 x + \alpha^6, \end{aligned}$$

and

$$\begin{aligned} \Lambda^{(21,3)}(x,y) &= \Lambda^{(20,3)}(x,y) \\ &= y^2 + \alpha^{10} xy + \alpha^{13} y + \alpha^{13} x + \alpha^{11}, \end{aligned}$$

which we abbreviate as follows:

$$\Lambda^{(21,1)} = \begin{vmatrix} \alpha^7 & 0 & \alpha^9 \\ \alpha^{13} & \alpha^5 & 1 & \alpha^5 & 1 \end{vmatrix} \qquad \Lambda^{(21,2)} = \begin{vmatrix} \alpha^3 & \alpha^{13} & 1 \\ \alpha^6 & \alpha^6 & \alpha^3 \end{vmatrix}$$

$$\Lambda^{(21,3)} = \begin{vmatrix} 1 \\ \alpha^{13} & \alpha^{10} \\ \alpha^{11} & \alpha^{13} \end{vmatrix}$$

Step (22) Set $r = 22 = (5,1)$. Syndrome $S_{5,1}$ is known and is equal to α^5. Two minimal connection polynomials $\Lambda^{(21,1)}(x,y)$ and $\Lambda^{(21,2)}(x,y)$ reach the point $(5,1)$. Using polynomial $\Lambda^{(21,1)}(x,y)$ and $r - s = (1,1)$, we compute the discrepancy $\delta_{22}^{(21,1)} = \alpha \neq 0$. Using polynomial $\Lambda^{(22,2)}(x,y)$ and $r - s = (3,0)$, we compute $\delta_{22}^{(21,2)} = \alpha \neq 0$.

Because $(1,1)$ and $(3,0)$ are already in the footprint, the footprint is not enlarged. The new minimal connection polynomials are

$$\Lambda^{(22,1)}(x,y) = \Lambda^{(21,1)}(x,y) + \alpha B^{(21,1)}(x,y)$$
$$= x^4 + \alpha^9 x^2 y + \alpha^3 x^3 + \alpha^{14} xy + \alpha^4 x^2 + \alpha^6 y + \alpha^5 x + 1,$$
$$\Lambda^{(22,2)}(x,y) = \Lambda^{(21,2)}(x,y) + \alpha B^{(21,2)}(x,y)$$
$$= x^2 y + \alpha^4 xy + \alpha^3 x^2 + \alpha y + \alpha^7 x + \alpha^4,$$

and

$$\Lambda^{(22,3)}(x,y) = \Lambda^{(21,3)}(x,y)$$
$$= y^2 + \alpha^{10} xy + \alpha^{13} y + \alpha^{13} x + \alpha^{11},$$

which we abbreviate as follows:

$$\Lambda^{(22,1)} = \begin{array}{|ccc}
\alpha^6 & \alpha^{14} & \alpha^9 \\
\hline
1 & \alpha^5 & \alpha^4 & \alpha^3 & 1
\end{array}
\qquad
\Lambda^{(22,2)} = \begin{array}{|ccc}
\alpha & \alpha^4 & 1 \\
\hline
\alpha^4 & \alpha^7 & \alpha^3
\end{array}$$

$$\Lambda^{(22,3)} = \begin{array}{|cc}
1 & \\
\alpha^{13} & \alpha^{10} \\
\hline
\alpha^{11} & \alpha^{13}
\end{array}$$

At this point, the situation changes. The next syndrome S_{23} is missing. In fact, the next three syndromes $S_{23}, S_{24},$ and S_{25} are missing, though other syndromes placed later in the sequence are known. Accordingly, the next three steps use the minimal connection polynomials to infer the missing syndromes.

Step (23) Set $r = 23 = (4,2)$. Syndrome $S_{4,2}$ is missing. All three polynomials $\Lambda^{(22,1)}(x,y)$, $\Lambda^{(22,2)}(x,y)$, and $\Lambda^{(23,3)}(x,y)$ reach the point $r = (4,2)$, so each can be used to estimate the unknown $S_{4,2}$. The three estimates are $S_{4,2}^{(1)} = \alpha^6$, $S_{4,2}^{(2)} = \alpha^6$, and $S_{4,2}^{(3)} = \alpha^6$. These estimates are all the same, so $S_{4,2}$ is now known to be α^6. Because the three estimates agree, the Sakata algorithm can continue. Because of the choice of $S_{4,2}$, the discrepancy for every minimal connection polynomial is zero. The set of minimal connection polynomials does not change during this iteration. If one of the three estimates were different, then that minimal connection polynomial would need to be updated.

Step (24) Set $r = 24 = (3,3)$. Syndrome $S_{3,3}$ is missing. Only two minimal connection polynomials reach the point $r = (3,3)$. Each can be used to estimate $S_{3,3}$. The estimates are $\widehat{S}_{3,3}^{(2)} = \alpha^3$ and $\widehat{S}_{3,3}^{(3)} = \alpha^3$. These estimates are the same, so $S_{3,3}$ is now known to be α^3. Because of this choice of $S_{3,3}$, the discrepancy for both connection polynomials is zero. The set of minimal connection polynomials does not change during this iteration.

Step (25) Set $r = 25 = (2, 4)$. Syndrome $S_{2,4}$ is missing. Only two minimal connection polynomials reach the point $r = (2, 4)$. Each can be used to estimate $S_{2,4}$. The two estimates are both α^6. Thus $S_{2,4}$ is now known to be α^6. The set of minimal connection polynomials does not change during this iteration.

Step (26) Set $r = 26 = (1, 5)$. Syndrome $S_{1,5}$ is known and is equal to α^4. Only one minimal connection polynomial $\Lambda^{(25,3)}(x, y)$ reaches the point $r = (1, 5)$ from the earlier syndromes. The discrepancy $\delta_{26}^{(25,3)}$ is zero. The set of minimal connection polynomials does not change during this iteration.

Step (27) Set $r = 27 = (0, 6)$. Syndrome $S_{0,6}$ is known and is equal to α^{11}. Only one minimal connection polynomial $\Lambda^{(26,3)}(x, y)$ reaches the point $r = (0, 6)$. The discrepancy $\delta_{27}^{(26,3)}$ is zero. The set of minimal connection polynomials does not change during this iteration.

Step (28) Set $r = 28 = (7, 0)$. Syndrome $S_{7,0}$ is known and is equal to α^8. Only one minimal connection polynomial $\Lambda^{(27,1)}(x, y)$ reaches the point $r = (7, 0)$. Using polynomial $\Lambda^{(27,1)}(x, y)$ and $r - s = (3, 0)$, we compute the discrepancy $\delta_{28}^{(27,1)} = \alpha^8$. Because $(3, 0)$ is already in the footprint, the footprint is not enlarged. The set of minimal connection polynomials changes to

$$\Lambda^{(28,1)}(x, y) = \Lambda^{(27,1)}(x, y) + \alpha^8 B^{(27,2)}(x, y)$$
$$= x^4 + \alpha^3 x^3 + \alpha^9 x^2 y + \alpha^4 x^2 + xy + \alpha x + \alpha^{11} y + \alpha,$$

$$\Lambda^{(28,2)}(x, y) = \Lambda^{(27,2)}(x, y)$$
$$= x^2 y + \alpha^4 xy + \alpha^3 x^2 + \alpha y + \alpha^7 x + \alpha^4,$$

and

$$\Lambda^{(28,3)}(x, y) = \Lambda^{(27,3)}(x, y)$$
$$= y^2 + \alpha^{10} xy + \alpha^{13} y + \alpha^{13} x + \alpha^{11},$$

which we abbreviate as follows:

$$\Lambda^{(28,1)} = \begin{vmatrix} \alpha^{11} & \alpha^0 & \alpha^9 \\ \alpha^1 & \alpha^1 & \alpha^4 & \alpha^3 & 1 \end{vmatrix} \qquad \Lambda^{(28,2)} = \begin{vmatrix} \alpha & \alpha^4 & 1 \\ \alpha^4 & \alpha^7 & \alpha^3 \end{vmatrix}$$

$$\Lambda^{(28,3)} = \begin{vmatrix} 1 \\ \alpha^{13} & \alpha^{10} \\ \alpha^{11} & \alpha^{13} \end{vmatrix}$$

Step (29) to step (35) There is no change to the connection set during these iterations; all missing syndromes are filled in turn.

Step (36) Set $r = 36 = (8, 0)$. Syndrome $S_{8,0}$ is known and is equal to α^{13}. Only one connection polynomial $\Lambda^{(35,1)}$ reaches the point $(8, 0)$. The discrepancy $\delta_{36}^{(35,1)}$ is α^7.

Because $(8,0) - (4,0) = (4,0)$, which was not in the footprint, the footprint is enlarged to include the point $(4,0)$. The new minimal connection polynomials are

$$\Lambda^{(36,1)}(x,y) = x\Lambda^{(35,1)}(x,y) + \alpha^7 B^{(35,2)}(x,y)$$
$$= x^5 + \alpha^3 x^4 + \alpha^9 x^3 y + \alpha^4 x^3 + \alpha^0 x^2 y$$
$$+ \alpha^1 x^2 + \alpha^9 xy + \alpha^0 y + \alpha^3,$$

$$\Lambda^{(36,2)}(x,y) = \Lambda^{(35,2)}(x,y)$$
$$= x^2 y + \alpha^4 xy + \alpha^3 x^2 + \alpha y + \alpha^7 x + \alpha^4,$$

and

$$\Lambda^{(35,3)}(x,y) = \Lambda^{(35,3)}(x,y)$$
$$= y^2 + \alpha^{10} xy + \alpha^{13} y + \alpha^{13} x + \alpha^{11},$$

which we abbreviate as follows:

$$\Lambda^{(36,1)} = \begin{vmatrix} \alpha^0 & \alpha^9 & \alpha^0 & \alpha^9 \\ \alpha^3 & 0 & \alpha^1 & \alpha^4 & \alpha^3 & 1 \end{vmatrix} \qquad \Lambda^{(36,2)} = \begin{vmatrix} \alpha & \alpha^4 & 1 \\ \alpha^4 & \alpha^7 & \alpha^3 \end{vmatrix}$$

$$\Lambda^{(36,3)} = \begin{vmatrix} 1 \\ \alpha^{13} & \alpha^{10} \\ \alpha^{11} & \alpha^{13} \end{vmatrix}$$

At this point we notice a concern. The current footprint has area 7, while the hyperbolic code has minimum distance 13, and so can only guarantee the correction of six errors. Thus we could now choose to declare the error pattern to be uncorrectable. Instead, we continue as before, filling missing syndromes as long as this process continues to work.

Step (37) to step (44) There is no change to the set of minimal connection polynomials during these iterations; all missing syndromes are filled in turn.
Step (45) Set $r = 45 = (9,0)$. Syndrome $S_{9,0}$ is known and is equal to α^{12}. The discrepancy $\delta_{45}^{(44,1)} = \alpha^{10}$. Only one connection polynomial $\Lambda^{(44,1)}(x,y)$ reaches the point $(9,0)$. Then

$$\Lambda^{(45,1)}(x,y) = \Lambda^{(44,1)}(x,y) + \alpha^{10} B^{(44,1)}(x,y)$$
$$= x^5 + \alpha^9 x^3 y + \alpha^{12} x^3 + \alpha^{11} x^2 y + \alpha^{14}$$
$$x^2 + \alpha^1 xy + \alpha^4 x + \alpha^3 y + \alpha^7,$$

$$\Lambda^{(45,2)}(x,y) = \Lambda^{(44,2)}(x,y)$$
$$= x^2 y + \alpha^4 xy + \alpha^3 x^2 + \alpha y + \alpha^7 x + \alpha^4,$$

and

$$\Lambda^{(45,3)}(x,y) = \Lambda^{(44,3)}(x,y)$$
$$= y^2 + \alpha^{10}xy + \alpha^{13}y + \alpha^{13}x + \alpha^{11},$$

which we abbreviate as follows:

$$\Lambda^{(45,1)} = \begin{vmatrix} \alpha^3 & \alpha^1 & \alpha^{11} & \alpha^9 \\ \alpha^7 & \alpha^4 & \alpha^{14} & \alpha^{12} & 0 & 1 \end{vmatrix} \qquad \Lambda^{(45,2)} = \begin{vmatrix} \alpha & \alpha^4 & 1 \\ \alpha^4 & \alpha^7 & \alpha^3 \end{vmatrix}$$

$$\Lambda^{(45,3)} = \begin{vmatrix} 1 & & \\ \alpha^{13} & \alpha^{10} \\ \alpha^{11} & \alpha^{13} \end{vmatrix}$$

At the end of this step, if the set of three minimal connection polynomials is tested, it will be found to have seven zeros. This is equal to the area of the connection footprint as shown in Figure 12.3.

Therefore the set of minimal connection polynomials generates the locator ideal for an error word with errors at the locations of these seven zeros. Indeed, if the senseword lies within the packing radius about the correct codeword, this must be the error pattern. Therefore the computation of the locator ideal is complete, and further iterations will not change the locator ideal, but will only fill missing syndromes. If desired, the process can be continued to fill missing syndromes until all known syndromes have been visited to ensure that the error pattern is indeed a correctable error pattern.

At this point, the locator ideal has been computed and the error correction can proceed by any of the several methods described in Section 12.5. The most straightforward of the methods described there is to continue the iterations described above to fill syndromes, producing a new syndrome at each iteration until the full bispectrum of the error pattern is known. An inverse two-dimensional Fourier transform then completes the computation of the error pattern.

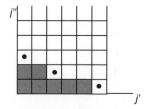

Figure 12.3. Final footprint for the hyperbolic code.

12.4   Decoding of hermitian codes

An epicyclic hermitian code of designed distance d was defined in Section 10.6 as a two-dimensional code shortened, or punctured, to lie on an hermitian curve in the bicyclic plane. The shortened code c is suitable for decoding, because all components $c_{i'i''}$ are zero for indices (i', i'') that do not correspond to points of the curve. This code can be decoded to the designed distance by the methods of two-dimensional locator decoding, such as the Sakata algorithm. However, embellishments to the Sakata algorithm are needed to account for missing and implied syndromes. The implied syndromes are found by using the knowledge that nonzero components of the codeword and the senseword can only occur for points on the curve. This works because the equation of the curve $G(x, y) = 0$ introduces dependencies among the Fourier transform components. Let $g_{i'i''} = (1/n^2)G(\omega^{-i'}, \omega^{-i''})$ and consider the array $v_{i'i''}g_{i'i''}$, where v is the senseword padded with zeros and embedded into a two-dimensional array. For every (i', i''), either $v_{i'i''}$ or $g_{i'i''}$ equals zero, so $v_{i'i''}g_{i'i''} = 0$. By the convolution theorem, $G(x, y)V(x, y) = 0$. This gives a relationship among the coefficients of $V(x, y)$. Likewise, $G(x, y)C(x, y) = 0$ and $G(x, y)E(x, y) = 0$. These equations provide relationships among the coefficients of $C(x, y)$, and among the coefficients of $E(x, y)$. In particular, because $G(x, y) = x^{q+1} = y^q + y$, we know that $x^{q+1}E(x, y) + y^q E(x, y) + yE(x, y) = 0$. This means that $E_{j'-q-1, j''} = E_{j', j''-q} + E_{j', j''-1}$.

For example, in the plane $GF(256)^2$, the hermitian curve is defined by the polynomial $G(x, y) = x^{17} + y^{16} + y$. The expression $G(x, y)C(x, y) = 0$ implies that

$$C_{j'+17, j''} + C_{j', j''+16} + C_{j', j''+1} = 0.$$

By replacing $j' + 17$ by j', this becomes

$$C_{j'j''} = C_{j'-17, j''+16} + C_{j'-17, j''+1}.$$

When these three terms are arranged in the total order, $C_{j'j''}$ is the last term of the three. If the other two terms are known, then $C_{j'j''}$ can be computed from them. We then say that $C_{j'j''}$ is given by the equation of the curve. If $(j' - 17, j'' + 16)$ and $(j' - 17, j'' + 1)$ are in the defining set of the code, then we say that the extra syndrome $S_{j'j''}$ is given by the equation of the curve, or that $S_{j'j''}$ is an *implied syndrome*. In particular, if the syndromes $S_{j'j''}$ are known for $j' = 0, \ldots, 16$ and for $j'' = 0, \ldots, 254$, then all other $S_{j'j''}$ are implied by the equation of the curve. Figure 12.4 shows the array of syndromes partitioned into known syndromes, implied syndromes, and missing syndromes. The implied syndromes can be inferred from the known syndromes by the equation of the curve. We shall see that the missing syndromes can then be inferred by using the Sakata–Massey theorem, provided the number of errors does not exceed the designed packing radius of the code.

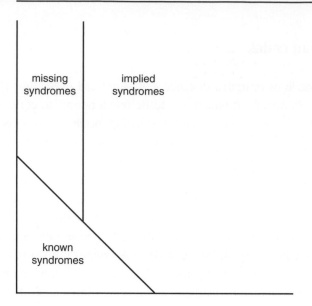

Figure 12.4. Syndromes for an hermitian code.

We summarize this observation as follows. Let v be a vector on *an* hermitian curve with weight at most t and $v \leftrightarrow V$. Suppose that r satisfies $J < r' + r'' < J + q$ and $r' \leq q$. There is exactly one way to extend the sequence $V_0, V_1, \ldots, V_{r-1}$ to a sequence $V_0, V_1, \ldots, V_r$ so that the locator ideal has a footprint with area at most t. This statement will be examined more fully in Section 12.7. For now, we will accept the statement as an unsupported assertion.

Figure 12.5 is a detailed algorithm for finding the connection set for an hermitian code. The algorithm uses the Sakata algorithm, augmented with two options for syndrome filling. If a syndrome is known, then it is used in the natural way to proceed with the current iteration. If a syndrome is implied, then it is computed from the equation of the curve before proceeding with the current iteration. If the syndrome is missing, then it is chosen as the unique value for which $\|\Delta_r\| \leq t$. To find the missing syndrome, there is no need to try every possible value for S_r; only those values produced by the most recent set of minimal connection polynomials need to be tried. We shall see that the majority of the connection polynomials must produce the correct syndrome.

As a simple example of a decoder based on the Sakata algorithm, we shall consider the $(60, 40, 15)$ epicyclic hermitian code over $GF(16)$, defined on the curve $x^5 + y^4 + y = 0$ over $GF(16)$. This code can correct seven errors. We will decode a senseword based on the set of syndromes shown in Figure 12.1. These syndromes are based on an error pattern with seven errors at the seven points $(i', i'') = (0, 1), (2, 3), (8, 3), (1, 7), (11, 3), (5, 3),$ and $(14, 3)$ as discussed in Section 12.5. These seven points all happen to be on the curve, because for each of these values of (i', i''), the point $(\alpha^{i'}, \alpha^{i''})$ is a zero of $x^5 + y^4 + y$. This fact was not a prior condition and was not

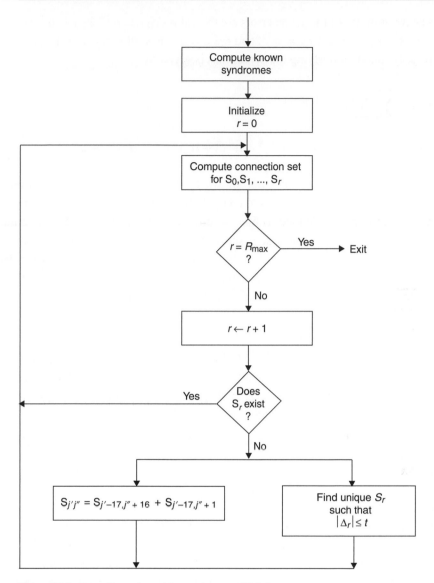

Figure 12.5. Decoding a hermitian code over $GF(16)$.

relevant to the decoder for the hyperbolic code in Section 12.3, and there that fact was completely incidental to the calculations of that decoder. The hermitian code, however, only exists on the points of the curve, and errors can only occur on the points of the curve, and the decoder can use this prior information. By choosing the same error pattern for both the hyperbolic code and the hermitian code, the calculations of the decoding algorithms are nearly the same, although some of the reasoning is different.

The magnitudes of the seven errors are the values $e_{0,1} = \alpha^{12}$, $e_{2,3} = \alpha^2$, $e_{8,3} = \alpha^{15}$, $e_{1,7} = \alpha^{11}$, $e_{11,3} = \alpha$, $e_{5,3} = \alpha^{12}$, and $e_{14,3} = \alpha^7$. For all other (i', i''), $e_{i'i''} = 0$. The error polynomial underlying the syndromes is given by

$$e(x, y) = \sum_{i'=0}^{14} \sum_{i''=0}^{14} e_{i'i''} x^{i'} y^{i''}$$

$$= (\alpha^{10} x^{14} + \alpha^0 x^{11} + \alpha^8 x^8 + \alpha^6 x^5 + \alpha^1 x^2) y^3 + \alpha^7 xy^7 + \alpha^6 y.$$

The error polynomial $e(x, y)$ is not known to the decoder; only the syndromes are known. The task of the decoder is to compute this polynomial from the syndromes.

The syndromes are defined as $S_{j'j''} = e(\alpha^{j'}, \alpha^{j''})$ for $j' + j'' \leq 5$. The pattern of syndromes for the hermitian code is shown in Figure 12.6. This is a much smaller set of syndromes than the set available to the hyperbolic code. However, now the decoder knows that errors can only occur at the points of the curve, and it uses this prior information. The decoder for the hyperbolic code had to deal with errors anywhere in the bicyclic plane.

The next two syndromes in sequence, $S_{6,0}$ and $S_{5,1}$, can be filled immediately by using the equation of the curve to write

$$S_{j'j''} = S_{j'-5,j''+4} + S_{j'-5,j''+1}.$$

Although this equation could also be used to compute syndrome $S_{5,0}$ from syndromes $S_{0,4}$ and $S_{0,1}$, this is not necessary because syndrome $S_{5,0}$ is already known. The two

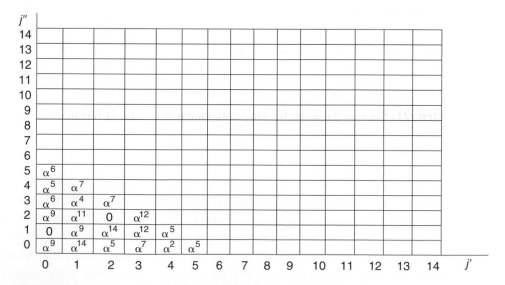

Figure 12.6. Syndromes for decoding a $(64, 44, 15)$ hermitian code.

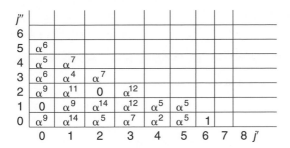

Figure 12.7. The start of syndrome filling.

implied syndromes are seen to have the values $S_{6,0} = S_{1,4} + S_{1,1} = 1$ and $S_{5,1} = S_{0,5} + S_{0,2} = \alpha^5$. The syndrome array, when filled with these two implied syndromes, $S_{6,0}$ and $S_{5,1}$, is given in Figure 12.7.

There are twenty-three syndromes here, twenty-one satisfying $j' + j'' \leq 5$, and the two that were computed as implied syndromes from the equation of the curve. Because we are using the graded order, with y larger than x, the two extra syndromes $S_{6,0} = 1$ and $S_{5,1} = \alpha^5$ come immediately after the twenty-one normal syndromes. In the graded order, the sequence of syndromes is as follows:

$$\alpha^9, \alpha^{14}, 0, \alpha^5, \alpha^9, \alpha^9, \alpha^7, \alpha^{14}, \alpha^{11}, \alpha^6, \alpha^2, \alpha^{12},$$

$$0, \alpha^4, \alpha^5, \alpha^5, \alpha^5, \alpha^{12}, \alpha^7, \alpha^7, \alpha^6, 1, \alpha^5, S_{4,2}, S_{3,3}, \ldots$$

Starting in position (23) and continuing, the missing syndromes $S_{4,2}, S_{3,3}, \ldots$ appear as unknowns. This is the same sequence of syndromes used in Section 12.3 for decoding a hyperbolic code. Hence the Sakata algorithm consists of the same steps insofar as the known sequence of syndromes allows. In particular, steps (21) and (22) are the same for decoding the senseword of the hermitian code as they were for the senseword of the hyperbolic code. We can simply repeat these steps from Section 12.3 because the syndromes are the same.

Step (21) Set $r = 21 = (6, 0)$. One polynomial $\Lambda^{(20,1)}(x, y)$ reaches the point $(6, 0)$. Using polynomial $\Lambda^{(20,1)}(x, y)$ and $r - s = (2, 0)$, we compute $\delta_{21}^{(20,1)}$ to be $\alpha^6 \neq 0$. Because $(2, 0)$ is already in the footprint, the footprint is not enlarged. The new minimal connection polynomials are

$$\Lambda^{(21,1)}(x, y) = \Lambda^{(20,1)}(x, y) + \alpha^6 x B^{(20,2)}(x, y)$$

$$= x^4 + \alpha^9 x^2 y + \alpha^5 x^3 + x^2 + \alpha^7 y + \alpha^5 x + \alpha^{13},$$

$$\Lambda^{(21,2)}(x, y) = \Lambda^{(20,2)}(x, y)$$

$$= x^2 y + \alpha^{13} xy + \alpha^3 x^2 + \alpha^3 y + \alpha^6 x + \alpha^6,$$

and

$$\Lambda^{(21,3)}(x,y) = \Lambda^{(20,3)}(x,y)$$
$$= y^2 + \alpha xy + y + \alpha^5 x + \alpha^9,$$

which we abbreviate as follows:

$$\Lambda^{(21,1)} = \begin{array}{|ccc}
\alpha^7 & 0 & \alpha^9 \\
\alpha^{13} & \alpha^5 & 1 & \alpha^5 & 1
\end{array} \qquad
\Lambda^{(21,2)} = \begin{array}{|ccc}
\alpha^3 & \alpha^{13} & 1 \\
\alpha^6 & \alpha^6 & \alpha^3
\end{array}$$

$$\Lambda^{(21,3)} = \begin{array}{|cc}
1 & \\
\alpha^{13} & \alpha^{10} \\
\alpha^{11} & \alpha^{13}
\end{array}$$

Step (22) Set $r = 22 = (5,1)$. Two polynomials $\Lambda^{(21,1)}(x,y)$ and $\Lambda^{(21,2)}(x,y)$ reach the point $(5,1)$. Using polynomial $\Lambda^{(21,1)}(x,y)$ and $r-s = (1,1)$, we compute $\delta_{22}^{(21,1)} = \alpha \neq 0$. Using polynomial $\Lambda^{(22,2)}(x,y)$ and $r-s = (3,0)$, we compute $\delta_{22}^{(21,2)} = \alpha \neq 0$. Because $(1,1)$ and $(3,0)$ are already in the footprint, the footprint is not enlarged. The new minimal connection polynomials are

$$\Lambda^{(22,1)}(x,y) = \Lambda^{(21,1)}(x,y) + \alpha B^{(21,1)}(x,y)$$
$$= x^4 + \alpha^9 x^2 y + \alpha^3 x^3 + \alpha^{14} xy + \alpha^4 x^2 + \alpha^6 y + \alpha^5 x + 1, \quad .$$
$$\Lambda^{(22,2)}(x,y) = \Lambda^{(21,2)}(x,y) + \alpha B^{(21,2)}(x,y)$$
$$= x^2 y + \alpha^4 xy + \alpha^3 x^2 + \alpha y + \alpha^7 x + \alpha^4,$$

and

$$\Lambda^{(22,3)}(x,y) = \Lambda^{(21,3)}(x,y)$$
$$= y^2 + \alpha xy + y + \alpha^5 x + \alpha^9,$$

which we abbreviate as follows:

$$\Lambda^{(22,1)} = \begin{array}{|cccc}
\alpha^6 & \alpha^{14} & \alpha^9 & \\
1 & \alpha^5 & \alpha^4 & \alpha^3 & 1
\end{array} \qquad
\Lambda^{(22,2)} = \begin{array}{|ccc}
\alpha & \alpha^4 & 1 \\
\alpha^4 & \alpha^7 & \alpha^3
\end{array}$$

$$\Lambda^{(22,3)} = \begin{array}{|cc}
1 & \\
\alpha^{13} & \alpha^{10} \\
\alpha^{11} & \alpha^{13}
\end{array}$$

Before continuing, we will make some remarks. All of the original syndromes have now been used. No more unused syndromes are known. The next syndromes, $S_{4,2}$, $S_{3,3}$, and $S_{2,4}$, are not known. Indeed, no other syndromes later in the sequence are known, though some are implied by the equation of the curve. Because $\|\Delta_{22}\| = 6$, it could be that Δ_{22} is the correct connection footprint for a pattern of six errors, but it cannot possibly be the connection footprint for a pattern of seven errors, which we know is the case. However, all syndromes have now been used, even the two extra syndromes, $S_{6,0}$ and $S_{5,1}$. There is no more information available to the decoder, even though it is not yet finished. Thus this example makes it evident that something more can be squeezed from the syndromes to enable the correction of seven errors. At this point, the notion of syndrome filling should become almost intuitively self-evident, though not proved. Indeed, how could the situation be otherwise?

The next syndrome is missing and will be filled by a majority decision among syndrome estimates. Subsequent syndromes will be found by this same method, or will be found from the equation of the curve. We shall see that the footprint of the connection set does not change until syndrome $S_{3,4}$ is reached at iteration number 32.

The sequence of intermediate steps is as follows.

Step (23) Set $r = 23 = (4, 2)$. All three polynomials reach the point $r = (4, 2)$, so each can be used to estimate the unknown $S_{4,2}$. The three estimates are $\widehat{S}_{4,2}^{(1)} = \alpha^6$, $\widehat{S}_{4,2}^{(2)} = \alpha^6$, and $\widehat{S}_{4,2}^{(3)} = \alpha^6$. The majority decision is that $S_{4,2} = \alpha^6$, and the Sakata algorithm can continue. Because the three estimates agree, the three discrepancies all are equal to zero, so the set of minimal connection polynomials does not change.

Step (24) Set $r = 24 = (3, 3)$. Only two polynomials reach the point $r = (3, 3)$. Each can be used to estimate $S_{3,3}$. The estimates are $\widehat{S}_{3,3}^{(2)} = \alpha^3$, and $\widehat{S}_{3,3}^{(3)} = \alpha^3$. Thus $S_{3,3} = \alpha^3$. Because the two estimates agree, both discrepancies are zero, so the set of the minimal connection polynomials does not change.

Step (25) Set $r = 25 = (2, 4)$. Only two polynomials reach the point $r = (2, 4)$. Each can be used to estimate $S_{2,4}$. The two estimates are both α^6. Thus $S_{2,4} = \alpha^6$. Again, the set of minimal connection polynomials does not change.

Step (26) Set $r = 26 = (1, 5)$. Only one polynomial reaches the point $r = (1, 5)$. Thus $S_{1,5} = \alpha^4$. The discrepancy is zero, and the set of minimal connection polynomials does not change.

Step (27) Set $r = 27 = (0, 6)$. Only one polynomial reaches the point $r = (0, 6)$. Thus $S_{0,6} = \alpha^{11}$. Again, the discrepancy is zero, and the set of minimal connection polynomials does not change.

The next three syndromes can be inferred from the equation of the curve.

Step (28) Set $r = 28 = (7, 0)$. Syndrome $S_{7,0}$ can be computed as an implied syndrome using the equation of the curve as $S_{7,0} = S_{2,4} + S_{2,1} = \alpha^8$. Only one minimal connection polynomial $\Lambda^{(27,1)}(x, y)$ reaches the point $r = (7, 0)$. Using polynomial $\Lambda^{(27,1)}(x, y)$ and $r - s = (3, 0)$, we compute that the discrepancy $\delta_{28}^{(27,1)}$ is α^8. Because $(3, 0)$ is already in the footprint, the footprint is not enlarged. The set of minimal connection polynomials changes to

$$\Lambda^{(28,1)}(x,y) = \Lambda^{(27,1)}(x,y) + \alpha^8 B^{(27,2)}(x,y)$$
$$= x^4 + \alpha^3 x^3 + \alpha^9 x^2 y + \alpha^4 x^2 + xy + \alpha x + \alpha^{11} y + \alpha,$$

$$\Lambda^{(28,2)}(x,y) = \Lambda^{(27,2)}(x,y)$$
$$= x^2 y + \alpha^4 xy + \alpha^3 x^2 + \alpha y + \alpha^7 x + \alpha^4,$$

and

$$\Lambda^{(28,3)}(x,y) = \Lambda^{(27,3)}(x,y)$$
$$= y^2 + \alpha^{10} xy + \alpha^{13} y + \alpha^{13} x + \alpha^{11},$$

which we abbreviate as follows:

$$\Lambda^{(28,1)} = \begin{vmatrix} \alpha^{11} & \alpha^0 & \alpha^9 \\ \alpha^1 & \alpha^1 & \alpha^4 & \alpha^3 & 1 \end{vmatrix} \qquad \Lambda^{(28,2)} = \begin{vmatrix} \alpha & \alpha^4 & 1 \\ \alpha^4 & \alpha^7 & \alpha^3 \end{vmatrix}$$

$$\Lambda^{(28,3)} = \begin{vmatrix} 1 & \\ \alpha^{13} & \alpha^{10} \\ \alpha^{11} & \alpha^{13} \end{vmatrix}$$

Step (29) Set $r = 29 = (6,1)$. Syndrome $S_{6,1}$ can be computed as an implied syndrome, using the equation of the curve, as $S_{6,1} = S_{1,5} + S_{1,2} = \alpha^4 + \alpha^{11} = \alpha^{13}$. Two polynomials $\Lambda^{(28,1)}$ and $\Lambda^{(28,2)}$ reach the point $(6,1)$. The discrepancy is zero for both. The set of minimal connection polynomials does not change.

Step (30) Set $r = 30 = (5,2)$. Syndrome $S_{5,2}$ can be computed as an implied syndrome, using the equation of the curve, as $S_{5,2} = S_{0,6} + S_{0,3} = \alpha^{11} + \alpha^6 = \alpha^1$. All three connection polynomials reach the point $(15,2)$. All three discrepancies are zero. The set of minimal connection polynomials does not change.

Step (31) Set $r = 31 = (4,3)$. Syndrome $S_{4,3}$ is missing. All three connection polynomials reach the point $(4,3)$ to give the three estimates $\widehat{S}_{4,3}^{(1)} = \alpha^{10}$, $\widehat{S}_{4,3}^{(2)} = \alpha^{10}$, and $\widehat{S}_{4,3}^{(3)} = \alpha^{10}$. Then, by syndrome voting, the value of $S_{4,3}$ is α^{10}. Because the three estimates of $S_{4,3}$ are the same, all discrepancies are zero. The set of minimal connection polynomials does not change.

In this way, one by one, the missing syndromes are filled. At the end of iteration (31), the array of syndromes has been filled, as shown in Figure 12.8.

Step (32) to step (35) There is no change to the connection set during these iterations. All missing syndromes are filled.

j''	0	1	2	3	4	5	6	7	8
6	α^{11}								
5	α^6	α^4							
4	α^5	α^7	α^6						
3	α^6	α^4	α^7	α^3	α^{10}				
2	α^9	α^{11}	0	α^{12}	α^6	α			
1	0	α^9	α^{14}	α^{12}	α^5	α^5	α^{13}		
0	α^9	α^{14}	α^5	α^7	α^2	α^5	1	α^8	j'

Figure 12.8. Continuation of syndrome filling.

Step (36) Set $r = 36 = (8,0)$. Syndrome $S_{8,0}$ is implied by the equation of the curve; it is equal to α^{13}. Only one connection polynomial reaches the point $(8,0)$. The discrepancy $\delta_{36}^{(35,1)}$ is α^7. Because $(8,0) - (4,0) = (4,0)$, which is not in the current footprint, the footprint must be enlarged to include this point. The new minimal connection polynomials are

$$\Lambda^{(36,1)}(x,y) = x\Lambda^{(35,1)}(x,y) + \alpha^7 B^{(35,2)}(x,y)$$
$$= x^5 + \alpha^3 x^4 + \alpha^9 x^3 y + \alpha^4 x^3$$
$$+ \alpha^0 x^2 y + \alpha^1 x^2 + \alpha^9 xy + \alpha^0 y + \alpha^3,$$

$$\Lambda^{(36,2)}(x,y) = \Lambda^{(35,2)}(x,y)$$
$$= x^2 y + \alpha^4 xy + \alpha^3 x^2 + \alpha y + \alpha^7 x + \alpha^4,$$

and

$$\Lambda^{(36,3)}(x,y) = \Lambda^{(26,3)}(x,y)$$
$$= y^2 + \alpha^{10} xy + \alpha^{13} y + \alpha^{13} x + \alpha^{11},$$

which we abbreviate as follows:

$$\Lambda^{(36,1)} = \begin{vmatrix} \alpha^0 & \alpha^9 & \alpha^0 & \alpha^9 \\ \alpha^3 & 0 & \alpha^1 & \alpha^4 & \alpha^3 & 1 \end{vmatrix} \qquad \Lambda^{(36,2)} = \begin{vmatrix} \alpha & \alpha^4 & 1 \\ \alpha^4 & \alpha^7 & \alpha^3 \end{vmatrix}$$

$$\Lambda^{(36,3)} = \begin{vmatrix} 1 & \\ \alpha^{13} & \alpha^{10} \\ \alpha^{11} & \alpha^{13} \end{vmatrix}$$

The footprint Δ_{36} is given in Figure 12.9. For this footprint, $\|\Delta_{36}\| = 7$. Because we know that the error pattern has seven errors, the footprint will not change further during subsequent iterations. However, further iterations are necessary to complete the formation of the minimal connection polynomials.

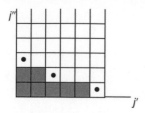

Figure 12.9. Final footprint for the hermitian code.

Step (37) to step (44) There is no change to the set of minimal connection polynomials during these steps.

Step (45) Set $r = 45 = (9, 0)$. The discrepancy $\delta_{45}^{(44,1)} = \alpha^{10}$. Then

$$\Lambda^{(45,1)}(x, y) = \Lambda^{(44,1)}(x, y) + \alpha^{10} B^{(44,1)}(x, y)$$
$$= x^5 + \alpha^9 x^3 y + \alpha^{12} x^3 + \alpha^{11} x^2 y + \alpha^{14} x^2$$
$$+ \alpha^1 xy + \alpha^4 x + \alpha^3 y + \alpha^7,$$

$$\Lambda^{(45,2)}(x, y) = \Lambda^{(44,2)}(x, y)$$
$$= x^2 y + \alpha^4 xy + \alpha^3 x^2 + \alpha y + \alpha^7 x + \alpha^4,$$

and

$$\Lambda^{(45,3)}(x, y) = \Lambda^{(44,3)}(x, y)$$
$$= y^2 + \alpha^{10} xy + \alpha^{13} y + \alpha^{13} x + \alpha^{11},$$

which we abbreviate as

$$\Lambda^{(45,1)} = \begin{vmatrix} \alpha^3 & \alpha^1 & \alpha^{11} & \alpha^9 \\ \alpha^7 & \alpha^4 & \alpha^{14} & \alpha^{12} & 0 & 1 \end{vmatrix} \qquad \Lambda^{(45,2)} = \begin{vmatrix} \alpha & \alpha^4 & 1 \\ \alpha^4 & \alpha^7 & \alpha^3 \end{vmatrix}$$

$$\Lambda^{(45,3)} = \begin{vmatrix} 1 \\ \alpha^{13} & \alpha^{10} \\ \alpha^{11} & \alpha^{13} \end{vmatrix}$$

At this point, the area of the footprint has area 7, and the three minimal connection polynomials have seven common zeros. The computation of the locator ideal is complete. Any further iterations will not change the connection polynomials, only fill missing syndromes. If desired, the process can be continued to fill missing syndromes until all syndromes have been visited. Because this is a correctable error pattern, the minimal connection polynomials become the minimal generator polynomials for the locator ideal.

At this point, the locator ideal has been computed and can be used to correct the errors by any of the several methods described in Section 12.5. The most straightforward of the methods, described in that section, is to continue the iterations, producing a new syndrome at each iteration until the full bispectrum of the error pattern is known. An inverse two-dimensional Fourier transform then completes the computation of the error pattern.

12.5 Computation of the error values

After the error-locator ideal has been computed, the location and the value of each error must be computed. There are several methods we may use to achieve this. We shall describe three methods of computing the errors from the locator ideal. These are the counterparts to the three methods of computing the error values that were described earlier, in Section 3.2, for codes on the line. Whichever method is used, the result of the computation is the array of errors, which is then subtracted from the senseword to obtain the corrected codeword.

The array of errors that underlies our running example is shown in Figure 12.10. The error pattern can be represented by the polynomial $e(x, y)$, given by

$$e(x, y) = (\alpha^{10}x^{14} + \alpha^0 x^{11} + \alpha^8 x^8 + \alpha^6 x^5 + \alpha^1 x^2)y^3 + \alpha^7 xy^7 + \alpha^6 y.$$

$$e = \begin{bmatrix}
0 & \alpha^6 & 0 & 0 & 0 & 0 & 0 & 0 & 0 & 0 & 0 & 0 & 0 & 0 & 0 \\
0 & 0 & 0 & 0 & 0 & 0 & 0 & \alpha^7 & 0 & 0 & 0 & 0 & 0 & 0 & 0 \\
0 & 0 & 0 & \alpha^1 & 0 & 0 & 0 & 0 & 0 & 0 & 0 & 0 & 0 & 0 & 0 \\
0 & 0 & 0 & 0 & 0 & 0 & 0 & 0 & 0 & 0 & 0 & 0 & 0 & 0 & 0 \\
0 & 0 & 0 & 0 & 0 & 0 & 0 & 0 & 0 & 0 & 0 & 0 & 0 & 0 & 0 \\
0 & 0 & 0 & \alpha^6 & 0 & 0 & 0 & 0 & 0 & 0 & 0 & 0 & 0 & 0 & 0 \\
0 & 0 & 0 & 0 & 0 & 0 & 0 & 0 & 0 & 0 & 0 & 0 & 0 & 0 & 0 \\
0 & 0 & 0 & 0 & 0 & 0 & 0 & 0 & 0 & 0 & 0 & 0 & 0 & 0 & 0 \\
0 & 0 & 0 & \alpha^8 & 0 & 0 & 0 & 0 & 0 & 0 & 0 & 0 & 0 & 0 & 0 \\
0 & 0 & 0 & 0 & 0 & 0 & 0 & 0 & 0 & 0 & 0 & 0 & 0 & 0 & 0 \\
0 & 0 & 0 & 0 & 0 & 0 & 0 & 0 & 0 & 0 & 0 & 0 & 0 & 0 & 0 \\
0 & 0 & 0 & \alpha^0 & 0 & 0 & 0 & 0 & 0 & 0 & 0 & 0 & 0 & 0 & 0 \\
0 & 0 & 0 & 0 & 0 & 0 & 0 & 0 & 0 & 0 & 0 & 0 & 0 & 0 & 0 \\
0 & 0 & 0 & 0 & 0 & 0 & 0 & 0 & 0 & 0 & 0 & 0 & 0 & 0 & 0 \\
0 & 0 & 0 & \alpha^{10} & 0 & 0 & 0 & 0 & 0 & 0 & 0 & 0 & 0 & 0 & 0
\end{bmatrix}$$

Figure 12.10. Error pattern for the running example.

The error bispectrum, shown in Figure 12.11, is the Fourier transform of this array of errors. It can be computed by evaluating the polynomial $e(x, y)$ at all points of the bicyclic plane. Because the errors lie on the hermitian curve, the first five columns of this array determine all others by the relationship $E_{j'j''} = E_{j'-5,j''+4} + E_{j'-5,j''+1}$. The syndromes that were used in Sections 12.3 and Section 12.4 are the appropriate components of Figure 12.8.

The locator ideal for the running example, as computed in both Section 12.3 and Section 12.4, is generated by the following minimal locator polynomials:

$$\mathbf{\Lambda}^{(1)} = \begin{array}{|cccc} \alpha^3 & \alpha^1 & \alpha^{11} & \alpha^9 \\ \alpha^7 & \alpha^4 & \alpha^{14} & \alpha^{12} \quad 0 \quad 1 \end{array} \qquad \mathbf{\Lambda}^{(2)} = \begin{array}{|ccc} \alpha & \alpha^4 & 1 \\ \alpha^4 & \alpha^7 & \alpha^3 \end{array} \qquad \mathbf{\Lambda}^{(3)} = \begin{array}{|cc|} 1 & \\ \alpha^{13} & \alpha^{10} \\ \alpha^{11} & \alpha^{13} \end{array}$$

To find the location of the errors, the common zeros of these three polynomials are found. For the hyperbolic code, all 225 points of the bicyclic plane $GF(16)^{*2}$ need to be tested. For the hermitian code, only the sixty points on the epicyclic curve need to be tested. (We have chosen the seven errors to be the same for the two examples, so the seven errors in the senseword of the hyperbolic code are all on the hermitian curve, but the senseword of the hyperbolic code could have had errors anywhere in the bicyclic plane.)

The first method uses recursive extension to compute all coefficients of the bispectrum polynomial $S(x, y)$. This amounts to a continuation of the steps of the embellished Sakata algorithm, with additional steps to fill all missing or implied syndromes. Each minimal polynomial $\Lambda^{(\ell)}(x, y)$ of the locator ideal satisfies the polynomial equation $\Lambda^{(\ell)}(x, y)S(x, y) = 0$, which gives the two-dimensional recursion. Every such equation with a single unknown component, $S_{j'j''}$, can be used to solve for that unknown component, which then becomes known. Eventually, by two-dimensional recursive extension, the array of bispectral components shown in Figure 12.11 is completely filled. The inverse Fourier transform of this array is the error pattern shown in Figure 12.10.

To compute the array $S = E$ by recursive extension, it is enough first to compute only the first five columns. Locator polynomials $\Lambda^{(2)}(x, y)$ and $\Lambda^{(3)}(x, y)$ are sufficient for this purpose. The other columns can be filled by using the equation of the curve or, better, the equation of the curve can be built into the computation of the inverse Fourier transform.

The second method uses a straightforward matrix inverse, in the manner used in the Peterson algorithm. First, the locations of the v errors are found by finding the common zeros of the three generators of the locator ideal. This gives the indices of the nonzero locations of the array e. This is a sparse array with at most t nonzero locations. Each syndrome is a linear combination of these nonzero error magnitudes. Then it is straightforward to set up a system of v linear equations in the v unknown error magnitudes.

The third method is based on a generalization of the Forney formula. The formal derivative of a polynomial in one variable is replaced by a formal derivative along a curve. The formula needed for a curve is more complicated and less attractive, however, than the original Forney formula. Moreover, a locator polynomial and its partial derivative may both be zero at the same point. In this case, the generalized formula becomes indeterminate. Then either a generalization of l'Hôpital's rule must be used, or a different locator polynomial must be chosen.

The generalization of the Forney formula has the following form:

$$
e_i = \begin{cases} \dfrac{\Gamma^{[\ell-1]}(\omega^{j'}, \omega^{j''})}{\Lambda^{[\ell]}(\omega^{j'}, \omega^{j''})} & \text{if } \Lambda(\omega^{j'}, \omega^{j''}) = 0 \\ 0 & \text{if } \Lambda(\omega^{j'}, \omega^{j''}) \neq 0, \end{cases}
$$

where ℓ is the smallest integer for which $\Lambda^{[\ell]}(\omega^{j'}, \omega^{j''}) \neq 0$, and $[\ell]$ denotes the ℓth Hasse derivative. The Hasse derivative along the hermitian curve is given by

$$
\Lambda^{[\ell]}(x', y'') = \Lambda^{[\ell,0]}(x', y'') + (x')^q \Lambda^{[\ell-1,1]}(x', y') + \cdots + (x')^{\ell q} \Lambda^{[0,\ell]}(x', y'),
$$

where the notation $\Lambda^{[\ell',\ell'']}(x, y)$ denotes the Hasse partial derivative combining the ℓ'th derivative in x and the ℓ''th derivative in y.

12.6 Supercodes of hermitian codes

Bicyclic codes on the plane were studied in Chapter 6. No bicyclic code (without puncturing or shortening) was found to be significant in terms of the minimum distance, though perhaps some may be practical if they are to be decoded beyond the minimum

$$
\begin{bmatrix}
\alpha^9 & \alpha^{14} & \alpha^5 & \alpha^7 & \alpha^2 & \alpha^5 & \alpha^0 & \alpha^8 & \alpha^{13} & \alpha^{12} & 0 & \alpha^2 & \alpha^{12} & \alpha^9 & \alpha^{10} \\
0 & \alpha^9 & \alpha^{14} & \alpha^{12} & \alpha^5 & \alpha^5 & \alpha^{13} & \alpha^{10} & 0 & \alpha^5 & \alpha^{13} & \alpha^6 & \alpha^8 & \alpha^2 & \alpha^7 \\
\alpha^9 & \alpha^{11} & 0 & \alpha^{12} & \alpha^6 & \alpha^1 & \alpha^9 & \alpha^3 & \alpha^6 & \alpha^3 & \alpha^{13} & \alpha^0 & \alpha^{13} & \alpha^5 & \alpha^1 \\
\alpha^6 & \alpha^4 & \alpha^7 & \alpha^3 & \alpha^{10} & \alpha^5 & \alpha^{14} & \alpha^6 & \alpha^5 & \alpha^3 & 0 & 0 & \alpha^{13} & \alpha^2 & \alpha^5 \\
\alpha^5 & \alpha^7 & \alpha^6 & \alpha^1 & \alpha^{14} & \alpha^5 & \alpha^{14} & \alpha^3 & \alpha^9 & \alpha^0 & \alpha^{10} & \alpha^8 & \alpha^4 & \alpha^{12} & \alpha^{12} \\
\alpha^6 & \alpha^4 & \alpha^{10} & \alpha^{12} & \alpha^9 & \alpha^{12} & \alpha^5 & \alpha^{13} & \alpha^3 & \alpha^4 & \alpha^{13} & \alpha^7 & \alpha^2 & \alpha^{14} & \alpha^{10} \\
\alpha^{11} & \alpha^{14} & \alpha^4 & \alpha^2 & \alpha^{12} & \alpha^7 & \alpha^3 & \alpha^0 & 0 & \alpha^{14} & \alpha^9 & \alpha^{11} & \alpha^{13} & \alpha^7 & \alpha^9 \\
0 & \alpha^1 & 0 & \alpha^2 & \alpha^1 & \alpha^5 & \alpha^{14} & \alpha^8 & \alpha^{11} & \alpha^6 & \alpha^7 & \alpha^5 & \alpha^3 & \alpha^{10} & \alpha^3 \\
\alpha^9 & \alpha^9 & \alpha^{12} & \alpha^8 & \alpha^7 & \alpha^3 & \alpha^4 & \alpha^{11} & \alpha^{10} & \alpha^7 & \alpha^7 & 0 & \alpha^3 & \alpha^7 & \alpha^{11} \\
1 & \alpha^{12} & \alpha^{11} & \alpha^6 & \alpha^6 & \alpha^5 & \alpha^4 & \alpha^8 & \alpha^{14} & \alpha^4 & \alpha^5 & \alpha^{13} & \alpha^9 & \alpha^2 & 0 \\
\alpha^7 & \alpha^9 & \alpha^0 & \alpha^2 & \alpha^7 & \alpha^6 & \alpha^{10} & \alpha^3 & \alpha^8 & \alpha^2 & \alpha^8 & \alpha^{12} & \alpha^7 & \alpha^1 & \alpha^{13} \\
\alpha^6 & \alpha^4 & \alpha^9 & \alpha^7 & \alpha^{10} & \alpha^4 & \alpha^8 & \alpha^5 & 0 & \alpha^4 & \alpha^7 & \alpha^1 & \alpha^3 & \alpha^{12} & \alpha^6 \\
\alpha^{14} & \alpha^6 & 0 & \alpha^7 & \alpha^{10} & \alpha^{13} & \alpha^4 & \alpha^{13} & \alpha^1 & \alpha^{11} & \alpha^6 & \alpha^{10} & \alpha^8 & \alpha^0 & \alpha^7 \\
\alpha^{13} & \alpha^{14} & \alpha^2 & \alpha^{13} & \alpha^3 & \alpha^9 & \alpha^9 & \alpha^1 & \alpha^0 & \alpha^{12} & \alpha^2 & 0 & \alpha^8 & \alpha^{12} & \alpha^7 \\
0 & \alpha^2 & \alpha^1 & \alpha^{11} & \alpha^4 & \alpha^5 & \alpha^9 & \alpha^{13} & \alpha^4 & \alpha^8 & 0 & \alpha^3 & \alpha^{14} & \alpha^7 & \alpha^2
\end{bmatrix}
$$

Figure 12.11. Error spectrum for the running example.

distance. In Chapters 10 and 11, certain bicyclic codes were punctured or shortened to lie on a plane curve to obtain codes that are attractive in terms of minimum distance. In this section, we show that some of these codes can be further improved by augmenting certain of their cosets to produce better codes.

The true minimum distance of a shortened hermitian code is at least as large as the Feng–Rao distance, while the designed distance of a shortened hermitian code is often defined to be the Goppa distance. For some hermitian codes, the Feng–Rao distance is larger than the Goppa distance. This means that the true minimum distance of the code is larger than the designed distance. For these codes, it is possible to enlarge the hermitian code – without reducing its designed distance – to a new linear code that contains the hermitian code. We will not follow this line of thought in this form further in this section. Instead, we will follow a different line of thought to a similar purpose. We will construct a code that merges the definition of a hyperbolic code with the definition of an hermitian code. This construction draws on the intuition developed in the decoding examples of Sections 12.3 and 12.4. In Section 12.3, to infer certain missing syndromes the extended Sakata algorithm used the fact that the code being decoded was a hyperbolic code. In Section 12.4, to infer certain missing syndromes the extended Sakata algorithm used the fact that the code being decoded is an hermitian code. We will merge the attributes of these two codes into one code so that both kinds of syndromes can be inferred.

The hyperbolic bound suggests how to construct this new code by forming the union of an hermitian code with certain of its cosets to increase its dimension. The original hermitian code is a subcode of the new code, and the new code is a supercode of the original hermitian code.

A codeword of the hermitian code is a vector with the components c_i for $i = 0, \ldots, n-1$, where i indexes the n points (i', i'') of the hermitian curve, those points at which $G(\omega^{-i'}, \omega^{-i''})$ is equal to zero. For the shortened hermitian code, the codeword spectral components satisfy $C_{j'j''} = 0$ if $j' + j'' \leq J$. Every other spectral component $C_{j'j''}$ is arbitrary, provided the array c is zero for those components that do not lie on the curve. The defining set of the shortened hermitian code is given by

$$\mathcal{A}_1 = \{(j',j'') \mid j' + j'' \leq J\}.$$

The designed distance of the shortened code is $d^* = mJ - 2g + 2$, and the dimension of this code is $k = n - mJ + g - 1$.

The defining set of the hyperbolic code is given by

$$\mathcal{A}_2 = \{(j',j'') \mid (j' + 1)(j'' + 1) \leq d^*\}.$$

The designed distance of the hyperbolic code is d^*.

For the hermitian supercode, the defining set is chosen to consist of those (j',j'') that are in both defining sets. The codeword spectra satisfies $C_{j'j''} = 0$ if $(j',j'') \in \mathcal{A}_1 \cap \mathcal{A}_2$.

Thus,

$$\mathcal{A} = \{(j',j'') \mid j'+j'' \le J\} \bigcap \{(j',j'') \mid (j'+1)(j''+1) \le d^*\},$$

and J satisfies $mJ = d^* + 2g - 2$. Consequently $C_{j'j''}$ is equal to zero if the two conditions $m(j'+j'') \le d^* + 2g - 2$ and $(j'+1)(j''+1) \le d^*$ are both satisfied. For all other (j', j''), the bispectrum component $C_{j'j''}$ is arbitrary, provided the constraint imposed by the curve is satisfied.

If the set $\mathcal{A}_2$ is not contained in the set $\mathcal{A}_1$, there will be fewer elements in the intersection $\mathcal{A} = \mathcal{A}_1 \cap \mathcal{A}_2$ than in the set $\mathcal{A}_2$. Because there are fewer such (j', j'') in $\mathcal{A}$ than in $\mathcal{A}_1$, the constraints of the defining set are looser. There will be more codewords satisfying the new constraints, so the dimension of the resulting code is larger.

Syndrome $S_{j'j''}$ will be known only if $m(j'+j'') \le d^*+2g-2$ and $(j'+1)(j''+1) \le d^*$ are both satisfied. It follows from the Sakata–Massey theorem that each unknown syndrome can be inferred by a subsidiary calculation, augmenting the Sakata algorithm just at the time that it is needed. Because the unknown syndromes that result from the new hyperbolic constraint can be inferred by the decoder, there is no reduction in the designed distance because of the redefinition of the defining set.

It remains to show that this discussion is fruitful by showing the existence of hermitian codes with defining set $\mathcal{A}_1$ such that $\mathcal{A}_2 \not\subset \mathcal{A}_1$. Only if this is so will there exist such codes better than hermitian codes. We will not provide general conditions under which this is so (see Problem 12.10). Instead, we will establish this fact by giving a simple example.

A bispectrum of an hermitian supercode with designed distance 27 over $GF(64)$ is shown in Figure 12.12. The hermitian polynomial $x^9 + y^8 + y$ over $GF(64)$ has genus

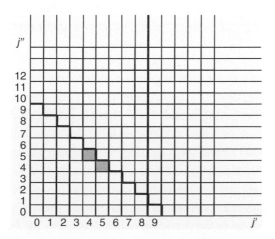

Figure 12.12. Bispectrum of an hermitian supercode over $GF(64)$.

28 and degree $m = 9$. For an hermitian code and a hyperbolic code of designed distance 27, the two defining sets are

$$A_1 = \{(j',j'') \mid 9(j' + j'') \leq 81\}$$

and

$$A_2 = \{(j',j'') \mid (j' + 1)(j'' + 1) \leq 27\}.$$

These sets are, in part,

$$A_1 = \{\ldots, (6,3), (5,4), (4,5), (3,6), \ldots\}$$

and

$$A_2 = \{\ldots, (6,3), (5,3), (4,4), (3,5), (3,6), \ldots\}.$$

In particular, the points $(5,4)$ and $(4,5)$ are not elements of $A_1 \cap A_2$. These points are shown as shaded in Figure 12.12. These shaded points correspond to components of the bispectrum that are constrained to zero in the hermitian code but are not so constrained in the hermitian supercode because they do not lie in the hyperbolic set $(j' + 1)(j'' + 1) \leq 27$. Accordingly, the supercode has a larger dimension than the hermitian code. The number of syndromes can be found to be fifty-two by counting the fifty-two white squares below $j' = 9$ and under the shaded region. Thus, the dimension is 452, so the code is a $(504, 452, 27)$ code over $GF(64)$. This code is superior to the $(504, 450, 27)$ hermitian code over $GF(64)$.

12.7   The Feng–Rao decoder

The Feng–Rao decoder is a procedure for inferring hidden syndromes from the given syndromes by using matrix rank arguments. The structure of the Feng–Rao decoder is closely related to the proof of the Feng–Rao bound. The Feng–Rao decoder is presented here because its conceptual structure provides valuable insight into the decoding problem. Indeed, it replaces the various strong tools of algebraic geometry with a rather straightforward decoding algorithm that is based on elementary matrix rank arguments. Nothing beyond linear algebra is needed to understand the computations of the decoder, but the proof of performance requires a statement of the Feng–Rao bound. Even more, the Feng–Rao decoder will decode up to the decoding radius defined by the Feng–Rao bound, whereas the Riemann–Roch theorem asserts only that the minimum distance is at least as large as the Goppa bound.

The Feng–Rao decoder requires that the ring $F[x, y]/\langle G(x, y)\rangle$ has a weight function, which we will denote by ρ. Let $\rho_1, \rho_2, \ldots$ be the weights corresponding to the monomials in the weighted graded order. Consider the bivariate syndromes arranged in the weighted graded order as follows:

$$S_{\rho_0}, S_{\rho_1}, S_{\rho_2}, S_{\rho_3}, \ldots, S_{\rho_i}, \ldots, S_{\rho_{r-1}}, S_{\rho_r}.$$

The next syndrome, $S_{\rho_{r+1}}$, is missing because φ_{r+1} is not a monomial corresponding to an element of the defining set. We will find $S_{\rho_{r+1}}$ by a majority vote of multiple estimates of it. Once $S_{\rho_{r+1}}$ is found, the same procedure can then be used to find $S_{\rho_{r+2}}$, and so, in turn, all missing syndromes. An inverse two-dimensional Fourier transform of the array of syndromes then gives the error pattern.

Because ρ is a weight function, we know that ρ forms a semigroup. Accordingly, for any i and j, there is a k such that[1] $\rho_k = \rho_i + \rho_j$. This allows us to define an array, $\mathbf{R}$, with the terms from the sequence of syndromes as elements according to the definition of terms as

$$R_{ij} = S_{\rho_i + \rho_j}.$$

Thus

$$\mathbf{R} = \begin{bmatrix} S_{\rho_0} & S_{\rho_1} & S_{\rho_2} & S_{\rho_3} & S_{\rho_4} & \cdots \\ S_{\rho_1} & S_{\rho_1+\rho_1} & S_{\rho_1+\rho_2} & S_{\rho_1+\rho_3} & & \cdots \\ S_{\rho_2} & S_{\rho_2+\rho_1} & S_{\rho_2+\rho_2} & S_{\rho_2+\rho_3} & & \cdots \\ S_{\rho_3} & S_{\rho_3+\rho_1} & S_{\rho_3+\rho_2} & & & \\ S_{\rho_4} & & & & & \\ \vdots & & & & & \end{bmatrix}.$$

Some of the elements of this array are known. Other elements of this array are not known because they come after the last known syndrome $S_{\rho_{r+1}}$. In each row, the initial elements are known, starting from the left side of the row over to the last known element in that row. After this last known element in a row, all subsequent elements of that row are unknown.

We have seen a matrix with a similar structure earlier, in Section 9.8, in connection with the proof of the Feng–Rao bound. In that section, we wrote the matrix $\mathbf{W}$

---

[1] Because the array $\mathbf{R}$ will be defined with elements $S_j = S_{j'j''}$ that themselves form a different two-dimensional array comprising the bispectrum of the error pattern, we will henceforth use i and j as the indices of $\mathbf{R}$.

suggestively as follows:

$$
W = \begin{bmatrix}
0 & & & & & & & & * \\
& & & & & & & * & \\
& & & & & & * & & \\
& & & & & * & & & \\
& & & & * & & & & \\
& & & * & & & & & \\
& & * & & & & & & \\
& * & & & & & & & \\
* & & & & & & & &
\end{bmatrix},
$$

where each element denoted by an asterisk is in a different row and a different column. Then, in the proof of the Feng–Rao bound, we saw that rank $W = $ wt $\boldsymbol{v}$. But wt $\boldsymbol{v} \le t$, so we know that rank $\boldsymbol{W} \le t$.

We can do a similar analysis of the matrix $\boldsymbol{R}$. The known syndromes appear in the matrix $\boldsymbol{R}$ where $\boldsymbol{W}$ has one of its constrained zeros. These are the positions above and to the left of the asterisk. First, we rewrite the matrix as follows:

$$
R = \begin{bmatrix}
S_{\rho_0} & S_{\rho_1} & S_{\rho_2} & & & & & * & \\
S_{\rho_1} & S_{\rho_1+\rho_1} & S_{\rho_1+\rho_2} & & & & * & & \\
& & & & & * & & & \\
& & & & * & & & & \\
& & & * & & & & & \\
& & * & & & & & & \\
& * & & & & & & & \\
* & & & & & & & &
\end{bmatrix},
$$

where each asterisk denotes the first syndrome in that row whose value is not known. If gaps are inserted within the rows and columns, as in Section 9.8, then the asterisks will lie on a straight line. As in the proof of the Feng–Rao bound in Section 9.8, this matrix of syndromes can be factored as follows:

$$
R = [\varphi_{\rho_i}(P_\ell)] \begin{bmatrix}
e_0 & 0 & \cdots & 0 \\
0 & e_1 & & 0 \\
\vdots & & \ddots & \\
0 & & & e_{n-1}
\end{bmatrix} [\varphi_{\rho_j}(P_\ell)]^{\mathrm{T}}.
$$

This equation has the form

$$
R = MEM^{\mathrm{T}}.
$$

Each of the two outer matrices M and M^T has full rank, and E is a diagonal matrix with rank equal to the weight of e. This means that the rank of R is equal to the number of errors. Thus a bounded distance decoder can presume that the rank of R is at most t. If the rank of R is larger than t, the decoder is not expected to find the correct codeword.

The proof of the next theorem uses the notion of a *pivot* or *sentinel* of a matrix. This notion is well known in the process of *gaussian elimination*, a popular method of solving a system of linear equations. Let $R(\imath, \jmath)$ denote the submatrix of R with index $(0, 0)$ as the upper left corner and index $(\imath, \jmath)$ as the lower right corner. The submatrix $R(\imath, \jmath)$ is obtained by cropping the rows of R after row $\imath$ and cropping the columns of R after column $\jmath$. In the language of gaussian elimination, a *pivot* of the matrix R is a matrix location $(\imath, \jmath)$ at which $R(\imath, \jmath)$ and $R(\imath - 1, \jmath - 1)$ have different ranks. Specifically, if $(\imath, \jmath)$ is a pivot, then

$$\operatorname{rank} R(\imath, \jmath) = \operatorname{rank} R(\imath - 1, \jmath - 1) + 1.$$

The number of pivots in the matrix R is equal to the rank of R. For our situation, the rank of the matrix R is at most t. This means that the number of pivots is at most t.

To see that there are ν pivots when ν is the rank of R, consider the sequence of submatrices formed from R by starting with a column matrix consisting of the first column, then appending columns one by one to obtain a sequence of submatrices. At some steps of this process, the rank of the submatrix does not change; at other steps, the rank increases by one. Because the rank of R is ν, there will be exactly ν steps of the process at which the rank of the submatrix increases by one. Mark those submatrices where the rank increases. There will be ν marked submatrices. Next, apply the same process to the rows of each marked submatrix. Start with the top row of the marked submatrix and append rows one by one until the submatrix reaches its maximum rank. The bottom right index $(\imath, \jmath)$ of this final submatrix is a pivot, and there will be ν pivots. Moreover, there can be only one pivot in each column because when a column is appended the rank of a submatrix can increase only by one. Further, by interchanging rows and columns of R, and then repeating this discussion, we can conclude that there can be only one pivot in each row.

We will argue in proving the next theorem that, for many of the appearances of the first missing syndrome $S_{\rho_{r+1}}$ as an element of R, that element can be estimated, and the majority of estimates of this missing syndrome will be the correct value of $S_{\rho_{r+1}}$. Hence the missing syndrome can be estimated by majority vote and the missing syndrome can be recovered. The process then can be repeated to recover the next missing syndrome, $S_{\rho_{r+2}}$ and so on. In this way, all syndromes can be found and the error spectrum can be fully recovered. An inverse Fourier transform then completes the decoding.

The first missing syndrome in the total order is $S_{\rho_{r+1}}$, which we will write more simply as S_{r+1}. If syndrome S_{r+1} were known, it would replace the asterisk in the expression for R. Let ν be the rank of R, which is initially unknown, except that ν cannot be larger than t for a bounded distance decoder. Then for any (ι, j) that indexes a matrix location of the first missing syndrome S_{r+1}, consider the four submatrices $\mathbf{R}(\iota - 1, j - 1)$, $\mathbf{R}(\iota, j - 1)$, $\mathbf{R}(\iota - 1, j)$, and $\mathbf{R}(\iota, j)$. If the first three of these submatrices have the same rank, then (ι, j) is declared to be a candidate for an estimate of S_r. This criteria ensures that a candidate cannot be in a row or a column that has a pivot in the known part of the matrix. The matrix $\mathbf{R}(\iota, j)$ has the unknown S_r in its lower right corner. This element can be chosen so that the rank of $\mathbf{R}(\iota, j)$ is equal to the rank of $\mathbf{R}(\iota - 1, j - 1)$. For every candidate (ι, j), an estimate $\hat{S}_{r+1}$ of S_{r+1} can be computed in this way. Some of these estimates may be correct and some may be wrong.

Theorem 12.7.1 *If the number of errors in a senseword is at most $(d_{FR}(r) - 1)/2$, then the majority of the estimates of the first missing syndrome S_r give the correct value.*

Proof: Let K denote the number of known pivots. These are the pivots in the known part of the matrix. Of the estimates of S_r, let T denote the number of correct estimates and let F denote the number of incorrect estimates. Each incorrect estimate is another pivot of the full matrix. This means that $K + F$ is not larger than the rank of the matrix $\mathbf{R}$, which is not larger than t. Thus

$$K + F \leq t = \lfloor (d_{FR}(r) - 1)/2 \rfloor.$$

There cannot be two pivots in the same row or column of matrix $\mathbf{R}$. Therefore, if index (ι, j) is a pivot in the known part of the matrix, then all entries (ι, j') in the ιth row with $j' > j$ are not pivots, and all entries (ι', j) in the jth column with $\iota' > \iota$ are not pivots. This means that at most $2K$ appearances of S_r can fail the rank condition on candidates. Therefore, because there are at least $d_{FR}(r)$ appearances of S_r in $\mathbf{R}$, there must be at least $d_{FR}(r) - 2K$ candidates. Because there are $T + K$ candidates, we conclude that

$$d_{FR}(r) - 2K \leq T + F.$$

Combining these inequalities yields

$$2(K + F) + 1 \leq d_{FR}(r) \leq T + F + 2K,$$

from which we conclude that $F < T$, as was to be proved. ∎

12.8   The theory of syndrome filling

All that remains to be discussed is the formal justification of syndrome filling. In Section 12.7, we discussed syndrome filling from the perspective of the Feng–Rao bound. In this section we will discuss the fact that, for the extended Sakata decoders that we have studied in Sections 12.3 and 12.4, the missing syndromes needed for decoding can always be filled, provided the number of errors is less than the packing radius of the code. This discussion is necessary in order to validate formally the Sakata decoder. This has already been discussed for hyperbolic codes in the introduction to Section 10.3. The proof there used the fact that the defining set is a hyperbolic set. In that case, the Sakata–Massey theorem asserts that there is only one possible value for the missing syndrome. We will summarize this discussion here for hyperbolic codes so as to prepare the way for a similar discussion on hermitian codes.

A hyperbolic code, with the designed distance $d^* = 2t + 1$, is defined in Section 6.4 as a two-dimensional cyclic code on the bicyclic plane $GF(q)^2$ with the defining set given by

$$\mathcal{A} = \{(j', j'') \mid (j' + 1)(j'' + 1) < 2t + 1\}.$$

The defining set $\mathcal{A}$ is described as the set of bi-indices bounded by a hyperbola. The syndromes are those components of the bispectrum indexed by each $(j', j'') \in \mathcal{A}$ of the two-dimensional Fourier transform of the noisy senseword.

Proposition 12.8.1 *For a senseword within the designed distance of a hyperbolic code, there is only one way to extend an ordered sequence of syndromes with a subsequent missing syndrome, not in the defining set, that will produce a connection footprint of area at most t, where $2t + 1 \leq d^*$.*

Proof: A corollary to the Sakata–Massey theorem (Corollary 8.3.3) states that if $\Lambda(x, y)$, a polynomial of degree s, produces the bi-index sequence $S_0, S_1, \ldots, S_{r-1}$, but does not produce the longer sequence $S_0, S_1, \ldots, S_{r-1}, S_r$, then the connection footprint Δ_{r-1} for $S_0, S_1, \ldots, S_{r-1}$ and the connection footprint Δ_r for $S_0, S_1, \ldots, S_r$ satisfy the following:

$$\|\Delta_{r-1} \cup (r - \Delta_r)\| \geq (r' + 1)(r'' + 1),$$

where $r = (r', r'')$. But if syndrome S_r is a missing syndrome, we know that $(r', r'') \notin \mathcal{A}$, so $(r' + 1)(r'' + 1) \geq 2t + 1$. Combining these statements leads to

$$\|\Delta_{r-1}\| + \|\Delta_r\| \geq \|\Delta_r \cup (r' - \Delta_{r-1})\|$$
$$\geq (r' + 1)(r'' + 1) \geq 2t + 1.$$

But since $\|\Delta_{r-1}\|$ is not larger than t, $\|\Delta_r\|$ must be larger than t. Because we are considering a decodable senseword for a code with designed distance at least $2t + 1$, the true footprint has area at most t. We conclude that Δ_r cannot be the true footprint because its area is too large, and so only the correct S_r can be the missing syndrome. ∎

Recall that the Sakata algorithm, when decoding a hyperbolic code, will develop a set of connection polynomials. The conclusion of the proposition regarding a missing syndrome will apply to each connection polynomial that reaches that missing syndrome.

We must also present a similar proof for the case of a shortened hermitian code. In that proof, however, the equation of the curve will become entangled with the Sakata–Massey theorem, so we must start the proof at a more basic level. We shall develop the notions for the hermitian code based on the polynomial

$$G(x, y) = x^{q+1} + y^q + y,$$

which gives an epicyclic code of blocklength $n = q^3 - q$ over the field $GF(q^2)$. The designed distance of the shortened code is given by

$$d^* = (q+1)J - 2g + 2.$$

Because $2g = q(q-1)$, this also can be written as follows:

$$d^* = (q+1)(J - q + 2).$$

We require that the algorithm decode up to this designed distance.

The equation of the curve $G(x, y) = 0$ creates a linear relationship among the syndromes given by

$$S_{j'+q+1,j''} + S_{j',j''+q} + S_{j',j''+1} = 0.$$

This expression mimics the equation of the curve, but expressed in terms of syndromes in the transform domain. If only one of these three syndromes is missing, it can be inferred by using this equation. If two of these syndromes were missing, then either or both can be estimated by using any minimal connection polynomial that reaches it. However, the case where two syndromes of this equation are both missing does not occur in the decoding of hermitian codes.

Suppose that the connection set has been computed for syndromes $S_0, S_1, \ldots, S_{r-1}$ and that syndrome S_r is missing. We will show first that only one choice of the missing S_r will give a connection set whose footprint has area at most t. This means that only one choice of the missing S_r corresponds to an error pattern of weight at most t. Therefore, in principle, a search over $GF(q^2)$, trying each possible syndrome in turn, will give the unknown syndrome S_r. A better procedure uses each minimal connection polynomial from the previous iteration to estimate a candidate value of that syndrome.

The first step is to define a subset of the set of syndrome indices where the required facts will be proved. Given any point $r = (r', r'')$, let

$$\mathcal{K}_1 = \{(j', j'') \mid 0 \le j' \le r'; 0 \le j'' \le r''\},$$
$$\mathcal{K}_2 = \{(j', j'') \mid 0 \le j' \le q; 0 \le j'' \le r'' - q\},$$

and

$$\mathcal{K} = \mathcal{K}_1 \cup \mathcal{K}_2.$$

This region of the j', j'' plane will be enough for our needs.

The following proposition bounds the cardinality of $\mathcal{K}$ provided $r' \leq q$. If $r' > q$, the proposition is not needed because syndromes with $r' > q$ are implied by the equation of the curve.

Proposition 12.8.2   *Suppose that $r = (r', r'')$ satisfies $J < r' + r'' < J + q$ and $r' \leq q$. Then $\|\mathcal{K}\|$ is at least as large as the designed distance d^*.*

Proof: If $\mathcal{K}_2$ is not empty, then

$$\|\mathcal{K}_1 \cup \mathcal{K}_2\| = (r' + 1)(r'' + 1) + (q - r')(r'' - q + 1).$$

If $\mathcal{K}_2$ is empty, then $r'' - q < 0$, so

$$\|\mathcal{K}_1 \cup \mathcal{K}_2\| = (r' + 1)(r'' + 1)$$
$$\geq (r' + 1)(r'' + 1) + (q - r')(r'' - q + 1).$$

Both cases can be combined, and we proceed as follows:

$$\begin{aligned}
\|\mathcal{K}\| &\geq (r' + 1)(r'' + 1) + (q - r')(r'' - q + 1) \\
&= (q + 1)(r' + r'') - r' - q^2 + q + 1 \\
&\geq (q + 1)(J + 1) - r' - q^2 + q + 1 \\
&\geq (q + 1)(J + 1) - q - q^2 + q + 1 \\
&= (q + 1)(J + 1) - q^2 + 1 \\
&= (q + 1)(J - q + 2) \\
&= d^*,
\end{aligned}$$

as was to be proved.   ■

The next proposition justifies the definition of the set $\mathcal{K}$. Before presenting this proposition, we motivate it by a simple example in the ring $R = GF(16)[x, y]/\langle x^5 + y^4 + y \rangle$. Suppose we are given that the point $(0, 4)$ is contained in the footprint $\Delta(I)$ of an ideal I of the ring R. Then we claim that $(4, 0)$ is also in the footprint $\Delta(I)$ of the ideal I. For if $(4, 0)$ is not in $\Delta(I)$, then there is a monic polynomial of I with leading monomial x^4. This polynomial, $p(x, y)$ must have the form

$$p(x, y) = x^4 + ax^3 + bx^2y + cxy^2 + dy^3 + ex^2 + fxy + gy^2 + hx + iy + j$$

for some constants $a, b, \ldots, j$. Therefore,

$$xp(x, y) + (x^5 + y^4 + y) = y^4 + ax^4 + bx^3 + \cdots + y + jx.$$

This polynomial is in the ideal I and has leading monomial y^4, so the point $(0, 4)$ is not in the footprint, contradicting the stated assumption. Thus $(4, 0)$ is in the footprint of I, as asserted.

Proposition 12.8.3 *In the ring $GF(q^2)[x, y]/\langle x^{q+1} + y^q + y\rangle$, if the point (r', r'') is in the footprint $\Delta(I)$ of ideal I, then $(q, r'' - q)$ is also in the footprint $\Delta(I)$.*

Proof: The statement is empty if $r'' - q$ is negative. Suppose $r'' - q$ is nonnegative and suppose that $\Delta(I)$ does not contain $(q, r'' - q)$. Then there is a polynomial $p(x, y)$ in I with leading term $x^q y^{r'' - q}$. Without loss of generality, we may assume that $p(x, y)$ contains no monomial $x^{j'} y^{j''}$ with $j' > q$ because such monomials may be canceled by adding an appropriate multiple of $x^{q+1} + y^q + y$ to $p(x, y)$. Now consider $p(x, y)$ with terms written in the graded order as follows:

$$p(x, y) = x^q y^{r'' - q} + \sum_{j'} \sum_{j''} p_{j'j''} x^{j'} y^{j''}.$$

Rewrite this by setting $j' + j'' = \ell$ and summing over ℓ and j'. This yields

$$p(x, y) = x^q y^{r'' - q} + \sum_{\ell=0}^{r'+r''} \sum_{j'=0}^{\ell} p_{j', \ell-j'} x^{j'} y^{\ell-j'}.$$

The only nonzero term in the sum with $\ell = r' + r''$ is the term $p_{r'r''} x^{r'} y^{r''}$. Multiply by $x^{r'+1}$ to write

$$x^{r'+1} p(x, y) = x^{r'+q+1} y^{r'' - q} + x^{r'} \sum_{\ell=0}^{r'+r''} \sum_{j'=0}^{\ell} p_{j', \ell-j'} x^{j'} y^{\ell-j'}.$$

Make the substitution $x^{q+1} = y^q + y$ in the first term on the right. Then

$$x^{r'+1} p(x, y) = x^{r'} y^{r''} + \sum_{\ell=0}^{r'+r''} \sum_{j'=0}^{\ell} p_{j', \ell-j'} x^{r'+j'} y^{\ell-j'} \pmod{x^{q+1} + y^q + y}.$$

It only remains to argue that the first term on the right is the leading term of the polynomial. This situation is illustrated in Figure 12.13. The square marked by a circle indicates the point (r', r''); the left square marked by an asterisk indicates the point $(q, r'' - q)$; and the right square marked by an asterisk indicates the point $(r' + q + 1, r'' - q)$. The substitution of $y^q + y$ for x^{q+1} deletes the point $(r' + q + 1, r'' - q)$, replacing it by the point (r', r''), which now is the leading monomial, as asserted. ∎

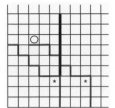

Figure 12.13. Geometric proof.

Proposition 12.8.4 *For a senseword within the designed distance of an hermitian code, there is only one way to extend an ordered sequence of syndromes with a subsequent missing syndrome, not in the defining set, that will produce a connection footprint of area at most t, where $2t + 1 \leq d^*$.*

Proof: A corollary to the Sakata–Massey theorem (Corollary 8.3.3) says that if $\Lambda(x, y)$, a polynomial of degree s, produces the bi-index sequence $S_0, S_1, \ldots, S_{r-1}$, but does not produce the larger sequence $S_0, S_1, \ldots, S_{r-1}, S_r$, then the connection footprint Δ_{r-1} for $S_0, S_1, \ldots, S_{r-1}$ and the connection footprint Δ_r for $S_0, S_1, \ldots, S_{r-1}, S_r$ satisfies

$$(r', r'') \in \Delta_{r-1} \cup (r - \Delta_r).$$

This statement applies even if the code were not restricted to lie on the hermitian curve. By requiring the code to lie on the hermitian curve, and appealing to Proposition 12.8.3, we also have

$$(q, r'' - q) \in \Delta_{r-1} \cup (r - \Delta_r)$$

Combining these statements with Proposition 12.8.2 gives

$$\|\Delta_r\| + \|\Delta_{r-1}\| \geq \|\Delta_r \cup \Delta_{r-1}\|$$
$$\geq \|\mathcal{K}\|$$
$$\geq d^* \geq 2t + 1.$$

But since $\|\Delta_{r-1}\|$ is not larger than t, $\|\Delta_r\|$ must be larger than t. Because we are considering a decodable senseword for a code with designed distance at least $2t + 1$, the true footprint has an area at most equal to t. We conclude that Δ_r cannot be the true footprint because its area is too large, and so only the correct S_r can be the missing syndrome.   ∎

Problems

12.1   The (49, 35, 7) hyperbolic code over $GF(8)$, with syndromes in the graded order, is to be decoded using the Sakata algorithm. Prove that the missing syndromes are actually not needed because the discrepancy must be zero on each iteration, corresponding to a missing syndrome, and from this condition the missing syndrome can be inferred.

12.2   For the hermitian code over $GF(16)$, prove that the syndromes satisfy

$$\sum_{\ell=0}^{4} S_{j',j''+3\ell} = 0.$$

12.3   Prove the following generalization of l'Hôpital's rule in an arbitrary field, F: if $P(x)$ and $Q(x)$ are polynomials over F, with a common zero at β, then $P'(x)/Q'(x) = p(x)/q(x)$, where $p(x)$ and $q(x)$ are defined by $P(x) = (x - \beta)p(x)$ and $Q(x) = (x - \beta)q(x)$.

12.4   Describe how the Sakata algorithm can be used to fill erasures in the absence of errors.

12.5   Derive a two-dimensional generalization of erasure and error decoding.

12.6   Derive a two-dimensional generalization of the Forney algorithm.

12.7   Formulate a halting condition for the Sakata algorithm that ensures that the set of connection polynomials is a minimal basis of a locator ideal wherever the error radius is within the packing radius. Can this condition be extended beyond the packing radius?

12.8   Estimate the complexity of the process of recursive extension for two-dimensional codes. Consider both codes on a plane and codes on curves. Is there a difference in complexity? Why?

12.9   Over the bicyclic plane $GF(16)^{*2}$, let C be the two-dimensional Fourier transform of a two-dimensional array c that is nonzero only on some or all of the points of the epicyclic hermitian curve, based on the polynomial $x^5 + y^4 + y$. Suppose that only the first five columns of C are given. Formulate an efficient computational procedure for computing the array c.

12.10   (a) Show that a proper supercode of an hermitian code over $GF(q^2)$ based on the polynomial $x^{q+1} + y^q + y$ with designed distance d^* cannot be formed based on the defining set

$$\mathcal{A} = \{(j' + 1)(j'' + 1) \ \leq \ d^*\} \bigcup \{m(j' + j'') \ \leq \ d^* + 2g - 2\},$$

where $m = g + 1$ if

$$q + 1 - \sqrt{q+1} < \sqrt{d^*} < q + 1 + \sqrt{q+1}.$$

(b) For what d^* do such codes exist over $GF(64)$?
(c) Find the dimension of such a code over $GF(64)$ for $d^* = 180$.

Notes

The history of how locator decoding was generalized to decode codes on curves is a fascinating case study in the development of ideas. We shall outline this history only for the decoding of codes on plane curves. The introduction by Goppa (1977, 1981) of codes on algebraic curves demanded the development of efficient decoding algorithms for these codes. The natural first step is to generalize the methods of locator decoding for one-dimensional cyclic codes to the decoding of two-dimensional codes defined on the plane. Accordingly, the Peterson algorithm for locator decoding was generalized to codes on plane curves by Justesen *et al.* (1989). This paper introduced the two-dimensional locator polynomial, a polynomial that has the error locations among its zeros. In retrospect, it is easy to see that the generalizations of both the Berlekamp–Massey algorithm and the Sugiyama algorithm should have been sought immediately, but this was not so obvious at the time. The Sakata algorithm (Sakata, 1990) was first developed as a two-dimensional generalization of the Berlekamp–Massey algorithm to bicyclic codes on the full plane, though not without a great deal of insight and effort. The Sakata algorithm was later applied to codes on curves by Justesen *et al.* (1992), but without reaching the designed distance of the code. Later, the Sakata algorithm was embellished to reach the designed distance of the code in a paper by Sakata *et al.* (1995), based on the notion of majority voting for missing syndromes. The method of majority voting was introduced by Feng and Rao (1993), and refined by Duursma (1993a).

In addition to the class of decoders that generalize the Berlekamp–Massey algorithm to two dimensions, there are the decoders that generalize the Sugiyama algorithm to two dimensions. Porter (1988), in his Ph.D. thesis, provided this generalization of the Sugiyama algorithm that corrects to within σ of the Goppa bound, where σ is a parameter known as the Clifford defect. Ehrhard provided an algorithm from this point of view that decodes to the Goppa radius.

The Sakata algorithm provides only the locations of the errors, not the magnitudes. Leonard (1995, 1996) provided an appropriate generalization of the Forney formula to codes on curves, and also generalized erasure and error decoding. The generalization of the Forney formula to codes on curves was also studied by Hansen, Jensen, and Koetter (1996), and by O'Sullivan (2002). The use of recursive extension to obtain the full bispectrum of the error pattern was discussed by Sakata, Jensen, and Høholdt (1995).

The construction of supercodes of hermitian codes by incorporating the hyperbolic bound was suggested by Blahut (1995). The method of syndrome filling by majority voting, developed from the point of view of matrix rank arguments and gaussian elimination, is due to Feng and Rao (1993). These ideas were made precise by Duursma (1993b). The complexity of the Feng–Rao decoder may not compare well with other methods of decoding codes on curves because it involves the computation of the rank for each of a series of matrices of increasing size. However, it has the virtue that the computations can be understood with only linear algebra. Perhaps future improvements or embellishments will make the Feng–Rao decoder more attractive, even practical. Syndrome filling was applied to hyperbolic codes by Saints and Heegard (1995). Many other aspects of decoding codes on curves were explored in the 1990s, such as can be found in O'Sullivan (1995, 1997, 2002).

The term "missing syndrome" is an oxymoron, since the term "syndrome" refers to a known piece of data. Although the term is not really satisfactory, it is used here because there does not seem to be a better alternative. What is needed is a good name for an individual component of a Fourier transform. One may hope that a satisfactory term will be suggested in the future.

Bibliography

S. Arimoto, Encoding and Decoding of p-ary Group Codes and the Correction System, *Information Processing in Japan*, vol. 2, pp. 320–325, 1961 (in Japanese).

E. F. Assmus, Jr. and H. F. Mattson, Jr., Coding and Combinatorics, *SIAM Review*, vol. 16, pp. 349–388, 1974.

E. F. Assmus, Jr., H. F. Mattson, Jr., and R. J. Turyn, *Cyclic Codes*, Report AFCRL-65-332, Air Force Cambridge Research Laboratories, Bedford, MA, 1965.

E. F. Assmus, Jr., H. F. Mattson, Jr., and R. J. Turyn, *Research to Develop the Algebraic Theory of Codes*, Report AFCRL-67-0365, Air Force Cambridge Research Laboratories, Bedford, MA, 1967.

S. Arimoto, Encoding and Decoding of p-ary Group Codes and the Correction System, *Information Processing in Japan*, (in Japanese), vol. 2, pp. 320–325, 1961.

E. R. Berlekamp, *Algebraic Coding Theory*, McGraw-Hill, New York, 1968.

E. R. Berlekamp, Coding Theory and the Mathieu Groups, *Information and Control*, vol. 18, pp. 40–64, 1971.

E. R. Berlekamp, Bounded Distance +1 Soft-Decision Reed–Solomon Decoding, *IEEE Transactions on Information Theory*, vol. IT-42, pp. 704–720, 1996.

E. R. Berlekamp, G. Seroussi, and P. Tong, A Hypersystolic Reed–Solomon Decoder, in *Reed–Solomon Codes and Their Applications*, S. B. Wicker and V. K. Bhargava, editors, IEEE Press, 1994.

E. Bézout, Sur le degré des équations résultantes de l'évanouissement des inconnus, *Histoire de l'Académie Royale des Sciences*, anneé 1764, pp. 288–338, Paris, 1767.

E. Bézout, *Théorie générale des équation algébriques*, Paris, 1770.

R. E. Blahut, A Universal Reed–Solomon Decoder, *IBM Journal of Research and Development*, vol. 28, pp. 150–158, 1984.

R. E. Blahut, *Algebraic Codes for Data Transmission*, Cambridge University Press, Cambridge, 2003.

R. E. Blahut, Algebraic Geometry Codes without Algebraic Geometry, *IEEE Information Theory Workshop*, Salvador, Bahia, Brazil, 1992.

R. E. Blahut, *Algebraic Methods for Signal Processing and Communications Coding*, Springer-Verlag, New York, 1992.

R. E. Blahut, *Fast Algorithms for Digital Signal Processing*, Addison-Wesley, Reading, MA, 1985.

R. E. Blahut, On Codes Containing Hermitian Codes, *Proceedings of the 1995 IEEE International Symposium on Information Theory*, Whistler, British Columbia, Canada, 1995.

R. E. Blahut, The Gleason–Prange Theorem, *IEEE Transactions on Information Theory*, vol. IT-37, pp. 1264–1273, 1991.

R. E. Blahut, *Theory and Practice of Error Control Codes*, Addison-Wesley, Reading, MA, 1983.

R. E. Blahut, Transform Techniques for Error-Control Codes, *IBM Journal of Research and Development*, vol. 23, pp. 299–315, 1979.

I. F. Blake, Codes over Certain Rings, *Information and Control*, vol. 20, pp. 396–404, 1972.

I. F. Blake, C. Heegard, T. Høholdt, and V. Wei, Algebraic Geometry Codes, *IEEE Transactions on Information Theory*, vol. IT-44, pp. 2596–2618, 1998.

E. L. Blokh and V. V. Zyablov, Coding of Generalized Concatenated Codes, *Problems of Information Transmission*, vol. 10, pp. 218–222, 1974.

R. C. Bose and D. K. Ray-Chaudhuri, On a Class of Error Correcting Binary Group Codes, *Information and Control*, vol. 3, pp. 68–79, 1960.

B. Buchberger, *Ein Algorithmus zum Auffinden der Basiselemente des Restklassenringes nach einem nulldimensionalen Polynomideal*, Ph.D. Thesis, University of Innsbruck, Austria, 1965.

B. Buchberger, Gröbner Bases: An Algorithmic Method in Polynomial Ideal Theory, in *Multidimensional Systems Theory*, N. K. Bose, editor, D. Reidel Publishing Company, pp. 184–232, 1985.

A. R. Calderbank and G. McGuire, Construction of a $(64, 2^{37}, 12)$ Code via Galois Rings, *Designs, Codes, and Cryptography*, vol. 10, pp. 157–165, 1997.

A. R. Calderbank, G. McGuire, P. V. Kumar, and T. Helleseth, Cyclic Codes over Z_4, Locator Polynomials, and Newton's Identities, *IEEE Transactions on Information Theory*, vol. IT-42, pp. 217–226, 1996.

T. K. Citron, *Algorithms and Architectures for Error Correcting Codes*, Ph.D. Dissertation, Stanford University, Stanford, CA, 1986.

J. H. Conway and N. J. A. Sloane, *Sphere Packings, Lattices, and Groups*, Second Edition, Springer-Verlag, New York, 1992.

D. Coppersmith and G. Seroussi, On the Minimum Distance of Some Quadratic Residue Codes, *IEEE Transactions on Information Theory*, vol. IT-30, pp. 407–411, 1984.

D. Cox, J. Little, and D. O'Shea, *Ideals, Varieties, and Algorithms*, Springer-Verlag, New York, 1992.

D. Dabiri, *Algorithms and Architectures for Error-Correction Codes*, Ph.D. Dissertation, University of Waterloo, Ontario, Canada, 1996.

D. Dabiri and I. F. Blake, Fast Parallel Algorithms for Decoding Reed–Solomon Codes Based on Remainder Polynomials, *IEEE Transactions on Information Theory*, vol. IT-41, pp. 873–885, 1995.

P. Delsarte, Four Fundamental Parameters of a Code and Their Combinatorial Significance, *Information and Control*, vol. 27, pp. 407–438, 1973.

P. Delsarte, On Subfield Subcodes of Modified Reed–Solomon Codes, *IEEE Transactions on Information Theory*, vol. IT-21, pp. 575–576, 1975.

L. E. Dickson, Finiteness of the Odd Perfect and Primitive Abundant Numbers with n Distinct Prime Factors, *American Journal of Mathematics*, vol. 35, pp. 413–422, 1913.

J. L. Dornstetter, On the Equivalence Between Berlekamp's and Euclid's Algorithms, *IEEE Transactions on Information Theory*, vol. IT-33, pp. 428–431, 1987.

V. G. Drinfeld and S. G. Vlădut, Number of Points on an Algebraic Curve, *Functional Analysis*, vol. 17, pp. 53–54, 1993.

I. M. Duursma, *Decoding Codes from Curves and Cyclic Codes*, Ph.D. Dissertation, Eindhoven University, Eindhoven, Netherlands, 1993a.

I. M. Duursma, Algebraic Decoding Using Special Divisors, *IEEE Transactions on Information Theory*, vol. IT-39, pp. 694–698, 1993b.

D. Ehrhard, *Über das Dekodieren Algebraisch-Geometrischer Coden*, Ph.D. Dissertation, Universität Düsseldorf, Düsseldorf, Germany, 1991.

D. Ehrhard, Achieving the Designed Error Capacity in Decoding Algebraic-Geometric Codes, *IEEE Transactions on Information Theory*, vol. IT-39, pp. 743–751, 1993.

S. V. Fedorenko, A Simple Algorithm for Decoding Reed–Solomon Codes and its Relation to the Welch–Berlekamp Algorithm, *IEEE Transactions on Information Theory*, vol. IT-51, pp. 1196–1198, 2005.

W. Feit, A Self Dual Even $(96, 48, 16)$ Code, *IEEE Transactions on Information Theory*, vol. IT-20, pp. 136–138, 1974.

G.-L. Feng and T. R. N. Rao, Decoding Algebraic-Geometric Codes up to the Designed Minimum Distance, *IEEE Transactions on Information Theory*, vol. IT-39, pp. 37–45, 1993.

G.-L. Feng and T. R. N. Rao, A Simple Approach for Construction of Algebraic-Geometric Codes From Affine Plane Curves, *IEEE Transactions on Information Theory*, vol. IT-40, pp. 1003–1012, 1994.

G.-L. Feng and T. R. N. Rao, Improved Geometric Goppa Codes, Part I, Basic Theory, *IEEE Transactions on Information Theory*, vol. IT-41, pp. 1678–1693, 1995.

G.-L. Feng, V. K. Wei, T. R. N. Rao, and K. K. Tzeng, Simplified Understanding and Effect Decoding of a Class of Algebraic-Geometric Codes, *IEEE Transactions on Information Theory*, vol. 40, pp. 981–1002, 1994.

W. Feng, *On Decoding Reed–Solomon Codes and Hermitian Codes*, Ph.D. Dissertation, University of Illinois, Urbana, Illinois, 1999.

W. Feng and R. E. Blahut, A Class of Codes that Contains the Klein Quartic Codes, *Proceedings of the 30th Conference on Information Sciences and Systems*, Princeton, NJ, 1996.

W. Feng and R. E. Blahut, Some Results on the Sudan Algorithm, *Proceedings of the 1998 IEEE International Symposium on Information Theory*, Cambridge, MA, 1998.

P. Fitzpatrick, On the Key Equation, *IEEE Transactions on Information Theory*, vol. IT-41, pp. 1–13, 1995.

G. D. Forney, Jr., On Decoding BCH Codes, *IEEE Transactions on Information Theory*, vol. IT-11, pp. 549–557, 1965.

G. D. Forney, Jr., *Concatenated Codes*, The M.I.T. Press, Cambridge, MA, 1966.

G. D. Forney, Jr., Transforms and Groups, in *Codes, Curves, and Signals: Common Threads in Communication*, A. Vardy, editor, Kluwer Academic, Norwell, MA, 1998.

G. D. Forney, Jr., N. J. A. Sloane, and M. D. Trott, The Nordstrom–Robinson code is the Binary Image of the Octacode, *Proceedings DIMACS/IEEE Workshop on Coding and Quantization*, DIMACS Series in Discrete Mathematics and Theoretical Computer Science, American Mathematical Society, 1993.

W. Fulton, *Algebraic Curves*, Benjamin-Cummings, 1969; reprinted in Advanced Book Classic Series, Addison-Wesley, Reading, MA, 1989.

A. Garcia, S. J. Kim, and R. F. Lax, Consecutive Weierstrass Gaps and Minimum Distance of Goppa Codes, *Journal of Pure and Applied Algebra*, vol. 84, pp. 199–207, 1993.

O. Geil and T. Høholdt, Footprints or Generalized Bézout's Theorem, *IEEE Transactions on Information Theory*, vol. IT-46, pp. 635–641, 2000.

O. Geil and T. Høholdt, On Hyperbolic Codes, *Proceedings of AAECC-14*, Melbourne, November 2001, Springer LNCS, 2002.

G. Goertzel, An Algorithm for the Evaluation of Finite Trigometric Series, *American Mathematical Monthly*, vol. 65, pp. 34–35, 1968.

J. M. Goethals, Nonlinear Codes Defined by Quadratic Forms Over $GF(2)$, *Information and Control*, vol. 31, pp. 43–74, 1976.

M. J. E. Golay, Notes on Digital Coding, *Proceedings of the IRE*, vol. 37, p. 657, 1949.

I. J. Good, The Interaction Algorithm and Practical Fourier Analysis, *Journal of the Royal Statistical Society*, vol. B20, pp. 361–375, 1958; addendum, vol. 22, pp. 372–375, 1960.

V. D. Goppa, A New Class of Linear Error-Correcting Codes, *Problemy Peredachi Informatsii*, vol. 6, pp. 207–212, 1970.

V. D. Goppa, Codes Associated with Divisors, *Problemy Peredachi Informatsii*, vol. 13, pp. 33–39, 1977; *Problems of Information Transmission*, vol. 13, pp. 22–26, 1977.

V. D. Goppa, Codes on Algebraic Curves, *Doklady Akad. Nauk SSSR*, vol. 259, pp. 1289–1290, 1981; *Soviet Math. Doklady*, vol. 24, pp. 170–172, 1981.

D. C. Gorenstein and N. Zierler, A Class of Error-Correcting Codes in p^m Symbols, *Journal of the Society of Industrial and Applied Mechanics*, vol. 9, pp. 207–214, 1961.

M. W. Green, Two Heuristic Techniques for Block Code Construction, *IEEE Transactions on Information Theory*, vol. IT-12, p. 273, 1966.

V. Guruswami and M. Sudan, Improved Decoding of Reed–Solomon Codes and Algebraic Geometry Codes, *IEEE Transactions on Information Theory*, vol. IT-45, pp. 1757–1767, 1999.

R. W. Hamming, Error Detecting and Error Correcting Codes, *Bell System Technical Journal*, vol. 29, pp. 147–160, 1950.

A. R. Hammons, Jr., P. V. Kumar, A. R. Calderbank, N. J. A. Sloane, and P. Solé, The Z_4-linearity of Kerdock, Preparata, Goethals, and Related Codes, *IEEE Transactions on Information Theory*, vol. IT-40, pp. 301–319, 1994.

J. P. Hansen, Codes from the Klein Quartic, Ideals and Decoding, *IEEE Transactions on Information Theory*, vol. IT-33, pp. 923–925, 1987.

J. P. Hansen and H. Stichtenoth, Group Codes on Certain Algebraic Curves with Many Rational Points, *Proceedings of Applied Algebra Engineering Communications Computing-1*, vol. 1, pp. 67–77, 1990.

J. P. Hansen, H. E. Jensen, and R. Koetter, Determination of Error Values for Algebraic-Geometric Codes and the Forney Formula, *IEEE Transactions on Information Theory*, vol. IT-42, pp. 1263–1269, 1996.

C. R. P. Hartmann, Decoding Beyond the BCH Bound, *IEEE Transactions on Information Theory*, vol. IT-18, pp. 441–444, 1972.

C. R. P. Hartmann and K. K. Tzeng, Generalizations of the BCH Bound, *Information and Control*, vol. 20, pp. 489–498, 1972.

H. Hasse, Theorie der høheren Differentiale in einem algebraischen Funktionenkørper mit vollkommenen Konstantenkørper bei beliebiger Charakteristik, *J. Reine & Angewandte Math.*, vol. 175, pp. 50–54, 1936.

C. Heegard, J. H. Little, and K. Saints, Systematic Encoding via Gröbner Bases for a Class of Algebraic Geometric Goppa Codes, *IEEE Transactions on Information Theory*, vol. IT-41, 1995.

H. H. Helgert, Alternant Codes, *Information and Control*, vol. 26, pp. 369–381, 1974.

T. Helleseth and T. Kløve, The Newton Radius of Codes, *IEEE Transactions on Information Theory*, vol. IT-43, pp. 1820–1831, 1997.

A. E. Heydtmann and J. M. Jensen, On the Equivalence of the Berlekamp–Massey and the Euclidian Algorithm for Decoding, *IEEE Transactions on Information Theory*, vol. IT-46, pp. 2614–2624, 2000.

D. Hilbert, Über die Theorie der Algebraischen Formen, *Mathematische Annalen*, vol. 36, pp. 473–534, 1890.

D. Hilbert, Über die Vollen Invarientensysteme, *Mathematische Annalen*, vol. 42, pp. 313–373, 1893.

D. Hilbert, *Gesammelte Abhandlungen*, vol. II (collected works), Springer, Berlin, 1933.

J. W. P. Hirschfeld, M. A. Tsfasman, and S. G. Vlâdut, The Weight Hierarchy of Higher Dimensional Hermitian Codes, *IEEE Transactions on Information Theory*, vol. IT-40, pp. 275–278, 1994.

A. Hocquenghem, Codes Correcteurs d'erreurs, *Chiffres*, vol. 2, pp. 147–156, 1959.

T. Høholdt, On (or in) the Blahut Footprint, in *Codes, Curves, and Signals: Common Threads in Communication*, A. Vardy, editor, Kluwer Academic, Norwell, MA, 1998.

T. Høholdt and R. Pellikaan, On the Decoding of Algebraic-Geometric Codes, *IEEE Transactions on Information Theory*, vol. IT-41, pp. 1589–1614, 1995.

T. Høholdt, J. H. van Lint, and R. Pellikaan, Algebraic Geometry Codes, *Handbook of Coding Theory*, V. S. Pless and W. C. Huffman, editors, Elsevier, Amsterdam, pp. 871–961, 1998.

T. Horiguchi, High Speed Decoding of BCH Codes Using a New Error Evaluation Algorithm, *Elektronics and Communications in Japan*, vol. 72, no. 12, part 3, 1989.

W. C. Huffman, The Automorphism Group of the Generalized Quadratic Residue Codes, *IEEE Transactions on Information Theory*, vol. IT-41, pp. 378–386, 1995.

T. W. Hungerford, *Algebra*, Springer-Verlag, 1974.

T. Ikai, H. Kosako, and Y. Kojima, On Two-Dimensional Cyclic Codes, *Transactions of the Institute of Electronic Communication of Japan*, vol. 57A, pp. 279–286, 1974.

H. Imai, A Theory of Two-Dimensional Cyclic Codes, *Information and Control*, vol. 34, pp. 1–21, 1977.

C. D. Jensen, *Codes and Geometry*, Ph.D. Dissertation, Denmarks Teknishe Højskole, Denmark, 1991.

J. Justesen, K. J. Larson, H. E. Jensen, A. Havemose, and T. Høholdt, Construction and Decoding of a Class of Algebraic Geometry Codes, *IEEE Transactions on Information Theory*, vol. IT-35, pp. 811–821, 1989.

J. Justesen, K. J. Larsen, H. E. Jensen, and T. Høholdt, Fast Decoding of Codes from Algebraic Plane Codes, *IEEE Transactions on Information Theory*, vol. IT-38, pp. 111–119, 1992.

W. M. Kantor, On the Inequivalence of Generalized Preparata Codes, *IEEE Transactions on Information Theory*, vol. IT-29, pp. 345–348, 1983.

T. Kasami, S. Lin, and W. W. Peterson, Some Results on Weight Distributions of BCH Codes, *IEEE Transactions on Information Theory*, vol. IT-12, p. 274, 1966.

A. M. Kerdock, A Class of Low-Rate Nonlinear Codes, *Information and Control*, vol. 20, pp. 182–187, 1972.

C. Kirfel and R. Pellikaan, The Minimum Distance of Codes in an Array Coming From Telescopic Semigroups, *Coding Theory and Algebraic Geometry: Proceedings of AGCT-4, Luminy, France*, 1993.

C. Kirfel and R. Pellikaan, The Minimum Distance of Codes in an Array Coming From Telescopic Semigroups, *IEEE Transactions on Information Theory*, vol. IT-41, pp. 1720–1732, 1995.

F. Klein, Über die Transformation Siebenter Ordnung Die Elliptischen Functionen, *Mathematics Annals*, vol. 14, pp. 428–471, 1879.

R. Koetter, A Unified Description of an Error Locating Procedure for Linear Codes, *Proceedings: Algebraic and Combinatorial Coding Theory*, Voneshta Voda, Bulgaria, 1992.

R. Koetter, A Fast Parallel Berlekamp–Massey Type Algorithm for Hermitian Codes, *Proceedings: Algebraic and Combinatorial Coding Theory*, pp. 125–128, Nogorad, Russia, 1994.

R. Koetter, *On Algebraic Decoding of Algebraic-Geometric and Cyclic Codes*, Ph.D. Dissertation, Linköping University, Linköping, Sweden, 1996.

R. Koetter, On the Determination of Error Values for Codes from a Class of Maximal Curves, *Proceedings of the 35th Allerton Conference on Communication, Control, and Computing*, University of Illinois, Monticello, Illinois, 1997.

R. Koetter, A Fast Parallel Implementation of a Berlekamp–Massey Algorithm for Algebraic-Geometric Codes, *IEEE Transactions on Information Theory*, vol. IT-44, pp. 1353–1368, 1998.

P. V. Kumar and K. Yang, On the True Minimum Distance of Hermitian Codes, *Coding Theory and Algebraic Geometry: Proceedings of AGCT-3*, Lecture Notes in Mathematics, vol. 158, pp. 99–107, Springer, Berlin, 1992.

N. Lauritzen, *Concrete Abstract Algebra*, Cambridge University Press, Cambridge, 2003.

D. A. Leonard, Error-Locator Ideals for Algebraic-Geometric Codes, *IEEE Transactions on Information Theory*, vol. IT-41, pp. 819–824, 1995.

D. A. Leonard, A Generalized Forney Formula for Algebraic-Geometric Codes, *IEEE Transactions on Information Theory*, vol. IT-42, pp. 1263–1269, 1996.

R. J. McEliece, The Guruswami-Sudan Decoding Algorithm for Reed–Solomon Codes, *IPN Progress Report*, pp. 42–153, 2003.

F. J. MacWilliams, A Theorem on the Distribution of Weights in a Systematic Code, *Bell System Technical Journal*, vol. 42, pp. 79–94, 1963.

D. M. Mandelbaum, Decoding of Erasures and Errors for Certain Reed–Solomon Codes by Decreased Redundancy, *IEEE Transactions on Information Theory*, vol. IT-28, pp. 330–336, 1982.

Yu. I. Manin, What is the Maximum Number of Points on a Curve Over F_2? *Journal of the Faculty of Science, University of Tokyo*, vol. 28, pp. 715–720, 1981.

J. L. Massey, Shift-Register Synthesis and BCH Decoding, *IEEE Transactions on Information Theory*, vol. IT-15, pp. 122–127, 1969.

J. L. Massey, Codes and Ciphers: Fourier and Blahut, in *Codes, Curves, and Signals: Common Threads in Communication*, A. Vardy, editor, Kluwer Academic, Norwell, MA, pp. 1998.

H. F. Mattson, Jr. and E. F. Assmus, Jr., Research Program to Extend the Theory of Weight Distribution and Related Problems for Cyclic Error-Correcting Codes, Report AFCRL-64-605, Air Force Cambridge Research Laboratories, Bedford, MA, July 1964.

H. F. Mattson, Jr. and G. Solomon, A New Treatment of Bose–Chaudhuri Codes, *Journal of the Society of Industrial and Applied Mathematics*, vol. 9, pp. 654–699, 1961.

M. Nadler, A 32-Point n equals 12, d equals 5 Code, *IRE Transactions on Information Theory*, vol. IT-8, p. 58, 1962.

A. W. Nordstrom and J. P. Robinson, An Optimum Linear Code, *Information and Control*, vol. 11, pp. 613–616, 1967.

M. E. O'Sullivan, Decoding of Codes Defined by a Single Point on a Curve, *IEEE Transactions on Information Theory*, vol. IT-41, pp. 1709–1719, 1995.

M. E. O'Sullivan, Decoding Hermitian Codes Beyond $(d_{min} - 1)/2$, *Proceedings of the IEEE International Symposium on Information Theory*, Ulm, Germany, 1997.

M. E. O'Sullivan, The Key Equation for One-Point Codes and Efficient Evaluation, *Journal of Pure and Applied Algebra*, vol. 169, pp. 295–320, 2002.

R. H. Paschburg, *Software Implementation of Error-Correcting Codes*, M.S. Thesis, University of Illinois, Urbana, Illinois, 1974.

R. Pellikaan, On the Decoding by Error Location and the Number of Dependent Error Positions, *Discrete Mathematics*, vols. 106–107, pp. 369–381, 1992.

R. Pellikaan, The Klein Quartic, the Fano Plane, and Curves Representing Designs, in *Codes, Curves, and Signals: Common Threads in Communication*, A. Vardy, editor, Kluwer Academic, Norwell, MA, pp. 1998.

W. W. Peterson, Encoding and Error-Correction Procedures for the Bose–Chaudhuri Codes, *IEEE Transactions on Information Theory*, vol. IT-6, pp. 459–470, 1960.

V. Pless, On the Uniqueness of the Golay Codes, *Journal of Combinatorial Theory*, vol. 5, pp. 215–228, 1968.

V. Pless and Z. Qian, Cyclic Codes and Quadratic Residue Codes over Z_4, *IEEE Transactions on Information Theory*, vol. IT-42, pp. 1594–1600, 1996.

A. Poli and L. Huguet, *Error Correcting Codes, Theory and Applications*, Mason, Paris, 1989.

S. C. Porter, *Decoding Codes Arising from Goppa's Construction on Algebraic Curves*, Ph.D. Dissertation, Yale University, New Haven, 1988.

S. C. Porter, B.-Z. Shen, and R. Pellikaan, Decoding Geometric Goppa Codes Using an Extra Place, *IEEE Transactions on Information Theory*, vol. IT-38, no. 6, pp. 1663–1676, 1992.

E. Prange, *Cyclic Error-Correcting Codes in Two Symbols*, Report AFCRC-TN-57-103, Air Force Cambridge Research Center, Cambridge, MA, 1957.

E. Prange, *Some Cyclic Error-Correcting Codes with Simple Decoding Algorithms*, Report AFCRC-TN-58-156, Air Force Cambridge Research Center, Bedford, MA, 1958.

F. P. Preparata, A Class of Optimum Nonlinear Double-Error-Correcting Codes, *Information and Control*, vol. 13, pp. 378–400, 1968.

Z. Qian, *Cyclic Codes over Z_4*, Ph.D. Dissertation, University of Illinois, Chicago, 1996.

C. M. Rader, Discrete Fourier Transforms When the Number of Data Samples is Prime, *Proceedings of the IEEE*, vol. 56, pp. 1107–1108, 1968.

K. Ranto, *Z_4-Goethals Codes, Decoding, and Designs*, Ph.D. Dissertation, University of Turku, Finland, 2002.

I. S. Reed and G. Solomon, Polynomial Codes Over Certain Finite Fields, *Journal of the Society of Industrial and Applied Mathematics*, vol. 8, pp. 300–304, 1960.

C. Roos, A New Lower Bound for the Minimum Distance of a Cyclic Code, *IEEE Transactions on Information Theory*, vol. IT-29, pp. 330–332, 1983.

R. M. Roth and A. Lempel, Application of Circulant Matrices to the Construction and Decoding of Linear Codes, *IEEE Transactions on Information Theory*, vol. IT-36, pp. 1157–1163, 1990.

R. M. Roth and G. Ruckenstein, Efficient Decoding of Reed–Solomon Codes Beyond Half the Minimum Distance, *IEEE Transactions on Information Theory*, vol. IT-46, pp. 246–257, 2000.

K. Saints and C. Heegard, On Hyperbolic Cascaded Reed–Solomon Codes, *Proceedings of Tenth International Symposium on Applied Algebra, Algebraic Algorithms, and Error-Correcting Codes*, San Juan, Puerto Rico, 1993.

K. Saints and C. Heegard, On Hyperbolic Cascade Reed–Solomon Codes, *Proceedings of AAECCC-10, Lecture Notes in Computer Science*, vol. 673, pp. 291–393, Springer, Berlin, 1993.

K. Saints and C. Heegard, Algebraic-Geometric Codes and Multidimensional Cyclic Codes: A Unified Theory Using Gröbner Bases, *IEEE Transactions on Information Theory*, vol. IT-41, pp. 1733–1751, 1995.

S. Sakata, Finding a Minimal Set of Linear Recurring Relations Capable of Generating a Given Finite Two-Dimensional Array, *Journal of Symbolic Computation*, vol. 5, pp. 321–337, 1988.

S. Sakata, Extension of the Berlekamp–Massey Algorithm to N Dimensions, *Information and Computation*, vol. 84, pp. 207–239, 1990.

S. Sakata, H. E. Jensen, and T. Høholdt, Generalized Berlekamp–Massey Decoding of Algebraic-Geometric Codes up to Half the Feng–Rao Bound, *IEEE Transactions on Information Theory*, vol. IT-41, pp. 1762–1768, 1995.

S. Sakata, J. Justesen, Y. Madelung, H. E. Jensen, and T. Høholdt, Fast Decoding of Algebraic-Geometric Codes up to the Designed Minimum Distance, *IEEE Transactions on Information Theory*, vol. IT-41, pp. 1672–1677, 1995.

D. V. Sarwate, Semi-fast Fourier Transforms over $GF(2^m)$, *IEEE Transactions on Computers*, vol. C-27, pp. 283–284, 1978.

T. Schaub, *A Linear Complexity Approach to Cyclic Codes*, Doctor of Technical Sciences Dissertation, ETH Swiss Federal Institute of Technology, 1988.

J. Schwartz, Fast Probabilistic Algorithms for Verification of Polynomial Identities, *Journal of the Association of Computing Machinery*, vol. 27, pp. 701–717, 1980.

J. P. Serre, Sur les nombres des points rationnels d'une courbe algébrique sur un corps fini, *Comptes Rendus Academy of Science, Paris*, vol. 297, serie I, pp. 397–401, 1983.

B.-Z. Shen, *Algebraic-Geometric Codes and Their Decoding Algorithms*, Ph.D. Thesis, Eindhoven University of Technology, 1992.

J. Simonis, The [23, 14, 5] Wagner Code is Unique, Report of the Faculty of Technical Mathematics and Informatics Delft University of Technology, Delft, pp. 96–166, 1996.

J. Simonis, The [23, 14, 5] Wagner Code Is Unique, *Discrete Mathematics*, vol. 213, pp. 269–282, 2000.

A. N. Skorobogatov and S. G. Vlăduţ, On the Decoding of Algebraic-Geometric Codes, *IEEE Transactions on Information Theory*, vol. IT-36, pp. 1461–1463, 1990.

P. Solé, A Quaternary Cyclic Code, and a Family of Quadriphase Sequences with Low Correlation Properties, *Lecture Notes in Computer Science*, vol. 388, pp. 193–201, 1989.

M. Srinivasan and D. V. Sarwate, Malfunction in the Peterson–Zierler–Gorenstein Decoder, *IEEE Transactions on Information Theory*, vol. IT-40, pp. 1649–1653, 1994.

H. Stichtenoth, A Note on Hermitian Codes over $GF(q^2)$, *IEEE Transactions on Information Theory*, vol. IT-34, pp. 1345–1348, 1988.

H. Stichtenoth, On the Dimension of Subfield Subcodes, *IEEE Transactions on Information Theory*, vol. IT-36, pp. 90–93, 1990.

H. Stichtenoth, *Algebraic Function Fields and Codes*, Springer-Verlag, Berlin, 1993.

M. Sudan, Decoding of Reed–Solomon Codes Beyond the Error-Correction Bound, *Journal of Complexity*, vol. 13, pp. 180–183, 1997.

M. Sudan, Decoding of Reed–Solomon Codes Beyond the Error-Correction Diameter, *Proceedings of the 35th Annual Allerton Conference on Communication, Control, and Computing*, University of Illinois at Urbana-Champaign, 1997.

Y. Sugiyama, M. Kasahara, S. Hirasawa, and T. Namekawa, A Method for Solving Key Equations for Decoding Goppa Codes, *Information and Control*, vol. 27, pp. 87–99, 1975.

L. H. Thomas, Using a Computer to Solve Problems in Physics, in *Applications of Digital Computers*, Ginn and Co., Boston, MA, 1963.

H. J. Tiersma, Remarks on Codes From Hermitian Curves, *IEEE Transactions on Information Theory*, vol. IT-33, pp. 605–609, 1987.

M. A. Tsfasman, S. G. Vlăduţ, and T. Zink, Modular Curves, Shimura Curves and Goppa Codes, Better Than Varshamov–Gilbert Bound, *Mathematische Nachrichten*, vol. 104, pp. 13–28, 1982.

B. L. van der Waerden, *Modern Algebra* (2 volumes), translated by F. Blum and T. J. Benac, Frederick Ungar, New York, 1950 and 1953.

J. H. van Lint and T. A. Springer, Generalized Reed–Solomon Codes From Algebraic Geometry, *IEEE Transactions on Information Theory*, vol. IT-33, pp. 305–309, 1987.

J. H. van Lint and R. M. Wilson, On the Minimum Distance of Cyclic Codes, *IEEE Transactions on Information Theory*, vol. IT-32, pp. 23–40, 1986.

T. J. Wagner, A Remark Concerning the Minimum Distance of Binary Group Codes, *IEEE Transactions on Information Theory*, vol. IT-11, p. 458, 1965.

T. J. Wagner, A Search Technique for Quasi-Perfect Codes, *Information and Control*, vol. 9, pp. 94–99, 1966.

R. J. Walker, *Algebraic Curves*, Dover, New York, 1962.

L. Welch and E. R. Berlekamp, *Error Correction for Algebraic Block Codes*, U.S. Patent 4 633 470, 1983.

J. K. Wolf, Adding Two Information Symbols to Certain Nonbinary BCH Codes and Some Applications, *Bell System Technical Journal*, vol. 48, pp. 2405–2424, 1969.

J. Wu and D. J. Costello, Jr., New Multi-Level Codes over $GF(q)$, *IEEE Transactions on Information Theory*, vol. IT-38, pp. 933–939, 1992.

C. Xing, On Automorphism Groups of the Hermitian Codes, *IEEE Transactions on Information Theory*, vol. IT-41, pp. 1629–1635, 1995.

T. Yaghoobian and I. F. Blake, Hermitian Codes as Generalized Reed–Solomon Codes, *Designs, Codes, and Cryptography*, vol. 2, pp. 15–18, 1992.

K. Yang and P. V. Kumar, On the True Minimum Distance of Hermitian Codes, in *Lecture Notes in Mathematics*, H. Stichtenoth and M. A. Tsfasman, editors, vol. 1518, pp. 99–107, Springer, Berlin, 1988.

Index

Page numbers in bold refer to the most important page, usually where the index entry is defined or explained in detail.